Computational Fluid Dynamics 2020

Computational Fluid Dynamics 2020

Editor

Mostafa Safdari Shadloo

MDPI • Basel • Beijing • Wuhan • Barcelona • Belgrade • Manchester • Tokyo • Cluj • Tianjin

Editor
Mostafa Safdari Shadloo
CORIA Lab., INSA Rouen/CNRS (UMR 6614)
France

Editorial Office
MDPI
St. Alban-Anlage 66
4052 Basel, Switzerland

This is a reprint of articles from the Special Issue published online in the open access journal *Mathematics* (ISSN 2227-7390) (available at: https://www.mdpi.com/si/mathematics/Computational_Fluid_Dynamics).

For citation purposes, cite each article independently as indicated on the article page online and as indicated below:

LastName, A.A.; LastName, B.B.; LastName, C.C. Article Title. *Journal Name* **Year**, *Volume Number*, Page Range.

ISBN 978-3-0365-2784-0 (Hbk)
ISBN 978-3-0365-2785-7 (PDF)

© 2022 by the authors. Articles in this book are Open Access and distributed under the Creative Commons Attribution (CC BY) license, which allows users to download, copy and build upon published articles, as long as the author and publisher are properly credited, which ensures maximum dissemination and a wider impact of our publications.

The book as a whole is distributed by MDPI under the terms and conditions of the Creative Commons license CC BY-NC-ND.

Contents

About the Editor ix

Preface to "Computational Fluid Dynamics 2020" xi

Nidal H. Abu-Hamdeh, Abdulmalik A. Aljinaidi, Mohamed A. Eltaher, Khalid H. Almitani, Khaled A. Alnefaie, Abdullah M. Abusorrah and Mohammad Reza Safaei
Implicit Finite Difference Simulation of Prandtl-Eyring Nanofluid over a Flat Plate with Variable Thermal Conductivity: A Tiwari and Das Model
Reprinted from: *Mathematics* **2021**, *9*, 3153, doi:10.3390/math9243153 1

Suresh Alapati
Simulation of Natural Convection in a Concentric Hexagonal Annulus Using the Lattice Boltzmann Method Combined with the Smoothed Profile Method
Reprinted from: *Mathematics* **2020**, *8*, 1043, doi:10.3390/math8061043 21

Muhammad Bilal Arain, Muhammad Mubashir Bhatti, Ahmad Zeeshan and Faris Saeed Alzahrani
Bioconvection Reiner-Rivlin Nanofluid Flow between Rotating Circular Plates with Induced Magnetic Effects, Activation Energy and Squeezing Phenomena
Reprinted from: *Mathematics* **2021**, *9*, 2139, doi:10.3390/math9172139 39

Umair Khan, Aurang Zaib and Anuar Ishak
Magnetic Field Effect on Sisko Fluid Flow Containing Gold Nanoparticles through a Porous Curved Surface in the Presence of Radiation and Partial Slip
Reprinted from: *Mathematics* **2021**, *9*, 921, doi:10.3390/math9090921 63

Nur Syazana Anuar and Norfifah Bachok
Double Solutions and Stability Analysis of Micropolar Hybrid Nanofluid with Thermal Radiation Impact on Unsteady Stagnation Point Flow
Reprinted from: *Mathematics* **2021**, *9*, 276, doi:10.3390/math9030276 81

Thanh Dam Mai and Jaiyoung Ryu
Effects of Leading-Edge Modification in Damaged Rotor Blades on Aerodynamic Characteristics of High-Pressure Gas Turbine
Reprinted from: *Mathematics* **2020**, *8*, 2191, doi:10.3390/math8122191 99

Anwar Khan, Muhammad Ashraf, Ahmed M. Rashad and Hossam A. Nabwey
Impact of Heat Generation on Magneto-Nanofluid Free Convection Flow about Sphere in the Plume Region
Reprinted from: *Mathematics* **2020**, *8*, 2010, doi:10.3390/math8112010 121

Thi Thanh Giang Le, Kyeong Sik Jang, Kwan-Sup Lee and Jaiyoung Ryu
Numerical Investigation of Aerodynamic Drag and Pressure Waves in Hyperloop Systems
Reprinted from: *Mathematics* **2020**, *8*, 1973, doi:10.3390/math8111973 139

Nurul Amira Zainal, Roslinda Nazar, Kohilavani Naganthran and Ioan Pop
Unsteady Stagnation Point Flow of Hybrid Nanofluid Past a Convectively Heated Stretching/Shrinking Sheet with Velocity Slip
Reprinted from: *Mathematics* **2020**, *8*, 1649, doi:10.3390/math8101649 163

Sivasankaran Sivanandam, Ali J. Chamkha, Fouad O. M. Mallawi, Metib S. Alghamdi and Aisha M. Alqahtani
Effects of Entropy Generation, Thermal Radiation and Moving-Wall Direction on Mixed Convective Flow of Nanofluid in an Enclosure
Reprinted from: *Mathematics* **2020**, *8*, 1471, doi:10.3390/math8091471 185

Mohammed M. Fayyadh, Kohilavani Naganthran, Md Faisal Md Basir, Ishak Hashim and Rozaini Roslan
Radiative MHD Sutterby Nanofluid Flow Past a Moving Sheet: Scaling Group Analysis
Reprinted from: *Mathematics* **2020**, *8*, 1430, doi:10.3390/math8091430 205

Firas A. Alwawi, Hamzeh T. Alkasasbeh, Ahmed M. Rashad and Ruwaidiah Idris
A Numerical Approach for the Heat Transfer Flow of Carboxymethyl Cellulose-Water Based Casson Nanofluid from a Solid Sphere Generated by Mixed Convection under the Influence of Lorentz Force
Reprinted from: *Mathematics* **2020**, *8*, 1094, doi:10.3390/math8071094 223

Hossam A. Nabwey, Waqar A. Khan and Ahmed M. Rashad
Lie Group Analysis of Unsteady Flow of Kerosene/Cobalt Ferrofluid Past A Radiated Stretching Surface with Navier Slip and Convective Heating
Reprinted from: *Mathematics* **2020**, *8*, 826, doi:10.3390/math8050826 243

Nurul Amira Zainal, Roslinda Nazar, Kohilavani Naganthran and Ioan Pop
Unsteady Three-Dimensional MHD Non-Axisymmetric Homann Stagnation Point Flow of a Hybrid Nanofluid with Stability Analysis
Reprinted from: *Mathematics* **2020**, *8*, 784, doi:10.3390/math8060784 257

Anwar Shahid, Hulin Huang, Muhammad Mubashir Bhatti, Lijun Zhang and Rahmat Ellahi
Numerical Investigation on the Swimming of Gyrotactic Microorganisms in Nanofluids through Porous Medium over a Stretched Surface
Reprinted from: *Mathematics* **2020**, *8*, 380, doi:10.3390/math8030380 281

Umair Khan, Aurang Zaib, Ilyas Khan, Kottakkaran Sooppy Nisar and Dumitru Baleanu
Insights into the Stability of Mixed Convective Darcy–Forchheimer Flows of Cross Liquids from a Vertical Plate with Consideration of the Significant Impact of Velocity and Thermal Slip Conditions
Reprinted from: *Mathematics* **2020**, *8*, 31, doi:10.3390/math8010031 299

Anum Shafiq, Ilyas Khan, Ghulam Rasool, Asiful H. Seikh and El-Sayed M. Sherif
Significance of Double Stratification in Stagnation Point Flow of Third-Grade Fluid towards a Radiative Stretching Cylinder
Reprinted from: *Mathematics* **2019**, *7*, 1103, doi:10.3390/math7111103 319

Anum Shafiq, Islam Zari, Ghulam Rasool, Iskander Tlili and Tahir Saeed Khan
On the MHD Casson Axisymmetric Marangoni Forced Convective Flow of Nanofluids
Reprinted from: *Mathematics* **2019**, *7*, 1087, doi:10.3390/math7111087 337

Naeem Faraz, Yasir Khan, Amna Anjum and Muhammad Kahshan
Three-Dimensional Hydro-Magnetic Flow Arising in a Long Porous Slider and a Circular Porous Slider with Velocity Slip
Reprinted from: *Mathematics* **2019**, *7*, 748, doi:10.3390/math7080748 353

Fahd Almutairi, S.M. Khaled and Abdelhalim Ebaid
MHD Flow of Nanofluid with Homogeneous-Heterogeneous Reactions in a Porous Medium under the Influence of Second-Order Velocity Slip
Reprinted from: *Mathematics* **2019**, *7*, 220, doi:10.3390/math7030220 . . . . **377**

Mohammad Yaghoub Abdollahzadeh Jamalabadi
Optimal Design of Isothermal Sloshing Vessels by Entropy Generation Minimization Method
Reprinted from: *Mathematics* **2019**, *7*, 380, doi:10.3390/math7050380 . . . . **389**

About the Editor

Mostafa Safdari Shadloo has been actively engaged in the fields of (i) (aero-)hydrodynamics, turbulence, and transitional boundary layers, as well as (ii) multiphase, multiphysics fluid flows and heat transfer for the last 10 years. His expertise is mainly in theoretical and computational fluid dynamics (CFD), but he has also been active in developing validation strategies and guidelines for CFDist. He aims to develop a new-generation, high-order, coupled algorithm for compressible/incompressible fluid flows with complex physical behaviors, relevant for industrial applications. In this framework, he uses high-performance computing (HPC), high-fidelity direct numerical simulations (DNS) and large-eddy simulations (LES) to decipher complex instabilities and flow behaviors caused mainly by multiphase and/or turbulent flows, with heat transfer and compressibility effects.

To summarize, at national, European and international levels, Dr. Shadloo has actively been the PI and Main Participant (MP) in numerous projects dealing with unsteady multi-physics flows, including multidisciplinary modeling, simulation and validation, with an overall budget of more than EUR 4 million. The main outcomes of Dr. Shadloo's research are published in 100 original scientific articles (plus book chapters) in highly prestigious, peer-reviewed journals, and 24 proceedings presented in international peer-reviewed conferences. His citations exceed 4600 and 3800, with an h-index of 41 and 38 at the time of writing, based on Google scholar and Scopus citation reports, respectively.

Preface to "Computational Fluid Dynamics 2020"

Hitherto, experimental approaches have been widely considered as the main source of information for predicting the physical behavior of fluid flow problems. However, in many applications, due to complexities in fluid behavior related to nonlinearity, experimental methods which are multiscale, multiphase, etc., are either extremely expensive or subjected to scaling issues, and in some cases are impossible. Under these constraints, scrutinizing the physical phenomena seems to be only possible through the alternative of numerical tools.

This Special Issue focuses on computational fluid dynamics (CFD) research, with an emphasis on its recent advancements and use in many industrial and academic applications. Papers on topics ranging from novel physical models and discoveries to the correct treatment of difficulties inherent to numerical modeling of fluid flow systems are invited for submission. These include, but are not limited to: (i) correct and effective modeling of the physical boundary conditions; (ii) mass and energy conservations; (iii) realistically treating complicated physical phenomena; (iv) extendibility to dealing with multiphysics phenomena such as those seen in magnetohydrodynamics (MHD), electrohydrodynamics (EHD), non-Newtonian flows, phase change, nanofluidics problems, etc.; and finally (v) the extension of the before-mentioned methodologies to three-dimensional modeling and massively parallel computing in order to handle real life problems of particular interest.

We are especially interested in the following manuscript topics: the use of conventional numerical methods such as finite difference (FDM), finite volume (FVM) and finite element (FEM) methods, elaborating on their differences, similarities, advantages and drawbacks; the development and validation of less established and novel, attractive numerical methodologies such as smoothed-particle hydrodynamics (SPH), moving particle semi-implicit (MPS), lattice Boltzmann (LBM) methods, etc. Manuscripts dealing with the benchmarking of new test cases, optimizing flow, fluid, and geometrical parameters, as well as using data-driven approaches such as reduced-order methods and machine learning (ML), are of particular interest. This Special Issue also welcomes related novel inter- or multi-disciplinary works in the emerging areas of mechanical, chemical, process and energy engineering.

Mostafa Safdari Shadloo
Editor

Article

Implicit Finite Difference Simulation of Prandtl-Eyring Nanofluid over a Flat Plate with Variable Thermal Conductivity: A Tiwari and Das Model

Nidal H. Abu-Hamdeh [1], Abdulmalik A. Aljinaidi [2], Mohamed A. Eltaher [2], Khalid H. Almitani [2], Khaled A. Alnefaie [2], Abdullah M. Abusorrah [3] and Mohammad Reza Safaei [4,5,*]

[1] Center of Research Excellence in Renewable Energy and Power Systems, and Department of Mechanical Engineering, Faculty of Engineering, K. A. CARE Energy Research and Innovation Center, King Abdulaziz University, Jeddah 21589, Saudi Arabia; nbuhamdeh@kau.edu.sa
[2] Mechanical Engineering Department, Faculty of Engineering, King Abdulaziz University, Jeddah 21511, Saudi Arabia; aljinaidi@kau.edu.sa (A.A.A.); meltaher@kau.edu.sa (M.A.E.); kalmettani@kau.edu.sa (K.H.A.); kalnefaie@kau.edu.sa (K.A.A.)
[3] Center of Research Excellence in Renewable Energy and Power Systems, Department of Electrical and Computer Engineering, Faculty of Engineering, King Abdulaziz University, Jeddah 21589, Saudi Arabia; aabusorrah@hotmail.com
[4] Department of Mechanical and Aeronautical Engineering, Clarkson University, Potsdam, NY 13699-5725, USA
[5] Department of Medical Research, China Medical University Hospital, China Medical University, Taichung 40402, Taiwan
* Correspondence: msafaei@clarkson.edu

Citation: Abu-Hamdeh, N.H.; Aljinaidi, A.A.; Eltaher, M.A.; Almitani, K.H.; Alnefaie, K.A.; Abusorrah, A.M.; Safaei, M.R. Implicit Finite Difference Simulation of Prandtl-Eyring Nanofluid over a Flat Plate with Variable Thermal Conductivity: A Tiwari and Das Model. *Mathematics* **2021**, *9*, 3153. https://doi.org/10.3390/math9243153

Academic Editor: Mostafa Safdari Shadloo

Received: 1 October 2021
Accepted: 1 December 2021
Published: 7 December 2021

Publisher's Note: MDPI stays neutral with regard to jurisdictional claims in published maps and institutional affiliations.

Copyright: © 2021 by the authors. Licensee MDPI, Basel, Switzerland. This article is an open access article distributed under the terms and conditions of the Creative Commons Attribution (CC BY) license (https://creativecommons.org/licenses/by/4.0/).

Abstract: The current article presents the entropy formation and heat transfer of the steady Prandtl-Eyring nanofluids (P-ENF). Heat transfer and flow of P-ENF are analyzed when nanofluid is passed to the hot and slippery surface. The study also investigates the effects of radiative heat flux, variable thermal conductivity, the material's porosity, and the morphologies of nano-solid particles. Flow equations are defined utilizing partial differential equations (PDEs). Necessary transformations are employed to convert the formulae into ordinary differential equations. The implicit finite difference method (I-FDM) is used to find approximate solutions to ordinary differential equations. Two types of nano-solid particles, aluminium oxide (Al_2O_3) and copper (Cu), are examined using engine oil (EO) as working fluid. Graphical plots are used to depict the crucial outcomes regarding drag force, entropy measurement, temperature, Nusselt number, and flow. According to the study, there is a solid and aggressive increase in the heat transfer rate of P-ENF Cu-EO than Al_2O_3-EO. An increment in the size of nanoparticles resulted in enhancing the entropy of the model. The Prandtl-Eyring parameter and modified radiative flow show the same impact on the radiative field.

Keywords: steady flow; Tiwari and Das model; Prandtl-Eyring nanofluid; entropy generation; implicit finite difference method

1. Introduction

Nanofluids, including nanomaterial dispersed in a pure fluid, are becoming applicable fluids in various systems due to their proved superior specification [1]. Augmented thermal conductivity is a remarkable property induced from nanofluid as compared with conventional fluids [2]. On the other hand, the viscosity of nanofluids is significantly varied depending on the type of nanoparticles, base fluid, and their interaction [3]. Some authors have observed Newtonian behaviour of nanofluids, while a non-Newtonian one has been widely revealed [4]. The non-Newtonian behaviours have practical implementations in wire and blade coating, molten plastic, dyeing of textile, some petroleum fluids, biological fluids movement, and food and slurries processing. In this regard, various kinds of rheological behaviours can be expected, defined by models such as power law, micropolar,

Reiner–Philippoff, viscoelastic, Casson, Carreau, Giesekus, Prandtl, Prandtl–Eyring, and Powell–Eyring [5]. These models introduce special impendence on the momentum conservation equations to be compatible with the targeted behaviour. Indeed, in mathematical language, the relationship between shear stress and deformation rate is described by each model. Prandtl and Prandtl–Eyring are a function of sine inverse and sine hyperbolic, respectively [6,7]. The power law model characterizes the relation as nonlinear [8]. Sajid et al. [9] studied a micropolar Prandtl fluid for a porous stretching sheet situation. They assumed that the heat source is related to temperature and a chemical reaction occurs into the medium. Maleki et al. [10] performed numerical research on power law nanofluid, which flows on a porous plate. They found that using Newtonian nanofluid has no improvement effect on heat transfer.

In contrast, the non-Newtonian one had an essential role in boosting heat transfer. Shankar and Naduvinamani [11] worked on the transport phenomena of a Prandtl–Eyring fluid through a sensor surface under magnetic force conditions. The observation showed that magnetic parameter augmentation causes a velocity field increment and temperature profile reduction. At the same time, heat transfer diminished by the Prandtl number rose. Temperature and concentration variations on Prandtl–Eyring fluid heat transfer were investigated by Al-Kaabi and Al-Khafajy [12] in a porous medium. Finally, Hayat et al. [13] evaluated the efficacy of Prandtl–Eyring nanofluid on gyrotactic microorganisms in a stretching sheet. The results indicated that higher melting parameters hike the velocity and pull down the temperature.

Stretching surface is a well-known and habitual process in industrial situations, i.e., extrusion, fiberglass, cooling of the metallic plate, glass blowing, hot rolling, etc. Boundary layer flow and heat transfer is the theory that helps better understand the scientific phenomena underlying it [14]. Nonlinear equations are expected from practical problems which are experienced in engineering applications. The Keller box method is an implicit finite difference method used to solve these types of equations [15]. Munjam et al. [16] proposed a new technique to solve the fluid flow of a Prandtl–Eyring fluid on a stretched sheet and compared their results with the Keller box method. The analytical outcomes indicated that as the fluid parameter rises, velocity enhances. In addition, they found that the Prandtl–Eyring fluid induces a grosser velocity value as compared to the viscous one.

Jamshed et al. [17] explored the entropy generation of Casson nanofluid by considering the Tiwari and Das model and the Keller box method to solve ODEs. Two methanol-based nanofluids were used by introducing Cu and TiO_2 nanoparticles; Cu nanofluids showed a better performance. In [18], they also used the same models and techniques for the same nanoparticles for engine oil base fluid. They concluded that entropy generation would enhance by Reynolds number and Brinkman numbers. Moreover, increasing nanofluid concentration led to shear rate enhancement. Abdelmalek et al. [19] discussed a Prandtl–Eyring nanofluid which influences Brownian motion and thermophoretic force through a stretched surface. It was proved that magnetic force is undesirable for the momentum. In contrast, Brownian motion and thermophoretic force raised the thermal energy.

In the viewpoint of heat transfer, thermal conductivity is a determinant parameter that is generally assumed to be unchanged. However, extensive studies emphasized that the efficacy of temperature changing the thermal conductivity would vary. Particularly, nanofluids have an intimate relation with temperature, which can considerably affect the heat transfer due to the higher aspect ratio that nanoparticles provide within the base fluid. It is proved that at higher temperatures, the thermal conductivity is typically more elevated. Thus, in a range of temperatures, the thermal conductivity is variable. Maleki et al. [20] studied the efficacy of different kinds of nanofluids on the heat transfer of a porous system. They claimed that the results are opposite with other researchers, i.e., adding more nanoparticles dwindled the heat transfer because it can alter radiation, viscous dissipation, and heat generation. In [21], they also surveyed the non-Newtonian nanofluids by considering the mixture of CMC and water as a base fluid. It was revealed that using non-Newtonian nanofluid in injection mode has a higher heat transfer efficiency as compared to the Newtonian one. Jamshed et al. [22] investigated Casson nanofluid in a stretching sheet system that included variable thermal conductivity. Keller box was the technique that solved ODEs. In this method, differential equations are solved numerically to reduce them into the 1st order differential equations. They used TiO_2 and Cu as nanoparticles in water. Cu/water nanofluid had better heat conduction performance. Carreau–Yasuda nanofluid was researched by Waqas et al. [5] by considering gyrotactic motile microorganisms. Velocity, thermal, and temperature fields were amended by decreasing the bioconvection Rayleigh number, increasing the thermal Biot number, and decreasing the Prandtl number. In addition, the concentration field improved by reducing Brownian motion. Xiong et al. [23] explained that variable thermal conductivity has a determinant role in field quantities. They scrutinized a fibre-reinforced generalized thermoelasticity system by considering temperature-dependent thermal conductivity. Ibrahim and Negera [24] inspected an MHD Williamson nanofluid effect within a stretching cylinder by considering chemical reaction conditions. They asserted that the higher the parameter of variable thermal conductivity, the higher the Sherwood number and skin friction, while the lower the Nusselt number. Dada and Onwubuoya [25] analysed an MHD Williamson fluid over a stretchable surface of variable thickness and thermal conductivity. The conclusion indicated that rising changeable thermal conductivity improves temperature. Hasona et al. [26] described the variable thermal conductivity of a non-Newtonian nanofluid in a special geometry channel. They reported that rising thermal and electrical conductivities enhance the temperature of working fluid, which in turn augments heat transfer performance within the system. Fatunmbi and Okoya [27] presented an investigation on hydromagnetic Casson nanofluid at the attendance of thermophoresis, ohmic heating, and a nonuniform heat source with variable thermal conductivity for a stretching sheet system. They demonstrated that driving up the Casson fluid parameter dwindled the fluid flow velocity, albeit, it augmented the viscous drag.

After a glance into the erstwhile studies, since most industrial fluids include non-Newtonian fluids in a situation like stretching surface, the significant concerns of the current project are to discuss the Prandtl–Eyring nanofluid flow over a stretching sheet under three conditions of variable thermal conductivity, thermal radiative flow, and porous material. In addition, their effects on the entropy formation were elaborated by considering the Tiwari and Das model. In this way, Al_2O_3/EO and Cu/EO nanofluids were analysed at volume concentrations of 3% to 20%. Furthermore, this study implemented the implicit finite difference method to solve the boundary layer equations nicely. Therefore, it can be said that the valuable outcomes of this research can be a guideline for practical applications because it was conducted to select parameters close to actual industrial conditions.

2. Flow Model Formulations

A nonregular stretching velocity was used in the flow analysis to characterize horizontal surface movement (for instance, Reference [28]):

$$U_w(x,0) = mx, \qquad (1)$$

where m is a pilot spreading ratio. The temperature at the surface is $¥_w(x,0) = ¥_\infty + m^* x$. For the sake of adaptability, it was fabricated as a perpetual at $x = 0$. m^*, $¥_w$ and $¥_\infty$ provided the temperature variation rate, thermal disparity rate, and ambient temperature congruently.

2.1. Prandtl–Eyring Fluid Stress Tensor

Prandtl–Eyring fluid stress tensor can be expressed as follows (Qureshi [29]),

$$\tau = \frac{A_p \, \mathrm{Sin}^{-1}\left\{\frac{1}{C}\left[\left(\frac{\partial G_1}{\partial y}\right)^2 + \left(\frac{\partial G_2}{\partial x}\right)^2\right]^{\frac{1}{2}}\right\}}{\left[\left(\frac{\partial G_1}{\partial y}\right)^2 + \left(\frac{\partial G_2}{\partial x}\right)^2\right]^{\frac{1}{2}}}\left(\frac{\partial G_1}{\partial y}\right). \qquad (2)$$

Here, τ signify extra stress tensor and $\overleftarrow{G} = [G_1(x,y,0), G_2(x,y,0), 0]$ indicates the flow velocity vector. A_w and C are fluid parameters. The complete derivation of this specific stress tensor and velocity field can found in Becker [30].

2.2. Model Assumptions and Restraints

The mathematical model was taken into account in the following presumptions and requirements:

- ✓ 2D laminar flow
- ✓ Boundary layer estimation
- ✓ Tiwari and Das nanofluid model
- ✓ Non-Newtonian Prandtl–Eyring nanofluid
- ✓ Copper (Cu) and aluminium oxide (Al_2O_3) nanoparticles
- ✓ Base fluid is engine oil (EO)
- ✓ Variable thermal conductivity
- ✓ Thermal radiation
- ✓ Permeable stretching surface
- ✓ Convective and slippery velocity conditions.

2.3. Geometry for Single-Phase Flow Model

Following is the geometric flow model: the flow goes over the sheet. Thermal leap was used to transfer heat from the fluid's surface to the fluid's inside as the velocity at the surface underwent the flow slip event. Alumina oxide nanoparticles and copper nanoparticles were mixed into the engine oil (see Figure 1).

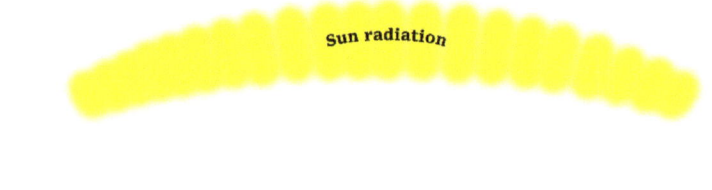

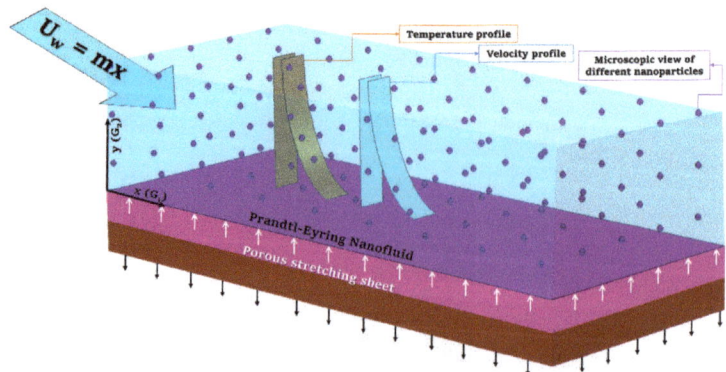

Figure 1. Diagram of the single-phase flow model.

2.4. Classical Equations

The flow formulae of viscous and steady Prandtl–Eyring nanofluid (P-ENF) in combination with variant thermal conductivity, radiation, and porous material are [31–33].

$$\frac{\partial G_1}{\partial x} + \frac{\partial G_2}{\partial y} = 0, \tag{3}$$

$$G_1 \frac{\partial G_1}{\partial x} + G_2 \frac{\partial G_1}{\partial y} = \frac{A_p}{C\rho_{nf}} \left(\frac{\partial^2 G_1}{\partial y^2}\right) - \frac{A_p}{2C^3 \rho_{nf}} \frac{\partial^2 G_1}{\partial y^2} \left[\left(\frac{\partial G_1}{\partial y}\right)^2\right] - \frac{\mu_{nf}}{\rho_{nf} k} G_1, \tag{4}$$

$$G_1 \frac{\partial \yen}{\partial x} + G_2 \frac{\partial \yen}{\partial y} = \frac{1}{(\rho C_p)_{\kappa_{nf}}} \left[\frac{\partial}{\partial y}\left(\kappa^*_{hnf}(\yen)\frac{\partial \yen}{\partial y}\right)\right] - \frac{1}{(\rho C_p)_{nf}} \left[\frac{\partial q_r}{\partial y}\right], \tag{5}$$

the appropriate connection conditions were as follows (Aziz et al. [34]):

$$G_1(x,0) = U_w + N_L\left(\frac{\partial G_1}{\partial y}\right), \quad G_2(x,0) = V_\pi, \quad -k_\pi\left(\frac{\partial \yen}{\partial y}\right) = h_\pi(\yen_w - \yen), \tag{6}$$

$$G_1 \to 0, \quad \yen \to \yen_\infty \text{ as } y \to \infty. \tag{7}$$

$\yen$ is the temperature of the nanofluid.

Other crucial parameters involved fluid parameters A_p, C, slip length N_L, surface permeability V_π, heat transfer coefficient h_π, and porosity (k), along with heat conductivity of firm k_π. It considered physical elements such as the thermal loss from a conventionally heated surface due to conduction and velocity at the surface as a function of the shear stress applied to it (slip condition). Because of the thickness of non-Newtonian P-ENF, just a short distance was covered by the radiative flow. Therefore, radiation heat flux estimation obtained through Rosseland [35] was applied in Equation (5).

$$q_r = -\frac{4\sigma^*}{3k^*}\frac{\partial \yen^4}{\partial y}, \tag{8}$$

herein, σ^* represents the Stefan–Boltzmann constant. Table 1 summarizes the equations of P-ENF material variables [36,37]:

Table 1. Formulae used for studied nanofluids [36,37].

Characteristics	Nanofluid
Dynamical viscosity (μ)	$\mu_{nf} = \mu_f (1-\phi)^{-2.5}$
Density (ρ)	$\rho_{nf} = (1-\phi)\rho_f - \phi\rho_s$
Heat capacity (ρC_p)	$(\rho C_p)_{nf} = (1-\phi)(\rho C_p)_f - \phi(\rho C_p)_s$
Thermal conductivity (κ)	$\frac{\kappa_{nf}}{\kappa_f} = \left[\frac{(\kappa_s + 2\kappa_f) - 2\phi(\kappa_f - \kappa_s)}{(\kappa_s + 2\kappa_f) + \phi(\kappa_f - \kappa_s)}\right]$
Variable thermal conductivity ($\kappa^*_{nf}(\yen)$)	$\kappa^*_{nf}(\yen) = \kappa_{nf}\left[1 + v^* \frac{\yen - \yen_\infty}{\yen_w - \yen_\infty}\right]$

ϕ represents the volume fraction coefficient of nanofluid. μ_f, ρ_f, κ_f and $(C_p)_f$ show dynamic viscosity, density, thermal conductivity, and functional heat capacity regarding the ideal fluid, respectively. The indice of "s" represents the solid nanoparticles. ($\kappa^*_{nf}(\yen)$) represents the temperature-reliant heat conductance of nanofluid.

The thermophysical properties of engine oil and studied nanoparticles are shown in Table 2 [38,39].

Table 2. Materials thermophysical properties [38,39].

Thermophysical	ρ (kg/m^3)	c_p (J/kgK)	k (W/mK)
Copper (Cu)	8933	385.0	401.00
Engine oil (EO)	884	1910	0.144
Aluminium oxide (Al$_2$O$_3$)	3970	765	40

3. Dimensionless Formulations Model

Similarity transformations that convert the governing PDEs into ODEs, and the BVP formulae (3)–(7) are modified. Familiarizing stream function ψ in the equation [28]

$$G_1 = \frac{\partial \psi}{\partial y}, \quad G_2 = -\frac{\partial \psi}{\partial x}. \tag{9}$$

The specified similarity quantities are ([28])

$$\gamma^*(x,y) = \sqrt{\frac{m}{\nu_f}} y, \quad \psi(x,y) = \sqrt{\nu_f m} x f(\gamma^*), \quad \theta(\gamma^*) = \frac{\yen - \yen_\infty}{\yen_w - \yen_\infty}. \tag{10}$$

into Equations (3)–(7). We get

$$\tau^* f''' \left(1 - \varsigma^* f''^2\right) + \phi_{\ddot{Y}_2} \left[f f'' - f'^2 \right] - \frac{1}{\phi_{\ddot{Y}_1}} F_\pi f' = 0, \tag{11}$$

$$\theta'' \left(1 + v^*\theta + \frac{1}{\phi_{\ddot{Y}_4}} P_r N_\pi \right) + v^* \theta'^2 + P_r \frac{\phi_{\ddot{Y}_3}}{\phi_{\ddot{Y}_4}} [f\theta' - f'\theta] = 0. \tag{12}$$

with

$$\left. \begin{array}{l} f(0) = S, \ f'(0) = 1 + \Lambda_\pi f''(0), \theta'(0) = -B_\pi(1 - \theta(0)) \\ f'(\gamma^*) \to 0, \ \theta(\gamma^*) \to 0, \ as \ \gamma^* \to \infty \end{array} \right\} \tag{13}$$

where $\phi'_{\ddot{Y}_i}$ is $1 \leq i \leq 4$ in formulae (11)–(12) signify the subsequent thermophysical structures for P-ENF [29].

$$\left. \begin{array}{l} \phi_{\ddot{Y}_1} = (1-\phi)^{2.5}, \quad \phi_{\ddot{Y}_2} = \left(1 - \phi + \phi \frac{\rho_s}{\rho_f}\right), \quad \phi_{\ddot{Y}_3} = \left(1 - \phi + \phi \frac{(\rho C_p)_s}{(\rho C_p)_f}\right) \\ \phi_{\ddot{Y}_4} = \left(\frac{(k_s + 2k_f) - 2\phi(k_f - k_s)}{(k_s + 2k_f) + \phi(k_f - k_s)}\right). \end{array} \right\} \tag{14}$$

Equation (2) is clearly shown to be valid. Table 3 shows the needed derivatives.

Table 3. Entrenched Control Constraints.

Symbols	Name		Default Value
τ^*	Prandtl–Eyring parameter-I	$\tau^* = \frac{A_p}{\mu_f C}$	1.3
ς^*	Prandtl–Eyring parameter-II	$\varsigma^* = \frac{m^3 x^2}{2C^2 \nu_f}$	0.3
P_r	Prandtl number	$P_r = \frac{\nu_f}{\alpha_f}$	6450
ϕ	Volume fraction coefficient	-	0.18
F_π	Porosity parameter	$F_\pi = \frac{\nu_f}{mk}$	0.6
S	Suction/Injection parameter	$S = -V_\pi \sqrt{\frac{1}{\nu_f m}}$	0.5
N_π	Thermal radiation parameter	$N_\pi = \frac{16}{3} \frac{\sigma^* \yen_\infty^3}{\kappa^* \nu_f (\rho C_p)_f}$	0.3
B_π	Biot number	$B_\pi = \frac{h_\pi}{k_\pi} \sqrt{\frac{\nu_f}{g}}$	0.2
Λ_π	Velocity slip	$\Lambda_\pi = \sqrt{\frac{m}{\nu_f}} N_L$	0.3

Other parameters like skin friction (C_f), Nusselt number (Nu_x) and entropy generation (N_G) can be expressed as [31,32]:

$$\left. \begin{array}{l} C_f Re_x^{\frac{1}{2}} = \tau^* f''(0) - \frac{1}{3}\tau^* \varsigma^* \left(f''(0) \right)^3, \\ Nu_x Re_x^{-\frac{1}{2}} = -\frac{k_{nf}}{k_f}(1+N_\pi)\theta'(0), \\ N_G = R_\pi \left[\phi_{\ddot{Y}_4}(1+N_\pi)\theta'^2 + \frac{1}{\phi_{\ddot{Y}_4}} \frac{B_\Gamma}{\Omega} \left(f''^2 + P_\xi f'^2 \right) \right]. \end{array} \right\} \tag{15}$$

4. Implicit Finite Difference Method

The implicit finite difference method (I-FDM) [40] was utilized to obtain the numerical solution for the equation set of models. I-FDM has an advantage as it has fast convergence. It is 2nd order convergent and inherently stable. I-FDM satisfies the Von Neumann stability test, which has the criterion of a real numerical solution for PDEs with the help of the stability and consistency of a numerical solution. I-FDM is employed to obtain the solution of Equations (11) and (12) using boundary conditions (13). This is a suitable method to obtain the approximated solution of boundary layer problems. I-FDM is widely applicable in the flow problems of the laminar boundary layer, and the obtained results are more effective than others.

To apply the implicit finite difference method [41], Equations (11) and (12) were written in the form of 1st order differential equations utilizing newly employed variables. Reduced equations are as follows [32]:

$$L_1 = f', \tag{16}$$

$$L_2 = L_1', \tag{17}$$

$$z_3 = \theta', \tag{18}$$

$$\tau^* L_2' \left(1 - \varsigma^* L_2^2 \right) + \phi_{\ddot{Y}_2} \left[fL_2 - L_1^2 \right] - \frac{1}{\phi_{\ddot{Y}_1}} F_\pi L_1 = 0, \tag{19}$$

$$L_3' \left(1 + v^*\theta + \frac{1}{\phi_{\ddot{Y}_4}} P_r N_\pi \right) + v^* L_3^2 + P_r \frac{\phi_{\ddot{Y}_3}}{\phi_{\ddot{Y}_4}} [fL_3 - L_1\theta] = 0. \tag{20}$$

With the presence of newly employed variables, boundary conditions eventually changed to [31]

$$f(0) = S, L_1(0) = 1 + \Lambda_\pi L_2(0), L_3(0) = -B_\pi(1 - \theta(0)), L_1(\infty) \to 0, \theta(\infty) \to 0. \tag{21}$$

The different formulae were calculated using central differencing, and average functions were replaced. Thus, the 1st ODEs (16) and (20) order decreases to the next series of nonlinear algebraic formulae.

$$\frac{(L_1)_j + (L_1)_{j-1}}{2} = \frac{f_j - f_{j-1}}{h}, \tag{22}$$

$$\frac{(L_2)_j + (L_2)_{j-1}}{2} = \frac{(L_1)_j - (L_1)_{j-1}}{h}, \tag{23}$$

$$\frac{(L_3)_j + (L_3)_{j-1}}{2} = \frac{\theta_j - \theta_{j-1}}{h}, \tag{24}$$

$$\tau^*\left(\frac{(L_2)_j-(L_2)_{j-1}}{h}\right)\left(1-\varsigma^*\left(\frac{(L_2)_j+(L_2)_{j-1}}{2}\right)^2\right)$$
$$+\left[\phi_{\ddot{Y}_2}\left(\left(\frac{f_j+f_{j-1}}{2}\right)\left(\frac{(L_2)_j+(L_2)_{j-1}}{2}\right)-\left(\frac{(L_1)_j+(L_1)_{j-1}}{2}\right)^2\right)-F_\pi\frac{1}{\phi_{\ddot{Y}_1}}\left(\frac{(L_1)_j+(L_1)_{j-1}}{2}\right)\right], \quad (25)$$

$$\left(\frac{(L_3)_j-(L_3)_{j-1}}{h}\right)\left(1+v^*\left(\frac{\theta_j+\theta_{j-1}}{2}\right)+\frac{1}{\phi_{\ddot{Y}_4}}P_rN_\pi\right)+v^*\left(\frac{(L_3)_j+(L_3)_{j-1}}{2}\right)^2$$
$$+P_r\frac{\phi_{\ddot{Y}_3}}{\phi_{\ddot{Y}_4}}\left[\left(\frac{f_j+f_{j-1}}{2}\right)\left(\frac{(L_3)_j+(L_3)_{j-1}}{2}\right)-\left(\frac{(L_1)_j+(L_1)_{j-1}}{2}\right)\left(\frac{\theta_j+\theta_{j-1}}{2}\right)\right]=0. \quad (26)$$

To linearize the resulting equations, Newton's technique was used. As an example, consider iteration $(i+1)$th

$$()_j^{(i+1)} = ()_j^{(i)} + \ddot{O}()_j^{(i)}. \quad (27)$$

under the substitutition of the linear tridiagonal equational system into Equations (22)–(26), disregarding the elevated $\ddot{O}_j^i$ components.

$$\ddot{O}f_j - \ddot{O}f_{j-1} - \frac{1}{2}h(\ddot{O}(L_1)_j + \ddot{O}(L_1)_{j-1}) = (d_1)_{j-\frac{1}{2}}, \quad (28)$$

$$\ddot{O}(L_1)_j - \ddot{O}(L_1)_{j-1} - \frac{1}{2}h(\ddot{O}(L_2)_j + \ddot{O}(L_2)_{j-1}) = (d_2)_{j-\frac{1}{2}}, \quad (29)$$

$$\ddot{O}\theta_j - \ddot{O}\theta_{j-1} - \frac{1}{2}h(\ddot{O}(L_3)_j + \ddot{O}(L_3)_{j-1}) = (d_3)_{j-\frac{1}{2}}, \quad (30)$$

$$(a_1)_j\ddot{O}f_j + (a_2)_j\ddot{O}f_{j-1} + (a_3)_j\ddot{O}L_{1j} + (a_4)_j\ddot{O}L_{j-1} + (a_5)_j\ddot{O}L_{2j} + (a_6)_j\ddot{O}L_{2j-1}$$
$$+(a_7)_j\ddot{O}\theta_j + (a_8)_j\ddot{O}\theta_{j-1} + (a_9)_j\ddot{O}(L_3)_j + (a_{10})_j\ddot{O}(L_3)_{j-1} = (d_4)_{j-\frac{1}{2}}, \quad (31)$$

$$(b_1)_j\ddot{O}f_j + (b_2)_j\ddot{O}f_{j-1} + (b_3)_j\ddot{O}L_{1j} + (b_4)_j\ddot{O}L_{1j-1} + (b_5)_j\ddot{O}L_{2j} + (b_6)_j\ddot{O}L_{2j-1}$$
$$+(b_7)_j\ddot{O}\theta_j + (b_8)_j\ddot{O}\theta_{j-1} + (b_9)_j\ddot{O}(L_3)_j + (b_{10})_j\ddot{O}(L_3)_{j-1} = (d_5)_{j-\frac{1}{2}}. \quad (32)$$

where

$$(d_1)_{j-\frac{1}{2}} = -f_j + f_{j-1} + \frac{h}{2}(L_1)_j + ((L_1)_{j-1}), \quad (33)$$

$$(d_2)_{j-\frac{1}{2}} = -(L_1)_j + (L_1)_{j-1} + \frac{h}{2}((L_2)_j + (L_2)_{j-1}), \quad (34)$$

$$(d_3)_{j-\frac{1}{2}} = -\theta_j + \theta_{j-1} + \frac{h}{2}((L_3)_j + (L_3)_{j-1}), \quad (35)$$

$$(d_4)_{j-\frac{1}{2}} = -h\left[\tau^*\left(\frac{(L_2)_j-(L_2)_{j-1}}{h}\right)\left(1-\varsigma^*\left(\frac{(L_2)_j+(L_2)_{j-1}}{2}\right)^2\right)\right]-h\left[\phi_b\left(\left(\frac{f_j+f_{j-1}}{2}\right)\left(\frac{(L_2)_j+(L_2)_{j-1}}{2}\right)-\left(\frac{(L_1)_j+(L_1)_{j-1}}{2}\right)^2\right)-F_\pi\frac{1}{\phi_a}\left(\frac{(L_1)_j+(L_1)_{j-1}}{2}\right)\right], \quad (36)$$

$$(d_5)_{j-\frac{1}{2}} = -h\left[\frac{((L_3)_j-(L_3)_{j-1})}{h}\left(1+v^*\left(\frac{\theta_j+\theta_{j-1}}{2}\right)+\frac{1}{\phi_{\ddot{Y}_4}}P_rN_\pi\right)\right]-h\left[v^*\left(\frac{(L_3)_j+(L_3)_{j-1}}{2}\right)^2\right]$$
$$-hP_r\frac{\phi_{\ddot{Y}_3}}{\phi_{\ddot{Y}_4}}\left[\left(\frac{(f_j+f_{j-1})((L_3)_j+(L_3)_{j-1})}{4}\right)\right]+hP_r\frac{\phi_{\ddot{Y}_3}}{\phi_{\ddot{Y}_4}}\left[\left(\frac{(\theta_j+\theta_{j-1})((L_1)_j+(L_1)_{j-1})}{4}\right)\right] \quad (37)$$

The boundary conditions become

$$\ddot{O}f_0 = 0, \ddot{O}(z_1)_0 = 0, \ddot{O}(z_3)_0 = 0, \ddot{O}(z_1)_J = 0, \ddot{O}\theta_J = 0. \quad (38)$$

The following are the formulae (33)–(37) that produce the bulk tridiagonal array,

$$R\ddot{O} = p, \quad (39)$$

where

$$H = \begin{bmatrix} \omega_1 & \omega_1 & & & & \\ \varphi_2 & \omega_2 & \varepsilon_2 & & & \\ & \ddots & \ddots & \ddots & & \\ & & \ddots & \ddots & \ddots & \\ & & & \varphi_{J-1} & \omega_{J-1} & \varepsilon_{J-1} \\ & & & & \varphi_J & \omega_J \end{bmatrix}, \ddot{O} = \begin{bmatrix} \ddot{O}_1 \\ \ddot{O}_2 \\ \vdots \\ \ddot{O}_{J-1} \\ \ddot{O} \end{bmatrix}, p = \begin{bmatrix} (d_1)_{j-\frac{1}{2}} \\ (d_2)_{j-\frac{1}{2}} \\ \vdots \\ (d_{J-1})_{j-\frac{1}{2}} \\ (d_J)_{j-\frac{1}{2}} \end{bmatrix}. \quad (40)$$

This matrix, H, resembles the generalized size of $J \times J$, whereas the $\ddot{O}$ and p indicate the column vectors order of $J \times 1$. Afterward, a unique LU factorization approach was employed to get the solution for $\ddot{O}$.

5. Code Validity

The technique's authenticity was assessed by comparing the thermal conveyance rate fallouts between the recent scheme and the previous results [42–45]. Table 4 summarises the consistency relationship found in all of the studies. Therefore, the findings of the present study are agreeable with previously published results and verified.

Table 4. Comparison of $-\theta'(0)$ with P_r, whenever $\phi = 0$, $N_\pi = 0$, $v^* = 0$, $\Lambda_\pi = 0$, $S = 0$, and $B_\pi \to \infty$.

P_r	Wang [41]	Gorla and Sidawi [42]	Khan and Pop [43]	Makinde and Aziz [44]	This Study
0.2	0.1691	0.1691	0.1691	0.1691	0.169
0.7	0.4539	0.4539	0.4539	0.4539	0.4537
2	0.9114	0.9114	0.9114	0.9114	0.9113
7	1.8954	1.8954	1.8954	1.8954	1.8958

6. Results and Discussion

The section discusses the numerical outcomes obtained on the model in consideration. The parameters involved in the results are ϕ, N_π, v^*, B_π, τ^*, ς^*, F_π, S, Λ_π, R_π, and B_Γ. Figures 2–21 display the physical behavior of the mentioned parameters regarding energy, entropy formation, and velocity on the nondimensional entities of the model. Results for non-Newtonian Al_2O_3-EO and Cu-EO P-ENFs were obtained. Temperature differences and coefficient of skin fraction are detailed in Table 5. The values used for the parameters are $\phi = 0.18$, $N_\pi = 0.3$, $v^* = 0.2$, $B_\pi = 0.3$, $\tau^* = 1.0$, $\varsigma^* = 0.2$, $F_\pi = 0.6$, $S = 0.5$, $\Lambda_\pi = 0.3$, $R_\pi = 5$, and $B_\Gamma = 5$. The power of Al_2O_3-EO and Cu-EO were decided with the fractional size of nanoparticles used in the working fluid. Flow stability of nanoparticles decreased when nanoparticles had a higher amount of fractional range. Al_2O_3-EO was favored more by fractional improvement than Cu-EO nanofluid. Figure 2 shows a lower flow of Cu-EO nanofluid than Al_2O_3-EO. As Al_2O_3 has a high heat transfer property, its primary purpose is to combine with EO. When the fractional volume of both fluids flows increases, thermal distribution transported to the domain from the surface is high, as shown in Figure 3. The increasing fractional volume also resulted in enhancing the fluctuations of the system entropy. Figure 4 shows the leading fluctuations of Cu-EO nanofluid, which settled down midway and increased further towards Al_2O_3-EO nanofluid. The thermal radiation parameter (N_π) models the radiation procedure used in enhancing the entropy rate and heat regarding induced temperature, as shown in Figures 5 and 6. Radiations had a negligible effect on the entropy variations caused by the prominent influence of flow conditions. Cu-EO had more control than the Al_2O_3-EO nanofluid. Regarding heat capability of Al_2O_3 and Cu-EO nanofluids, there was a dominant effect on entropy and thermal aspects of individual variations in v^*, i.e., thermal conductivity. Figures 7 and 8 represent these effects. When the variation parameter tries to increase the ranges of entropy and heat, the nominal impact of v^* is proved by close variations of entropy and thin layers of heat. In both behaviors of parameters, Al$_2$O$_3$-EO underestimated the Cu-EO nanofluid. Figures 9 and 10 clearly show the increment in convective heat on thermal as well as entropy states from lower surfaces in the domain. Parametric values of Biot number represent the ordinary heating procedure B_π. An increase in B_π resulted in enhancing the thermal state in the flow domain, but this only had a negligible impact on the entropy generation. The entropy profile is smaller than the thermal boundary layer, which proves the above statement. According to the study, Al$_2$O$_3$-EO is better than Cu-EO nanofluid. Figures 11–13 demonstrate the impact on the power, entropy, and velocity distributions of Prandtl–Eyring nanofluid τ^*. Figure 11 shows the speed (f') corresponding to τ^*. The velocity of both fluids increased with amplification in τ^*. However, the velocity of Al$_2$O$_3$-EO velocity was more incredible as compared to Cu-EO. Figure 12 shows the temperature curve concerning the Prandtl–Eyring parameter τ^*. An increment in τ^* resulted in reducing the temperatures of both fluids. However, the temperature profile of Cu-EO nanofluid is more critical than Al$_2$O$_3$-EO nanofluid. Figure 13 represents the entropy fluctuation of P-ENF caused by τ^*. An increase of τ^* resulted in lowering the entropy formation. The lower value of entropy of the Al$_2$O$_3$-EO fluid was used to represent Cu-EO nanofluid when both nanofluids were at the end of the graph. τ^* is strongly related to the profile of P-ENF. However, an increase of τ^* resulted in decreasing the entropy and temperature. Figures 14–16 illustrate the efficacy of the Prandtl–Eyring parameter ς^* on the profiles of temperature, velocity, and entropy formation. The velocity change regarding ς^* was displayed in Figure 16. A decrease in the velocity profile was the result of an increment in Cu-EO while increasing Al$_2$O$_3$-EO and a high rate in ς^*. Figure 15 shows the fluctuations in the profile of temperature concerning ς^*. The temperature grows as ς^* is increased, and Cu-EO obtains a quick temperature. Figure 16 highlights the difference in entropy caused by the Prandtl–Eyring parameter ς^*. An increment in entropy is obtained with increasing ς^*. Results obtained from modifying the slip conditions on the nature of the flow, heat, and generation of entropy, respectively, are shown in Figures 17–19. Viscous behavior was focused on the flow conditions in the combinations of the Prandtle-Eyring fluid. Variations in velocity, entropy formation, and thermal distributions have an essential role in slip conditions. The situation for fluidity becomes difficult when slip conditions of Prandtl–Eyring fluid flow are increased. Fluidity was reduced for Cu-EO than Al$_2$O$_3$-EO P-ENF. Such hierarchy mainly occurs in thermal distributions, i.e., Cu-EO has a higher thermal state than Al$_2$O$_3$-EO nanofluid, as depicted in Figure 18. Greater values of slip parameter Λ_π resulted in decreasing the entropy generation. It was caused by slip flow, which acted opposite to entropy generation, as shown in Figure 19. Figure 20 shows the performed estimations for F_π = 0.6, 1.6, and 2.6; meanwhile, parametric values of ς^* are 0.2, 0.4, and 0.6. An increment in the material parameter resulted in enhancing the coefficient of skin friction. Flow velocity was decreased due to an increase in skin friction as resistance in fluid increased. In Figure 21, calculations for N_π = 0.1, 0.3, and 0.5 were employed while Prandtl number P_r was kept fixed on 1.0, 6.2, and 7.38. The convective heat transfer rate rose whenever the radiation parameter N_π, is increased. The heat transfer rate was augmented when heat flux was increased.

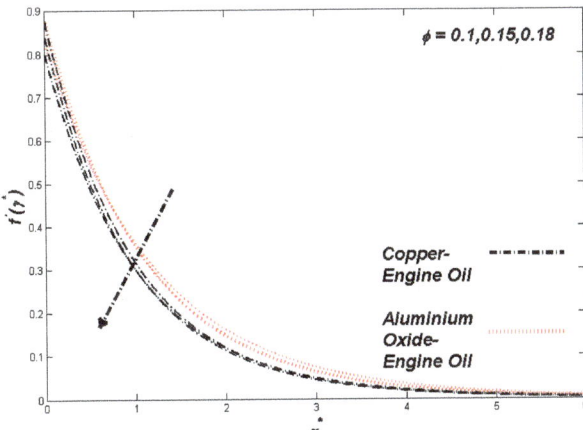

Figure 2. Velocity variation with ϕ.

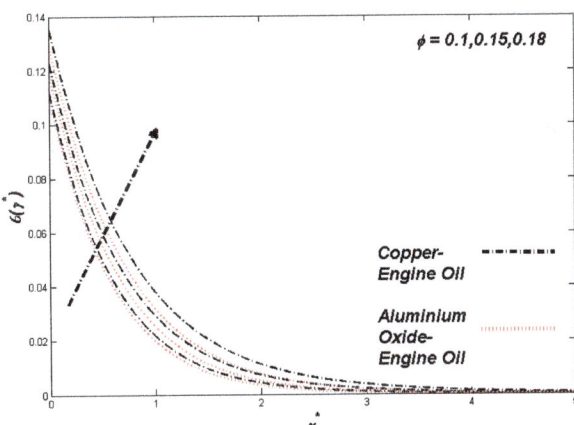

Figure 3. Temperature variation with ϕ.

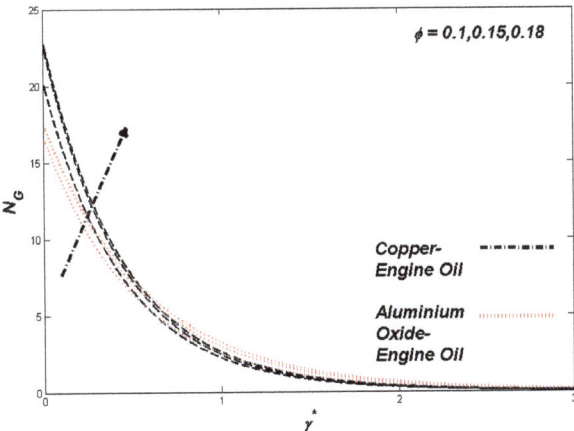

Figure 4. Entropy variation with ϕ.

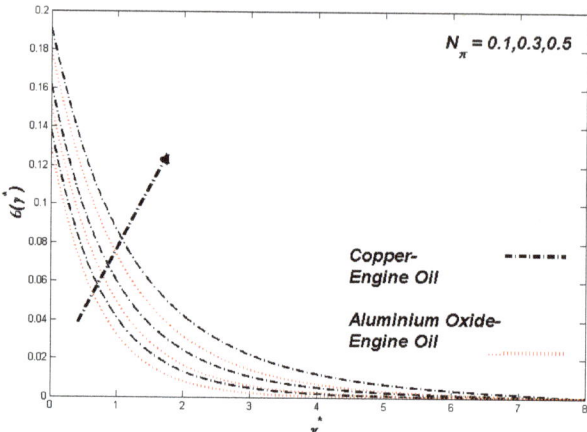

Figure 5. Temperature variation with N_π.

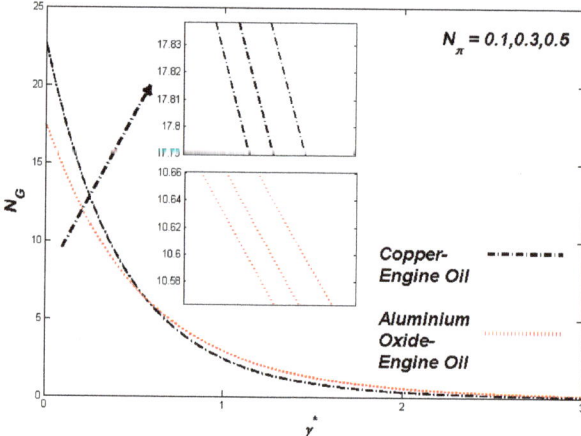

Figure 6. Entropy variation with N_π.

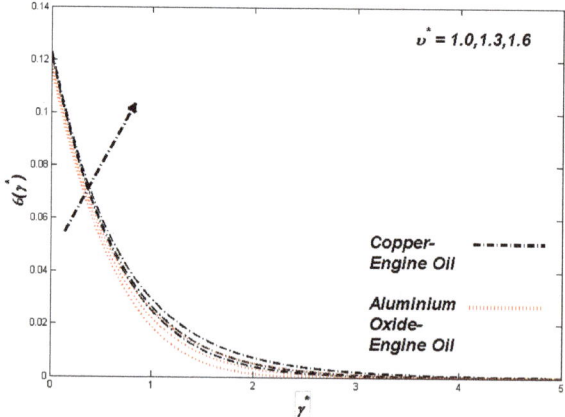

Figure 7. Temperature variation with v^*.

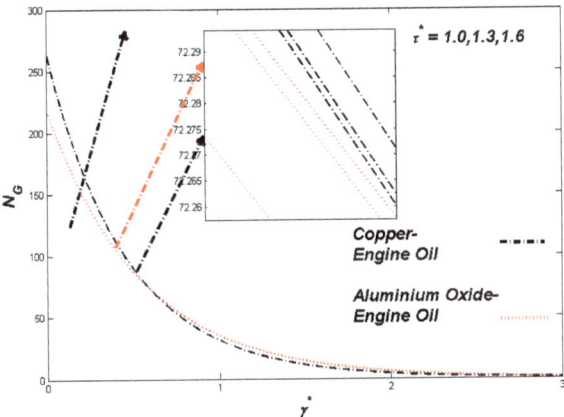

Figure 8. Entropy variation with v^*.

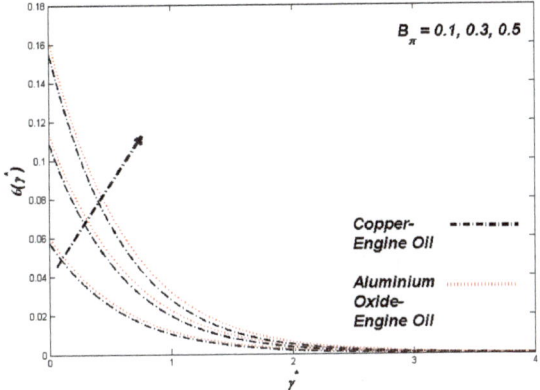

Figure 9. Temperature variation with B_π.

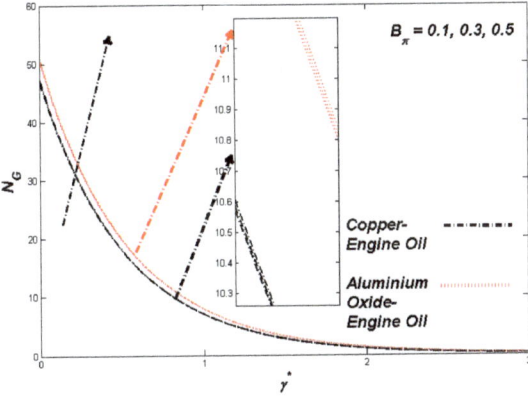

Figure 10. Entropy variation with B_π.

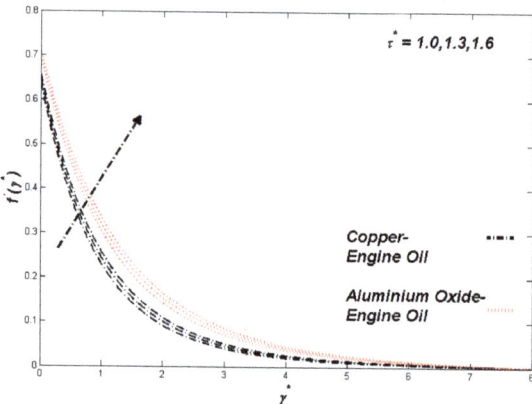

Figure 11. Velocity variation with τ^*.

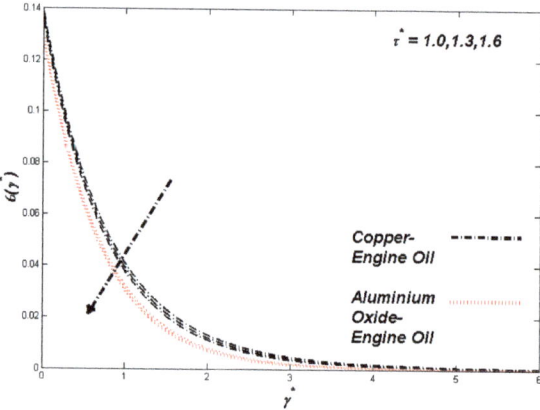

Figure 12. Temperature variation with τ^*.

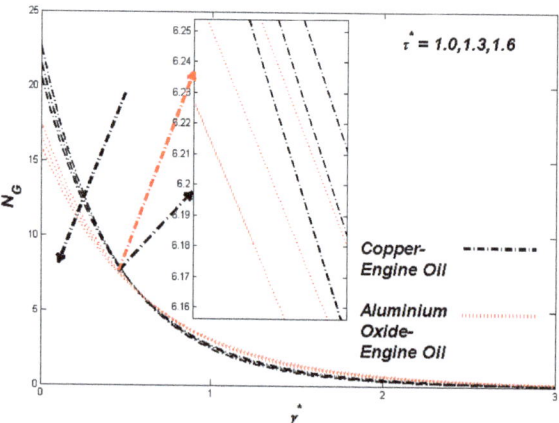

Figure 13. Entropy variation with τ^*.

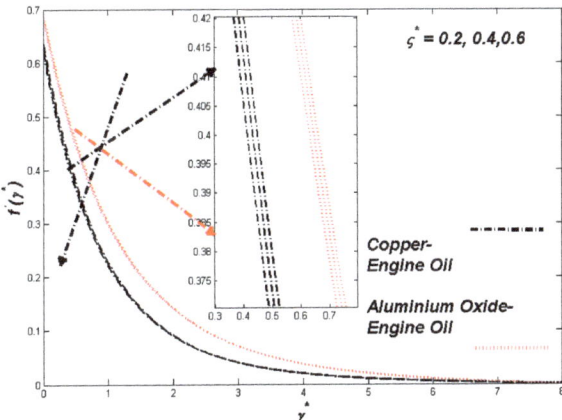

Figure 14. Velocity variation with ς^*.

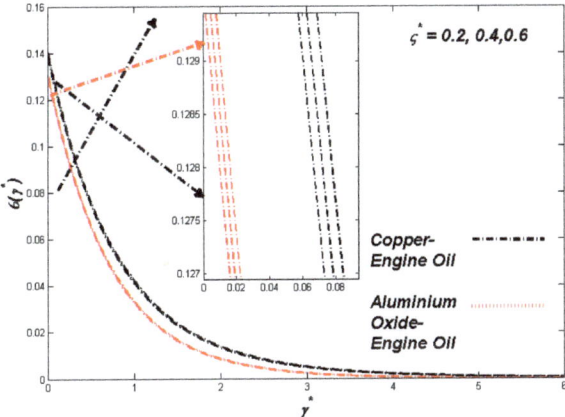

Figure 15. Temperature variation with ς^*.

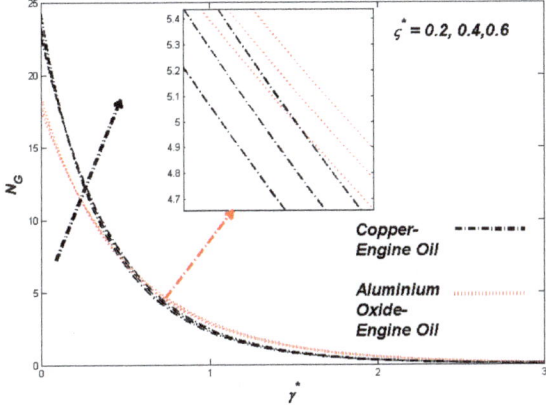

Figure 16. Entropy variation with ς^*.

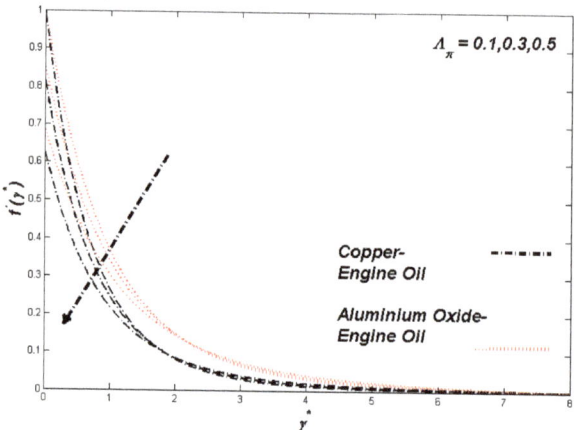

Figure 17. Velocity variation with Λ_π.

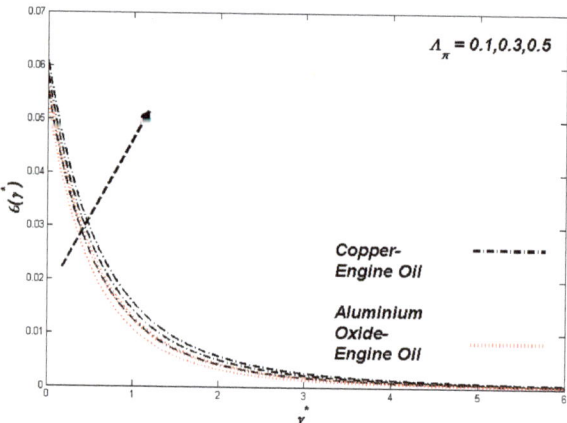

Figure 18. Temperature variation with Λ_π.

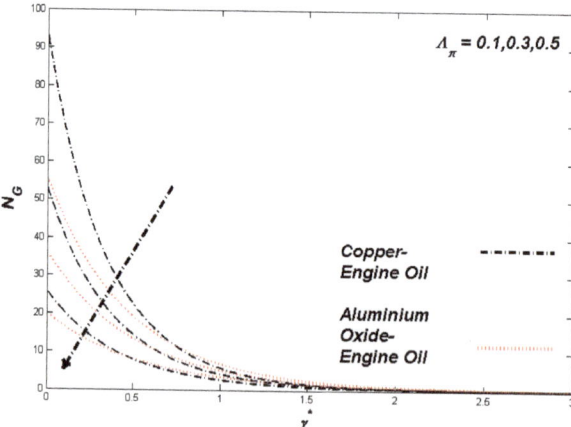

Figure 19. Entropy variation with Λ_π.

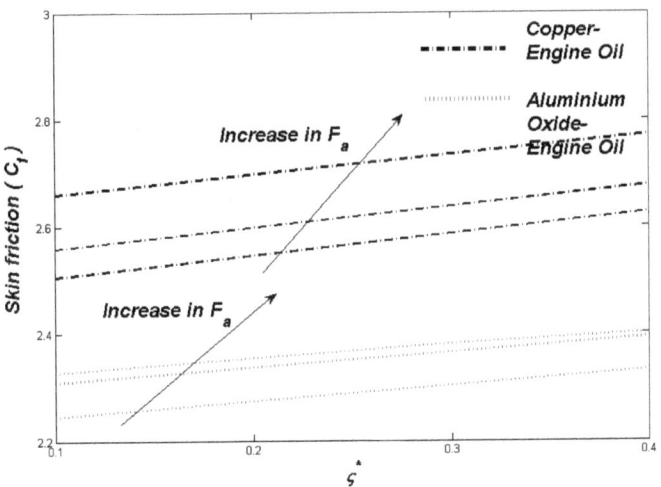

Figure 20. Skin friction C_f against the parameter ς^*.

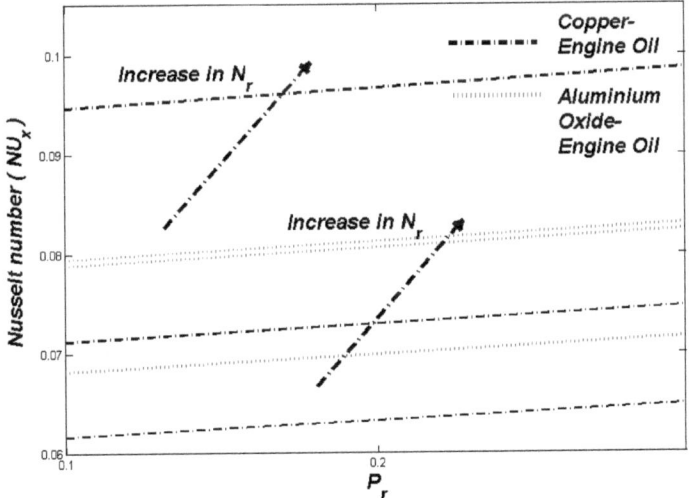

Figure 21. Nusselt number Nu_x against the parameter P_r.

Table 5. Values of $C_f Re_x^{1/2}$ and $Nu_x Re_x^{-1/2}$ for $P_r = 6450$.

$\tau*\tau*$	$\varsigma*\varsigma*$	F_π	ϕ	Λ_π	S	N_π	$v*$	B_π	$C_f Re_x^{\frac{1}{2}}$ Cu-EO	$C_f Re_x^{\frac{1}{2}}$ Al$_2$O$_3$-EO	$N_u Re_x^{\frac{-1}{2}}$ Cu-EO	$N_u Re_x^{\frac{-1}{2}}$ Al$_2$O$_3$-EO
1	0.2	0.6	0.18	0.3	0.5	0.3	0.2	0.3	5.5179	4.2061	3.5457	3.1859
1.3									5.5322	4.234	3.5785	3.216
1.6									5.5716	4.2695	3.6018	3.2474
	0.2								5.5179	4.2061	3.5457	3.1859
	0.4								5.4921	4.1875	3.5014	3.1522
	0.6								5.4513	4.1645	3.4787	3.139
		0.6							5.5179	4.2061	3.5457	3.1859
		1.6							5.5415	4.2431	3.5251	3.1633
		2.6							5.5823	4.2789	3.5075	3.141
			0.09						5.4565	4.1331	3.4865	3.1236
			0.15						5.484	4.1702	3.5186	3.1542
			0.18						5.5179	4.2061	3.5457	3.1859
				0.1					5.592	4.2609	3.5972	3.2328
				0.2					5.5405	4.2337	3.5649	3.2011
				0.3					5.5179	4.2061	3.5457	3.1859
					0.3				5.4932	4.1722	3.5143	3.1598
					0.5				5.5179	4.2061	3.5457	3.1859
					0.7				5.5416	4.2334	3.5766	3.2145
						0.1			5.5179	4.2061	5.5271	3.1604
						0.3			5.5179	4.2061	3.5457	3.1859
						0.5			5.5179	4.2061	3.5637	3.2194
							0.1		5.5179	4.2061	3.5950	3.2239
							0.2		5.5179	4.2061	3.5457	3.1859
							0.3		5.5179	4.2061	3.5121	3.1565
								0.1	5.5179	4.2061	3.5109	3.1718
								0.3	5.5179	4.2061	3.5457	3.1859
								0.5	5.5179	4.2061	3.5735	3.2274

7. Final Remarks

Investigations were made on HT properties and the entropy formation of P-ENF using a stretchable sheet. The single-phase method was employed to construct a computational model. Various physical parameters extract the results with the variations in energy, entropy, and velocity. The impacts of the thermal conductivity parameter $v*$, the thermal radiative parameter N_π, Prandtl–Eyring parameters $\tau*$ and $\varsigma*$, the velocity slip parameter Λ_π, Biot number B_π, B_Γ, and R_π, as well as nanomolecular size ϕ and porous media parameter F_π were examined in the study. Some of the main developments from the study were: The increment in the size of nanoparticles resulted in amplifying the heat transfer rate in engine oil. According to the analysis, copper nanofluid is a better heat conductor than aluminium oxide nanofluid. Increasing the porous media parameter F_π, thermal radiative flow N_π, size parameter ϕ, and Brinkman number B_Γ, the entropy was also enhanced. However, entropy was diminished with a rise in velocity slip parameter Λ_π. An increment in the porous media parameter resulted in increasing the velocity. At the same time, it decreased with the nanoparticles' size augmentation.

The results obtained from the present study can help future researchers improve the heat effect. Heating systems can be formed using various non-Newtonian nanofluids, including Casson, Carreau, second-grade, Maxwell, micropolar, etc. The efficacy of time-dependent porosity and viscosity along with magneto slip flow can be represented by expanding the study.

Author Contributions: Conceptualization, N.H.A.-H. and M.R.S.; methodology, N.H.A.-H.; software, K.H.A. and A.A.A.; validation, M.A.E.; formal analysis, K.A.A. and M.A.E.; investigation, resources, K.H.A.; visualization, A.A.A. and K.H.A.; data curation, K.A.A., writing—original draft preparation, A.M.A.; writing—review and editing, M.R.S. and K.A.A.; supervision, N.H.A.-H. and M.R.S.; project administration, A.A.A. and N.H.A.-H.; funding acquisition, M.R.S. and N.H.A.-H. All authors have read and agreed to the published version of the manuscript.

Funding: The authors extend their appreciation to the Deputyship for Research & Innovation, Ministry of Education in Saudi Arabia for funding this research work through the project number "IFPNC-006-135-2020" and King Abdulaziz University, DSR, Jeddah, Saudi Arabia.

Institutional Review Board Statement: Not applicable.

Informed Consent Statement: Not applicable.

Data Availability Statement: Not applicable.

Acknowledgments: The great help from King Abdulaziz University's faculty members, including Radi A. Alsulami, Muhyaddin J. H. Rawa, Mashhour A. Alazwari, and Hatem F. Sindi are really appreciated.

Conflicts of Interest: The authors declare no conflict of interest.

References

1. Zhang, X.; Tang, Y.; Zhang, F.; Lee, C.S. A novel aluminum–graphite dual-ion battery. *Adv. Energy Mater.* **2016**, *6*, 1502588. [CrossRef]
2. Hosseini, S.M.; Safaei, M.R.; Goodarzi, M.; Alrashed, A.A.; Nguyen, T.K. New temperature, interfacial shell dependent dimensionless model for thermal conductivity of nanofluids. *Int. J. Heat Mass Transf.* **2017**, *114*, 207–210. [CrossRef]
3. Ahmadi, M.H.; Mohseni-Gharyehsafa, B.; Ghazvini, M.; Goodarzi, M.; Jilte, R.D.; Kumar, R. Comparing various machine learning approaches in modeling the dynamic viscosity of CuO/water nanofluid. *J. Therm. Anal. Calorim.* **2020**, *139*, 2585–2599. [CrossRef]
4. Bahiraei, M.; Salmi, H.K.; Safaei, M.R. Effect of employing a new biological nanofluid containing functionalized graphene nanoplatelets on thermal and hydraulic characteristics of a spiral heat exchanger. *Energy Convers. Manag.* **2019**, *180*, 72–82. [CrossRef]
5. Waqas, H.; Farooq, U.; Khan, S.A.; Alshehri, H.M.; Goodarzi, M. Numerical analysis of dual variable of conductivity in bioconvection flow of Carreau–Yasuda nanofluid containing gyrotactic motile microorganisms over a porous medium. *J. Therm. Anal. Calorim.* **2021**, *145*, 2033–2044. [CrossRef]
6. Wang, X.; Li, C.; Zhang, Y.; Said, Z.; Debnath, S.; Sharma, S.; Yang, M.; Gao, T. Influence of texture shape and arrangement on nanofluid minimum quantity lubrication turning. *Int. J. Adv. Manuf. Technol.* **2021**, 1–16. [CrossRef]
7. Xie, Y.; Meng, X.; Mao, D.; Qin, Z.; Wan, L.; Huang, Y. Homogeneously dispersed graphene nanoplatelets as long-term corrosion inhibitors for aluminum matrix composites. *ACS Appl. Mater. Interfaces* **2021**, *13*, 32161–32174. [CrossRef]
8. Abu-Hamdeh, N.H.; Alsulami, R.A.; Rawa, M.J.; Alazwari, M.A.; Goodarzi, M.; Safaei, M.R. A Significant Solar Energy Note on Powell-Eyring Nanofluid with Thermal Jump Conditions: Implementing Cattaneo-Christov Heat Flux Model. *Mathematics* **2021**, *9*, 2669. [CrossRef]
9. Sajid, T.; Jamshed, W.; Shahzad, F.; Eid, M.R.; Alshehri, H.M.; Goodarzi, M.; Akgül, E.K.; Nisar, K.S. Micropolar fluid past a convectively heated surface embedded with nth order chemical reaction and heat source/sink. *Phys. Scr.* **2021**, *96*, 104010. [CrossRef]
10. Maleki, H.; Safaei, M.R.; Togun, H.; Dahari, M. Heat transfer and fluid flow of pseudo-plastic nanofluid over a moving permeable plate with viscous dissipation and heat absorption/generation. *J. Therm. Anal. Calorim.* **2019**, *135*, 1643–1654. [CrossRef]
11. Shankar, U.; Naduvinamani, N. Magnetized squeezed flow of time-dependent Prandtl-Eyring fluid past a sensor surface. *Heat Transf. Asian Res.* **2019**, *48*, 2237–2261. [CrossRef]
12. Al-Kaabi, W.; Al-Khafajy, D.G.S. Radiation and Mass Transfer Effects on Inclined MHD Oscillatory Flow for Prandtl-Eyring Fluid through a Porous Channel. *Al-Qadisiyah J. Pure Sci.* **2021**, *26*, 347–363. [CrossRef]
13. Hayat, T.; Ullah, I.; Muhammad, K.; Alsaedi, A. Gyrotactic microorganism and bio-convection during flow of Prandtl-Eyring nanomaterial. *Nonlinear Eng.* **2021**, *10*, 201–212. [CrossRef]
14. Waqas, H.; Farooq, U.; Alshehri, H.M.; Goodarzi, M. Marangoni-bioconvectional flow of Reiner–Philippoff nanofluid with melting phenomenon and nonuniform heat source/sink in the presence of a swimming microorganisms. *Math. Methods Appl. Sci.* **2021**. [CrossRef]
15. Alazwari, M.A.; Abu-Hamdeh, N.H.; Goodarzi, M. Entropy Optimization of First-Grade Viscoelastic Nanofluid Flow over a Stretching Sheet by Using Classical Keller-Box Scheme. *Mathematics* **2021**, *9*, 2563. [CrossRef]
16. Munjam, S.R.; Gangadhar, K.; Seshadri, R.; Rajeswar, M. Novel technique MDDIM solutions of MHD flow and radiative Prandtl-Eyring fluid over a stretching sheet with convective heating. *Int. J. Ambient. Energy* **2021**, 1–10. [CrossRef]
17. Jamshed, W.; Kumar, V.; Kumar, V. Computational examination of Casson nanofluid due to a non-linear stretching sheet subjected to particle shape factor: Tiwari and Das model. *Numer. Methods Partial. Differ. Equ.* **2020**. [CrossRef]
18. Jamshed, W.; Mishra, S.; Pattnaik, P.; Nisar, K.S.; Devi, S.S.U.; Prakash, M.; Shahzad, F.; Hussain, M.; Vijayakumar, V. Features of entropy optimization on viscous second grade nanofluid streamed with thermal radiation: A Tiwari and Das model. *Case Stud. Therm. Eng.* **2021**, *27*, 101291. [CrossRef]

19. Abdelmalek, Z.; Hussain, A.; Bilal, S.; Sherif, E.-S.M.; Thounthong, P. Brownian motion and thermophoretic diffusion influence on thermophysical aspects of electrically conducting viscoinelastic nanofluid flow over a stretched surface. *J. Mater. Res. Technol.* **2020**, *9*, 11948–11957. [CrossRef]
20. Maleki, H.; Alsarraf, J.; Moghanizadeh, A.; Hajabdollahi, H.; Safaei, M.R. Heat transfer and nanofluid flow over a porous plate with radiation and slip boundary conditions. *J. Cent. South Univ.* **2019**, *26*, 1099–1115. [CrossRef]
21. Maleki, H.; Safaei, M.R.; Alrashed, A.A.; Kasaeian, A. Flow and heat transfer in non-Newtonian nanofluids over porous surfaces. *J. Therm. Anal. Calorim.* **2019**, *135*, 1655–1666. [CrossRef]
22. Jamshed, W.; Goodarzi, M.; Prakash, M.; Nisar, K.S.; Zakarya, M.; Abdel-Aty, A.-H. Evaluating the unsteady casson nanofluid over a stretching sheet with solar thermal radiation: An optimal case study. *Case Stud. Therm. Eng.* **2021**, *26*, 101160. [CrossRef]
23. Xiong, C.-B.; Yu, L.-N.; Niu, Y.-B. Effect of variable thermal conductivity on the generalized thermoelasticity problems in a fiber-reinforced anisotropic half-space. *Adv. Mater. Sci. Eng.* **2019**, *2019*, 8625371. [CrossRef]
24. Ibrahim, W.; Negera, M. Viscous dissipation effect on mixed convective heat transfer of MHD flow of Williamson nanofluid over a stretching cylinder in the presence of variable thermal conductivity and chemical reaction. *Heat Transf.* **2021**, *50*, 2427–2453. [CrossRef]
25. Dada, M.S.; Onwubuoya, C. Variable viscosity and thermal conductivity effects on Williamson fluid flow over a slendering stretching sheet. *World J. Eng.* **2020**, *17*, 357–371. [CrossRef]
26. Hasona, W.; Almalki, N.; ElShekhipy, A.; Ibrahim, M. Combined effects of variable thermal conductivity and electrical conductivity on peristaltic flow of pseudoplastic nanofluid in an inclined non-Uniform asymmetric channel: Applications to solar collectors. *J. Therm. Sci. Eng. Appl.* **2020**, *12*, 021018. [CrossRef]
27. Fatunmbi, E.O.; Okoya, S.S. Quadratic Mixed Convection Stagnation-Point Flow in Hydromagnetic Casson Nanofluid over a Nonlinear Stretching Sheet with Variable Thermal Conductivity. In *Defect and Diffusion Forum*; Trans Tech Publications Ltd.: Bäch, Switzerland, 2021; pp. 95–109.
28. Aziz, A.; Shams, M. Entropy generation in MHD Maxwell nanofluid flow with variable thermal conductivity, thermal radiation, slip conditions, and heat source. *AIP Adv.* **2020**, *10*, 015038. [CrossRef]
29. Qureshi, M.A. A case study of MHD driven Prandtl-Eyring hybrid nanofluid flow over a stretching sheet with thermal jump conditions. *Case Stud. Therm. Eng.* **2021**, *28*, 101581. [CrossRef]
30. Becker, E. Simple non-Newtonian fluid flows. *Adv. Appl. Mech.* **1980**, *20*, 177–226.
31. Jamshed, W.; Aziz, A. A comparative entropy based analysis of Cu and Fe_3O_4/methanol Powell-Eyring nanofluid in solar thermal collectors subjected to thermal radiation, variable thermal conductivity and impact of different nanoparticles shape. *Results Phys.* **2018**, *9*, 195–205. [CrossRef]
32. Jamshed, W.; Nasir, N.A.A.M.; Isa, S.S.P.M.; Safdar, R.; Shahzad, F.; Nisar, K.S.; Eid, M.R.; Abdel-Aty, A.-H.; Yahia, I. Thermal growth in solar water pump using Prandtl–Eyring hybrid nanofluid: A solar energy application. *Sci. Rep.* **2021**, *11*, 1–21. [CrossRef] [PubMed]
33. Khan, M.I.; Khan, S.A.; Hayat, T.; Khan, M.I.; Alsaedi, A. Nanomaterial based flow of Prandtl-Eyring (non-Newtonian) fluid using Brownian and thermophoretic diffusion with entropy generation. *Comput. Methods Programs Biomed.* **2019**, *180*, 105017. [CrossRef] [PubMed]
34. Aziz, A.; Jamshed, W.; Aziz, T. Mathematical model for thermal and entropy analysis of thermal solar collectors by using Maxwell nanofluids with slip conditions, thermal radiation and variable thermal conductivity. *Open Phys.* **2018**, *16*, 123–136. [CrossRef]
35. Brewster, M.Q. *Thermal Radiative Transfer and Properties*; John Wiley & Sons: Hoboken, NJ, USA, 1992.
36. Jamshed, W. Numerical investigation of MHD impact on Maxwell nanofluid. *Int. Commun. Heat Mass Transf.* **2021**, *120*, 104973. [CrossRef]
37. Waqas, H.; Hussain, M.; Alqarni, M.; Eid, M.R.; Muhammad, T. Numerical simulation for magnetic dipole in bioconvection flow of Jeffrey nanofluid with swimming motile microorganisms. *Waves Random Complex Media* **2021**, 1–18. [CrossRef]
38. Iqbal, Z.; Azhar, E.; Maraj, E. Performance of nano-powders SiO_2 and SiC in the flow of engine oil over a rotating disk influenced by thermal jump conditions. *Phys. A Stat. Mech. Appl.* **2021**, *565*, 125570. [CrossRef]
39. Mohamed, M.K.A.; Ong, H.R.; Alkasasbeh, H.T.; Salleh, M.Z. Heat Transfer of Ag-Al_2O_3/Water Hybrid Nanofluid on a Stagnation Point Flow over a Stretching Sheet with Newtonian Heating. In *Journal of Physics: Conference Series*; IOP Publishing: Bristol, UK, 2020; p. 042085. [CrossRef]
40. Keller, H.B. A new difference scheme for parabolic problems. In *Numerical Solution of Partial Differential Equations–II*; Elsevier: Amsterdam, The Netherlands, 1971; pp. 327–350.
41. Hussain, S.M.; Jamshed, W. A comparative entropy based analysis of tangent hyperbolic hybrid nanofluid flow: Implementing finite difference method. *Int. Commun. Heat Mass Transf.* **2021**, *129*, 105671. [CrossRef]
42. Wang, C. Free convection on a vertical stretching surface. *ZAMM-J. Appl. Math. Mech./Z. Angew. Math. Und Mech.* **1989**, *69*, 418–420. [CrossRef]
43. Gorla, R.S.R.; Sidawi, I. Free convection on a vertical stretching surface with suction and blowing. *Appl. Sci. Res.* **1994**, *52*, 247–257. [CrossRef]
44. Khan, W.; Pop, I. Boundary-layer flow of a nanofluid past a stretching sheet. *Int. J. Heat Mass Transf.* **2010**, *53*, 2477–2483. [CrossRef]
45. Makinde, O.D.; Aziz, A. Boundary layer flow of a nanofluid past a stretching sheet with a convective boundary condition. *Int. J. Therm. Sci.* **2011**, *50*, 1326–1332. [CrossRef]

Article

Simulation of Natural Convection in a Concentric Hexagonal Annulus Using the Lattice Boltzmann Method Combined with the Smoothed Profile Method

Suresh Alapati

School of Mechanical & Mechatronics Engineering, Kyungsung University, 309, Suyeong-ro (Daeyeon-dong), Nam-gu, Busan 48434, Korea; sureshalapatimech@gmail.com or suresh@ks.ac.kr; Tel.: +82-51-663-4690

Received: 3 June 2020; Accepted: 23 June 2020; Published: 26 June 2020

Abstract: This research work presents results obtained from the simulation of natural convection inside a concentric hexagonal annulus by using the lattice Boltzmann method (LBM). The fluid flow (pressure and velocity fields) inside the annulus is evaluated by LBM and a finite difference method (FDM) is used to get the temperature filed. The isothermal and no-slip boundary conditions (BC) on the hexagonal edges are treated with a smooth profile method (SPM). At first, for validating the present simulation technique, a standard benchmarking problem of natural convection inside a cold square cavity with a hot circular cylinder is simulated. Later, natural convection simulations inside the hexagonal annulus are carried out for different values of the aspect ratio, AR (ratio of the inner and outer hexagon sizes), and the Rayleigh number, Ra. The simulation results are presented in terms of isotherms (temperature contours), streamlines, temperature, and velocity distributions inside the annulus. The results show that the fluid flow intensity and the size and number of vortex pairs formed inside the annulus strongly depend on AR and Ra values. Based on the concentric isotherms and weak fluid flow intensity at the low Ra, it is observed that the heat transfer inside the annulus is dominated by the conduction mode. However, multiple circulation zones and distorted isotherms are observed at the high Ra due to the strong convective flow. To further access the accuracy and robustness of the present scheme, the present simulation results are compared with the results given by the commercial software, ANSYS-Fluent®. For all combinations of AR and Ra values, the simulation results of streamlines and isotherms patterns, and temperature and velocity distributions inside the annulus are in very good agreement with those of the Fluent software.

Keywords: lattice Boltzmann method; smoothed profile method; hybrid method; natural convection simulation; concentric hexagonal annulus

1. Introduction

Natural convection heat transfer in an annular space between two concentric cylinders (also known as concentric annuli) is one of the most studied problems in the field of heat transfer. This fundamental problem has attracted many researchers because of its significance in many engineering applications such as the design of heat exchanger devices, solar energy collectors, cooling of electric power cables, nuclear and chemical reactors, food processing devices, aircraft cabin insulation, etc. [1,2]. For early research works of theoretical and/or experimental investigations on natural convection in an annular space between cold outer and hot inner cylinders, one can find the literature [1–8]. Studying the behavior of natural convection flow in an annulus with irregular geometries (other than the square or rectangular such as circular, elliptical, triangular, and hexagonal) by using the numerical simulations is highly challenging due to complex irregular boundaries. The irregular boundary problems are generally treated with unstructured (body-fitted) grid methods, also known as conforming-mesh methods, which are very complicated and are computationally intensive. In the past two decades, many researchers have paid attention to develop non-conforming-mesh methods, which use a fixed Cartesian grid to simulate the fluid flow in complex geometries. Some of those numerical methods are immersed boundary method (IBM) [9], distributed Lagrange multiplier method or fictitious domain method [10], and smoothed profile method (SPM) [11–13].

In the past three decades, the lattice Boltzmann method (LBM) has evolved as a powerful computational technique for solving fluid flow and heat transfer problems. In LBM, one gets the solution for the particle density distribution functions (PDF) (by solving Boltzmann kinetic equation on a discrete lattice mesh) instead of directly solving the pressure and velocity fields. The macroscopic variables (such as pressure, velocity, and temperature) are obtained by calculating the hydrodynamic moments of PDF [14]. Because of its many advantages [15] compared to the classical Navier–Stokes equations solvers, LBM has been successfully used to simulate various Multiphysics problems such as multiphase flows [15–17], magnetohydrodynamic (MHD) flows [18,19], micro- and nano-flows [20–22] and fluid-solid interactions [13,23–26]. LBM has also been successfully implemented to predict the behavior of the fluid flow due to natural convection in complex geometries [27–37]. In the above-mentioned research works on natural convection, the following techniques have been used to handle the no-slip and constant temperature BC on complex irregular surfaces: the bounce back (BB) scheme [27–32], IBM [33–35], and SPM [36,37].

BB rule was first proposed by the Ladd [38,39] to impose the no-slip BC at curved surfaces of solid particles. In this scheme, the irregular surface of a solid body is imagined as a flat edge that lies in-between two neighboring solid and fluid grid points. The no-slip BC can then be achieved with the help of the standard mid-plane BB scheme which bounces back the missing distribution functions coming from the solid nodes to the fluid nodes. Later, Bouzidi et al. [40] and Yu et al. [41] developed an improved version of the Ladd scheme, known as the interpolated bounce back (IBB) scheme to achieve the second-order accuracy for the fluid velocity and temperature. Sheikholeslami et al. [27,28] investigated MHD flow and heat transfer in an annular space between a heated inner circular and a cold outer square cylinder and they reported the fluid flow and heat transfer results at various Ra and AR values. Lin et al. [29] performed simulation of natural convection flow in an annulus between a heated inner circular cylinder, which is located eccentrically, and a cold square enclosure. Bararnia et al. [30] simulated the natural convection between a heated inner elliptical cylinder and a square outer cylinder. They reported the fluid flow and heat transfer characteristics for various combinations of the vertical positions of the inner cylinder and Ra. Sheikholeslami et al. [31] studied the effect of a magnetic field on the fluid flow and heat transfer characteristics of a nanofluid inside a circular cylinder with an inner triangular cylinder. Moutaouakil et al. [32] conducted lattice Boltzmann simulations of natural convection in an annulus between an inner hexagonal cylinder and an outer square cavity. In all the above-mentioned articles [27–32], IBB scheme was used to treat the complex boundaries of circular, triangular, elliptical, and hexagonal geometries. Even though IBB scheme can effectively be used for treating the complex curved boundary problems, the main drawback is that there may be fluctuations in the velocity and temperature fields at fluid-solid interfaces especially when the solid boundary is moving with a certain velocity.

In IBM, the complex irregular boundaries of solid bodies are represented with a set of Lagrangian nodes while the evaluation of the fluid flow is considered on a fixed Eulerian grid. To enforce the no-slip and constant temperature BC on the solid nodes, artificial body force and heat source terms are added to fluid momentum and energy equations, respectively. On can refer to the review article by Mittal and Iaccarino [42] for a clear discussion on different approaches for calculating the body force terms. Hu et al. [33] simulated natural convection in a concentric annulus of circular cylinders using LBM and they used IBM to treat the no-slip and isothermal BC on the curved boundaries. Hu et al. [34] developed an immersed boundary lattice Boltzmann method (IBLBM) for simulating fluid flow due to natural convection in a cold square cavity with a heated inner circular cylinder covered by a porous layer. They investigated the effects of thermal conductivity ratio, Ra, and Darcy number on the behavior of fluid flow and heat transfer. Khazaeli et al. [35] used IBLBM to simulate the natural convection due to a hot circular cylinder inside a square and circular enclosures (cold) for different Ra values. IBM could resolve the problem of fluctuations in the velocity field of IBB scheme. However, the main disadvantage of IBM is that it requires complex interpolation functions, and needs a lot of data exchange between the fluid (Eulerian) and solid (Lagrangian) nodes. Therefore, the parallel computational performance of the scheme based on the LBM and IBM becomes lower as the global data communication between the neighboring grid points increases [13].

In SPM, a smoothed profile function is used to recognize the complex surfaces of a solid body, and the same grid system is used for fluid and solid. The no-slip and isothermal BC at the complex surfaces are implemented by adding a hydrodynamic force and heat source terms to the fluid momentum and energy equations, respectively. The main advantage of SPM over the other non-conforming-mesh methods is that all operations are completely local to a grid point as both fluid and solid are represented with the same grid system; so, the implementation of this scheme to parallel computing applications is easier [13]. Also, SPM does not need any complex interpolation function as needed by IBM. Although SPM is computationally more efficient and easier to apply than IBB and IBM schemes, till now, only a few researchers have used the method based on LBM and SPM to study the fluid flow and heat transfer behavior in complex boundaries. Hu et al. [36] used the LBM combined with SPM for simulating natural convection in complex irregular geometries. They reported the simulation results for the velocity and temperature inside a square enclosure with a hot circular cylinder for different values of Ra and AR. All the above-mentioned research works [27–36] considered the double populations model (DPM) (where two sets of PDF are used: one set for solving the velocity field and another one for the temperature field). Recently, Alapati et al. [37] developed a numerical technique based on the combination of LBM and SPM to simulate particulate flows with heat transfer. They used a hybrid method (HM), which solves the fluid flow by LBM with a set of PDF and the temperature field by FDM and concluded that LBM-SPM method based on HM is computationally more efficient than the method based on DPM.

Fluid flow and heat transfer thorough or over hexagonal-shaped geometries is a ubiquitous problem in many engineering applications such as solar energy collectors [43,44], nuclear power plants [45], microfluidic heat sinks [46], lamella type compact heat exchangers [47], air-conditioning applications [48], etc. In solar energy collectors, to minimize the radiation and convection losses to the surrounding atmosphere, an array of transparent tubes, arranged in a hexagonal honeycomb pattern, is used in-between the absorbing surface and cover plate. Marshall et al. [43] and Buchberg et al. [44] found that the thermal efficiencies of honeycomb solar collectors were higher compared to the collectors without the honeycomb layer. The fuel rods of a nuclear reactor core are stacked in the form of a hexagonal lattice and are located inside a circular or hexagonal channel to pass a coolant longitudinally over them [45]. In a ministered heat sink used for electronic systems cooling, an array of pin-fins of various cross-sectional shapes are attached to a microchannel wall. Aliabadia et al. [44] found that hydrothermal (hydraulic and thermal) performance was best for the pin-fins with the circular and hexagonal cross-sections. Hexagonal duct shape is a commonly used shape in lamella type compact heat exchangers, which is used in many industries such as pulp and paper, alcohol, petrochemical, and other chemical industries [47]. A desiccant disk used in air-conditioning applications consists of an array of several ducts packed in the form of honeycomb pattern. Zhang [48] found that the heat and mass transfer efficiencies of ducts with hexagonal cross-section were higher compared to circular and rectangular ducts as the hexagonal duct walls are more uniformly placed in the desiccant wheel.

Even though it has great significance and applications only a few researchers have investigated the behavior of natural convection flow in an annulus with a heated hexagonal cylinder. Boyd [3] experimentally investigated natural convection in an annulus with a heated hexagonal inner cylinder and cold outer circular cylinder. Raithby et al. [6] simulated the natural convection in an annulus bounded by a circular cylinder outside and horizontal hexagonal cylinder inside by using an orthogonal curvilinear coordinates system (a body-fitted grid system). Galkape and Asfaw [7] employed a non-orthogonal coordinate system to study the same problem of Raithby et al. [6]. More recently,

Moutaouakil et al. [32] studied the natural convection due to a hot hexagonal cylinder inside a square enclosure (cold) with LBM. They presented the fluid flow and heat transfer characteristics for different combinations of AR and Ra by considering two orientations for the hexagonal cylinder. As mentioned earlier, they used DPM with IBB scheme to treat BC on hexagonal surfaces.

The literature review showed that the problem of natural convection in an annulus space between two hexagonal cylinders (concentric with each other) has not been studied. The main objective of this work is to solve the problem of two-dimensional natural convection flow caused by a hot hexagonal cylinder placed concentrically inside a cold hexagonal cylinder. The numerical technique that combines LBM, SPM, and FDM methods is employed because it offers many advantages over the other methods. In the present work, an equation for smoothed profile function that identifies the hexagonal boundaries is proposed. Assessing the accuracy and robustness of the present simulation technique by comparing results given by the present method with ANSYS-Fluent® results is also another purpose of this work. The remainder of this paper is arranged as follows. A brief description of the simulation technique is given in Section 2. A discussion on numerical results is presented in Section 3. At first, the validation results, by applying the present simulation scheme to a standard benchmarking test, are provided. Later, the simulation results of streamlines, isotherms, and temperature and velocity distributions inside the concentric hexagonal annulus are presented. The concluding remarks of the present study are provided in Section 4.

2. Numerical Method

Figure 1 shows the simulation set-up, which consists of an annulus region formed by two concentric horizontal hexagonal cylinders of different sizes, considered in the present work. Simulations are performed inside a square enclosure and 251 × 251 lattice grid points are used to divide the computational domain. The center positions of the two hexagonal cylinders are fixed at the center of the enclosure. L_{out} in the figure represents the distance between two opposite sides of the outer hexagonal cylinder (the size of the outer cylinder). Throughout the simulations, L_{out} is kept constant at $L_{out} = 212$ and the size for the inner hexagonal cylinder, L_{in} in Figure 1, is varied based on the aspect ratio, which is defined as: $AR = L_{in}/L_{out}$. g in the figure denotes the gravitational acceleration constant.

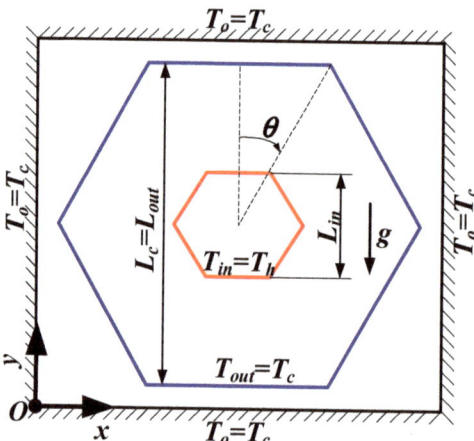

Figure 1. Simulation set-up for investigating the natural convection in an annulus region between two concentric hexagonal cylinders. The origin of the lattice grid is mentioned by 'O' in the figure.

The initial values for the fluid velocity and temperature inside the domain are set to zero. The no-slip and constant temperature BC ($T_o = T_c \equiv 0$) are applied at the enclosure walls for the flow field and temperature field, respectively, and the standard mid-plane BB scheme is used to treat the no-slip boundary condition. The temperature value for all edges/sides of the outer hexagonal cylinder is kept constant at, $T_{out} \equiv T_c = 0$ (cold surface), and that for the inner one is fixed at $T_{in} \equiv T_h = 1$ (hot surface). The constant temperature and no-slip BC on all sides of the inner and outer cylinder are treated with SPM. In below, the formulations for LBM for solving the fluid flow field, FDM for solving the temperature field, and SPM for treating BC on hexagonal edges are provided.

2.1. Solving Fluid Flow Using LBM

In this work, as mentioned earlier, LBM is used to obtain the flow field due to natural convection. In LBM, the macroscopic variables such as fluid pressure and the velocity, are computed from the fluid-PDF, $f_n(\mathbf{x}, t)$, which are evaluated by solving the Boltzmann kinetic equation on a discrete lattice mesh. Here, $f_n(\mathbf{x}, t)$ is the probability of finding a fluid particle at a lattice position, $\mathbf{x}$, and at a time, t, moving with a discrete velocity, $\mathbf{c}_n$ (the subscript n indicates the PDF number), which is selected in such a way that after time step Δt, the particle arrives at the nth neighboring grid point [21]. The single-relaxation-time lattice Boltzmann equation (LBE) with external body force term is given by [37]

$$f_n(\mathbf{x} + \mathbf{c}_n \Delta t, t + \Delta t) = f_n(\mathbf{x}, t) - \frac{1}{\lambda}\big(f_n(\mathbf{x}, t) - f_n^{eq}(\mathbf{x}, t)\big) + \frac{w_n \Delta t}{c_s^2}\big(\big(\mathbf{f}^{th}(\mathbf{x}, t) + \mathbf{f}^{fl}(\mathbf{x}, t)\big) \cdot \mathbf{c}_n\big) \quad (1)$$

where λ is the relaxation time, $f_n^{eq}(\mathbf{x}, t)$ is the equilibrium distribution functions, w_n is the weighing function, and c_s is the sound speed. $\mathbf{f}^{th}(\mathbf{x}, t)$ and $\mathbf{f}^{fl}(\mathbf{x}, t)$ in Equation (1) represent the buoyancy force and the hydrodynamic force (due to the no-slip BC on the hexagonal surfaces) source terms, respectively. Through the Chapman–Enskog analysis, the above equation recovers the Navier–Stokes equations in the low Mach number limit, $|\mathbf{u}|/c_s \ll 1$ [14]. The relation between the fluid kinematic viscosity, ν, and relaxation time, λ, is given by

$$\nu = c_s^2 \Delta t \left(\lambda - \frac{1}{2}\right). \quad (2)$$

To model the buoyancy force, $\mathbf{f}^{th}(\mathbf{x}, t)$, the Boussinesq approximation is used as follows

$$\mathbf{f}^{th} = \rho_0 \beta (T - T_0) g \, \hat{\mathbf{j}}, \quad (3)$$

where ρ_0 is the initial value for fluid density, T_0 is the initial fluid temperature, β is the fluid thermal expansion coefficient at T_0, T is the fluid temperature field, and $\hat{\mathbf{j}}$ is the unit vector in the vertical direction (y – direction). The hydrodynamic force term, $\mathbf{f}^{fl}(\mathbf{x}, t)$ of Equation (1) is obtained with SPM. After solving for $f_n(\mathbf{x}, t)$, the fluid density and the velocity fields, $\rho(\mathbf{x}, t)$ and $\mathbf{u}(\mathbf{x}, t)$, are obtained from

$$\rho(\mathbf{x}, t) = \sum_{n=0}^{b} f_n, \quad \mathbf{u}(\mathbf{x}, t) = \frac{1}{\rho} \sum_{n=0}^{b} f_n \mathbf{c}_n. \quad (4)$$

2.2. Solving Temperature Distribution with FDM

The temperature distribution inside the computational domain is obtained by discretizing the energy equation with FDM by using the standard central difference scheme in space and the forward difference scheme in time [37]. After discretization, the equation for finding the temperature value at a grid point in new time level is (q in the below expression represents the heat source term due to the constant temperature BC on the hexagonal edges)

$$T_{i,j}^{new} = T_{i,j}^{old} + (\text{RHS}^{old} + q_{i,j}^{old}) \Delta t, \quad (5)$$

with

$$\text{RHS}^{\text{old}} = -4\alpha T^{\text{old}}_{i,j} + \left(\alpha - \tfrac{1}{2}u^{\text{old}}_{i+1,j}\right)T^{\text{old}}_{i+1,j} + \left(\alpha + \tfrac{1}{2}u^{\text{old}}_{i-1,j}\right)T^{\text{old}}_{i-1,j} \\ + \left(\alpha - \tfrac{1}{2}v^{\text{old}}_{i,j+1}\right)T^{\text{old}}_{i,j+1} + \left(\alpha + \tfrac{1}{2}v^{\text{old}}_{i,j-1}\right)T^{\text{old}}_{i,j-1} \quad (6)$$

In the above equation, α is the thermal diffusivity, and the subscripts i & j represent lattice grid indices in $x-$ & $y-$ directions, respectively.

2.3. Evaluation of $\mathbf{f}^{fl}(\mathbf{x},t)$ and $q(\mathbf{x},t)$ with SPM

In SPM, a smoothed profile function (also termed as concentration function or indicator function), $\phi_k(\mathbf{x},t)$, is used to identify the solid regions [13] (here, k is the index value for the hexagonal cylinders; $k = 1$ for the inner hexagon and $k = 2$ outer one). The equation for $\phi_k(\mathbf{x},t)$ is defined in such a way that $\phi_k = 0$ in the fluid region, $\phi_k = 1$ in the solid region, and ϕ_k smoothly varies from 0 to 1 at the fluid-solid interface. Here, the following equation is used to evaluate $\phi_k(\mathbf{x},t)$ of each hexagon

$$\phi_k(\mathbf{x},t) = f(\mathbf{d}_k(\mathbf{x},t)), \quad (7a)$$

$$f(\mathbf{d}_k(\mathbf{x},t)) = \begin{cases} 0 & r < -\xi_k/2 \\ \tfrac{1}{2}\left(\sin\left(\pi\tfrac{d_k(\mathbf{x},t)}{\xi_k}\right) + 1\right) & |r| < \xi_k/2 \\ 1 & r > \xi_k/2 \end{cases}, \quad (7b)$$

where $\mathbf{d}_k(\mathbf{x},t)$ and ξ_k are the signed normal distance function to the solid surface (here, edges of two hexagons) and the interface thickness of each hexagon, respectively. Unless otherwise mentioned, throughout this work, the values for the interface thickness for the two hexagons are chosen as, $\xi_1 = \xi_2 \equiv 0.5$. The following equation is used for finding $\mathbf{d}_k(\mathbf{x},t)$ of each hexagon

$$\mathbf{d}_k(\mathbf{x},t) = \max\left\{\frac{L_{x,k}}{2}, \max\left[\left(|x - X_{c,k}|\sin(30°) + |y - Y_{c,k}|\cos(30°)\right), \left(|y - Y_{c,k}| - \frac{L_{y,k}}{2}\right)\right]\right\} \quad (8)$$

$L_{x,k}$ and $L_{y,k}$ in the above equation are the distance between two corners and two opposite sides of each hexagon, respectively ($L_{y,1} = L_{in}$ and $L_{y,2} = L_{out}$ are the sizes of the two hexagons), and $X_{c,k}$ and $Y_{c,k}$ are the center positions of hexagons, in $x-$ and $y-$ directions, respectively. The smoothly distributed concentration field of two hexagons, $\phi(\mathbf{x},t)$, is obtained by adding the $\phi_k(\mathbf{x},t)$ values of two hexagons

$$\phi(\mathbf{x},t) = \sum_{k=1}^{2} \phi_k(\mathbf{x},t). \quad (9)$$

The body force term for enforcing no-slip BC on hexagonal edges, $\mathbf{f}^{fl}(\mathbf{x},t)$ in Equation (1), is evaluated by using

$$\mathbf{f}^{fl}(\mathbf{x},t) = [\mathbf{u}_P(\mathbf{x},t) - \mathbf{u}(\mathbf{x},t)]\phi(\mathbf{x},t)/\Delta t, \quad (10)$$

$\mathbf{u}_P(\mathbf{x},t)$ in the above equation is the velocity field for the solid regions, which is zero as the two hexagonal cylinders are stationary. Similarly, the heat source term for treating the constant temperature BC, $q(\mathbf{x},t)$ of Equation (5), can be obtained by

$$q(\mathbf{x},t) = [T_P(\mathbf{x},t) - T(\mathbf{x},t)]\phi(\mathbf{x},t)/\Delta t, \quad (11)$$

where $T_P(\mathbf{x},t)$ is the hexagonal cylinders temperature field, which is evaluated by using

$$\phi(\mathbf{x},t)T_P(\mathbf{x},t) = \sum_{k=1}^{2} \phi_k(\mathbf{x},t)T_k(t), \quad (12)$$

where $T_k(t)$ is the temperature of each hexagon; $T_1(t) = T_{in} \equiv 1$ for hot inner cylinder and $T_2(t) = T_{out} \equiv 0$ the cold outer cylinder.

3. Simulation Results

3.1. Validation

The standard benchmarking problem of natural convection due to the hot circular cylinder inside a square enclosure [36,49] is chosen to validate the numerical code developed based on the present simulation technique (see Figure 2 for simulation set-up). The following equation is considered to fix the kinematic viscosity, ν [37]

$$\nu = \sqrt{\frac{Pr}{Ra}} U_c L_c, \tag{13}$$

where Pr, U_c, and L_c are the Prandtl number, the characteristic velocity, and the characteristic length, respectively, and the definitions for Ra, Pr, and U_c are given by

$$\text{Ra} = \frac{g\beta\Delta T L_c^3}{\nu\alpha}, \quad \text{Pr} = \frac{\nu}{\alpha}, \text{ and } U_c = \sqrt{g\beta\Delta T L_c}, \tag{14}$$

ΔT in the above equation is the temperature difference between the inner circular cylinder (hot, $T_{in} = T_h \equiv 1$) and outer square cavity (cold, $T_{out} = T_h \equiv 1$). The computational domain is divided into 201 × 201 lattice grid points. The simulations are performed for, Pr = 0.71 (i.e., heat transferring medium is air), $U_c = 0.1$, and $L_c = L_{out}$ (size of the outer square cavity). Three different combinations for Rayleigh number, Ra = 10^4, 10^5, and 10^6, and the aspect ratio, $AR = L_{in}/L_{out} \equiv 0.2, 0.4,$ and 0.6 are considered. Figure 3 shows the isotherms (left side) and streamlines (right side) inside the enclosure when Ra = 10^6 and $AR = 0.2$. Two symmetrical vortices appear in the upper region of the enclosure, as the natural convection flow intensity is predominant in the upper region of the cavity due to Ra value is very high. The details of the surface-averaged Nusselt number, $\overline{\text{Nu}}$, on the inner cylinder at all combinations of Ra and AR considered in the present work are provided in Table 1. The corresponding results obtained by the previous works [36,49] are also given in Table 1. $\overline{\text{Nu}}$ increases with Ra for all values of AR. The streamlines and the isotherms patterns, and $\overline{\text{Nu}}$ values of all Ra and AR combinations are in excellent agreement with the previous results [33,35]. After this validation, simulation of fluid flow and heat transfer due to natural convection inside the hexagonal annulus is performed and the corresponding results are discussed in the following section.

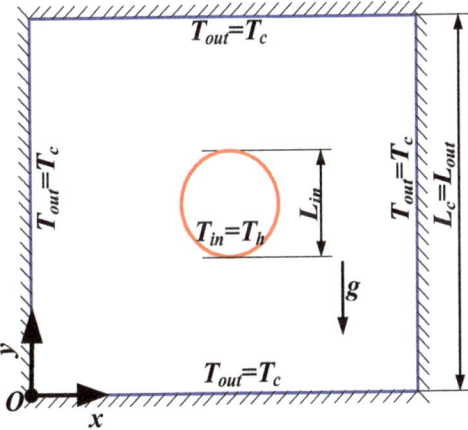

Figure 2. Set-up, consists of an annular space between an outer square enclosure (cold) and an inner circular cylinder (hot), considered for validating the present numerical method.

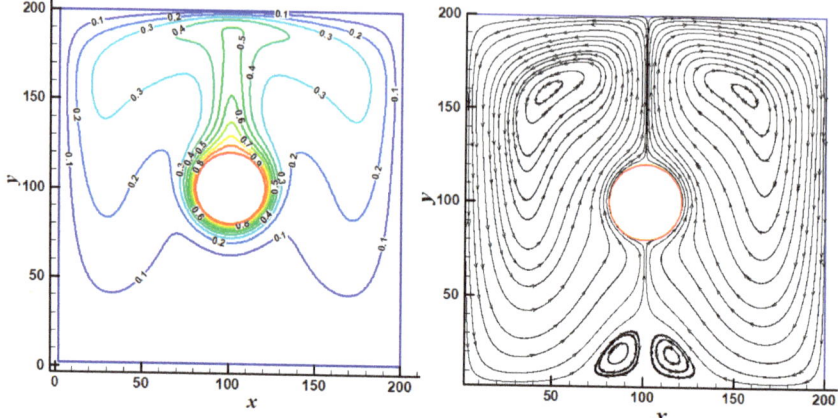

Figure 3. Isotherms (left side) and streamlines (right side) patterns inside the square enclosure when Rayleigh number is, Ra = 10^6, and aspect ratio, AR = 0.2.

Table 1. The details of the surface average Nusselt number, $\overline{Nu}$, on the inner cylinder at different combinations of Ra and AR.

Ra	AR	Present	Previous [36]	Previous [49]
10^4	0.2	2.042	2.035	2.071
	0.4	3.202	3.173	3.331
	0.6	5.349	5.266	5.826
10^5	0.2	3.714	3.751	3.825
	0.4	4.843	4.893	5.080
	0.6	6.182	6.175	6.212
10^6	0.2	5.959	6.115	6.107
	0.4	8.718	8.897	9.374
	0.6	11.662	11.940	11.620

3.2. Natural Convection in the Concentric Hexagonal Annulus

In this session, the results obtained by the simulation of the fluid flow and heat transfer in the annulus bounded by two horizontal concentric hexagonal cylinders are presented (set-up is shown in Figure 1). Simulations are performed for different values of AR and Ra, by varying AR in the range, AR = 0.2~0.6, and Ra in the range, Ra = $10^3 \sim 10^6$. The characteristic length in Equation (14) is set as, $L_c = L_{out}$ (the size of the outer hexagon).

All the results obtained by the present simulation technique are compared with those given by commercial software, ANSYS-Fluent®. Fluent 18.2 is used to simulate a steady laminar flow and heat transfer inside a two-dimensional annular space bounded by two concentric hexagonal cylinders. The size of the outer cylinder is fixed at L_{out} = 212 m and that of the inner cylinder is varied as per the AR. The values for the temperatures at the walls of the inner and outer cylinder are set at T_{in} = 289 K and T_{out} = 288 K, respectively. Constant temperature and no-slip BC are used for heat transfer and fluid flow, respectively. The initial value for density is taken as ρ_0 = 1.225 kg/m^3 (air density value at temperature 288 K) and the Boussinesq model is used to model the variation of the density as a function of temperature. Dry air properties at temperature 288 K are used to set the values for specific heat, viscosity, thermal conductivity, and thermal expansion coefficient. The value for the y– directional gravitational acceleration constant is varied corresponds to Ra. SIMPLE (Semi-Implicit Method for Pressure Linked Equation) scheme has opted for the pressure-velocity coupling. The governing equations are discretized using the least square cell-based method and a second-order upwind scheme is chosen to solve momentum and energy equations. The Gauss–Seidal iterative method with default under-relaxation factors is selected to solve the system of algebraic equations. The convergence criterion for the residuals of all continuity, momentum, and energy equations is set as 10^{-9}.

3.2.1. Streamlines and Isotherms Patterns Inside the Annulus

It is observed from the simulation results that irrespective of AR and Ra values, the fluid flow patterns (streamlines) and temperature contours (isotherms) are symmetrical about the vertical centerline of the annulus. Figure 4 shows the isotherms (left side) and streamlines (right side) pattern inside the annulus for the two values of $AR = 0.2$ (Figure 4a) and $AR = 0.6$ (Figure 4b), when Ra $= 10^3$. As Ra is very low, the strength of the buoyancy force (strength of the gravitational acceleration in this case) that causes the convective flow is very low. Therefore, the heat transfer process inside the annulus is mainly dominated by the conduction mode and the isotherms are very smooth (no distortion of isotherms takes place due to very weak fluid flow) and are almost concentric to the inner and outer hexagonal cylinders. Both isothermal and streamlines are symmetrical concerning the vertical as well as horizontal centerlines of the annulus. When $AR = 0.2$, the isotherms are almost circular and the spacing between them increases with the distance from the inner hexagon as the available space between inner and outer hexagons is more than that when $AR = 0.6$. On the other hand, when $AR = 0.6$, the isotherms in the vicinity of both inner and outer cylinders are in the form of the hexagon and the spacing between them is less as they get squeezed due to constricted space between inner and outer hexagons. The streamlines pattern for both $AR = 0.2$ and $AR = 0.6$ show that two symmetrical recirculating eddies (kidney-shaped cells) are formed inside the annulus and the location of cell centers is almost close to the horizontal centerline of the annulus as the fluid flow intensity in the upward direction is almost negligible because of very low Ra. When we observe Figure 4b carefully, we can see that there is slight penetration of the streamline into the solid edges, which is a slight drawback of the present scheme. The main reason for this phenomenon is that as SPM is a non-conforming-mesh method, the same grid system is used for the solid and fluid regions and simulations are also performed inside the solid regions (even though it is enough to consider the boundary effects on the hexagonal edges). However, performing the simulations inside the solid does not affect the flow field in the fluid domain and the overall behavior of the fluid flow and heat transfer is well captured.

Figure 5 shows the isotherms (left side) and streamlines (right side) pattern inside the annulus for the values of $AR = 0.2$ (Figure 5a) and $AR = 0.6$ (Figure 5b), and when Ra $= 10^6$. As the Ra is very high, the effect of buoyancy-driven flow is significant and hence the heat transfer in the upper region of the annulus is mainly dominated by the convection mode. Isotherms and streamlines are no longer symmetrical about the horizontal median of the annulus. For $AR = 0.2$, it is concluded from the isotherms and streamlines pattern that the fluid near the inner hexagonal cylinder surface gets heated and moves upwards along the upper inclined edges of the hexagon due to the buoyancy effect. Because of strong convection currents, a thermal plume is formed on the top of the inner cylinder and thermal boundary layer thickness at the top flat edge of the outer cylinder is very thin (indicated by close clustering of isothermal lines) as continuous impingement of fluid flow in the upper region of the annulus. The thermal boundary layer thickness at the bottom of the inner hexagonal cylinder is also found to be very low and the fluid temperature below the inner cylinder is almost uniform and is equal to that of the outer cylinder as heat transfer in this region is dominated by conduction. The centers of the symmetrical recirculating eddies are located well above the horizontal median as the fluid flow is dominant in the upper half of the annulus. The streamline pattern for $AR = 0.2$ also reveals that two symmetric secondary vortices are formed at the bottom wall of the outer cylinder due to the separation of the momentum boundary layer as a result of strong upward convective flow.

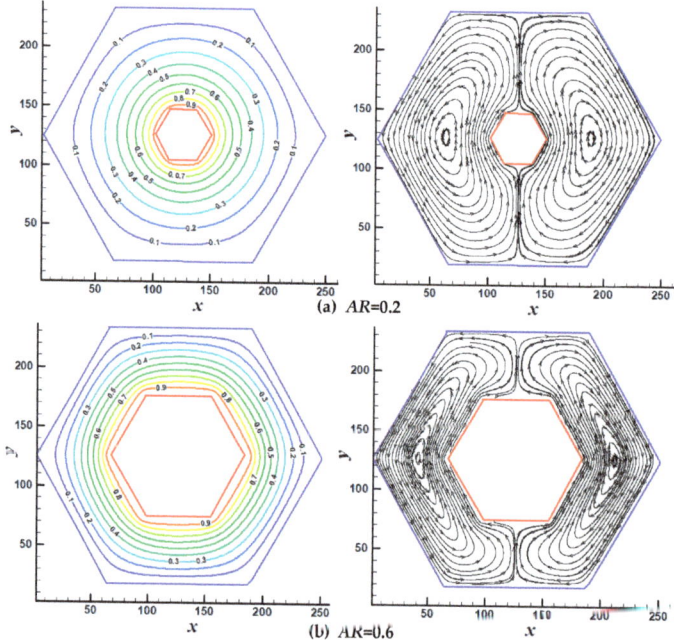

Figure 4. Isotherms (left side) and streamlines (right side) patterns inside the hexagonal annulus when the Rayleigh number is, Ra $= 10^3$, and for $AR = 0.2$ (**a**), and $AR = 0.6$ (**b**).

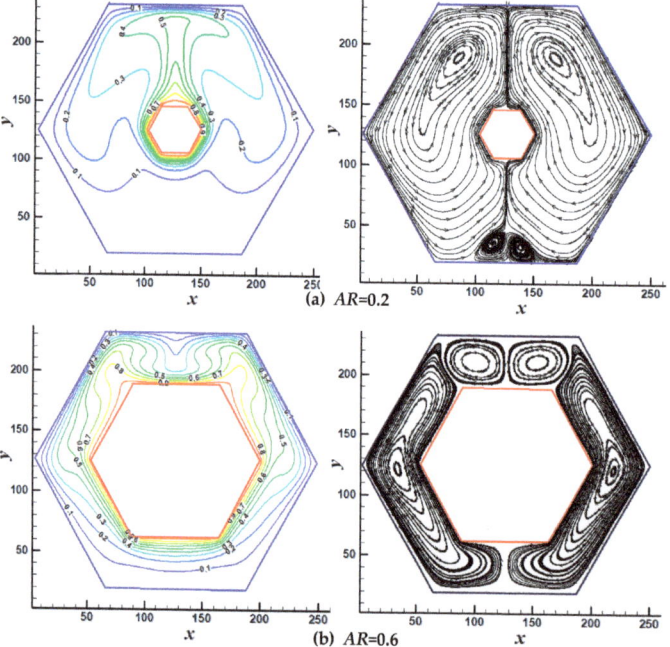

Figure 5. Isotherms (left side) and streamlines (right side) patterns inside the hexagonal annulus when the Rayleigh number is, Ra $= 10^6$, and for $AR = 0.2$ (**a**), and $AR = 0.6$ (**b**).

A completely different phenomenon is observed when $AR = 0.6$. Since there is limited available space for convection on the top of the inner cylinder, two separate thermal plumes (due to the buoyancy-driven fluid flow along the upper inclined edges of the hexagon) are formed along each upper corner of the hexagon. A third thermal plume is also seen on the top flat edge of the inner cylinder in the reverse direction as the uppermost corner of the inner hexagonal cylinder separates the fluid flow and generates two secondary vortices. The fluid flow separation phenomenon can be confirmed by noticing the two counter-rotating cells over the top flat edge of the inner cylinder from the streamline pattern of $AR = 0.6$. This type of flow separation phenomena at a high AR value was also observed by Raithby et al. [6], Bararnia et al. [30], Moutaouakil et al. [32], and Hu et al. [36], and even though their simulation domains were completely different from the present study.

To assess the capability of the present simulation method for predicting the behavior of natural convection flow in the concentric hexagonal annulus, the results obtained from the present method are compared with ANSYS-Fluent® results. Figure 6 shows the simulation results of isotherms and streamlines patterns obtained from Fluent for the values of $AR = 0.2$ (Figure 6a) and $AR = 0.6$ (Figure 6b), and the case when $Ra = 10^6$. By comparing the isotherms and streamlines patterns of Figures 5 and 6, we can say that the present simulation results are successfully reproduced the Fluent results.

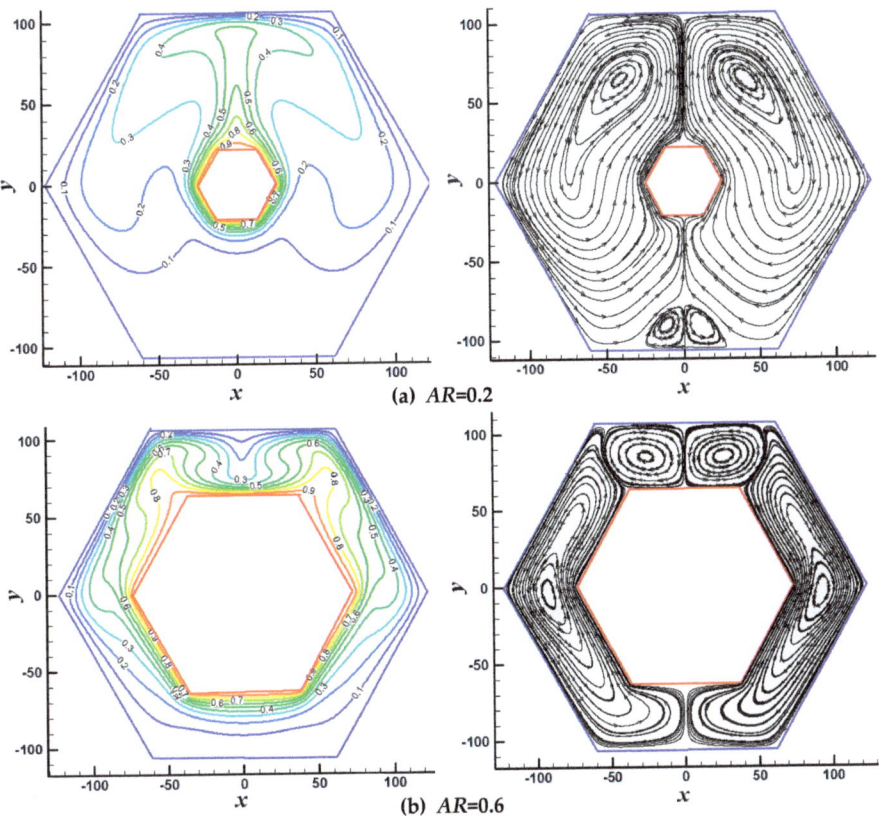

Figure 6. Isotherms (left side) and streamlines (right side) patterns obtained from ANSYS-Fluent® software for the values of $AR = 0.2$ (**a**) and $AR = 0.6$ (**b**) when $Ra = 10^6$.

3.2.2. Temperature and Velocity Profiles

Figure 7 shows the temperature distribution along the gap between the inner and outer hexagonal cylinders at three different angular directions, $\theta = 0°$ (along the line passing through the vertical median), $\theta = 30°$ (along the line passing through the right uppermost corners of the two hexagons), and at $\theta = 90°$ (along the line passing through the horizontal median) when $AR = 0.2$. R_i and R_o in the figure denote the radius of circles that pass through the edges of inner and outer hexagonal cylinders, respectively. The temperature distribution profiles are plotted for two Rayleigh numbers: $Ra = 10^3$ and $Ra = 10^6$. The symbols in the figure indicate the corresponding results obtained from the Fluent, which are in good agreement with the present simulation results. When $Ra = 10^3$, the temperature profiles, along all three θ directions, show a quasi-linear pattern (in other words, the temperature gradients along the gap are almost constant) with the gap as the conduction is the primary mode of the heat transfer process in the annulus. The temperature profiles along $\theta = 30°$ and $\theta = 90°$ are almost the same as the isothermal lines are almost concentric and R_i values at $\theta = 30°$ and $\theta = 90°$ are the same. When $Ra = 10^6$, as the strong convective fluid flow disturbs the uniform temperature distribution over the inner hexagon, high-temperature gradients are observed near the inner and outer hexagonal cylinder edges. As mentioned earlier, a thermal plume is formed in the direction of $\theta = 0°$ for $AR = 0.2$, the temperature gradients closer to the outer cylinder wall are very steep for the temperature profile along $\theta = 0°$ as the thermal boundary layer thickness is very thin due to continuous impingement thermal plume against the top flat edge of the outer hexagonal cylinder. The slope of the temperature profile near the outer cylinder wall is steeper for $\theta = 30°$ compared to that for $\theta = 90°$ as the convective fluid flow intensity is weaker at $\theta = 90°$.

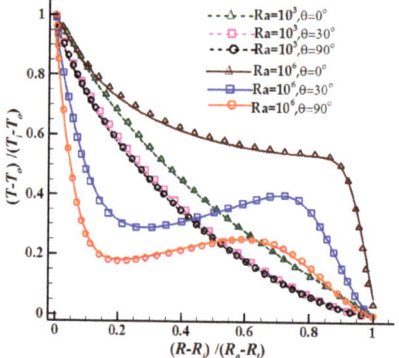

Figure 7. Temperature distribution along the gap between the inner and outer hexagonal cylinders at different angular directions, $\theta = 0°$, $\theta = 30°$, and at $\theta = 90°$ for $AR = 0.2$. The symbols represent corresponding data from Fluent software.

The curves for the temperature distributions along the gap when $AR = 0.6$ are provided in Figure 8 for Rayleigh numbers, $Ra = 10^3$ and $Ra = 10^6$. The temperature data obtained from Fluent software is also provided (the symbols in the figure) for comparison purposes. The agreement between the two results is excellent, implying the capability of the present simulation technique in the simulation of fluid flow and heat transfer in the hexagonal annuals. In this case, also the slope of the temperature profiles in each θ direction is almost constant when $Ra = 10^3$. For $Ra = 10^6$ case, the temperature profile in the direction of $\theta = 30°$ is in a similar trend with that of $\theta = 0°$ of Figure 7 data (for the case of $AR = 0.2$) as the formation of thermal plume for $AR = 0.6$ is along the direction of $\theta = 30°$. The slope of the temperature profile near the inner cylinder wall is very steep for $\theta = 0°$ as thermal boundary layer thickness over the flat top edge of the inner cylinder is very small due to the formation of the thermal plume in the reverse direction.

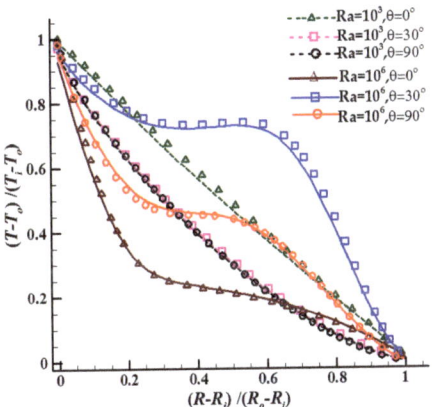

Figure 8. Temperature distribution along the gap between the inner and outer hexagonal cylinders at angular directions, $\theta = 0°$, $\theta = 30°$, and at $\theta = 90°$ for $AR = 0.6$. The symbols represent corresponding data from Fluent software.

The heat transfer between the outer and inner cylinders is enhanced by convection mode through fluid circulation. The rotational velocity (tangential velocity), u_θ, can be used as a good indication for the intensity of the convective fluid flow. Figure 9 shows the variations of tangential velocity, u_θ, along the gap between the inner and outer hexagonal cylinders for Rayleigh numbers, $Ra = 10^3$ and $Ra = 10^6$, and for aspect ratio, $AR = 0.2$. The profiles are plotted for $\theta = 30°$ and $\theta = 90°$. The reference velocity, $\alpha/(R_o - R_i)$, has chosen to normalize u_θ. It is noted that the magnitudes of u_θ when $Ra = 10^3$ are very small (almost zero) compared to those when $Ra = 10^6$ because of a weak fluid flow intensity at low Ra. For the velocity profile at $\theta = 90°$, the location of the flow reversal point is exactly at the center of the gap (at the halfway between the corners of two hexagons). However, the flow inversion point for the profile at $\theta = 30°$ is located a bit away from the gap center towards the outer cylinder. The velocity gradients for the profile at $\theta = 90°$ are steeper near the inner cylinder than those at the outer cylinder as the convective currents in the region adjacent to the inner cylinder (where fluid gets heated) are stronger than those near the outer cylinder.

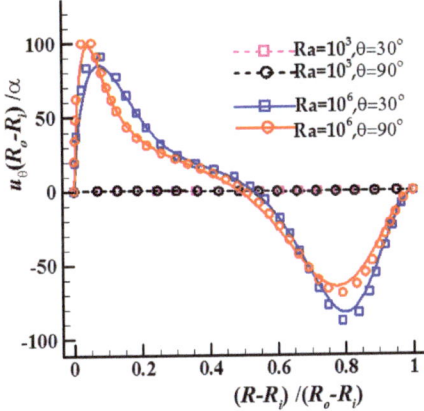

Figure 9. Variation of tangential velocity, u_θ, along the gap between the inner and outer hexagonal cylinders at angular directions, $\theta = 30°$, and $\theta = 90°$ for $AR = 0.2$. The symbols represent the corresponding data from Fluent software.

The profiles for the tangential velocity distributions when $AR = 0.6$ are provided in Figure 10 for Rayleigh numbers, $Ra = 10^3$ and $Ra = 10^6$. In this case, as well the magnitudes of u_θ when $Ra = 10^3$ are very small compared to those when $Ra = 10^6$. When $AR = 0.6$, as the thermal plume is formed along the $\theta = 30°$ direction, the tangential velocity in that direction is very low (fluid flow radially outwards along $\theta = 30°$). Therefore, the magnitude of u_θ along $\theta = 30°$ very low compared to that along $\theta = 90°$. The locations for the flow reversal points of the two velocity profiles along $\theta = 30°$ and $\theta = 90°$ are the same and are located away from the gap center and are towards the inner cylinder. The magnitudes of u_θ obtained for $AR = 0.6$ case are lower compared to those obtained for $AR = 0.2$ ($Ra = 10^6$ data of Figure 9) as available space for convection flow is constricted at $AR = 0.6$. The tangential velocity profiles data obtained from the present simulation technique, for cases of $AR = 0.2$ and $AR = 0.6$, and $Ra = 10^3$ and $Ra = 10^6$, are compared with those of Fluent software (the symbols in Figures 9 and 10). The present simulation results show good agreement with the Fluent data.

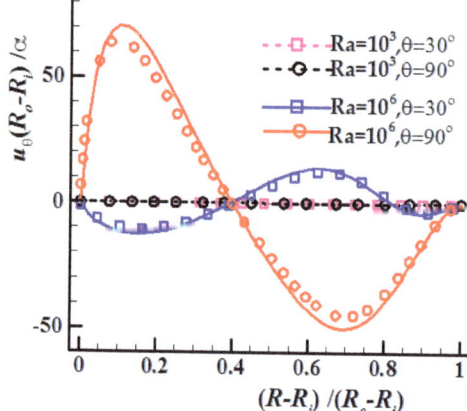

Figure 10. Variation of tangential velocity, u_θ, along the gap between the inner and outer hexagonal cylinders at angles, $\theta = 30°$, and $\theta = 90°$ for $AR = 0.6$. The symbols represent the corresponding data from Fluent software.

4. Conclusions

In this work, a FORTRAN code based on a hybrid method (which uses a combination of LBM, SPM, and FDM) has been developed to simulate the natural convection inside an annulus between two concentric hexagonal cylinders. After validating the numerical code by applying it to a standard benchmarking problem, natural convection simulations inside the hexagonal annulus have been performed by considering different combinations of Rayleigh number, Ra, and aspect ratio, AR. When $AR = 0.6$, two separate thermal plumes are formed due to the separation of convective flow at the upper corner of inner hexagon, which is in accordance with the previous studies. To verify the accuracy and robustness of the present method for simulating natural convection flow inside the hexagonal annulus, all the simulation results obtained from the present technique have been compared with the Fluent results. The simulation results of isotherms and streamlines patterns, temperature, and velocity distributions inside the annulus show good agreement with those obtained from Fluent software.

Funding: This research was funded by Kyungsung University Research Grants in 2019 and the APC was funded by the National Research Foundation of Korea (NRF) grant funded by the Korea government (Ministry of Science and ICT) (2020R1G1A1010247).

Acknowledgments: This research was supported by Kyungsung University Research Grants in 2019 and this research was supported by the National Research Foundation of Korea (NRF) grant funded by the Korea government (Ministry of Science and ICT) (2020R1G1A1010247).

Conflicts of Interest: The author declares no conflict of interest.

References

1. Kuehn, T.H.; Goldstein, R.J. An experimental and theoretical study of natural convection in the annulus between horizontal concentric cylinders. *J. Fluid Mech.* **1976**, *4*, 695–719. [CrossRef]
2. Kuehn, T.H.; Goldstein, R.J. An experimental study of natural convection heat transfer in concentric and eccentric horizontal cylindrical annuli. *J. Heat Trans.* **1978**, *100*, 635–640. [CrossRef]
3. Boyd, R.D. An Experimental Study on the steady natural convection in a horizontal annulus with irregular boundaries. *ASME-HTD* **1980**, *8*, 89–95.
4. Boyd, R.D. A unified theory for correlating steady laminar natural convective heat transfer data for horizontal annuli. *Int. J. Heat Mass Transf.* **1981**, *24*, 1545–1548. [CrossRef]
5. Chang, K.S.; Won, Y.H.; Cho, C.H. Patterns of natural convection around a square cylinder placed concentrically in a horizontal circular cylinder. *J. Heat Trans.* **1983**, *105*, 273–280. [CrossRef]
6. Raithby, G.D.; Galpin, P.F.; Van Doormaal, J.P. Prediction of heat and fluid flow in complex geometries using general orthogonal coordinates. *Numer. Heat Trans.* **1986**, *9*, 125–142. [CrossRef]
7. Glakpe, E.K.; Asfaw, A. Prediction of two-dimensional natural convection in enclosures with inner bodies of arbitrary shapes. *Numer. Heat Trans. A* **1991**, *20*, 279–296. [CrossRef]
8. Zhang, H.L.; Wu, Q.J.; Tao, W.Q. Experimental study of natural convection heat transfer between a cylindrical envelope and an internal concentric heated octagonal cylinder with or without slots. *ASME J. Heat Trans.* **1991**, *113*, 116–121. [CrossRef]
9. Peskin, C.S. Flow patterns around heart valves: A numerical method. *J. Comp. Phys.* **1972**, *10*, 252–271. [CrossRef]
10. Glowinski, R.; Pan, T.-W.; Helsa, T.; Joseph, D.D. A distributed lagrange multiplier/fictitious domain method for particulate flows. *Int. J. Multiph. Flow.* **1999**, *25*, 755–794. [CrossRef]
11. Nakayama, Y.; Yamamoto, R. Simulation method to resolve hydrodynamic interactions in colloidal dispersions. *Phys. Rev. E.* **2005**, *71*, 036707-1–036707-7. [CrossRef] [PubMed]
12. Luo, X.; Maxey, R.M.; Karniadakis, G.E. Smoothed profile method for particulate flows: Error analysis and simulations. *J. Comp. Phys.* **2009**, *228*, 1750–1769. [CrossRef]
13. Alapati, S.; Che, W.S.; Suh, Y.K. Simulation of sedimentation of a sphere in a viscous fluid using the lattice Boltzmann method combined with the smoothed profile method. *Adv. Mech. Eng.* **2015**, *7*, 794198-1–794198-12. [CrossRef]
14. Succi, S. *The Lattice Boltzmann Equation for Fluid Dynamics and Beyond*; Oxford University Press: Oxford, UK, 2001.
15. Alapati, S.; Kang, S.M.; Suh, Y.K. Parallel computation of two-phase flow in a microchannel using the lattice Boltzmann method. *J. Mech. Sci. Technol.* **2009**, *23*, 2492–2501. [CrossRef]
16. Di Palma, P.R.; Huber, C.; Viotti, P. A new lattice Boltzmann model for interface reactions between immiscible fluids. *Adv. Water Resour.* **2015**, *82*, 139–149. [CrossRef]
17. Chen, G.; Huang, X.; Wang, S.; Kang, Y. Study on the bubble growth and departure with a lattice Boltzmann method. *China Ocean Eng.* **2020**, *34*, 69–79. [CrossRef]
18. Dellar, P. Lattice Kinetic Schemes for Magnetohydrodynamics. *J. Comp. Phys.* **2002**, *179*, 95–126. [CrossRef]
19. Dellar, P. Lattice Boltzmann formulation for Braginskii magnetohydrodynamics. *Comput. Fluids* **2011**, *46*, 201–205. [CrossRef]
20. Fyta, M.; Melchionna, S.; Succi, S.; Kaxiras, E. Hydrodynamic correlations in the translocation of biopolymer through a nanopore: Theory and multiscale simulations. *Phys. Rev. E* **2008**, *78*, 036704-1–036704-7. [CrossRef]
21. Alapati, S.; Fernandes, D.V.; Suh, Y.K. Numerical simulation of the electrophoretic transport of a biopolymer through a synthetic nanopore. *Mol. Simulat.* **2011**, *37*, 466–477. [CrossRef]
22. Alapati, S.; Fernandes, D.V.; Suh, Y.K. Numerical and theoretical study on the mechanism of biopolymer translocation process through a nanopore. *J. Chem. Phys.* **2011**, *135*, 055103-1–055103-11. [CrossRef] [PubMed]
23. Hu, Y.; Li, D.C.; Shu, S.; Niu, X.D. Modified momentum exchange method for fluid-particle interactions in the lattice Boltzmann method. *Phys. Rev. E* **2015**, *91*, 033301-1–033301-14. [CrossRef] [PubMed]
24. Yuan, H.Z.; Niu, X.D.; Shu, S.; Li, M.J.; Yamaguchi, H. A momentum exchange-based immersed boundary-lattice Boltzmann method for simulating a flexible filament in an incompressible flow. *Comput. Math. Appl.* **2014**, *67*, 1039–1056. [CrossRef]

25. Kohestani, A.; Rahnama, M.; Jafari, S.; Javaran, E.J. Non-circular particle treatment in smoothed profile method: A case study of elliptical particles sedimentation using lattice Boltzmann method. *J. Dispers. Sci. Technol.* **2020**, *41*, 315–329. [CrossRef]
26. Zhang, Y.; Pan, G.; Zhang, Y.; Haeri, S. A relaxed multi-direct-forcing immersed boundary-cascaded lattice Boltzmann method accelerated on GPU. *Comput. Phys. Commun.* **2020**, *248*, 106980. [CrossRef]
27. Sheikholeslami, M.; Bandpy, M.G.; Ganji, D.D. Magnetic field effects on natural convection around a horizontal circular cylinder inside a square enclosure filled with nanofluid. *Int. Commun. Heat Mass.* **2012**, *39*, 978–986. [CrossRef]
28. Sheikholeslami, M.; Bandpy, M.G.; Ganji, D.D. Lattice Boltzmann method for MHD natural convection heat transfer using nanofluid. *Powder Technol.* **2014**, *254*, 82–93. [CrossRef]
29. Lin, K.-H.; Liao, C.-C.; Lien, S.-Y.; Lin, C.-A. Thermal lattice Boltzmann simulations of natural convection with complex geometry. *Comput. Fluids* **2012**, *69*, 35–44. [CrossRef]
30. Bararnia, H.; Soleimani, S.; Ganji, D.D. Lattice Boltzmann simulation of natural convection around a horizontal elliptic cylinder inside a square enclosure. *Int. Commun. Heat Mass.* **2011**, *38*, 1436–1442. [CrossRef]
31. Sheikholeslami, M.; Bandpy, M.G.; Vajravelu, K. Lattice Boltzmann simulation of magnetohydrodynamic natural convection heat transfer of Al_2O_3-water nanofluid in a horizontal cylindrical enclosure with an inner triangular cylinder. *Int. J. Heat Mass Transf.* **2015**, *80*, 16–25. [CrossRef]
32. Moutaouakil, L.E.; Zrikem, Z.; Abdelbaki, A. Lattice Boltzmann simulation of natural convection in an annulus between a hexagonal cylinder and a square enclosure. *Math. Probl. Eng.* **2017**, *2017*, 3834170-1–3834170-11. [CrossRef]
33. Hu, Y.; Niu, X.-D.; Shu, S.; Yuan, H.; Li, M. Natural convection in a concentric annulus: A lattice Boltzmann method study with boundary condition-enforced immersed boundary method. *Adv. Appl. Math. Mech.* **2013**, *5*, 321–336. [CrossRef]
34. Hu, Y.; Li, D.; Shu, S.; Niu, X. Immersed boundary-lattice Boltzmann simulation of natural convection in a square enclosure with a cylinder covered by porous layer. *Int. J. Heat Mass Transf.* **2016**, *92*, 1166–1170. [CrossRef]
35. Khazaeli, R.; Mortazavi, S.; Ashrafizaadeh, M. Application of an immersed boundary treatment in simulation of natural convection problems with complex geometry via the lattice Boltzmann method. *J. Appl. Fluid Mech.* **2015**, *8*, 309–321. [CrossRef]
36. Hu, Y.; Li, D.; Shu, S.; Niu, X. An efficient smoothed profile-lattice Boltzmann method for the simulation of forced and natural convection flows in complex geometries. *Int. Commun. Heat Mass.* **2015**, *68*, 188–199. [CrossRef]
37. Alapati, S.; Che, W.S.; Ahn, J.-W. Lattice Boltzmann method combined with the smoothed profile method for the simulation of particulate flows with heat transfer. *Heat Transfer Eng.* **2019**, *40*, 166–183. [CrossRef]
38. Ladd, A.J.C. Numerical simulations of particulate suspensions via a discretized Boltzmann equation. Part 1. Theoretical foundation. *J. Fluid Mech.* **1994**, *271*, 285–309. [CrossRef]
39. Ladd, A.J.C. Numerical simulations of particulate suspensions via a discretized Boltzmann equation. Part 2. Numerical results. *J. Fluid Mech.* **1994**, *271*, 311–339. [CrossRef]
40. Bouzidi, M.; Firdaouss, M.; Lallemand, P. Momentum transfer of a Boltzmann-lattice fluid with boundaries. *Phys. Fluids* **2001**, *13*, 3452–3459. [CrossRef]
41. Yu, D.; Mei, R.; Luo, L.-S.; Shyy, W. Viscous flow computations with the method of lattice Boltzmann equation. *Prog. Aerosp. Sci.* **2003**, *39*, 329–367.
42. Mittal, R.; Iaccarino, G. Immersed boundary methods. *Ann. Rev. Fluid Mech.* **2005**, *37*, 239–261. [CrossRef]
43. Marshall, K.N.; Wedel, R.K.; Dammann, R.E. *Development of Plastic Honeycomb Flat-Plate Solar Collectors*; Lockheed Missiles & Space Company Inc.: Palo Alto, CA, USA, 1976.
44. Buchberg, H.; Edwards, D.K.; Mackenzie, J.D. *Transparent Glass Honeycombs for Energy Loss Control*; School of Engineering and Applied Science, University of California: Los Angeles, CA, USA, 1977.
45. Marin, O.; Vinuesa, R.; Obabko, A.V.; Schlatter, P. Characterization of the secondary flow in hexagonal ducts. *Phys. Fluids* **2016**, *28*, 125101. [CrossRef]
46. Aliabadia, M.K.; Deldarb, S.; Hassanic, S.M. Effects of pin-fins geometry and nanofluid on the performance of a pin-fin miniature heat sink (PFMHS). *Int. J. Mech. Sci.* **2018**, *148*, 442–458. [CrossRef]
47. Sadasivam, R.; Manglik, R.M.; Jog, M.A. Fully developed forced convection through trapezoidal and hexagonal ducts. *Int. J. Heat Mass Transf.* **1999**, *42*, 4321–4331. [CrossRef]

48. Zhang, L.Z. Transient and conjugate heat and mass transfer in hexagonal ducts with adsorbent walls. *Int. J. Heat Mass Transf.* **2015**, *84*, 271–281. [CrossRef]
49. Moukalled, F.; Acharya, S. Natural convection in the annulus between concentric horizontal circular and square cylinders. *J. Thermophys. Heat Trans.* **1996**, *10*, 524–531. [CrossRef]

© 2020 by the author. Licensee MDPI, Basel, Switzerland. This article is an open access article distributed under the terms and conditions of the Creative Commons Attribution (CC BY) license (http://creativecommons.org/licenses/by/4.0/).

Article

Bioconvection Reiner-Rivlin Nanofluid Flow between Rotating Circular Plates with Induced Magnetic Effects, Activation Energy and Squeezing Phenomena

Muhammad Bilal Arain [1], Muhammad Mubashir Bhatti [2,*], Ahmad Zeeshan [1] and Faris Saeed Alzahrani [3]

[1] Department of Mathematics and Statistics, International Islamic University, Islamabad 44000, Pakistan; muhammad.phdma79@iiu.edu.pk (M.B.A.); ahmad.zeeshan@iiu.edu.pk (A.Z.)
[2] College of Mathematics and Systems Science, Shandong University of Science & Technology, Qingdao 266590, China
[3] Department of Mathematics, Faculty of Sciences, King Abdul-Aziz University, P.O. Box 80203, Jeddah 21589, Saudi Arabia; falzahrani1@kau.edu.sa
* Correspondence: mmbhatti@sdust.edu.cn or mubashirme@yahoo.com

Citation: Arain, M.B.; Bhatti, M.M.; Zeeshan, A.; Alzahrani, F.S. Bioconvection Reiner-Rivlin Nanofluid Flow between Rotating Circular Plates with Induced Magnetic Effects, Activation Energy and Squeezing Phenomena. *Mathematics* **2021**, *9*, 2139. https://doi.org/10.3390/math9172139

Academic Editor: Mostafa Safdari Shadloo

Received: 29 July 2021
Accepted: 31 August 2021
Published: 2 September 2021

Publisher's Note: MDPI stays neutral with regard to jurisdictional claims in published maps and institutional affiliations.

Copyright: © 2021 by the authors. Licensee MDPI, Basel, Switzerland. This article is an open access article distributed under the terms and conditions of the Creative Commons Attribution (CC BY) license (https://creativecommons.org/licenses/by/4.0/).

Abstract: This article deals with the unsteady flow in rotating circular plates located at a finite distance filled with Reiner-Rivlin nanofluid. The Reiner-Rivlin nanofluid is electrically conducting and incompressible. Furthermore, the nanofluid also accommodates motile gyrotactic microorganisms under the effect of activation energy and thermal radiation. The mathematical formulation is performed by employing the transformation variables. The finalized formulated equations are solved using a semi-numerical technique entitled Differential Transformation Method (DTM). Padé approximation is also used with DTM to present the solution of nonlinear coupled ordinary differential equations. Padé approximation helps to improve the accuracy and convergence of the obtained results. The impact of several physical parameters is discussed and gives analysis on velocity (axial and tangential), magnetic, temperature, concentration field, and motile gyrotactic microorganism functions. The impact of torque on the lower and upper plates are deliberated and presented through the tabular method. Furthermore, numerical values of Nusselt number, motile density number, and Sherwood number are given through tabular forms. It is worth mentioning here that the DTM-Padé is found to be a stable and accurate method. From a practical point of view, these flows can model cases arising in geophysics, oceanography, and in many industrial applications like turbomachinery.

Keywords: Reiner-Rivlin nanofluid; circular plates; induced magnetic effects; activation energy; bioconvection nanofluid

1. Introduction

Nanofluids were first explained by Choi [1] in 1995. Nanofluids are a composition of nanoparticles and a base fluid including oil, water, ethylene-glycol, kerosene, polymeric solutions, bio-fluids, lubricants, oil, etc. The material of the nanoparticles [2] involves chemically stable metals, carbon in multiple forms, oxide ceramics, metal oxides, metal carbides, etc. The magnitude of the nanoparticles is substantially smaller (approx. less than 100 nm). Nanofluids have multitudinous applications in engineering and industry [3,4], such as smart fluids, nuclear reactors, industrial cooling, geothermal power extract, and distant energy resources, nanofluid coolant, nanofluid detergents, cooling of microchips, brake and distant vehicular nanofluids, and nano-drug delivery. In the light of these applications, numerical researchers discussed the nanofluids in different geometrical configurations. For instance, Gourarzi et al. [5] scrutinized the impact of thermophoretic force and Brownian motion on hybrid nanofluid. They concluded with the excellent point that nanoparticle formation on cold walls is more essential due to thermophoresis migration. Ghalandari et al. [6] used CFD to model silver/water nanofluid flow towards a root canal. The effects of injection height, nanofluid concentration, and the rate of volumetric flow were explored and addressed. Sheikholeslami and Vajravelu [7] studied the control volume-based finite element approach to determine magnetite nanofluid flow into the same heat flux in the whole cavity. The impact of Rayleigh number, Hartmann number, and volume friction of nanofluid flow magnetite (an iron oxide) and heat transfer features were discussed. Sheikholeslami and Ganji [8] addressed hydrothermal nanofluid in the existence of magnetohydrodynamics by using DTM. They discussed the impact of squeezing number and nanofluid volume fraction on heat transfer and fluid flow. Biswal et al. [9] deliberated fluid flow in a semi-permeable channel with the influence of a transverse magnetic field. Zhang et al. [10] considered the outcome of thermal diffusivity and conductivity of numerous nanofluids utilizing the transient short-hot-wire technique. Fakour et al. [11] inquired the laminar nanofluid flow in the channel using the least square approach with porous walls. This study shows that by enhancing Hartman and Reynolds number, the velocity of the nanofluid flow in the channel declines and an extreme amount of temperature is enhanced. More, enhancing the Prandtl number along with the Eckert number also increases the temperature distribution. Zhu et al. [12] inquired the second-order slip and migration of nanoparticles from a magnetically influenced annulus. They applied a well-known HAM technique for solving the equations, and a h-curve was drawn to validate the exactness of the obtained solution. Ellahi et al. [13] revealed the impact of Poiseuille nanofluid flow with Stefan blowing and second-order slip. The accuracy of the analytical solution is obtained by the HAM and verified by h-curve and residual error norm for each case. They claim that the ratio of buoyancy forces in the existence of a magnetic field played a vital role in velocity distribution.

Magnetohydrodynamic (MHD) has grabbed different researchers' attention because of its multitudinous applications in the agricultural, physics, medicine, engineering, and petroleum industries, etc. For instance, applications of MHD involve bearing sand boundary layer control, MHD generators, rotating machines, viscometry, electronic storing components, turbomachines, lubrications, oceanographically processes, reactor chemical vapor deposition, and pumps. The magnetic field plays an essential role in controlling the boundary layer of momentum and heat transfer. The presence of magnetics is beneficial to control fluid movement. It is worthwhile to mention that the magnetic essential modified the outcomes of heat transfer in the flow by maneuvering the suspended nanoparticles and reorganized the fluid concentration. Khan et al. [14] studied the magnetohydrodynamic nanofluid flow between the pair of rotating plates. Zangooee et al. [15] analyzed the hydrothermal magnetized nanofluid flow between a pair of radiative rotating disks. From their studies, it is perceived that concentration decreases while increasing in Reynolds number, but on the other hand, the temperature is increasing for Reynolds number. By enhancing the value of the stretching parameter, the Reynolds number increases at the upper disc and decreases at the lower plates. Hatami et al. [16] analytically inquired the magnetized nanofluid flow in the porous medium. These results showed that the magnetic field opposes fluid flow in all directions. In addition, they claimed that the action of thermophoresis increases temperature and reduces the flow of heat from the disc. Nanoparticles shape effect on magnetized nanofluid flow over a rotating disc embedded in porous medium investigated by Rashid and Liang [17]. Abbas et al. [18] studied a fully developed flow of nanofluid with activation energy and MHD. The study's main findings demonstrate that flow field and entropy rate are highly affected by a magnetic field. The results indicate that both the flow and entropy rates of the magnetic field are significantly affected. Rashidi et al. [19] inquired steady MHD nanofluid flow with entropy generation and due to permeable rotating plates. Alsaedi et al. [20] inquired the flow of copper-water nanofluid with MHD and partial slip due to a rotating disc. They contemplated water as a base fluid and copper nanoparticles. They concluded with the remark that for greater values of a nanoparticle volume fraction, the magnitude of skin friction coefficient had been increased both for radial and azimuthal profiles. Asma et al. [21] numerically discussed the MHD nanofluid flow over a rotating disk under the impact of activation energy.

They observed that the concentration and temperature both show a growing tendency by increasing Hartman numbers. Aziz et al. [22] inquired the three-dimensional motion of viscous nanoparticles over rotating plates with slip effects. They showed that concentration profile and temperature distribution show enhancing behaviors for increasing values of Hartmann number. Hayat et al. [23] numerically inquired the nanofluid flow because of rotating disks with slip effects and magnetic field. These studies showed that more significant levels of the magnetic parameter indicate reduced velocity distribution behavior, whereas temperature and concentration distribution show opposite behavior. The hydromagnetic fluid flow of nanofluid due to stretchable/shrinkable disk with non-uniform heat generation/absorption is inquired by Naqvi et al. [24]. The graphical results of the studies showed that the higher values of the Prandtl number give an improved temperature, but when thermophoresis and Brownian motion parameters are reduced, the temperature distribution reduces.

Svante Arrhenius, a Swedish physicist, used the phrase energy for the first time in 1889. Activation energy is measured in KJ/mol and denoted by E_a, which means the minimum energy achieved by molecules/atoms to initiate the chemical process. For various chemical processes, the amount of energy activation is varying, even sometimes zero. The activation energy in heat transfer and mass transfer has its usages in chemical engineering, emulsions of different suspensions, food processing, geothermal reservoirs, etc. Bestman [25] published the first paper on activation energy with a binary chemical process. Discussion on the inclusion of chemical reaction into nanofluids flow and Arrhenius activation energy was determined by Khan et al. [26]. Zeeshan et al. [27] studied the Couette-Poiseuille flow with activation energy and analyzed convective boundary conditions. Bhatti and Michaelides [28] discussed the influence of activation energy on a Riga plate with gyrotactic microorganisms. Khan et al. [29] reveal that the impact of activation energy on the flow of nanofluid against stagnation point flow by considering it nonlinear with activation energy. Their investigation revealed that activation energy decline for the mass transfer phenomena. Hamid et al. [30] inquired about the effects of activation energy inflow of Williamson nanofluid with the influence of chemical reactions. The study concluded that the heat transfer rate in cylindrical surfaces declines when increasing the reaction rate parameter. Azam et al. [31] inquired about the impact of activation energy in the axisymmetric nanofluid flow. Waqas et al. [32] inquired the flow of Oldroyd-B bioconvection nanofluid numerically with nonlinear radiation through a rotating disc with activation energy.

Bioconvection characterizes the hydrodynamic instabilities and the forms of suspended biased swimming microorganisms. The hydrodynamics instabilities occur due to the coupling between the cell's swimming performance and physical features of the cell, i.e., fluid flows and density. For example, a combination of gravitational and viscous torques tend to swim the cells in the direction of down welling fluid. A gyrotactic instability ensues if the fluid is less dense than the cells. Bioconvection portrays a classical structure where a macroscopic mechanism occurs due to the microscopic cellular ensuing in relatively dilute structures. There is also the ecological impact for bioconvection and its mechanisms, which is promising for industrial development. In the recent era, many scientists have discussed the mechanism of bioconvection using nanofluid models. For instance, Makinde et al. [33] examined the nanofluid flow due to rotating disk and thermal radiation with titanium and aluminum nanoparticles. They showed that the base liquid thermal efficiency is remarkable when the nanoparticles of titanium alloy are introduced in contrast to the nanoparticles of aluminum alloy. Reddy et al. [34] studied the Maxwell thermally radiative nanofluid flow on a double rotating disk. Waqas et al. [35] examined the effect of thermally bioconvection Sutterby nanofluid flow between two rotating disks along with microorganisms. The fluid speed with mixed convection parameters grew quicker but delayed the magnetic field parameter and the Rayleigh number bioconvection. Some important studies on the bioconvection mechanism can be found from the list of references [36–39].

For many industrial applications such as the production of glass, furnaces, space technologies, comic aircraft, space vehicles, propulsion systems, plasma physics, and reentry aerodynamics in the field of aero-structure flows, combustion processes, and other spacecraft applications, the role of thermal radiation is significant. Raju et al. [40] examined the flow of convective magnesium oxide nanoparticles with nonlinear thermal convective over a rotating disk. Sheikholeslami et al. [41] presented the analysis of thermally radiative MHD nanofluid through the porous cavity. Muhammad et al. [42] analyzed the characteristics of thermal radiation for Powell-Eyring nanofluid flow with additional effects of activation energy. Aziz et al. [43] numerically analyzed hybrid nanofluid with entropy analysis, thermal radiation, and viscous dissipation. Mahanthesh et al. [44] investigated the significance of radiation effects of the two-phase flow of nanoparticles over a vertical plate. Jawad et al. [45] investigated the bio-convection nanofluid flow of Darcy law through a channel (Horizontal) with magnetic field effects and thermal radiation. Majeed et al. [46] thermally analyzed magnetized bioconvection flow with additional effects of activation energy. Numerous fresh developments on this topic can be envisaged through [47–52].

After studying the preexistent literature, it is noticed that there is no addition to the research of Reiner-Rivlin fluid flow between rotating circular plates filled with microorganisms and nanoparticles. In the present study, we assume that the flow in the tangential and axial direction. The Reiner-Rivlin nanofluid with motile gyrotactic microorganisms is filled between the pair of rotating plates. The thermally radiative Reiner-Rivlin fluid is electrically conducted under the existence of activation energy. The famous Differential Transform scheme is used to obtain the solution of the ordinary differential equations. Padé approximation is also applied to enhance the convergence rate of the solution obtained by the Differential Transform Method. The impact of various parameters in nanoparticle concentration, velocity, temperature, and motile microorganism function is analyzed thoroughly using graphs and tabular forms.

2. Physical and Mathematical Structure of Three-Dimensional Flow

Let us anticipate incompressible three-dimensional, unsteady, axisymmetric squeezed film flow of Reiner-Rivlin nanofluid between a circular rotating parallel plate. The height of both plates is taken as $\widehat{\Gamma}(t) \left[= D(-\beta t + 1)^{1/2} \right]$ at time t. Let (r, θ, z) be the cylindrical polar coordinates with velocity field $V = [v_r, v_\theta, v_z]$. The lower circular plate is fixed while the upper circular plate is considered as moving towards the lower plate. The moving plate velocity is represented by $\widehat{\Gamma}'(t)$. Both plates are rotating at a symmetric axis, which is characterized by Z-axis. The components of the magnetic field applied $\mathbf{H}$ on the moving plate in axial and azimuthal direction are:

$$\widehat{H}_\theta = \frac{rN_0}{\mu_2}\sqrt{\frac{D}{\widehat{\Gamma}(t)}}, \quad \widehat{H}_z = -\frac{\beta M_0 D}{\mu_1 \widehat{\Gamma}(t)}, \tag{1}$$

Here N_0, M_0 in Equation (1) denotes the dimensionless quantities, which results $\widehat{H}_\theta$, $\widehat{H}_z$ in dimensionless, and the magnetized permeability of medium inside and outside of both plates are characterized by μ_2 and μ_1, respectively. In the case of liquid metals, $\mu_2 = \mu_\ell$ where μ_ℓ indicates the free space permeability. H_θ, H_z on a fixed plate is expected to be zero. The extrinsic applied magnetic field $\mathbf{H}$ tends to generate an induced magnetic field $\mathbf{B}(r, \theta, z)$ having components $\widehat{B}_r, \widehat{B}_\theta, \widehat{B}_z$ between the two plates (see Figure 1). The temperature and the concentration at the lower plate is denoted as (T_0, C_0) while at the upper plate is taken as (T_1, C_1).

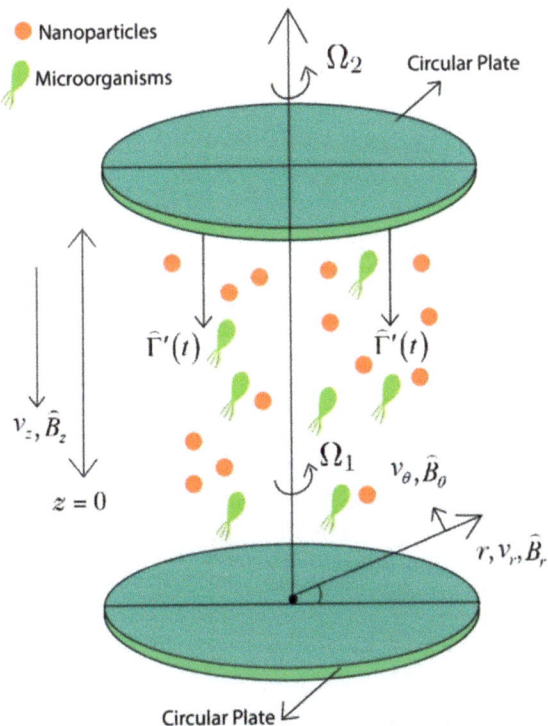

Figure 1. A physical structure for nanofluid flow between parallel circular plates in the existence of motile gyrotactic microorganisms and induced MHD.

2.1. Mathematical Modeling of Reiner-Rivlin Fluid

The constitutive equation of Reiner-Rivlin fluid flow is defined as [53]:

$$\tau_{ij} = -p\delta_{ij} + \mu e_{ij} + \mu_c e_{ik} e_{kj}, \; e_{jj} = 0, \tag{2}$$

where τ_{ij} represents stress tensor, p denotes pressure, μ denotes the viscosity coefficient, μ_c denotes cross-viscosity coefficient, δ_{ij} denotes Kronecker symbol, and deformation rate tensor is represented by $e_{ij} = (\partial u_i/\partial x_j) + (\partial u_j/\partial x_i)$. Components of deformation rate tensor are:

$$\begin{array}{c} e_{rr} = 2D_2 v_r, \; e_{\theta\theta} = 2\frac{v_r}{r}, \; e_{zz} = 2D_4 v_z, \; e_{r\theta} = e_{\theta r} = rD_2\left(\frac{v_\theta}{r}\right) = D_2 v_\theta - \frac{v_\theta}{r}, \\ e_{z\theta} = e_{\theta z} = D_4 v_\theta, \; e_{rz} = e_{zr} = D_4 v_r + D_2 v_z \end{array} \tag{3}$$

with the help of Equation (2), components of stress tensor are attained as

$$\tau_{rr} = -p + \mu e_{rr} + \mu_c \left(e_{rr}^2 + e_{r\theta}^2 + e_{rz}^2 \right), \tag{4}$$

$$\tau_{rr} = -p + 2\mu D_2 v_r + \mu_c \left[4(D_2 v_r)^2 + \left(D_2 v_\theta - \frac{v_\theta}{r} \right)^2 + (D_4 v_r + D_2 v_z)^2 \right], \tag{5}$$

$$\tau_{r\theta} = \tau_{\theta r} = 0 + \mu e_{r\theta} + \mu_c (e_{rr} e_{r\theta} + e_{r\theta} e_{\theta\theta} + e_{rz} e_{z\theta}), \tag{6}$$

$$\tau_{r\theta} = \mu\left(D_2 v_\theta - \frac{v_\theta}{r}\right) + \frac{\mu_c \left[2(D_2 v_r)\left(D_2 v_\theta - \frac{v_\theta}{r}\right) \right.}{\left. + \left(D_2 v_\theta - \frac{v_\theta}{r}\right)\left(2\frac{v_r}{r}\right) + (D_4 v_\theta)(D_4 v_r + D_2 v_z) \right]} \tag{7}$$

$$\tau_{rz} = \mu e_{rz} + \mu_c (e_{rr} e_{rz} + e_{r\theta} e_{\theta z} + e_{rz} e_{zz}), \tag{8}$$

$$\tau_{rz} = \mu(D_4 v_r + D_2 v_z) + \begin{matrix} \mu_c[2(D_2 v_r)(D_4 v_r + D_2 v_z) \\ + (D_2 v_\theta - \frac{v_\theta}{r})(D_4 v_\theta) + 2(D_4 v_z)(D_4 v_r + D_2 v_z)] \end{matrix}, \quad (9)$$

$$\tau_{\theta\theta} = -p + \mu e_{\theta\theta} + \mu_c \left(e_{r\theta}^2 + e_{\theta\theta}^2 + e_{z\theta}^2 \right), \quad (10)$$

$$\tau_{\theta\theta} = -p + \mu \left(2\frac{v_r}{r} \right) + \mu_c \left[\left(D_2 v_\theta - \frac{v_\theta}{r} \right)^2 + 4\left(\frac{v_r}{r} \right)^2 + (D_4 v_\theta)^2 \right], \quad (11)$$

$$\tau_{\theta z} = \mu e_{\theta z} + \mu_c (e_{\theta r} e_{rz} + e_{\theta\theta} e_{\theta z} + e_{\theta z} e_{zz}), \quad (12)$$

$$\tau_{\theta z} = \tau_{z\theta} = \mu(D_4 v_\theta) + \mu_c \left[\left(D_2 v_\theta - \frac{v_\theta}{r} \right)(D_4 v_r + D_2 v_z) + 2(D_4 v_\theta)\left(\frac{v_r}{r} + D_4 v_z \right) \right], \quad (13)$$

where $D_1 = \frac{\partial}{\partial t}$, $D_2 = \frac{\partial}{\partial r}$, $D_3 = \frac{\partial}{\partial \theta}$, $D_4 = \frac{\partial}{\partial z}$.

2.2. Proposed Governing Equations

Assuming the above-mentioned Reiner-Rivlin fluid model, the proposed governing equations for continuity and momentum in the direction of r, θ, z read as

$$\frac{1}{r} D_2(r v_r) + \frac{1}{r} D_3(v_\theta) + D_4(v_z) = 0, \quad (14)$$

$$\begin{aligned} \rho \left(D_1(v_r) + v_r D_2(v_r) + \frac{v_\theta}{r} D_3(v_r) + v_z D_4(v_r) - \frac{v_\theta}{r} \right) &= -D_2 p \\ + \mu \left[\frac{1}{r} D_2(v_r) + \frac{1}{r^2} D_3^2(v_r) + D_4{}^2(v_r) - \frac{2}{r} D_3(v_\theta) - \frac{v_r}{r^2} \right] & \\ + \frac{1}{r} \frac{\partial}{\partial r}(r \tau_{rr}) + \frac{1}{r} D_3(\tau_{r\theta}) - \frac{\tau_{\theta\theta}}{r} + D_4(\tau_{rz}) - D_4(B_r) R_n - D_4(B_\theta) B_\theta & \end{aligned} \quad (15)$$

$$\begin{aligned} \rho \left(D_1(v_\theta) + v_r D_2(v_\theta) + \frac{v_\theta}{r} D_3(v_\theta) + v_z D_4(v_\theta) - \frac{v_r v_\theta}{r} \right) &= -\frac{1}{r} D_3 p \\ + \mu \left[\frac{1}{r} D_2(r D_2(v_\theta)) + \frac{1}{r^2} D_3^2(v_\theta) + D_4{}^2(v_\theta) + \frac{2}{r^2} D_3(v_r) - \frac{v_\theta}{r^2} \right] & \\ + \frac{1}{r} D_3(\tau_{\theta\theta}) + \frac{1}{r^2} D_2(r^2 \tau_{r\theta}) + D_4(\tau_{\theta z}) - D_4(B_\theta) B_z - D_2(B_\theta) B_r & \end{aligned} \quad (16)$$

$$\begin{aligned} \rho \left(D_1(v_z) + v_r D_2(v_z) + \frac{v_\theta}{r} D_3(v_z) + v_z D_4(v_z) \right) &= -D_4 p \\ + \mu \left[\frac{1}{r} D_2(r D_2(v_z)) + \frac{1}{r^2} D_3^2(v_z) + D_4{}^2(v_z) \right] & \\ + D_4(\tau_{zz}) + \frac{1}{r} D_2(r \tau_{rz}) + \frac{1}{r} D_3(\tau_{\theta z}) - D_4(B_\theta) B_\theta + D_4(B_r) B_r & \end{aligned} \quad (17)$$

where p represents pressure, ρ represents fluid density, stress tensor is denoted by τ, and μ represents fluid viscosity. The equation of the magnetic field is

$$\frac{1}{r} D_2 r B_r + \frac{1}{r} D_3 B_\theta + D_4 B_z = 0, \quad (18)$$

$$D_1 B_r + v_r D_2 B_r + v_\theta D_3 B_r + v_z D_4 B_r = -D_4(v_r B_z - v_z B_r) + \frac{1}{\delta \mu_2}\left(D_4{}^2 B_r \right), \quad (19)$$

$$\begin{aligned} D_1 B_\theta + v_r D_2 B_\theta + v_\theta D_3 B_\theta + v_z D_4 B_\theta &= D_2(v_r B_\theta - v_\theta B_r) \\ &\quad - D_4(v_\theta B_z - B_\theta v_z) + \frac{1}{\delta \mu_2}\left(D_4{}^2 B_\theta \right) \end{aligned}, \quad (20)$$

$$D_1 B_z + v_r D_2 B_z + v_\theta D_3 B_z + v_z D_4 B_z = D_2(v_r B_z - v_z B_r) + \frac{1}{\delta \mu_2}\left(D_4{}^2 B_z \right), \quad (21)$$

where δ is the electrical conductivity.

The energy equation reads as:

$$\begin{aligned} D_1 \breve{T} + v_r D_2 \breve{T} + v_z D_4 \breve{T} &= \frac{k}{(\rho c)_f} D_4{}^2 \breve{T} - \frac{1}{(\rho c)_f}\left(\frac{\partial q_r}{\partial r} \right) \\ &\quad + \frac{(\rho c)_p}{(\rho c)_f}\left[D_B \left(D_2 \breve{T} \cdot D_2 \breve{C} + D_4 \breve{T} \cdot D_4 \breve{C} \right) + \frac{D_T}{T_u}\left[\left(D_2 \breve{T} \right)^2 + \left(D_4 \breve{T} \right)^2 \right] \right] \end{aligned}, \quad (22)$$

where $\breve{T}$ represents temperature, $\bar{k}$ the thermal conductivity, $\breve{C}$ represents concentration, mean fluid temperature is represented by $\breve{T}_m$, the specific heat capacity of nanofluid $(\rho c)_p$,

$(\rho c)_f$ the specific heat capacity of the base fluid, Brownian diffusivity is represented by D_B, thermophoretic diffusion coefficient is represented by D_T. In accordance with Rosseland approximation radiation heat flux, which is uni-directional (acting axially) takes the form, $q_r = -\frac{4\sigma_e}{3\beta_r}\frac{\partial \breve{T}^4}{\partial r}$, in which σ_e represents the Stefan–Boltzmann constant and β_r represents the mean absorption coefficient, respectively. Rosseland's model applies for optically thick nanofluids and yields a reasonable estimate for radiative transfer effects, although it neglects non-gray effects.

The equation of nanoparticle concentration reads as [54]

$$D_1\breve{C} + v_r D_2\breve{C} + v_z D_4\breve{C} = D_B D_4^2 \breve{C} + \frac{D_T}{\breve{T}_u} D_4^2 \breve{T} - k_r^2\left(\breve{C} - \breve{C}_u\right)\left(\frac{\breve{T}}{\breve{T}_u}\right)^n e^{-\frac{E_a}{\kappa \breve{T}}}, \quad (23)$$

where k_r^2 is the reaction rate, n is the rate constant, κ is the Boltzmann constant, and E_a is the activation energy.

The microorganism conservation equation reads as

$$D_1 n + v_r D_2 n + v_\theta D_3 n + v_z D_4 n + \frac{bW_{mo}}{\breve{C}_l - \breve{C}_u}\left[D_4\left(n D_4 \breve{C}\right)\right] = D_{mo}\left(D_4^2 n\right). \quad (24)$$

Here bW_{mo} is considered constant, where b are chemotaxis constant, cell swimming maximal speed is denoted by W_{mo}, and D_{mo} denotes diffusivity of microorganism. The corresponding boundary conditions are [54].

$$v_r = 0, v_\theta = \Omega_1 r\frac{D^2}{\widehat{\Gamma}^2(t)}, v_z = 0, B_z = B_\theta = 0, n = n_l, \breve{T} = \breve{T}_l, \breve{C} = \breve{C}_l, \text{ at } z = 0, \quad (25)$$

$$\left.\begin{array}{l}v_r = 0, v_\theta = \Omega_2 r\frac{D^2}{\widehat{\Gamma}^2(t)}, B_\theta = N_0 r\frac{D^2}{\widehat{\Gamma}^2(t)}, B_z = -\frac{\beta DM_0}{\widehat{\Gamma}(t)},\\ \breve{C} = \breve{C}_u, \breve{T} = \breve{T}_u, n = n_u, v_z = -\frac{\beta D^2}{2\widehat{\Gamma}(t)},\end{array}\right\} \text{ at } z = \widehat{\Gamma}(t), \quad (26)$$

3. Similarity Transformations

Introducing the subsequent similarity variables satisfying the continuity equation, for instance:

$$\left\{\begin{array}{l} v_r = r\frac{\partial F}{\partial z} = \frac{\beta r}{2}\frac{D^2}{\widehat{\Gamma}^2(t)} f'(\lambda), v_\theta = G(z,t)r = r\Omega_1\frac{D^2}{\widehat{\Gamma}^2(t)} g(\lambda),\\ v_z = -2F(z,t) = -\frac{\beta D^2 f(\lambda)}{\widehat{\Gamma}(t)},\\ B_r = r\frac{\partial M}{\partial z} = \frac{\beta r DM_0}{2\widehat{\Gamma}^2(t)} m'(\lambda), B_\theta = rN(z,t) = rN_0 \frac{D^2}{\widehat{\Gamma}^2(t)} n(\lambda),\\ B_z = -2M(z,t) = -\frac{\beta DM_0 m(\lambda)}{\widehat{\Gamma}(t)},\\ \phi(\lambda) = \frac{\breve{C}-\breve{C}_u}{\breve{C}_l - \breve{C}_u}, \chi(\lambda) = \frac{n-n_u}{n_l - n_u}, \theta = \frac{\breve{T}-\breve{T}_u}{\breve{T}_l - \breve{T}_u}, \lambda = \frac{z}{\widehat{\Gamma}(t)}.\end{array}\right. \quad (27)$$

where similarity variable is λ and $f(\lambda), g(\lambda), m(\lambda), n(\lambda), \theta(\lambda), \phi(\lambda)$ and $\chi(\lambda)$ are non-dimensional velocity in axial and tangential direction, the magnetic field in axial and tangential direction, temperature, concentration, and motile density function, respectively.

Now substituting the above-mentioned similarity transformation in Equations (6)–(16), following coupled, nonlinear ODE's with independent variable (λ) obtained as,

$$f^{(iv)}(\eta) = 4R_Q\left[3f'' - 2\left(\frac{R_\Omega}{S_Q}\right)^2 gg' + 2F_T^2(mm''' + m'm'') - (2f - \lambda)f''' + 2F_A^2\left(\frac{R_\Omega}{S_Q}\right)^2 nn'\\ -4K\left[\frac{2R_\Omega}{R_Q}g'g'' + \frac{R_Q}{R_\Omega}[2f''f''' + 2(f''f''' + f'f^{iv})]\right]\right], \quad (28)$$

$$g''(\eta) = 2S_Q^2\left[2g + \lambda g' + 2gf' - fg' + 2F_AF_T(mn' + nm')\right] - 2K\left[g'(\eta)f''(\eta) - f'(\eta)g''(\eta)\right], \quad (29)$$

$$m'' = \text{Re}_M\left[m + \lambda m' + 2mf' - 2fm'\right], \quad (30)$$

$$n'' = \text{Re}_M\left[2n - fn' + \lambda n' + 2\left(\frac{F_A}{F_T}\right)mg'\right], \quad (31)$$

$$\left(1 + \frac{4}{3}R_d(1 + (T_r - 1)\breve{\theta})^3\right)\breve{\theta}'' + 4Rd(T_r - 1)(1 + (T_r - 1)\breve{\theta})^2\breve{\theta}'^2 + S_QP_tf\breve{\theta}' + T_t\breve{\theta}'^2 + T_b\breve{\theta}'\phi' = 0, \quad (32)$$

$$\phi'' + \frac{T_t}{T_b}\breve{\theta}'' + S_QS_Mf\phi' - S_M\sigma(1 + \widetilde{\delta}\breve{\theta})^n\exp\left(-\frac{E}{1 + \widetilde{\delta}\breve{\theta}}\right)\phi = 0, \quad (33)$$

$$\chi'' - S_QB_s\left(\frac{\lambda}{2}\right)\chi' + B_sS_Qf\chi' - P_l\left[\chi'\phi' + (\chi + \Phi)\phi''\right] = 0. \quad (34)$$

where S_Q represents the squeezed Reynolds number, R_Ω the rotational Reynolds number, F_A, F_T, denote the strength of the magnetic field in axial and azimuthal direction, Re_M the magnetic Reynolds number, K the material parameter of Reiner-Rivlin fluid, T_b the Brownian motion, P_t the Prandtl number, T_t the Thermophoresis parameter, E the non-dimensional form of Arrhenius activation energy, S_M the Schmidt number, B_s the bioconvection Schmidt number, σ the rate of chemical reaction, P_l the Peclet number, $\widetilde{\delta}$ represents the temperature ratio, T_r the temperature ratio parameter, R_d the radiation parameter, and Φ the constant number, respectively. They can be written as

$$\begin{cases} S_Q = \frac{\beta D^2}{2\nu}, R_\Omega = \frac{\Omega_1 D^2}{\nu}, F_T = \frac{M_0}{D\sqrt{\mu_2\rho}}, F_A = \frac{N_0}{\Omega_1\sqrt{\mu_2\rho}}, K = \frac{\mu_c\Omega}{\mu}, \\ T_b = \frac{\tau D_B(\breve{C}_l - \breve{C}_u)}{\widetilde{\alpha}}, T_t = \frac{\tau D_T(\breve{T}_l - \breve{T}_u)}{\widetilde{\alpha}\breve{T}_u}, P_t = \frac{\nu}{\widetilde{\alpha}}, \widetilde{\alpha} = \frac{k}{(\rho c)_p}, S_M = \frac{\nu}{D_B}, \\ B_s = \frac{\nu}{D_n}, P_l = \frac{bW_{mo}}{D_{mo}}, \Phi = \frac{n_u}{n_l - n_u}, Bt = \delta\mu_2\nu, \text{Re}_M = R_QBt, R_d = \frac{4\breve{T}_u^3\sigma_e}{\beta_r k}, \\ E = \frac{E_a}{\kappa T_u}, \sigma = \frac{k_r^2\widehat{\Gamma}(t)^2}{\nu}, \widetilde{\delta} = \frac{\breve{T}_l - \breve{T}_u}{\breve{T}_u}, \tau = \frac{(\rho c)_p}{(\rho c)_f}, T_r = \frac{T_l}{T_u} \end{cases} \quad (35)$$

where Bt represents Batchelor number.

The boundary conditions said in Equations (25) and (26) reduced as

$$\begin{cases} f'(0) = 0, f(0) = 0, m(0) = 0, g(0) = 1, n(0) = 1, \breve{\theta}(0) = 1, \chi(0) = 1, \phi(0) = 1, \\ f(1) = \frac{1}{2}, m(1) = 1, g(1) = \dot{\xi}, n(1) = 1, \breve{\theta}(1) = 0, \phi(1) = 0, \chi(1) = 0 \end{cases} \quad (36)$$

where $f, g, n, m, \theta, \phi, \chi$ denotes axial velocity and tangential velocity, magnetic field components in the tangential and axial direction, temperature distribution, nanoparticles concentration, motile gyrotactic microorganism profile, $\dot{\xi}(= \Omega_2/\Omega_1)$ represents the angular velocity, and its range is in between the rotating plates $-1 \leq \dot{\xi} \leq 1$. It is beneficial to investigate various revolving flow attributes of rotating plates in the reverse or same direction.

On the upper (moving) plate, the dimensionless torque can be calculated as

$$\widehat{T}_{up} = 2\pi\rho\int_0^b\left(\frac{\partial v}{\partial z}\right)_{z=\widehat{\Gamma}(t)}dr, \quad (37)$$

where the plate radius is signified by b.

Using Equation (27) in Equation (37), it becomes

$$\widehat{T}_{up} = \frac{dg(1)}{d\lambda}, \quad (38)$$

where the upper plate torque is designated by $\hat{T}_{up}$, and the tangential velocity gradient on the upper (moving) plate is $dg(1)/d\lambda$.

In the same fashion, the lower plate torque in dimensionless form is achieved by similar calculation and it becomes for $\lambda = 0$ as

$$\hat{T}_{lp} = \frac{dg(0)}{d\lambda}. \tag{39}$$

4. Solution of the Problem by DTM-Padé

DTM was first introduced by Zhou [55] in an engineering analysis for electric circuit theory for linear and nonlinear problems. It is an extremely powerful method for finding the solutions of magnetohydrodynamics and complex material flow problem. The Differential Transform Method (DTM) is distinct from the conventional higher-order Taylor series scheme. It was also used in combination with Padé approximants very successfully. The purpose of applying Padé-approximation is to improve the convergence rate of series solutions. The reason behind this is that sometimes the DTM fails to converge. That is why most of the researchers' merge DTM and Padé approximation to deal with the high order nonlinear differential equations. The Padé approximation is a rational function that can be thought of as a generalization of a Taylor polynomial. A rational function is the ratio of polynomials. Because these functions only use the elementary arithmetic operations, they are very easy to evaluate numerically. The polynomial in the denominator allows one to approximate functions that have rational singularities All the codes are developed on Mathematica software. The dimensionless Equations (28)–(36) are attained with the help of similar transformations stated in Equation (27), which are solved by virtue of the Differential Transform Method. To proceed further with the DTM technique, let us define q^{th} derivative as:

$$F(\lambda) = \frac{1}{q!}\left[\frac{d^q f}{d\lambda^q}\right]_{\lambda=\lambda_0}, \tag{40}$$

where $f(\lambda)$ are original and $F(\lambda)$ represent transformed functions. Now the differential inverse transform $F(\lambda)$ can be defined as

$$f(\lambda) = \sum_{q=0}^{\infty} F(\lambda)(\lambda - \lambda_0)^q, \tag{41}$$

The objective of differential transformation has been achieved by the Taylor extension series, and in terms of the finite series, the function $f(\lambda)$ can be defined as

$$f(\lambda) \cong \sum_{q=0}^{k} F(\lambda)(\lambda - \lambda_0)^q, \tag{42}$$

The rate of convergence depends upon the value of k. Each BVP can be converted to IVP with the replacement of unknown initial conditions. Taking differential transformation of the separate term by term of Equations (28)–(36), the following transformations are attained:

$$\left.\begin{array}{r}
f'' \to (1+\lambda)(2+\lambda)f(\lambda+2),\\
f'''^3 \to \left[\sum_{\tilde{v}=0}^{\lambda}\left(\begin{array}{c}\sum_{r=0}^{\omega}(\omega+1)(\omega+2)(-\omega+\tilde{v}+1)(-\tilde{v}+\lambda+1)(-\tilde{v}+\lambda+2)\\ f(-\tilde{v}+\lambda+2)f(2+\omega)f(-\omega+2+\lambda)\end{array}\right)\right],\\
f'f''f''' \to \left[\sum_{\tilde{v}=0}^{\lambda}\left(\begin{array}{c}\sum_{\tilde{v}=0}^{\lambda-\omega}(\omega+1)(1+\omega)(2+\omega)(-\omega+\lambda-\tilde{v}+1)(-\omega+2+\lambda-\tilde{v})\\ (-\tilde{v}+\lambda-\omega+3)f(1+\omega)f(2+\tilde{v})f(-\omega+\lambda-\tilde{v}+3)\end{array}\right)\right],\\
f''f'''^2 \to \left[\sum_{\tilde{v}=0}^{\lambda}\left(\begin{array}{c}\sum_{\tilde{v}=0}^{\lambda-\omega}(1+\omega)(\omega+2)(3+\omega)(\tilde{v}+1)(2+\tilde{v})(-\tilde{v}+\lambda+1-\omega)\\ (-\tilde{v}+\lambda+2-\omega)(-\tilde{v}+\lambda-\omega+3)f(3+\omega)f(2+\tilde{v})f(-\tilde{v}+\lambda-\omega+3)\end{array}\right)\right],
\end{array}\right\} \tag{43}$$

$$
\left.\begin{aligned}
g &\to g(l), \\
\lambda g' &\to \sum_{\omega=0}^{\lambda} ((-\omega+1+\lambda)\varepsilon(\omega)g(-\omega+1+\lambda)),
\end{aligned}\right\} \quad (44)
$$

$$
\left.\begin{aligned}
fg' &\to \sum_{\omega=0}^{\lambda} (-\omega+1+\lambda)f(\omega)g(-\omega+1+\lambda), \quad gf' \to \sum_{\omega=0}^{\lambda} (-\omega+1+\lambda)g(\omega)f(-\omega+1+\lambda), \\
g'f'g'' &\to \sum_{\omega=0}^{\lambda} (\omega+1)(-\omega+1+\lambda)(-\omega+2+\lambda)f(1+\omega)g(1+\omega)g(-\omega+2+\lambda), \\
g''g'f''' &\to \sum_{\omega=0}^{\lambda} (1+\omega)(\omega+2)(-\omega+1+\lambda)(-\omega+2+\lambda)(-\omega+\lambda+3)g(1+\omega)g(2+\omega) \\
&\qquad g(-\omega+\lambda+3), \\
g'f'f'' &\to \sum_{m=0}^{\lambda} (1+\omega)(-\omega+\lambda+1)(-\omega+2+\lambda)f(1+\omega)g(1+\omega)f(-\omega+2+\lambda), \\
f'''g'f'' &\to \sum_{\omega=0}^{\lambda} (1+\omega)(2+\omega)(-\omega+1+\lambda)(-\omega+2+\lambda)(-\omega+\lambda+3)g(\omega+1)f(2+\omega) \\
&\qquad f(-\omega+3+\lambda), \\
f''g'^2 &\to \sum_{\omega=0}^{\lambda} \left(\begin{aligned}\sum_{\tilde{v}=0}^{\lambda} (1+\omega)(2+\omega)(1-\omega+\tilde{v})(-\tilde{v}+1+\lambda)g(-\tilde{v}+1+\lambda)f(2+\omega) \\ g(-\omega+1+\lambda)\end{aligned}\right), \\
g''^2 f'' &\to \sum_{\omega=0}^{\lambda} \left(\begin{aligned}\sum_{\tilde{v}=0}^{\lambda-\omega} (\omega+1)(2+\omega)(\tilde{v}+1)(\tilde{v}+2)(-\tilde{v}-1-\omega+\lambda)(-\omega+2+\lambda-\tilde{v}) \\ g(2+\omega)f(q+2)g(-\omega+2+\lambda-\tilde{v})\end{aligned}\right), \\
f'^2 g'' &\to \sum_{\tilde{v}=0}^{\lambda} \left(\begin{aligned}\sum_{\omega=0}^{\tilde{v}} (1+\omega)(2+\omega)(-\omega+1+\tilde{v})(-\tilde{v}+1+\lambda)f(-\tilde{v}+1+\lambda)g(2+\omega) \\ f(-\omega+\lambda+1)\end{aligned}\right),
\end{aligned}\right\} \quad (45)
$$

$$
\left.\begin{aligned}
m'm'' &\to \sum_{\omega=0}^{\lambda} (\omega+1)(2+\omega)(-\omega+1+\lambda)(-\omega+2+\lambda)m(\omega+1)m(-\omega+2+\lambda), \\
\lambda m' &\to \sum_{\omega=0}^{\lambda} ((-\omega+1+\lambda)\varepsilon(\omega)m(-\omega+\lambda+1)), \\
mf' &\to \sum_{\omega=0}^{\lambda} ((-\omega+1+\lambda)m(\omega)f(-\omega+1+\lambda)), \\
fm' &\to \sum_{\omega=0}^{\lambda} ((-\omega+1+\lambda)f(\omega)m(-\omega+1+\lambda)), \\
mg' &\to \sum_{\omega=0}^{\lambda} ((-\omega+1+\lambda)m(\omega)g(-\omega+1+\lambda)),
\end{aligned}\right\} \quad (46)
$$

$$
\left.\begin{aligned}
nn' &\to \sum_{m=0}^{\lambda} ((-\omega+1+\lambda)n(\omega)n(-\omega+1+\lambda)), \\
fn' &\to \sum_{\omega=0}^{\lambda} ((-\omega+1+\lambda)f(\omega)n(-\omega+1+\lambda)), \\
\lambda n' &\to \sum_{\omega=0}^{\lambda} ((-\omega+1+\lambda)\varepsilon(\omega)n(-\omega+1+\lambda)),
\end{aligned}\right\} \quad (47)
$$

$$
\left.\begin{aligned}
f\breve{\theta}' &\to \sum_{\omega=0}^{\lambda} \left((-\omega+1+\lambda)f(\omega)\breve{\theta}(-\omega+1+\lambda)\right), \\
\breve{\theta}'^2 &\to \sum_{\omega=0}^{\lambda} \left((1+\omega)(-\omega+1+\lambda)\breve{\theta}(1+\omega)\breve{\theta}(1-\omega+\lambda)\right),
\end{aligned}\right\} \quad (48)
$$

$$
\left.\begin{aligned}
\breve{\theta}'\phi' &\to \sum_{\omega=0}^{\lambda} \left((1+\omega)(-\omega+1+\lambda)\breve{\theta}(1+\omega)\phi(-\omega+1+\lambda)\right), \\
f\phi' &\to \sum_{\omega=0}^{\lambda} ((-\omega+1+\lambda)f(\omega)\phi(-\omega+1+\lambda)),
\end{aligned}\right\} \quad (49)
$$

$$\left.\begin{aligned}
\lambda\chi' &\to \sum_{\omega=0}^{\lambda}((-\omega+1+\lambda)\varepsilon(\omega)\chi(-\omega+1+\lambda)), \\
f\chi' &\to \sum_{\omega=0}^{\lambda}((-\omega+1+\lambda)f(\omega)\chi(-\omega+1+\lambda)), \\
\chi'\phi' &\to \sum_{\omega=0}^{\lambda}((1+\omega)(-\omega+1+\lambda)\chi(\omega+1)\phi(-\omega+1+\lambda)), \\
\chi\phi'' &\to \sum_{\omega=0}^{\lambda}((-\omega+1+\lambda)(-\omega+2+\lambda)\chi(\omega)\phi(-\omega+2+\lambda)),
\end{aligned}\right\} \quad (50)$$

where $f(l), g(l), m(l), n(l), \breve{\theta}(l), \phi(l)$ and $\chi(l)$ are the transformed function of $f(\lambda), g(\lambda), m(\lambda), n(\lambda), \theta(\lambda), \phi(\lambda)$ and $\chi(\lambda)$, respectively, and are expressed as

$$f(\lambda) = \sum_{l=0}^{\infty} f(l)\lambda^l, \qquad (51)$$

$$g(\lambda) = \sum_{l=0}^{\infty} g(l)\lambda^l, \qquad (52)$$

$$m(\lambda) = \sum_{l=0}^{\infty} m(l)\lambda^l, \qquad (53)$$

$$n(\lambda) = \sum_{l=0}^{\infty} n(l)\lambda^l, \qquad (54)$$

$$\breve{\theta}(\lambda) = \sum_{l=0}^{\infty} \breve{\theta}(l)\lambda^l, \qquad (55)$$

$$\phi(\lambda) = \sum_{l=0}^{\infty} \phi(l)\lambda^l, \qquad (56)$$

$$\chi(\lambda) = \sum_{l=0}^{\infty} \chi(l)\lambda^l. \qquad (57)$$

By applying differential transform on corresponding boundary conditions, we obtained

$$\left.\begin{aligned}
&f(0) = 0, \quad f(1) = \tfrac{1}{2}, \quad g(0) = 1, \quad m(0) = 0, \quad n(0) = 0, \\
&\breve{\theta}(0) = 1, \quad \phi(0) = 0, \quad \chi(0) = 0, \quad f(2) = \Pi_1, \quad f(3) = \Pi_2, \\
&g(1) = \Pi_3, \quad m(1) = \Pi_4, \quad n(1) = \Pi_5, \quad \breve{\theta}(1) = \Pi_6, \quad \phi(1) = \Pi_6, \\
&\chi(1) = \Pi_8
\end{aligned}\right\}, \quad (58)$$

where Π_e ($e = 1, \ldots, 8$) are the constants. Substituting transformations given in Equations (43)–(50) into Equations (30)–(36), and solved with support of associated boundary conditions shown in Equation (58), the resulting solutions in the form of the series are:

$$f(\lambda) = \dot{f}_1\lambda^2 + \dot{f}_2\lambda^3 + \dot{f}_3\lambda^4 + \dot{f}_4\lambda^5 + \ldots, \qquad (59)$$

$$g(\lambda) = 1 - \dot{g}_1\lambda + \dot{g}_2\lambda^2 + \dot{g}_3\lambda^3 + \dot{g}_4\lambda^4 + \ldots, \qquad (60)$$

$$m(\lambda) = \dot{m}_1\lambda + \dot{m}_2\lambda^3 + \dot{m}_3\lambda^4 + \dot{m}_4\lambda^5 + \ldots, \qquad (61)$$

$$n(\lambda) = \dot{n}_1\lambda + \dot{n}_2\lambda^3 + \dot{n}_3\lambda^4 + \dot{n}_4\lambda^5 + \ldots, \qquad (62)$$

$$\breve{\theta}(\lambda) = 1 + \dot{\theta}_1\lambda + \dot{\theta}_2\lambda^2 + \dot{\theta}_3\lambda^3 + \dot{\theta}_4\lambda^4 + \ldots, \qquad (63)$$

$$\phi(\lambda) = 1 + \dot{\phi}_1\lambda + \dot{\phi}_2\lambda^2 + \dot{\phi}_3\lambda^3 + \dot{\phi}_4\lambda^4 + \ldots, \qquad (64)$$

$$\chi(\lambda) = 1 + \dot{\chi}_1\lambda + \dot{\chi}_2\lambda^2 + \dot{\chi}_3\lambda^3 + \dot{\chi}_4\lambda^4 + \ldots, \qquad (65)$$

where $\dot{f}_i, \dot{g}_i, \dot{m}_i, \dot{n}_i, \dot{\theta}_i, \dot{\phi}_i, \dot{\chi}_i$; where $i = (1, 2, 3, \ldots)$ are constants. It is not easy to express them here because of their complex and long numerical values. With the assistance of Mathematica computational software, the equation as mentioned above is solved with 30 iterations. However, it failed to obtain a reasonable rate of convergence. The convergence rate of certain sequences can be improved with certain techniques. Many researchers used the Padé technique, which was used in the form of a rational fraction, i.e., ratio of two polynomials. The results obtained by DTM, owing to the non-linearity on the governing equations, do not satisfy the boundary conditions at infinity without applying the Padé approximation. The obtained solution by DTM must then be merged with Padé-approximation, which gives a substantial rate of convergence at infinity. According to one's desired exactness, a higher order of approximation is required. Here, $[5 \times 5]$ order approximation is applied to Equations (59)–(65), the Padé approximants are as follows.

$$f(\lambda) = \frac{1.744240\lambda^2 - 6.384709\lambda^3 + 7.800949\lambda^4 - 2.873131\lambda^5 + \ldots}{1 - 2.775474\lambda + 1.798969\lambda^2 + 0.612641\lambda^3 - 0.047549\lambda^4 - 0.001808\lambda^5 + \ldots}, \quad (66)$$

$$g(\lambda) = \frac{1 - 0.461931\lambda - 0.480814\lambda^2 - 0.030027\lambda^3 - 0.036507\lambda^4 - 0.008722\lambda^5 + \ldots}{1 + 0.580516\lambda + 0.110708\lambda^2 + 0.014909\lambda^3 + 0.003980\lambda^4 + 0.003613\lambda^5 + \ldots}, \quad (67)$$

$$m(\lambda) = \frac{0.706586\lambda - 0.052291\lambda^2 - 0.0453936\lambda^3 + 0.272736\lambda^4 - 0.322963\lambda^5 + \ldots}{1 - 0.074006\lambda - 0.397576\lambda^2 + 0.119954\lambda^3 - 0.060980\lambda^4 - 0.027780\lambda^5 + \ldots}, \quad (68)$$

$$n(\lambda) = \frac{0.767837\lambda + 1.046017\lambda^2 + 0.365143\lambda^3 + 0.429179\lambda^4 + 0.171075\lambda^5 + \ldots}{1 + 1.362290\lambda + 0.039499\lambda^2 + 0.254792\lambda^3 + 0.340033\lambda^4 - 0.212403\lambda^5 + }, \quad (69)$$

$$\breve{\theta}(\lambda) = \frac{1 - 0.794515\lambda - 0.240481\lambda^2 + 0.053229\lambda^3 - 0.032680\lambda^4 + 0.014409\lambda^5 + \ldots}{1.0 + 0.038878\lambda - 0.035237\lambda^2 + 0.050992\lambda^3 - 0.033598\lambda^4 + 0.00129\lambda^5 + \ldots}, \quad (70)$$

$$\phi(\lambda) = \frac{1 - 1.715367\lambda + 0.560355\lambda^2 + 0.391456\lambda^3 - 0.354482\lambda^4 + 0.119370\lambda^5 + \ldots}{1 + 0.217143\lambda - 0.068712\lambda^2 + 0.047085\lambda^3 - 0.043507\lambda^4 - 0.008306\lambda^5 + \ldots}, \quad (71)$$

$$\chi(\lambda) = \frac{1 - 0.776897\lambda + 0.662042\lambda^2 - 0.785269\lambda^3 + 0.099751\lambda^4 - 0.193734\lambda^5 + \ldots}{1 + 2.4876949\lambda + 2.462925\lambda^2 + 1.100656\lambda^3 + 0.134289\lambda^4 - 0.0376572\lambda^5 + \ldots}, \quad (72)$$

5. Graphical and Numerical Analysis

In this segment, graphical and numerical analysis is made on the solutions of resulting nonlinear ordinary differential equations mentioned in Equations (28)–(36). The differential transformation scheme is applied to present the solutions of the foregoing equations. Our principal focus is to inspect the physical characteristics of numerous physical parameters in the momentum equation, induced MHD equations, temperature distribution, motile microorganism density function, and mass transfer equation. For instance, the influence of squeezing and Rotational Reynolds number S_Q, R_Ω, Reiner-Rivlin fluid parameter K, Brownian motion T_b, magnetic Reynolds number Re_M, Prandtl number P_t, thermophoresis parameter T_t, Schmidt number S_M, Bioconvection number B_s, and Peclet number P_l are examined.

Table 1 shows the numerical comparison with previous results [56] against the torque values at the upper and the lower plate by taking $K = 0, R_d = 0, \sigma = 0$ in the present results. It is found that the results obtained in the present study are not only correct but also converge rapidly. Furthermore, we can also say that the proposed methodology, i.e., DTM-Padé shows promising results against the coupled nonlinear different equations.

Tables 2–4 shows the different physical parameters developed against Sherwood number, Nusselt number, and motile density function $[\phi'(0), \theta'(0), \chi'(0)]$. Moreover, the torque values at the lower plate $dg(0)/d\lambda$, and upper plate $dg(1)/d\lambda$ are also calculated numerically in Tables 5 and 6.

Table 1. Comparison of the torque values at the lower and upper plate with previous results [56] when the fluid behaves as a Newtonian model ($K = 0$) and the remaining values are $R_\Omega = 0.3$, $F_T = 0.5$, $Bt = 0.6$, $K = 0$, $R_d = 0$, $\sigma = 0$ for various values of S_Q and R_Ω.

S_Q	$\frac{dg(0)}{d\lambda}$		$\frac{dg(1)}{d\lambda}$	
	Zhang et al. [56]	Present Results	Zhang et al. [56]	Present Results
0.1	−1.0929372214309236	−1.0929372214309236	−0.948663684660318	−0.948663684660318
0.2	−1.180889912821983	−1.180889912821983	−0.9013607839508947	−0.9013607839508947
R_Ω				
0.1	−1.265492575299778	−1.265492575299778	−0.8533000683642988	−0.8533000683642988
0.2	−1.2652748717875888	−1.2652748717875888	−0.8549052425970227	−0.8549052425970227

Table 2. Analysis of Nusselt number $\theta'(0)$, for multiple values T_t, T_b, P_t, S_Q by DTM-Padé [5 × 5].

				$K = 0$	$K = 0.1$
T_t	T_b	P_t	S_Q	DTM-Padé	
0.03	0.01	6.8	0.05	−0.8944762272711257	−0.8944824906474336
0.06				−0.8063777250952755	−0.806383306245722
0.09				−0.7253079370829076	−0.7253128728932438
0.05	0.2			−0.37080900827594065	−0.37081139690286175
	0.3			−0.2316716738652042	−0.23167310222667004
	0.4			−0.1411942252519443	−0.14119505897571566
	0.01	4		−0.8998893435594736	−0.899893033465222
		7		−0.8304624112857932	−0.8304683534204742
		10		−0.7656313391774631	−0.7656391365026328
		6.8	−0.01	−0.8068399501595973	−0.8068414709109244
			0.05	−0.8334251602591399	−0.8334257278827505
			0.10	−0.8553872157158188	−0.8553910187144096

Table 3. Analysis of Sherwood number $\phi'(0)$ for various values T_t, T_b, S_Q, S_M, E, σ by DTM-Padé [5 × 5].

						$K = 0$	$K = 0.1$
T_t	T_b	S_Q	S_M	E	σ	DTM-Padé	DTM-Padé
0.03	0.01	0.05	1	1	1	−1.423723583572087	−1.4237062656685955
0.06						−2.242251830648612	−2.24222066465715
0.09						−3.52021034762737	−3.5201695528198456
0.05	0.01					−1.915474205493601	−1.9154471661812198
	0.02					−1.5921974868312598	−1.5921851742637334
	0.03					−1.4831389820481062	−1.4831315689992728
	0.01	−0.01				−2.0517295527621053	−2.0517273504948057
		0.05				−1.9367842616882913	−1.9268295351904132
		0.10				−1.832060050336652	−1.8320099157048728
		0.05	2			−2.0146398467429014	−2.0241644922138637
			4			−2.2012470588694386	−2.2102937724970664
			6			−2.3744735466373146	−2.3831035839166357
			1	2		−1.8531063430623982	−1.8630765987997622

Table 3. Cont.

						K = 0	K = 0.1
T_t	T_b	S_Q	S_M	E	σ	DTM-Padé	DTM-Padé
				3		−1.829718893888726	−1.8397573168178392
				4		−1.821048746320453	−1.8311125640363681
				1	2	−2.01093920918129	−2.020462840246665
					4	−2.19121593549506	−2.2002581288517207
					6	−2.359245916149752	−2.367866612938209

Table 4. Analysis of $\chi'(0)$ for various values of S_Q, B_s, P_l by DTM-Padé [5 × 5].

			K = 0	K = 0.1
S_Q	B_s	P_l	DTM-Padé	DTM-Padé
−0.01	1	1	−2.5484740630681886	−2.5484721652641165
0.05			−2.4422867164206274	−2.442278141197654
0.10			−2.3326945059751685	−2.3545141881576237
0.05	5		−2.4315407355757515	−2.4315225967036653
	10		−2.4300912988489185	−2.430083525231539
	15		−2.4286390039889302	−2.4286422222719586
	1	0.5	−1.6814808605966678	−1.6814680934795112
		1.0	−2.432698678642556	−2.4326722455225793
		1.5	−3.2408290233020116	−3.2407890961974375

Table 5. Numerical computations of Torque at a fix circular and upper circular plates by DTM-Padé [5 × 5] for various values of Squeezing Reynolds Number S_Q.

	$\frac{dg(0)}{d\lambda}$		$\frac{dg(1)}{d\lambda}$	
S_Q	K = 0	K = 0.1	K = 0	K = 0.1
0.1	−1.0944523632334688	−1.0885716574469078	−0.9499690408077309	−0.9876461173721226
0.2	−1.1811908734455248	−1.1746346120145237	−0.9071760445081409	−0.9777866647632717

Table 6. Numerical computations of Torque at a fix circular and upper circular plates by DTM-Padé [5 × 5] for multiple values of Rotational Reynolds Number R_Ω.

	$\frac{dg(0)}{d\lambda}$		$\frac{dg(1)}{d\lambda}$	
R_Ω	K = 0	K = 0.1	K = 0	K = 0.1
0.1	−1.0470698634685973	−1.0416647863739605	−0.9735621738611226	−0.9805868144292514
0.2	−1.047887492344034	−1.0424472521086106	−0.9736078836468506	−0.9885069828641508

Figure 2 illustrates the influence of the velocity profile in the axial direction f' because of the squeezed Reynolds number S_Q, rotational Reynolds number R_Ω, and the material parameter of Reiner-Rivlin fluid K. From Figure 2 one can perceive that increasing the squeezed Reynolds number S_Q axial velocity decreases, but increasing rotational Reynolds number R_Ω, the axial velocity profile increases. The physical reason behind this is that when we increase the value of Squeezing Reynolds number S_Q, the distance between the plates increases, the fluid velocity decreases, and the fluid accelerates by rotation of the plate when we increase the values of rotational Reynolds number R_Ω. Figure 3 depicts that increasing the values of the material parameter of the Reiner-Rivlin fluid increases the velocity distribution against axial direction f'.

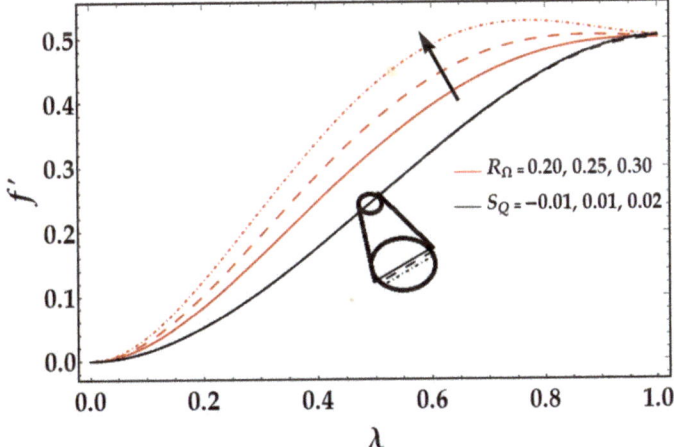

Figure 2. Implications of S_Q and R_Ω on velocity distribution (axial) $f'(\lambda)$.

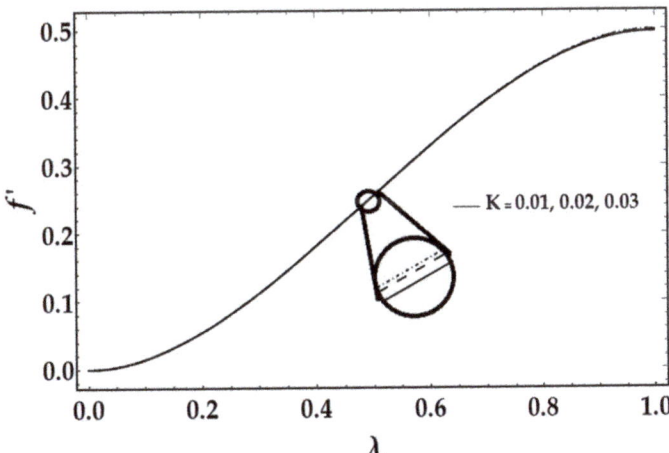

Figure 3. Implications of K on velocity distribution (axial) $f'(\lambda)$.

Figure 4 depicts the influence of squeezing Reynolds number S_Q and Rotational Reynolds Number R_Ω against tangential velocity distribution g'. From Figure 4, it can be ascertained that by enhancing the values of the squeezed Reynolds number S_Q, the tangential velocity distribution decreases. Similar phenomena are observed in Figure 5, i.e., by increasing the values of the rotational Reynolds number, the tangential velocity profile declines.

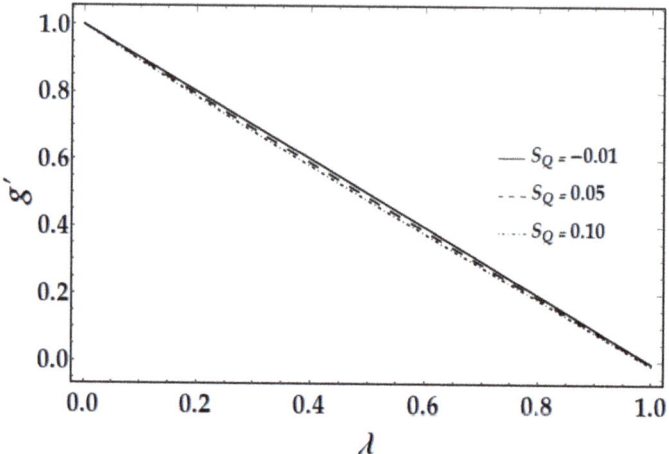

Figure 4. Implications of S_Q on velocity distribution (tangential) $g'(\lambda)$.

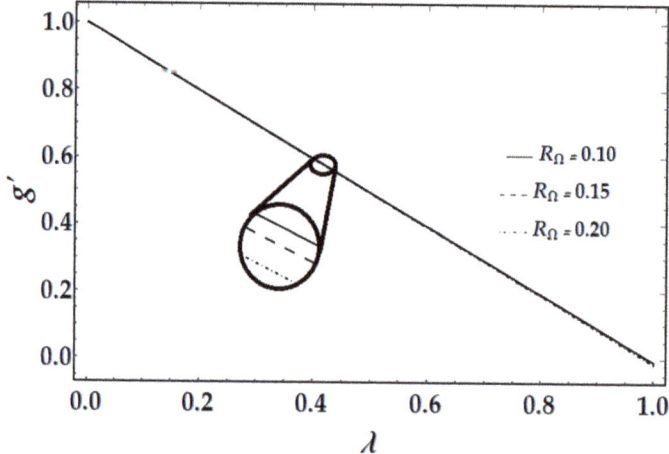

Figure 5. Implications of R_Ω on velocity distribution (tangential) $g'(\lambda)$.

From Figure 6, it can be seen that by increasing the values of magnetic Reynolds number Re_M, the tangential and axial magnetic field decreases, as the magnetic Reynolds number is the ratio of fluid flux to the mass diffusivity. So, by increasing the magnetic Reynolds number, a decrease in mass diffusivity and increase in fluid flux is seen. This decline in mass diffusivity disrupts the diffusion of the magnetic field and resulting, a decline in axial and tangential induced magnetic fields is observed.

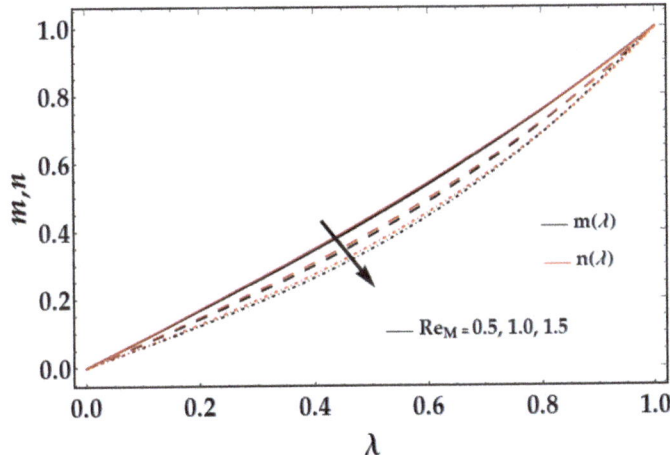

Figure 6. Implications of Re_M on $m(\lambda)$, $n(\lambda)$ in axial and tangential direction.

Figure 7 elucidates the consequences of the Brownian motion parameter and thermophoresis parameter T_b, T_t on the temperature field $\tilde{\theta}$. The graph shows that intensifying the values of thermophoresis, Brownian motion parameter T_t, T_b increases the temperature profile. The physical reason is that the fluid temperature increases due to strengthening the kinetic energy of nanoparticles. The effects of squeezing Reynolds number S_Q and Prandtl number P_t on temperature profile $\tilde{\theta}$ is displayed in Figure 8. One can notice that by enhancing the Prandtl number P_t and the squeezing Reynolds number S_Q, the temperature profile $\tilde{\theta}$ diminishes. When the thermal conductivity reduces by intensifying the values of the Prandtl number P_t then the temperature profile $\tilde{\theta}$ declines. The effects of radiation parameter R_d on temperature profile $\tilde{\theta}$ are shown in Figure 9. It is observed that by enhancing the radiation parameter R_d the temperature profile $\tilde{\theta}$ increases. The physical reason behind this is that an increase in radiation releases the heat energy from flow; hence there is an increase in temperature.

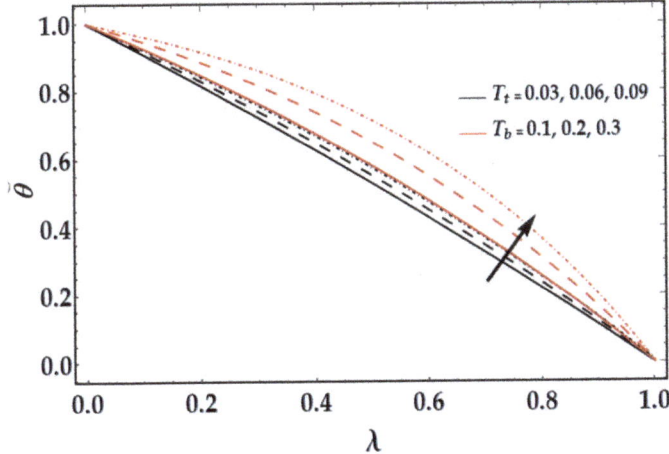

Figure 7. Implications of T_t and T_b on temperature function $\tilde{\theta}(\lambda)$.

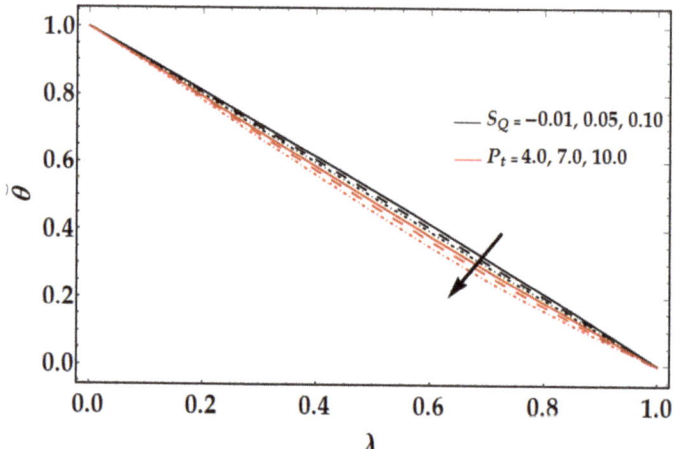

Figure 8. Implications of S_Q and P_t on temperature function $\tilde{\theta}(\lambda)$.

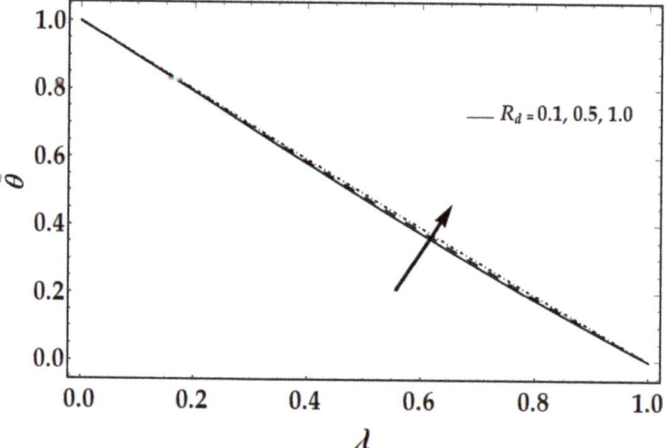

Figure 9. Implications of R_d, on temperature function $\tilde{\theta}(\lambda)$.

Figure 10 shows the consequences of thermophoresis parameter T_t and Brownian motion T_b on nanoparticle concentration ϕ. It is perceived that nanoparticle concentration declines by increasing the values of Brownian motion T_b, and concentration of nanoparticle intensifies by increasing values of thermophoresis parameter T_t. In fact, gradual growth in T_b increases the random motion and collision among nanoparticles of the fluid, which produces more heat and eventually it results in a decrease in the concentration field. Due to increasing values of T_t, more nanoparticles are pulled towards the cold surface from the hot one, which ultimately results in increasing the concentration distributions. Figure 11 shows the consequences of Schmidt number S_M and squeezed Reynolds number S_Q on nanoparticle concentration. By enlarging the values of squeezed Reynolds number S_Q, nanoparticle concentration ϕ increases, on the other hand, converse phenomena are noticed by enhancing the values of Schmidt number S_M. Figure 12 deliberates the influence of reaction rate σ and activation energy E on the nanoparticle concentration ϕ. It may be observed that nanoparticle concentration displays a substantial rise by increasing values of E. Since high energy activation and low temperatures impart to a constant reaction rate, the resulting chemical reaction is therefore slowed down. Consequently, the concentration of the solute rises. On the other side, by increasing values of σ, the nanoparticle concentration decreases.

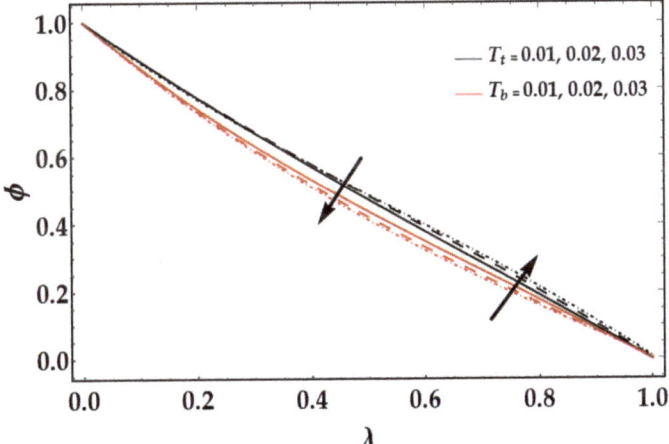

Figure 10. Implications of T_t and T_b on concentration function $\phi(\lambda)$.

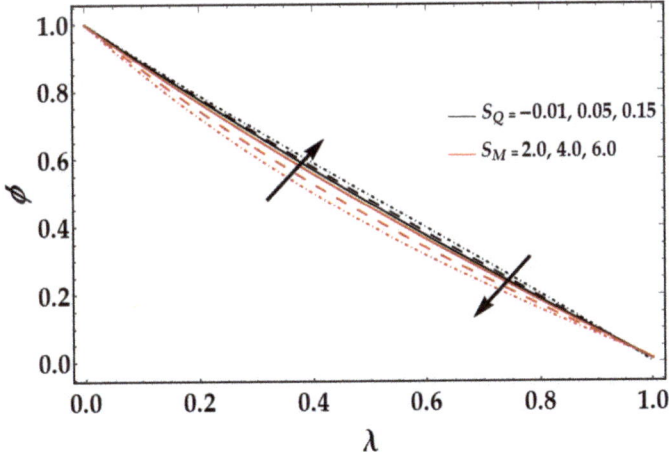

Figure 11. Implications of S_Q, S_M on concentration function $\phi(\lambda)$.

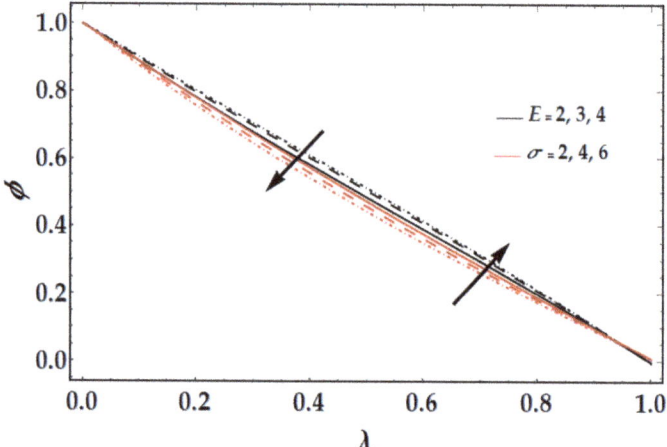

Figure 12. Implications of S_Q, S_M on concentration function $\phi(\lambda)$.

Figure 13 portrays the consequences of Peclet number P_l and squeezed Reynolds number S_Q on motile microorganism density function χ. One can experience that enhancing values of squeezed Reynolds number S_Q tends to boost the microorganism density function, while increasing the values of Peclet number P_l, the motile microorganism density function diminishes. The reason behind this is that the diffusivity of the microorganism reduces, then the speed of the microorganism also decreases. This is the physical fact and resulting in the microorganism density function decreasing while increasing the value of Peclet number P_l. Figure 14 is plotted to see the physical performance of the Bioconvection Schmidt number B_s. It is apparent that by enhancing values of bioconvection Schmidt number B_s the motile microorganism density function rises, but the consequences are negligible.

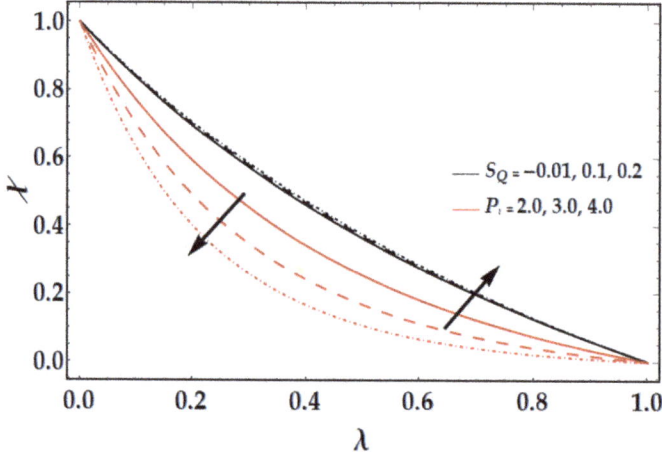

Figure 13. Implications of S_Q, P_l on motile microorganism density function $\chi(\lambda)$.

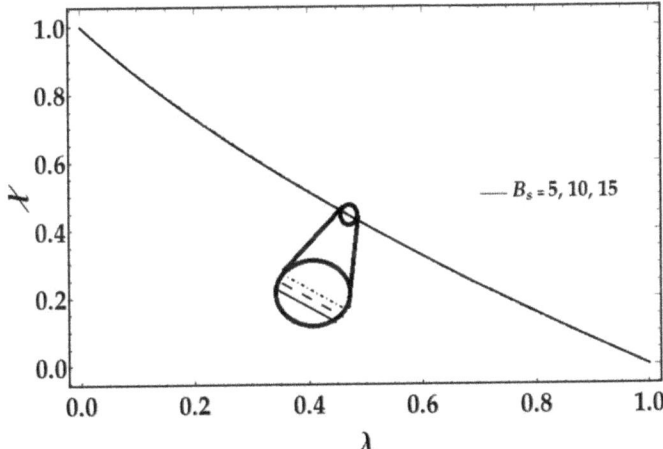

Figure 14. Implications of B_s on motile microorganism density function $\chi(\lambda)$.

6. Conclusions

In this study, we determined incompressible three-dimensional, unsteady, axisymmetric squeezed film flow of Reiner-Rivlin nanofluid between parallel circular plates. The impact of an induced magnetic field, the suspension of motile gyrotactic microorganisms, activation energy, and thermal radiation are also contemplated. DTM-Padé is applied to present the solutions of the ordinary differential equations after employing the similarity transformations. Padé approximant is applied because it provides a good rate of convergence and gives reliable results. Comparison is made for the values of toque on the lower and upper plates. The main findings are accomplished below:

i. The opposite behavior is experienced for the rotational Reynolds number on tangential and axial velocity distribution.
ii. Enhancing the value of squeezing Reynolds number, the tangential and axial velocity distribution decreases.
iii. By enlarging the value of the magnetic Reynolds number, the magnetic field (induced) in tangential and axial directions decreases.
iv. The Reiner-Rivlin opposes the fluid motion; however, the impact is negligible.
v. By increasing the Brownian motion and thermophoresis parameter, the temperature distribution rises.
vi. Temperature distribution decays due to the effects of the Prandtl number-like phenomenon is observed for enlarging the values of squeezed Reynolds number.
vii. The thermal radiation parameter enhances the temperature distribution.
viii. Nanoparticle concentration and motile density increase by enhancing the in value of squeezing Reynolds number.
ix. Nanoparticle concentration shows opposed phenomena for Brownian motion parameter compared with thermophoresis parameter.
x. Increasing values of activation energy tends to intensify the nanoparticle concentration profile.
xi. The microorganism profile declines by increasing the values of Peclet number, but the microorganism profile rises by enlarging the bioconvection number.

Future Work: The present study shows perfect accuracy of the proposed methodology; however, attention has been given to non-Newtonian fluid with induced magnetic. Future studies may generalize the present study to consider applied magnetics effects, porosity effects, slip effects, entropy generation, and other non-Newtonian fluid models, etc., which are beneficial to bioreactor configurations and lubrication regimes and will be presented soon.

Author Contributions: Conceptualization, M.M.B.; investigation, M.B.A. and A.Z.; methodology, M.B.A. and M.M.B.; validation, F.S.A.; writing-review and editing, M.B.A., A.Z and M.M.B. All authors have read and agreed to the published version of the manuscript.

Funding: This research received no external funding.

Institutional Review Board Statement: Not applicable.

Informed Consent Statement: Not applicable.

Data Availability Statement: Not applicable.

Conflicts of Interest: The authors declare no conflict of interest.

Nomenclature

H_θ	Axial components
H_z	Azimuthal components
μ_1	Magnetic permeability inside the plate
μ_2	Magnetic permeability outside the plate
μ_ℓ	Free space permeability (NA^{-2})
$\widehat{\mathbf{B}}(r,\theta,z)$	Induced magnetic field
T_0	Lower plates constant temperature (K)
T_l	Upper plates constant temperature (K)
C_l	Concentration at lower plate
C	Concentration at upper plate
p	Pressure (Pa)
ρ	Fluid density (Kg/m^3)
μ	Fluid viscosity (Ns/m^2)
δ	Electrical conductivity ($S \cdot m^{-1}$)
$\widetilde{T}$	Temperature (K)
$\widetilde{C}$	Concentration
$\widetilde{T}_m$	Mean fluid temperature (K)
c_p	Specific heat ($Jkg^{-1}K^{-1}$)
D_B	Brownian diffusivity
D_T	Thermophoretic diffusion coefficient
F_A	Magnetic force strength in the axial direction
F_T	Magnetic force in the tangential direction
Re_M	Magnetic Reynolds number
T_b	Brownian motion parameter
T_t	Thermophoresis parameter
P_t	Prandtl number
S_M	Schmidt number
B_s	bioconvection Schmidt number
f	Axial velocity (m/s)
g	Tangential velocity (m/s)
$\widetilde{\theta}$	Temperature profile of nanofluids (K)
ϕ	Concentration profile of nanofluids
χ	Motile density microorganism
$\dot{\zeta}$	Angular velocity (m/s)
b	Radius of the disk
$\widehat{T}_{up}$	Dimensionless torque applied on the upper plate
$\widehat{T}_{lp}$	Dsimensionless torque applied on the lower plate
b	Chemotaxis constant
W_{mo}	Maximal speed
S_Q	Squeezed Reynold number
D_{mo}	Diffusivity of micro-organisms

References

1. Choi, S.U.; Eastman, J.A. *Enhancing Thermal Conductivity of Fluids with Nanoparticles* (No. ANL/MSD/CP-84938; CONF-951135-29); Argonne National Lab.: Lemont, IL, USA, 1995.
2. Sarkar, J. A critical review on convective heat transfer correlations of nanofluids. *Renew. Sustain. Energy Rev.* **2011**, *15*, 3271–3277. [CrossRef]
3. Das, S.K.; Choi, S.U.; Patel, H.E. Heat transfer in nanofluids—A review. *Heat Trans. Eng.* **2006**, *27*, 3–19. [CrossRef]
4. Wong, K.V.; De Leon, O. Applications of nanofluids: Current and future. *Adv. Mechanic. Eng.* **2017**, *2*, 105–132. [CrossRef]
5. Goudarzi, S.; Shekaramiz, M.; Omidvar, A.; Golab, E.; Karimipour, A.; Karimipour, A. Nanoparticles migration due to thermophoresis and Brownian motion and its impact on Ag-MgO/Water hybrid nanofluid natural convection. *Powder Technol.* **2020**, *375*, 493–503. [CrossRef]
6. Ghalandari, M.; Koohshahi, E.M.; Mohamadian, F.; Shamshirband, S.; Chau, K.W. Numerical simulation of nanofluid flow inside a root canal. *Eng. Appl. Comput. Fluid Mech.* **2019**, *13*, 254–264. [CrossRef]
7. Sheikholeslami, M.; Vajravelu, K. Nanofluid flow and heat transfer in a cavity with variable magnetic field. *Appl. Math. Comput.* **2016**, *298*, 272–282. [CrossRef]
8. Sheikholeslami, M.; Ganji, D.D. Nanofluid flow and heat transfer between parallel plates considering Brownian motion using DTM. *Comput. Methods Appl. Mech. Eng.* **2014**, *283*, 651–663. [CrossRef]
9. Biswal, U.; Chakraverty, S.; Ojha, B.K.; Hussein, A.K. Numerical simulation of magnetohydrodynamics nanofluid flow in a semi-porous channel with a new approach in the least square method. *Int. Commun. Heat Mass Transf.* **2020**, *121*, 105085. [CrossRef]
10. Zhang, X.; Gu, H.; Fujii, M. Effective thermal conductivity and thermal diffusivity of nanofluids containing spherical and cylindrical nanoparticles. *Exp. Therm. Fluid Sci.* **2007**, *31*, 593–599. [CrossRef]
11. Fakour, M.; Vahabzadeh, A.; Ganji, D. Study of heat transfer and flow of nanofluid in permeable channel in the presence of magnetic field. *Propuls. Power Res.* **2015**, *4*, 50–62. [CrossRef]
12. Zhu, J.; Wang, S.; Zheng, L.; Zhang, X. Heat transfer of nanofluids considering nanoparticle migration and second-order slip velocity. *Appl. Math. Mech.* **2016**, *38*, 125–136. [CrossRef]
13. Alamri, S.Z.; Ellahi, R.; Shehzad, N.; Zeeshan, A. Convective radiative plane Poiseuille flow of nanofluid through porous medium with slip: An application of Stefan blowing. *J. Mol. Liq.* **2018**, *273*, 292–304. [CrossRef]
14. Sheikholeslami, M.; Hatami, M.; Ganji, D.D.G.-D. Analytical investigation of MHD nanofluid flow in a semi-porous channel. *Powder Technol.* **2013**, *246*, 327–336. [CrossRef]
15. Zangooee, M.; Hosseinzadeh, K.; Ganji, D. Hydrothermal analysis of MHD nanofluid (TiO_2-GO) flow between two radiative stretchable rotating disks using AGM. *Case Stud. Therm. Eng.* **2019**, *14*, 100460. [CrossRef]
16. Khan, J.A.; Mustafa, M.; Hayat, T.; Alsaedi, A. A revised model to study the MHD nanofluid flow and heat trans-fer due to rotating disk: Numerical solutions. *Neural Comput. Appl.* **2018**, *30*, 957–964. [CrossRef]
17. Rashid, U.; Liang, H. Investigation of nanoparticles shape effects on MHD nanofluid flow and heat transfer over a rotating stretching disk through porous medium. *Int. J. Numer. Methods Heat Fluid Flow* **2020**, *30*, 5169–5189. [CrossRef]
18. Abbas, S.Z.; Khan, M.I.; Kadry, S.; Khan, W.A.; Israr-Ur-Rehman, M.; Waqas, M. Fully developed entropy optimized second order velocity slip MHD nanofluid flow with activation energy. *Comput. Methods Programs Biomed.* **2020**, *190*, 105362. [CrossRef] [PubMed]
19. Rashidi, M.M.; Abelman, S.; Mehr, N.F. Entropy generation in steady MHD flow due to a rotating porous disk in a nanofluid. *Int. J. Heat Mass Transf.* **2013**, *62*, 515–525. [CrossRef]
20. Hayat, T.; Rashid, M.; Imtiaz, M.; Alsaedi, A. Magnetohydrodynamic (MHD) flow of Cu-water nanofluid due to a rotating disk with partial slip. *AIP Adv.* **2015**, *5*, 067169. [CrossRef]
21. Asma, M.; Othman, W.; Muhammad, T.; Mallawi, F.; Wong, B. Numerical study for magnetohydrodynamic flow of nanofluid due to a rotating disk with binary chemical reaction and arrhenius activation energy. *Symmetry* **2019**, *11*, 1282. [CrossRef]
22. Aziz, A.; Alsaedi, A.; Muhammad, T.; Hayat, T. Numerical study for heat generation/absorption in flow of nanofluid by a rotating disk. *Results Phys.* **2018**, *8*, 785–792. [CrossRef]
23. Hayat, T.; Muhammad, T.; Shehzad, S.A.; Alsaedi, A. On magnetohydrodynamic flow of nanofluid due to a rotating disk with slip effect: A numerical study. *Comput. Methods Appl. Mech. Eng.* **2017**, *315*, 467–477. [CrossRef]
24. Naqvi, S.M.R.S.; Muhammad, T.; Saleem, S.; Kim, H.M. Significance of non-uniform heat generation/absorption in hydromagnetic flow of nanofluid due to stretching/shrinking disk. *Phys. A Stat. Mech. Its Appl.* **2019**, *553*, 123970. [CrossRef]
25. Bestman, A.R. Natural convection boundary layer with suction and mass transfer in a porous medium. *Int. J. Energy Res.* **1990**, *14*, 389–396. [CrossRef]
26. Khan, N.S.; Kumam, P.; Thounthong, P. Second law analysis with effects of Arrhenius activation energy and binary chemical reaction on nanofluid flow. *Sci. Rep.* **2020**, *10*, 1–16. [CrossRef]
27. Zeeshan, A.; Shehzad, N.; Ellahi, R. Analysis of activation energy in Couette-Poiseuille flow of nanofluid in the presence of chemical reaction and convective boundary conditions. *Results Phys.* **2018**, *8*, 502–512. [CrossRef]
28. Bhatti, M.M.; Michaelides, E.E. Study of Arrhenius activation energy on the thermo-bioconvection nanofluid flow over a Riga plate. *J. Therm. Anal. Calorim.* **2021**, *143*, 2029–2038. [CrossRef]
29. Khan, M.I.; Hayat, T.; Alsaedi, A. Activation energy impact in nonlinear radiative stagnation point flow of Cross nanofluid. *Int. Commun. Heat Mass Transf.* **2018**, *91*, 216–224. [CrossRef]

30. Hamid, A.; Hashim; Khan, M. Impacts of binary chemical reaction with activation energy on unsteady flow of magneto-Williamson nanofluid. *J. Mol. Liq.* **2018**, *262*, 435–442. [CrossRef]
31. Azam, M.; Xu, T.; Shakoor, A.; Khan, M. Effects of Arrhenius activation energy in development of covalent bond-ing in axisymmetric flow of radiative-cross nanofluid. *Int. Comm. Heat Mass Trans.* **2020**, *113*, 104547. [CrossRef]
32. Waqas, H.; Imran, M.; Muhammad, T.; Sait, S.M.; Ellahi, R. Numerical investigation on bioconvection flow of Ol-droyd-B nanofluid with nonlinear thermal radiation and motile microorganisms over rotating disk. *J. Therm. Anal. Calorim.* **2021**, *145*, 523–539. [CrossRef]
33. Makinde, O.D.; Mahanthesh, B.; Gireesha, B.J.; Shashikumar, N.; Monaledi, R.; Tshehla, M. MHD nanofluid flow past a rotating disk with thermal radiation in the presence of aluminum and titanium alloy nanoparticles. *Defect Diffus. Forum* **2018**, *384*, 69–79. [CrossRef]
34. Reddy, P.S.; Jyothi, K.; Reddy, M.S. Flow and heat transfer analysis of carbon nanotubes-based Maxwell nanofluid flow driven by rotating stretchable disks with thermal radiation. *J. Braz. Soc. Mech. Sci. Eng.* **2018**, *40*, 576. [CrossRef]
35. Waqas, H.; Farooq, U.; Muhammad, T.; Hussain, S.; Khan, I. Thermal effect on bioconvection flow of Sutterby nanofluid between two rotating disks with motile microorganisms. *Case Stud. Therm. Eng.* **2021**, *26*, 101136. [CrossRef]
36. Latiff, N.; Uddin, J.; Ismail, A.M. Stefan blowing effect on bioconvective flow of nanofluid over a solid rotating stretchable disk. *Propuls. Power Res.* **2016**, *5*, 267–278. [CrossRef]
37. Bég, O.A.; Kabir, M.N.; Uddin, J.; Ismail, A.I.M.; Alginahi, Y.M. Numerical investigation of Von Karman swirling bioconvective nanofluid transport from a rotating disk in a porous medium with Stefan blowing and anisotropic slip effects. *Proc. Inst. Mech. Eng. Part C J. Mech. Eng. Sci.* **2020**. [CrossRef]
38. Zohra, F.T.; Uddin, M.J.; Basir, F.; Ismail, A.I.M. Magnetohydrodynamic bio-nano-convective slip flow with Stefan blowing effects over a rotating disc. *Proc. Inst. Mech. Eng. Part N J. Nanomater. Nanoeng. Nanosyst.* **2020**, *234*, 83–97. [CrossRef]
39. Zohra, F.T.; Uddin, M.J.; Ismail, A.I.M. Magnetohydrodynamic bio-nanoconvective Naiver slip flow of micropolar fluid in a stretchable horizontal channel. *Heat Trans. Asian Res.* **2019**, *48*, 3636–3656. [CrossRef]
40. Khan, S.A.; Nie, Y.; Ali, B. Multiple slip effects on MHD unsteady viscoelastic nano-fluid flow over a permeable stretching sheet with radiation using the finite element method. *SN Appl. Sci.* **2020**, *2*, 66. [CrossRef]
41. Raju, C.S.K.; Mamatha, S.U.; Rajadurai, P.; Khan, I. Nonlinear mixed thermal convective flow over a rotating disk in suspension of magnesium oxide nanoparticles with water and EG. *Eur. Phys. J. Plus* **2019**, *134*, 196. [CrossRef]
42. Muhammad, T.; Waqas, H.; Khan, S.A.; Ellahi, R.; Sait, S.M. Significance of nonlinear thermal radiation in 3D Eyring–Powell nanofluid flow with Arrhenius activation energy. *J. Therm. Anal. Calorim.* **2021**, *143*, 929–944. [CrossRef]
43. Aziz, A.; Jamshed, W.; Aziz, T.; Bahaidarah, H.M.S.; Rehman, K.U. Entropy analysis of Powell–Eyring hybrid nanofluid including effect of linear thermal radiation and viscous dissipation. *J. Therm. Anal. Calorim.* **2021**, *143*, 1331–1343. [CrossRef]
44. Mahanthesh, B.; Mackolil, J.; Radhika, M.; Al-Kouz, W. Siddabasappa Significance of quadratic thermal radiation and quadratic convection on boundary layer two-phase flow of a dusty nanoliquid past a vertical plate. *Int. Commun. Heat Mass Transf.* **2021**, *120*, 105029. [CrossRef]
45. Jawad, M.; Saeed, A.; Khan, A.; Gul, T.; Zubair, M.; Shah, S.A.A. Unsteady bioconvection Darcy-Forchheimer nanofluid flow through a horizontal channel with impact of magnetic field and thermal radiation. *Heat Transf.* **2021**, *50*, 3240–3264. [CrossRef]
46. Majeed, A.; Zeeshan, A.; Amin, N.; Ijaz, N.; Saeed, T. Thermal analysis of radiative bioconvection magnetohydro-dynamic flow comprising gyrotactic microorganism with activation energy. *J. Therm. Anal. Calorim.* **2021**, *143*, 2545–2556. [CrossRef]
47. Amanulla, C.; Wakif, A.; Boulahia, Z.; Reddy, M.S.; Nagendra, N. Numerical investigations on magnetic field modeling for Carreau non-Newtonian fluid flow past an isothermal sphere. *J. Braz. Soc. Mech. Sci. Eng.* **2018**, *40*, 462. [CrossRef]
48. Kalaivanan, R.; Ganesh, N.V.; Al-Mdallal, Q.M. An investigation on Arrhenius activation energy of second grade nanofluid flow with active and passive control of nanomaterials. *Case Stud. Therm. Eng.* **2020**, *22*, 100774. [CrossRef]
49. Shah, Z.; Kumam, P.; Deebani, W. Radiative MHD Casson Nanofluid Flow with Activation energy and chemical reaction over past nonlinearly stretching surface through Entropy generation. *Sci. Rep.* **2020**, *10*, 1–14. [CrossRef] [PubMed]
50. Reddy, S.R.R.; Reddy, P.B.A.; Rashad, A.M. Activation energy impact on chemically reacting eyring–powell nanofluid flow over a stretching cylinder. *Arab. J. Sci. Eng.* **2020**, *45*, 5227–5242. [CrossRef]
51. Kotresh, M.J.; Ramesh, G.K.; Shashikala, V.K.R.; Prasannakumara, B.C. Assessment of Arrhenius activation energy in stretched flow of nanofluid over a rotating disc. *Heat Transf.* **2021**, *50*, 2807–2828. [CrossRef]
52. Abdelmalek, Z.; Khan, S.U.; Waqas, H.; Nabwey, H.A.; Tlili, I. Utilization of second order slip, activation energy and viscous dissipation consequences in thermally developed flow of third grade nanofluid with gyrotactic microorganisms. *Symmetry* **2020**, *12*, 309. [CrossRef]
53. Naz, R.; Noor, M.; Shah, Z.; Sohail, M.; Kumam, P.; Thounthong, P. Entropy generation optimization in MHD pseudoplastic fluid comprising motile microorganisms with stratification effect. *Alex. Eng. J.* **2020**, *59*, 485–496. [CrossRef]
54. Rashidi, M.M.; Freidoonimehr, N.; Momoniat, E.; Rostami, B. Study of nonlinear MHD tribological squeeze film at generalized magnetic reynolds numbers using DTM. *PLoS ONE* **2015**, *10*, e0135004. [CrossRef] [PubMed]
55. Zhou, J.K. *Differential Transformation and Its Applications for Electrical Circuits*; Huazhong University Press: Wuhan, China, 1986.
56. Zhang, L.; Arain, M.B.; Bhatti, M.M.; Zeeshan, A.; Hal-Sulami, H. Effects of magnetic Reynolds number on swimming of gyrotactic microorganisms between rotating circular plates filled with nanofluids. *Appl. Math. Mech.* **2020**, *41*, 637–654. [CrossRef]

Article

Magnetic Field Effect on Sisko Fluid Flow Containing Gold Nanoparticles through a Porous Curved Surface in the Presence of Radiation and Partial Slip

Umair Khan [1,2], Aurang Zaib [3] and Anuar Ishak [1,*]

1. Department of Mathematical Sciences, Faculty of Science and Technology, Universiti Kebangsaan Malaysia, UKM Bangi 43600, Malaysia; umairkhan@iba-suk.edu.pk
2. Department of Mathematics and Social Sciences, Sukkur IBA University, Sukkur 65200, Pakistan
3. Department of Mathematical Sciences, Federal Urdu University of Arts, Science & Technology, Gulshan-e-Iqbal Karachi 75300, Pakistan; aurangzaib@fuuast.edu.pk
* Correspondence: anuar_mi@ukm.edu.my

Abstract: The radiation and magnetic field effects of nanofluids play a significant role in biomedical engineering and medical treatment. This study investigated the performance of gold particles in blood flow (Sisko fluid flow) over a porous, slippery, curved surface. The partial slip effect was considered to examine the characteristics of nanofluid flow in depth. The foremost partial differential equations of the Sisko model were reduced to ordinary differential equations by using suitable variables, and the boundary value problem of the fourth-order (bvp4c) procedure was applied to plot the results. In addition, the effects of the parameters involved on temperature and velocity were presented in light of the parametric investigation. A comparison with published results showed excellent agreement. The velocity distribution was enhanced due to the magnetic field, while the temperature increased due to the effects of a magnetic field and radiation, which are effective in therapeutic hyperthermia. In addition, the nanoparticle suspension showed increased temperature and decelerated velocity.

Keywords: Sisko fluid flow; gold particles; magnetohydrodynamics (MHD); radiation effect; slip effect; curved surface

1. Introduction

Nanofluids are a prominent topic of research. They have a wide range of applications in engineering and technology fields. Nanofluids have potential benefits in cancer therapy, drug delivery, nuclear reactors, and solar energy. Growth enrichment and convection thermal conductivity are needed during fluid flow when an outside source is essential. Nanofluids are synthesized by scattering nanoparticles in regular fluids. In addition to regular fluids, such as lubricant, oils, water, and polymer solutions, biological fluids can also be used as base fluids. A notable development in this area was investigated after the initial research by Choi [1]. Eastman et al. [2] experimentally analyzed heat transport in the presence of water-based CuO particles and ethylene-glycol-based Al_2O_3 particles. Since then, different researchers have discussed the features of nanofluids [3–12]. However, the use of gold nanoparticles (Au-NPs) in biomedical science is also important, and Au-NPs can be used as therapeutic agents. They are currently used as contrast and photovoltaic agents and as drug transporters. In addition, Au-NPs have many characteristics that make them suitable for use in cancer treatment. Furthermore, owing to the high atomic number of gold, Au-NPs engender heat, which can be used for photothermal therapy of tumorous glands [13,14].

Non-Newtonian fluids play an imperative role in numerous manufacturing and engineering processes, such as food processing, petroleum digging, and chemical and biological treatment. Previously, blood was treated as a Newtonian fluid [15]; however, Thurston [16] clarified that visco-elasticity is considered a basic property of rheological blood, which indicates that human blood is non-Newtonian, depending on the visco-elastic performance of red blood cells. Several non-Newtonian fluids are treated as blood, e.g., Sisko fluid. Khan and Shahzad [17] inspected Sisko fluid flow over a stretching sheet. Munir et al. [18] extended this research by considering the bidirectional Sisko fluid flow over a stretching surface. Khan et al. explored the effects of a magnetic field and radiation on Sisko fluid flow over a bidirectional stretching sheet [19]. Eid et al. [20] used gold nanoparticles to investigate the effects of radiation on Sisko biofluid flow over a nonlinear stretching sheet. Ahmad et al. [21] numerically investigated the significance of Sisko fluid flow over a stretching curved sheet using a nanofluid and a magnetic field. Khan et al. [22] investigated 2D Sisko fluid flow impeding nanoparticles via a radially stretching/shrinking sheet under zero-flux conditions.

The effect of radiation on blood flow is considered important in biomedical science and other medical treatment techniques, especially in thermal therapeutic procedures. One effective technique commonly used for heat treatment of different body parts is infrared radiation. This method is favored in heat therapy because it is applied directly to blood capillaries in the affected regions. In addition, it is used in the treatment of bursitis, which is inflammation of the fluid-filled sacs (bursae) that lie between bone and tendon or between skin and tendon. Inoue and Kabaya [23], Kobu [24], and Nishimoto et al. [25] experimentally investigated the effects of infrared radiation on blood flow. He et al. [26] used laser irradiation to analyze oxygen transport, temperature, and blood flow in breast tumors. Prakash and Makinde [27] explored the effects of radiation on blood flow with heat transport through an artery with stenosis. Misra and Sinha investigated the effects of a magnetic field and radiation on time-dependent blood flows with heat transfer through a porous capillary in a stretching motion [28]. Khan et al. [29] used gold nanoparticles to investigate the effects of a magnetic field on radiative blood flow over a slippery surface and obtained multiple solutions. In addition, Zaib et al. investigated the effects of a magnetic field and radiation on the mixed convective flow of a tangent hyperbolic fluid over a flat, non-isothermal vertical plate [30].

The present study investigated radiative blood flow with heat transport using a Sisko fluid containing gold nanoparticles, over a porous, curved surface with a magnetic field and partial slip. The governing partial differential equations were converted to a system of ordinary differential equations before they were solved numerically via the boundary value problem of the fourth-order (bvp4c) function available in MATLAB software, which is based on the Lobatto IIIA technique. To analyze the capability of the numerical solution process, the skin friction coefficient was compared with published results. Graphical results were presented for the velocity profile, temperature distribution, skin friction, and heat transfer rate for different values of the parameters involved. To the best of our knowledge, no study has investigated flow situations with gold nanoparticles using the Sisko model with similarity solutions. The results have implications for clinical sciences, especially in thermal therapy.

2. Problem Definition and Modeling

2.1. Problem Definition

The steady flow of a non-Newtonian fluid containing gold nanoparticles over a porous curved surface was examined. According to Chen [31], blood is treated as an electric conducting fluid. Thus, blood flow in nature is magnetohydrodynamic. Blood flow is due to the movement of the stretching surface along the s direction and suction along the r direction. In addition, the stretching curve is coiled in a circle of radius R and center O and contains 12–85 nm gold nanoparticles, as shown in Figure 1. A magnetic field B_0 is applied normal to the curved surface. A larger R signifies a vaguely curved surface. Moreover, the stretching and shrinking sheet of the curved surface depends upon the arbitrary constant ($c > 0$ for stretching and $c < 0$ for shrinking), where velocity is represented as $U_w(s) = cs$, with $c > 0$, which moves along the s direction, and suction velocity is represented as v_o. In addition, the characteristics of nanoparticles and the carrier-based fluid are assumed to be constant. The temperature and ambient temperature of the surface are represented, respectively, as T_w and T_∞. The radiation and partial slip effects are also incorporated.

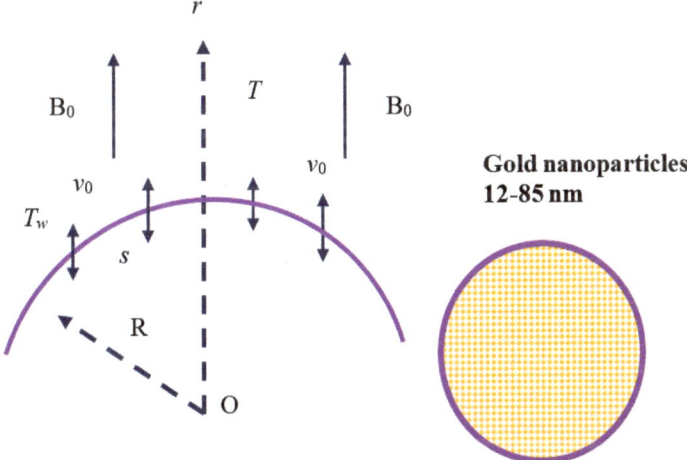

Figure 1. Physical diagram of the problem.

2.2. Flow Analysis

Using boundary layer approximations, the governing equations involving the magnetic effect on Sisko blood flow with gold nanoparticles over a porous, curved surface with radiation and partial slip are

$$\frac{\partial}{\partial r}[(r+R)v] + R\frac{\partial u}{\partial s} = 0, \tag{1}$$

$$\frac{1}{(r+R)}u^2 = \frac{1}{\rho_{nf}}\frac{\partial p}{\partial r}, \tag{2}$$

$$v\frac{\partial u}{\partial r} + \frac{b}{\rho_{nf}(r+R)^2}\frac{\partial}{\partial r}\left[(r+R)^2\left(-\left(\frac{\partial u}{\partial r} - \frac{u}{(r+R)}\right)\right)^n\right] + \left(\frac{Ru}{r+R}\right)\frac{\partial u}{\partial s} + \frac{uv}{(r+R)} = -\frac{1}{\rho_{nf}}\left(\frac{R}{r+R}\right)\frac{\partial p}{\partial s} - \frac{\sigma_{nf}B_0^2}{\rho_{nf}}u + \frac{\mu_{nf}}{\rho_{nf}(r+R)^2}\frac{\partial}{\partial r}\left[(r+R)^2\left(\frac{\partial u}{\partial r} - \frac{u}{(r+R)}\right)\right], \tag{3}$$

$$v\frac{\partial T}{\partial r} + \frac{1}{(c_p\rho)_{nf}(r+R)}\frac{\partial}{\partial r}[(r+R)q_r^*] + \left(\frac{Ru}{r+R}\right)\frac{\partial T}{\partial s} = \frac{k_{nf}}{(c_p\rho)_{nf}}\left(\frac{\partial^2 T}{\partial r^2} + \frac{1}{(r+R)}\frac{\partial T}{\partial r}\right), \tag{4}$$

with the boundary conditions

$$\begin{cases} u = \lambda U_w + L\left[\frac{\partial u}{\partial r} - \frac{u}{r+R}\right], \ v = v_o, \ T = T_w \text{ at } r = 0, \\ u \to 0, \ \frac{\partial u}{\partial r} \to 0, \ T \to T_\infty \text{ as } r \to \infty. \end{cases} \tag{5}$$

Here, u and v are components of the velocity, such that the corresponding stretching velocity moves along the axial s direction and suction along the radial r direction, respectively. In addition, n and b are material constants, L constant slip parameter, p the pressure, R the radius of curvature, and T the nanofluid temperature. In Equation (4), the relative heat flux is represented by q_r^* and can be articulated by employing the Rosseland approximation as

$$q_r^* = -4(3k_1)^{-1}\sigma_1\frac{\partial T^4}{\partial r}, \tag{6}$$

where σ_1 is the mean proportion coefficient and k_1 the Stefan–Boltzmann constant. Thus, the term T^4 at point T_∞ is exercised by the Taylor series. Avoiding the highest-order terms, we get

$$T^4 \approx 4TT_\infty^3 - 3T_\infty^4 \tag{7}$$

The thermo-physical quantities of the gold particle nanofluid introduced in the governing equations are given by

$$\begin{aligned}
&\frac{\mu_{nf}}{\mu_f} = 1 + 7.3\phi + 123\phi^2 \text{ for } \phi > 0.02, \quad \frac{\rho_{nf}}{\rho_f} = (1-\phi) + \frac{\rho_{s_1}}{\rho_f}\phi, \\
&\frac{(c_p\rho)_{nf}}{(c_p\rho)_f} = (1-\phi) + \frac{(c_p\rho)_{s_1}}{(c_p\rho)_f}\phi, \quad \frac{k_{nf}}{k_f} = \left[\frac{(k_{s_1}+2k_f)-2(\phi k_f - \phi k_{s_1})}{(k_{s_1}+2k_f)+\phi(k_f-k_{s_1})}\right], \\
&\frac{\sigma_{nf}}{\sigma_f} = \left(1 + \frac{3\phi\left(\frac{\sigma_{s_1}}{\sigma_f}-1\right)}{\left(\frac{\sigma_{s_1}}{\sigma_f}+2\right)-\left(\frac{\sigma_{s_1}}{\sigma_f}-1\right)\phi}\right),
\end{aligned} \tag{8}$$

where k_f, ρ_f, μ_f and σ_f are the thermal conductivity, density, viscosity, and electrical conductivity of the carrier-based fluid, respectively; $k_{nf}, \rho_{nf}, \mu_{nf}$ and σ_{nf} are the corresponding quantities of the nanofluid, respectively; c_p is the heat capacity; and subscripts f, nf, and s_1 are quantities of the carrier-based fluid, the nanofluid, and the solid volume fraction of the nanoparticles, respectively.

Upon applying the following similarity transformation:

$$\begin{aligned}
&u = csF'(\eta), \; v = \frac{-csR}{1+R}\text{Re}_b^{-\frac{1}{n+1}}\frac{1}{n+1}[2n\Gamma(\eta) + (1-n)\eta F'(\eta)], \\
&\psi = cs^2\text{Re}_b^{-\frac{1}{n+1}}F(\eta), \eta = \left(\frac{r}{s}\right)\text{Re}_b^{\frac{1}{n+1}}, p = \rho_f c^2 s^2 P(\eta), \theta(\eta) = \frac{T-T_\infty}{T_w-T_\infty}.
\end{aligned} \tag{9}$$

Equation (1) is satisfied identically, whereas Equations (2)–(5) become

$$\frac{\frac{\partial P}{\partial \eta}}{\frac{\rho_{nf}}{\rho_f}} = \frac{F'^2}{\eta + B} \tag{10}$$

$$\begin{aligned}
\frac{2BP}{(\eta+B)\frac{\rho_{nf}}{\rho_f}} &= \frac{B}{\eta+B}\left(\frac{2n}{n+1}\right)\left(FF'' + \frac{FF'}{\eta+B}\right) - \frac{B}{\eta+B}F'^2 + \frac{\frac{\mu_{nf}}{\mu_f}}{\frac{\rho_{nf}}{\rho_f}}B_1\left(F''' + \frac{F'}{(\eta+B)^2} + \frac{F''}{\eta+B}\right) + \\
&\frac{n}{\frac{\rho_{nf}}{\rho_f}}\left(-\left(F'' \frac{F'}{\eta+B}\right)\right)^{n-1}\left(F'' + \frac{F'}{(\eta+B)^2} - \frac{F''}{\eta+B}\right) - \frac{2}{(\eta+B)\frac{\rho_{nf}}{\rho_f}}\left(-\left(F''\frac{F'}{\eta+B}\right)\right)^n - M\frac{\frac{\sigma_{nf}}{\sigma_f}}{\frac{\rho_{nf}}{\rho_f}}F',
\end{aligned} \tag{11}$$

By removing the pressure term from Equations (10) and (11), we obtain the following equations, along with the dimensionless form of the energy equation:

$$\begin{aligned}
&\frac{\Sigma_1}{\Sigma_2}B_1\left(F'''' + \frac{2F'''}{\eta+B} + \frac{F'}{(\eta+B)^3} - \frac{F''}{(\eta+B)^2}\right) + \left(\frac{2n}{n+1}\right)\begin{bmatrix}\left(\frac{B}{\eta+B}\right)(FF''' + F'F'') + \\ \left(\frac{B}{(\eta+B)^2}\right)(FF'' + F'^2) \\ -\frac{BFF'}{(\eta+B)^3}\end{bmatrix} \\
&- \frac{2BF'F''}{\eta+B} - \frac{2BF'^2}{(\eta+B)^2} + \frac{n}{\Sigma_2}\left(-\left(F''\frac{F'}{\eta+B}\right)\right)^{n-1}\left(F'''' + \frac{2F'''}{\eta+B} - \frac{F''}{(\eta+B)^2} + \frac{F'}{(\eta+B)^3}\right) - \\
&\frac{n(n-1)}{\Sigma_2}\left(-\left(F''\frac{F'}{\eta+B}\right)\right)^{n-2}\left(F''' + \frac{F'}{(\eta+B)^2} - \frac{F''}{\eta+B}\right)^2 - \frac{\Sigma_3}{\Sigma_2}M\left(\frac{F'}{\eta+B} + F''\right) = 0,
\end{aligned} \tag{12}$$

$$\left(\Sigma_4 + \frac{4}{3}R_d\right)\left(\theta'' + \frac{\theta'}{\eta+B}\right) + \Pr\Sigma_5 \frac{B}{\eta+B}\left(\frac{2n}{n+1}\right)F\theta' = 0, \tag{13}$$

in which:

$$\frac{\mu_{nf}}{\mu_f} = \Sigma_1, \frac{\rho_{nf}}{\rho_f} = \Sigma_2, \frac{\sigma_{nf}}{\sigma_f} = \Sigma_3, \frac{k_{nf}}{k_f} = \Sigma_4, \frac{(\rho c_p)_{nf}}{(\rho c_p)_f} = \Sigma_5.$$

2.3. Outcome of the Parameterization

The physical parameters in Equations (12) and (13) are radiation R_d; the Prandtl number Pr; the local Reynolds numbers Re_s and Re_b; the material parameter of the Sisko fluid, B_1; the magnetic parameter M; the radius of curvature B; the thermal conductivity α_f; the slip parameter B_2; and the suction parameter S, which are obtained as follows:

$$R_d = \frac{4\sigma_1 T_\infty^3}{k_1 k_f}, \Pr = \frac{su_w}{\alpha_f} Re_b^{\frac{-2}{n+1}}, Re_s = \frac{u_w s}{\nu_f}, Re_b = \frac{u_w^{2-n} s^n \rho_f}{b}, B_1 = \frac{Re_b^{\frac{2}{n+1}}}{Re_s}, M = \frac{\sigma_f B_0^2}{\rho_f c},$$

$$B = \frac{R}{s} Re_b^{\frac{1}{n+1}}, \alpha_f = \frac{k_f}{(c_p \rho)_f}, B_2 = \frac{L}{s} Re_b^{\frac{1}{n+1}}, S = \frac{-v_0}{u_w}\left(\frac{n+1}{2n}\right) Re_b^{\frac{1}{n+1}}.$$

The boundary conditions are

$$\begin{cases} F'(0) = \lambda + B_2\left(F''(0) - \frac{F'(0)}{B}\right), F(0) = S, \theta(0) = 1 \text{ at } \eta = 0, \\ F'(\infty) \to 0, F''(\infty) \to 0, \theta(\infty) \to 0 \text{ as } \eta \to \infty. \end{cases} \quad (14)$$

The pressure term can be obtained from Equation (11), which becomes:

$$P = \frac{(\eta+B)\frac{\rho_{nf}}{\rho_f}}{2B}\begin{bmatrix} \frac{B}{\eta+B}\left(\frac{2n}{n+1}\right)\left(FF'' + \frac{FF'}{\eta+B}\right) - \frac{B}{\eta+B}F'^2 + \frac{\frac{\mu_{nf}}{\mu_f}}{\frac{\rho_{nf}}{\rho_f}} B_1 \left(F''' \frac{F'}{(\eta+B)^2} + \frac{F'}{\eta+B}\right) + \\ \frac{n}{\frac{\rho_{nf}}{\rho_f}}\left(-\left(F'' - \frac{F'}{\eta+B}\right)\right)^{n-1}\left(F''' - \frac{F''}{\eta+B} + \frac{F'}{(\eta+B)^2}\right) - \\ \frac{2}{(\eta+B)\frac{\rho_{nf}}{\rho_f}}\left(-\left(F'' - \frac{F'}{\eta+B}\right)\right)^n - M\frac{\frac{\sigma_{nf}}{\sigma_f}}{\frac{\rho_{nf}}{\rho_f}} F' \end{bmatrix} \quad (15)$$

2.4. Physical Parameters

Following an engineering approach, the quantities C_F (local skin friction coefficient) and Nu_s (Nusselt number) are expressed as follows:

$$C_F = \frac{2\tau_{rs}}{\rho_f u_w^2}, \quad (16)$$

where

$$\tau_{rs} = \mu_{nf}\left(\frac{\partial u}{\partial r} - \frac{u}{r+R}\right) - b\left[-\left(\frac{\partial u}{\partial r} - \frac{u}{r+R}\right)\right]^n\bigg|_{r=0}, \quad (17)$$

After transformation, the reduced skin friction coefficient becomes

$$\frac{1}{2} Re_b^{\frac{1}{n+1}} C_F = \Sigma_1 B_1\left(F''(0) - \frac{F'(0)}{B}\right) - \left(-\left(F''(0) - \frac{F'(0)}{B}\right)\right)^n. \quad (18)$$

The heat transfer rate at the surface, Nu_s (Nusselt number), is

$$Nu_s = \frac{sq_w}{k_f(T_w - T_\infty)}, \quad (19)$$

where

$$q_w = -k_{nf}\left(\frac{\partial T}{\partial r}\right)\bigg|_{r=0}. \quad (20)$$

After transformation, the reduced Nusselt number becomes

$$\text{Re}_b^{\frac{-1}{n+1}} Nu_s = -\Sigma_4 \theta'(0). \tag{21}$$

3. Results and Discussion

The nonlinear ordinary differential Equations (12) and (13) subject to boundary restrictions (Equation (14)) were solved numerically using the bvp4c function in the MATLAB software. The effects of various physical parameters (the magnetic parameter M, curvature parameter B, slip parameter B_2, nanoparticle volume fraction ϕ, radiation parameter R_d, suction S, and stretching/shrinking parameter λ) on the velocity profile, temperature distribution, skin friction coefficient, and heat transfer rate are illustrated in Figures 2–19 for both the carrier-based fluid ($\phi = 0$) and the gold particle nanofluid ($\phi = 0.035$). The parameters fixed throughout computation were $B_1 = 1.1$, $S = 3.5$, $M = 0.5$, $B_2 = 0.2$, $\lambda = 1.5, n = R_d = 2$, and $B = 1.5$.

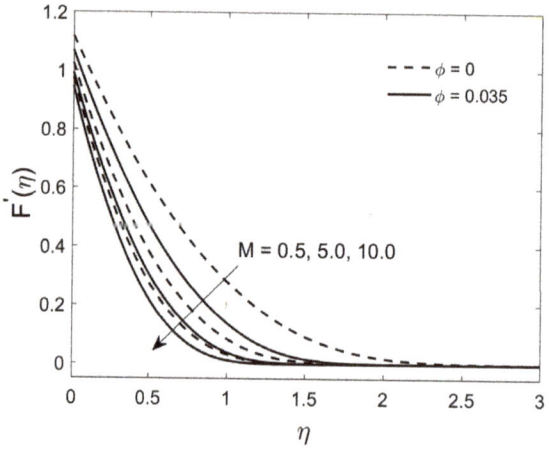

Figure 2. Effect of a magnetic field on the velocity profile.

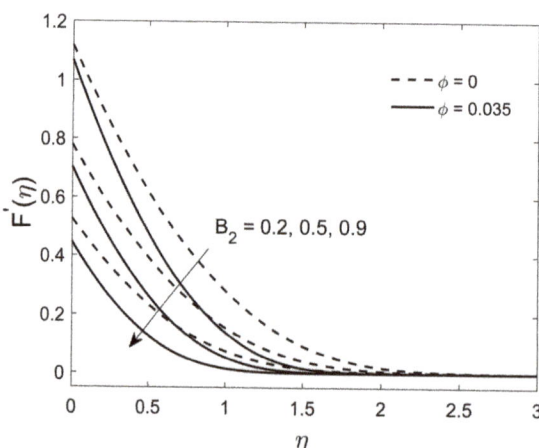

Figure 3. Effect of a partial slip on the velocity profile.

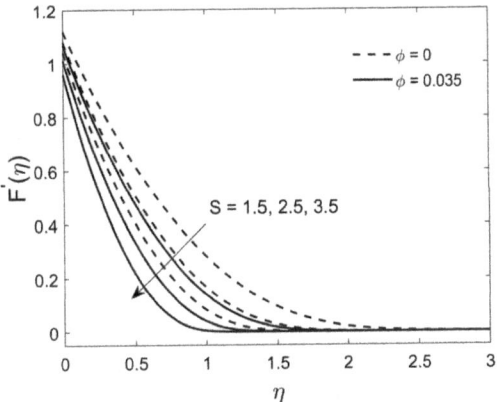

Figure 4. Effect of suction on the velocity profile.

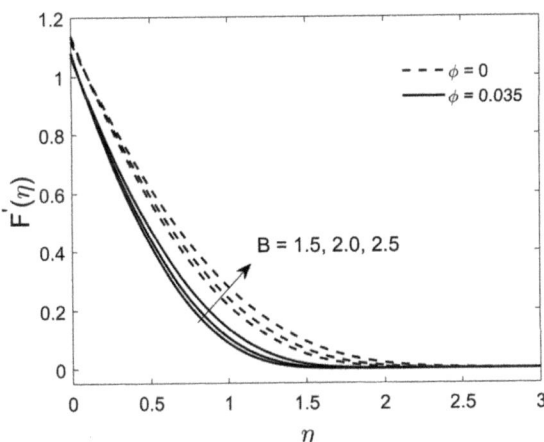

Figure 5. Effect of the radius of curvature on the velocity profile.

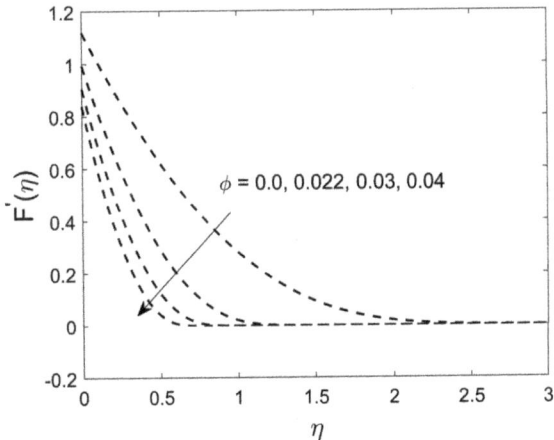

Figure 6. Effect of the nanoparticle volume fraction on the velocity profile.

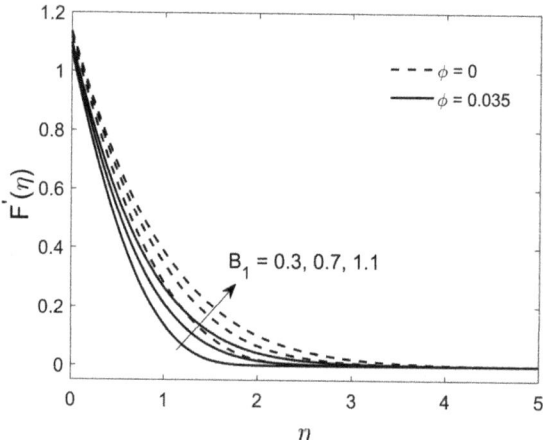

Figure 7. Effect of the material parameter on the velocity profile.

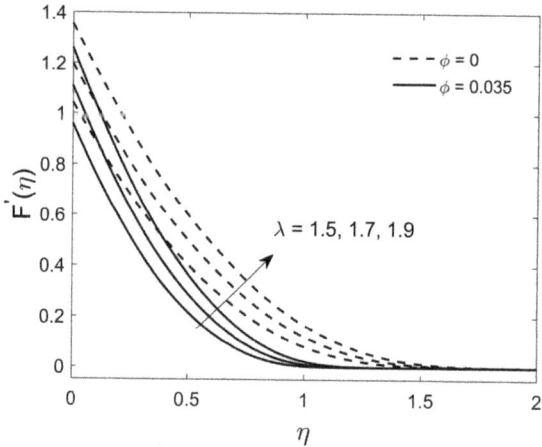

Figure 8. Effect of stretching/shrinking on the velocity profile.

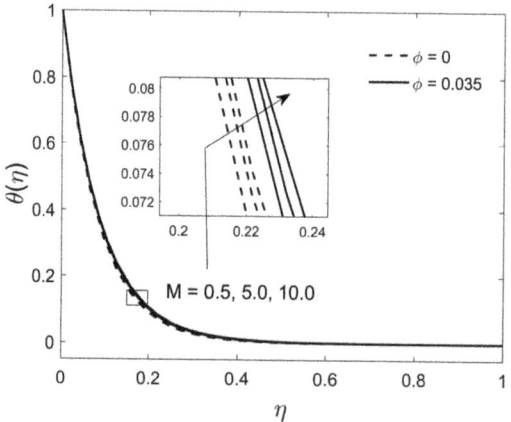

Figure 9. Effect of the magnetic field on the temperature distribution.

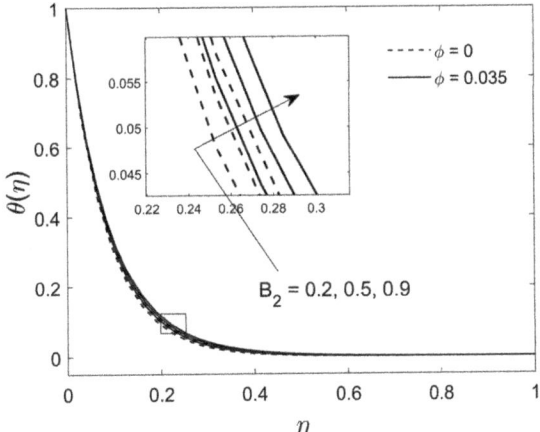

Figure 10. Effect of the slip on the temperature distribution.

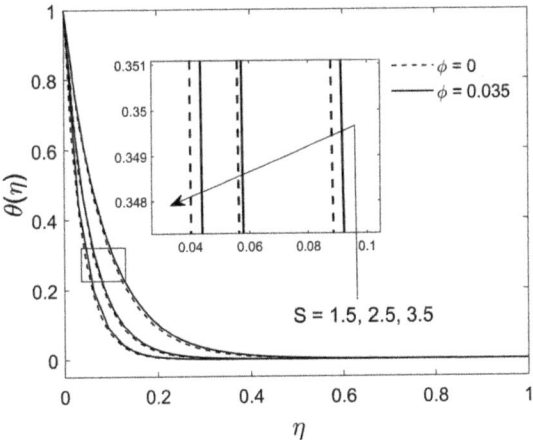

Figure 11. Effect of suction on the temperature distribution.

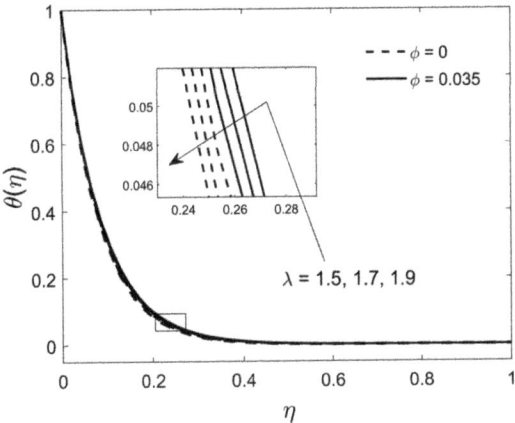

Figure 12. Effect of stretching/shrinking on the temperature distribution.

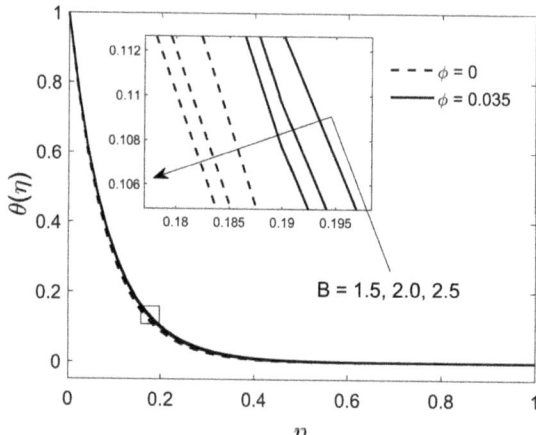

Figure 13. Effect of curvature on the temperature distribution.

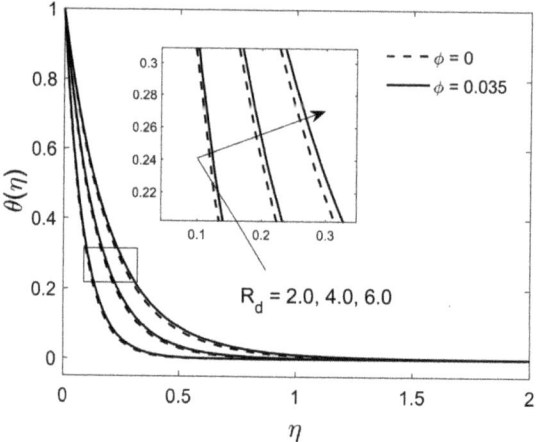

Figure 14. Effect of radiation on the temperature distribution.

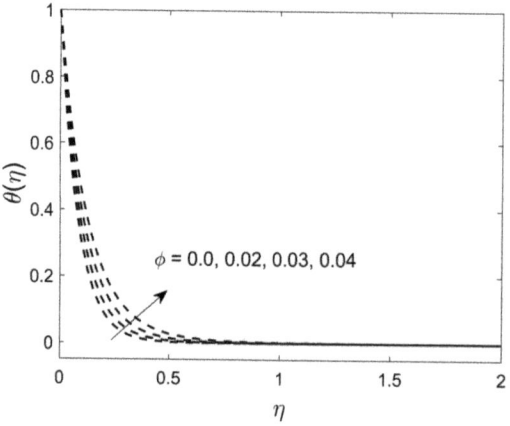

Figure 15. Effect of the nanoparticle volume fraction on the temperature distribution.

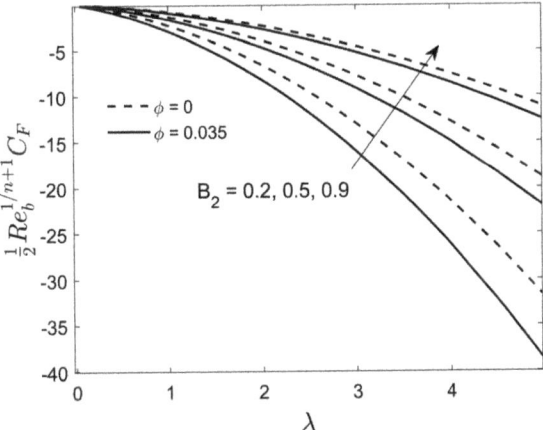

Figure 16. Effect of the slip on the skin friction coefficient against stretching/shrinking.

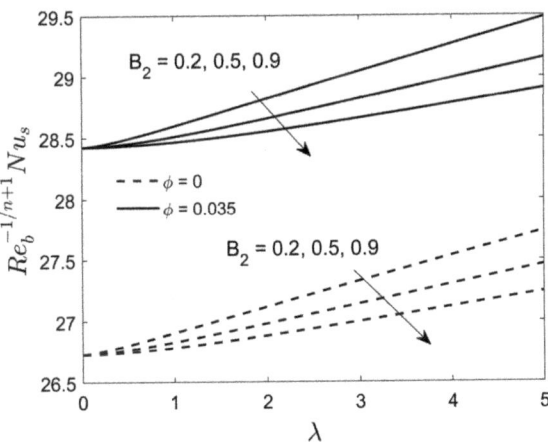

Figure 17. Effect of the slip on the Nusselt number against stretching/shrinking.

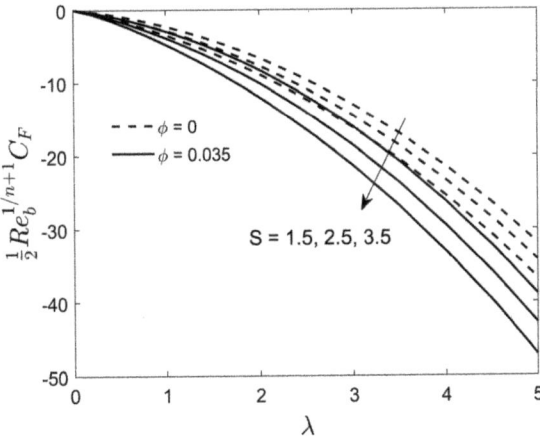

Figure 18. Effect of suction on the skin friction coefficient against stretching/shrinking.

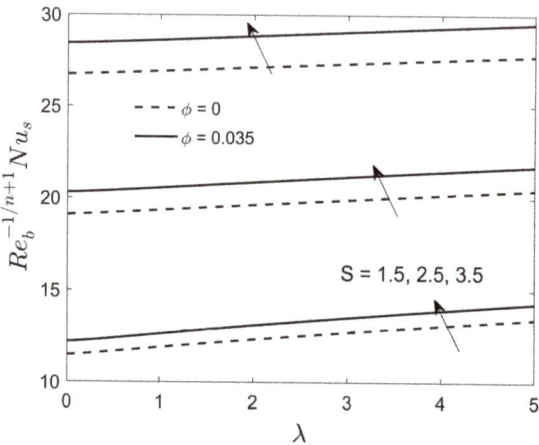

Figure 19. Effect of suction on the Nusselt number against stretching/shrinking.

3.1. Effect of Physical Parameters on the Velocity Profile

Figures 2–6 demonstrate the behavior of the velocity profile at different values of M, B_2, S, B, and nanoparticle volume fraction ϕ. The velocity decreases due to the magnetic field for the carrier-based fluid and the gold particle nanofluid (Figure 2). Initially, the velocity and thickness of the momentum boundary layer for the carrier-based fluid are larger compared to those for the gold particle nanofluid. In addition, the velocity behavior of the gold particle nanofluid reduces more in the presence of gold nanoparticles, because gold nanoparticles generate friction in the fluid. Physically, increasing values of M augment the Lorentz force, which ultimately decreases velocity. Figure 2 also shows that the velocity of blood decreases. Figure 3 shows the effect of a partial slip on the velocity. The velocity of the fluid decreases for both the carrier-based fluid and the gold particle nanofluid. The velocity decreases in both fluids ($\phi = 0$) and ($\phi = 0.035$) with increasing B_2. The velocity reduction at the curve surface shows that the fluid flow occurs at the stretching curve surface; thus, any increase in the velocity slip parameter of the fluid at the stretching surface decreases the velocity field. Moreover, the higher values of B_2 signify that the friction connecting the blood and the surface is removed. The impact of suction on the velocity is shown in Figure 4 for both the carrier-based fluid and the gold particle nanofluid. Physically, the resistance in the blood flow occurs because of viscosity, which can be handled by using suction. Increasing suction decreases the drag force at the sheet. Consequently, the momentum boundary layer thickness also reduces in both cases.

Figure 5 shows the effect of the curvature on the velocity profile, which increases the flow for both the carrier-based fluid and the gold particle nanofluid. The momentum boundary layer thickness and the magnitude of the velocity also improve with increasing curvature. Furthermore, the distance between the solution curves for the carrier-based fluid is slightly similar compared to that between the outcome curves of the gold particle nanofluid. Generally, this behavior of a fluid means that the bend of the curved stretching surface enhances fluid flow over it. This rise in the velocity gradient is slightly more in the carrier-based fluid compared to the gold particle nanofluid.

Figure 6 shows that the velocity of blood decreases as ϕ increases, which is responsible for the reduction in the velocity of the boundary layer. Physically, a higher ϕ enhances the blood viscosity, which consequently decreases the magnitude of the boundary layer thickness. Figures 7 and 8 show that the velocity increases for both the carrier-based fluid and the gold particle nanofluid due to the material and stretching/shrinking parameters. In addition, the magnitude of the velocity and the momentum boundary layer thickness increase with increasing material and stretching parameters. Both graphs are plotted for both fluids, where the increase in velocity is more for the carrier-based fluid compared to the gold particle nanofluid.

3.2. Effect of Physical Parameters on the Temperature Distribution

Figures 9–15 show the effects of the magnetic, slip, suction, stretching/shrinking, curvature, and radiation parameters on the temperature for the carrier-based fluid and the gold particle nanofluid. Figure 9 shows that the temperature increases due to the magnetic field for both fluids. This is because inclusion of the transverse strength of a magnetic field in an electrically conducting fluid increases the Lorentz force. This strength bears the potential to increase the blood temperature distribution.

Figure 10 shows that the velocity slip increases the temperature of the blood. The temperature distribution and the thermal boundary layer thickness increase with increasing slip for both the carrier-based fluid and the gold particle nanofluid because the slip slows down the fluid motion and ultimately affects the temperature. It is also evident from this figure that the increase in temperature is more for the nanofluid as compared to the regular fluid for the larger impacts of slip constraint.

Figure 11 shows that for larger values of suction, the blood temperature at any point of the flow is moderate for both the carrier-based fluid and the gold particle nanofluid. The thermal boundary layer thickness and the temperature distribution decrease at higher suction for both fluids. Stretching/shrinking reduces the temperature, as shown in Figure 12, for both fluids. A similar behavior is seen in Figure 12 for both fluids due to larger stretching compared to Figures 11 and 13, and shows that the temperature reduces due to the curvature for both fluids. The thermal boundary layer thickness and the temperature distribution decrease for both fluids due to the higher curvature. The figure insert shows the blood temperature distribution in terms of the significant effect of curvature, where the thermal conductivity is more for the gold particle nanofluid compared with the carrier-based fluid. From a physical point of view, this happens because the stretching curved surface increases the fluid flow for the velocity profiles and indirectly contributes to reducing the temperature distribution in terms of magnitude.

Radiation increases the temperature for both the carrier-based fluid and the gold particle nanofluid (Figure 14), which consequently increases the significant boundary layer thickness. Thus, the temperature of the boundary layer increases significantly. Moreover, the temperature distribution is greater for the gold particle nanofluid compared with the carrier-based fluid because the presence of gold nanoparticles produces more energy in the form of heat and, consequently, the temperature rises. The temperature increases due to the nanoparticle volume fraction in both fluids (Figure 15), because the inclusion of gold nanoparticles increases the thermal conductivity of blood, which increases the blood temperature.

3.3. Effect of Physical Parameters on the Skin Friction Coefficient and the Nusselt Number

The effects of the slip on skin friction coefficient and the Nusselt number against stretching/shrinking (λ) for both the carrier-based fluid and the gold particle nanofluid are shown in Figures 16 and 17, respectively. The velocity slip enhances the skin friction coefficient but reduces the Nusselt number in both fluids. In addition, the skin friction coefficient significantly shrinks due to stretching/shrinking, whereas heat transfer increases. Moreover, the Nusselt number is higher in the gold particle nanofluid compared to the carrier-based fluid, which ultimately enhances thermal conductivity. The effects of suction on shear stress and heat transfer against λ are shown in Figures 18 and 19, respectively. The skin friction coefficient shrinks due to suction, whereas the Nusselt number increases. The thermophysical data of the base fluid (blood) and gold nanoparticles are listed in Table 1.

Table 1. Thermophysical properties of blood and gold nanoparticles (Koriko et al. [32]).

Thermophysical Properties	Blood	Gold
ρ (kg/m^3)	1050	19,300
c_p (J/kgK)	3617	129
σ (S/m)	1090	4.1×10^6
k (W/mK)	0.52	318

Finally, Table 2 presents a comparison of the current outcomes of the skin friction coefficient for B when $B_1 = R_d = S = M = B_2 = \phi = 0$, $\lambda = 1$, and $n = 1$, which shows favorable agreement. For more details about the current technique, Table 3 shows a comparison of the current computational outcomes of the shear stress or friction factor for distinct values of the shrinking parameter when $B \to \infty$, $B_2 = M = \phi = B_1 = 0, n = 1$, and $S = 2$ with the results of Roşca et al. [33]. The values show excellent agreement, proving the feasibility of the present numerical scheme. In addition, the numerical computational values of the skin friction coefficient and the heat transfer rate for the various constraints are given in Table 4 for both the carrier-based fluid ($\phi = 0$) and the gold particle nanofluid ($\phi = 0.035$), while the rest of the fixed parameters are Pr = 21, $n = 2, S = 3.5$, and $B_2 = 0.2$. For the carrier-based fluid, the skin friction coefficient increases by 18.815%, 27.626%, 6.231%, and 1.552×10^{-4}% due to the impact of λ, M, B_1, and R_d, respectively, while it decreases by 1.774% due to B. By contrast, for the gold particle nanofluid, the skin friction coefficient increases by 18.311%, 15.034%, 4.541%, and 3.207×10^{-5}% due to the effect of λ, M, B_1, and R_d, respectively, while it decreases by approximately 0.751% due to the curvature parameter. Moreover, due to λ and B_1, the heat transport rate increases by 0.157% and 0.031%, respectively, for the carrier-based fluid and by 0.156% and 0.027%, respectively, for the gold particle nanofluid. Due to the effect of the magnetic, curvature, and radiation parameters, the heat transport rate decreases by 0.078%, 6.42×10^{-3}%, and 41.705%, respectively, for the carrier-based fluid and by 0.053%, 8.39×10^{-3}%, and 41.045%, respectively, for the gold particle nanofluid. Generally, the increasing skin friction coefficient is better for both fluids due to the magnetic and stretching parameters, while it is lower (approximately 1.552×10^{-4}% and 3.207×10^{-5}%, respectively) for the radiation parameter. Alternatively, the heat transport rate is maximum for the stretching parameter for both fluids when the parameter increases and minimum (about 6.42×10^{-3}% and 8.39×10^{-3}%, respectively) for the curvature parameter. Finally, these numerically calculated values show that the skin friction coefficient and the heat transfer rate are largely found in the carrier-based fluid compared to the gold particle nanofluid.

Table 2. Comparison of the skin friction coefficient $\left(-\frac{1}{2}\operatorname{Re}_b^{\frac{1}{n+1}} C_F\right)$ for different values of B with the results of [34–36].

B	$-\frac{1}{2}\operatorname{Re}_b^{\frac{1}{n+1}} C_F$			
	Saleh et al. [34]	Abbas et al. [35]	Waini et al. [36]	Current Outcome
5	1.15076	1.15763	1.15077	1.15756
10	1.07172	1.07349	1.07173	1.07347
20	1.03501	1.03561	1.03501	1.03566
30	1.02315	1.02353	1.02316	1.02353
40	1.01729	1.01759	1.01730	1.01704
50	1.01380	1.01405	1.01381	1.01440
100	1.00687	1.00704	1.00687	1.00703
200	1.00342	1.00356	1.00343	1.00354
1000	1.00068	1.00079	1.00068	1.00069

Table 3. Comparison of the skin friction coefficient $\left(-\frac{1}{2}\text{Re}_b^{\frac{1}{n+1}} C_F\right)$ for different values of λ with the results of Roşca et al. [33].

	$-\frac{1}{2}\text{Re}_b^{\frac{1}{n+1}} C_F$		
λ	Roşca et al. [33]		Current Outcome
	Numerical Outcome	Analytical Outcome	
-0.5	0.85289	0.85355	0.85365
-0.6	0.9786	0.97947	0.97966
-0.7	1.08255	1.08340	1.08382
-0.75	1.12366	1.12500	1.12563
-0.8	1.15619	1.15777	1.15879
-0.9	1.18214	1.18460	1.18796
-0.95	1.15876	1.16242	1.16340
-0.99	1.08018	1.08900	1.09448

Table 4. Skin friction coefficient $\left(\frac{1}{2}\right)\text{Re}_b^{\frac{1}{n+1}} C_F$ and Nusselt number $\text{Re}_b^{\frac{-1}{n+1}} Nu_s$ for different parameters at $Pr = 21$, $n = 2$, $S = 3.5$, and $B_2 = 0.2$.

λ	M	B	B_1	R_d	$(1/2)\text{Re}_b^{\frac{1}{n+1}} C_F$		$\text{Re}_b^{\frac{-1}{n+1}} Nu_s$	
					$\phi=0$	$\phi=0.035$	$\phi=0$	$\phi=0.035$
1.5	0.5	1.5	0.3	2.0	5.92616	8.10476	27.00936	28.70528
1.7	-	-	-	-	7.04118	9.58888	27.05197	28.75033
1.9	-	-	-	-	8.23084	11.15728	27.09459	28.79556
1.5	0.5	1.5	0.3	2.0	5.92616	8.10476	27.00936	28.70528
-	5.0	-	-	-	7.56336	9.32326	26.98816	28.68989
-	10.0	-	-	-	8.74439	10.33223	26.97428	28.67789
1.5	0.5	1.5	0.3	2.0	5.92616	8.10476	27.00936	28.70528
-	-	2.0	-	-	5.82099	8.04389	27.00762	28.70287
-	-	2.5	-	-	5.76707	8.01154	27.00647	28.70138
1.5	0.5	1.5	0.3	2.0	5.92616	8.10476	27.00936	28.70528
-	-	-	0.7	-	6.29542	8.47280	27.01654	28.71433
-	-	-	1.1	-	6.70957	8.85348	27.02234	28.72237
1.5	0.5	1.5	0.3	2.0	5.92616	8.10476	27.00936	28.70528
-	-	-	-	4.0	5.92617	8.10476	15.74498	16.92315
-	-	-	-	6.0	5.92617	8.10476	11.14985	12.03202

4. Conclusions

This study discussed the effects of a magnetic field on the flow of Sisko fluid containing gold nanoparticles over a porous, curved surface in the presence of radiation and partial slip. The key outcomes were as follows:

- The velocity of blood containing gold nanoparticles decreases with increasing intensity of an external magnetic field.
- The effect of thermal radiation can significantly modify the blood temperature. With increasing thermal radiation, the thickness of the thermal boundary layer increases significantly.
- An increase in the erythrocyte slip at the curved wall surface increases the temperature of the boundary layer.

- When suction increases in the curved wall surface, both the temperature and the velocity of blood decrease.

The current theoretical estimates will be significant for the more precise treatment of patients with regard to better outcomes of thermal therapy for reducing pain. In addition, this investigation will improve the understanding of thermal processes that occur during blood flow in arterial microvessels. Clinicians involved in tumor and cancer treatment will begin using the electromagnetic hyperthermia technique, which involves overheating target tissues to about 42 °C. The results show that the flow velocity of blood can be managed by suitably adjusting (decreasing/increasing) the magnetic field intensity. This finding should help surgeons who generally want to maintain blood flow at a preferred level throughout surgery.

Author Contributions: Conceptualization, A.Z. and A.I.; Formal analysis, A.Z. and A.I.; Methodology, U.K., A.Z. and A.I.; Software, U.K.; Supervision, A.Z. and A.I. All authors have read and agreed to the published version of the manuscript.

Funding: This research was funded by Universiti Kebangsaan Malaysia (project code: DIP-2020-001).

Institutional Review Board Statement: Not applicable.

Informed Consent Statement: Not applicable.

Data Availability Statement: Not applicable.

Acknowledgments: We gratefully acknowledge the financial support received from the Universiti Kebangsaan Malaysia (project code: DIP-2020-001).

Conflicts of Interest: The authors declare no conflict of interest

References

1. Choi, S.U.S.; Eastman, J.A. Enhancing thermal conductivity of fluids with nanoparticles. *ASME Fluids Eng. Div.* **1995**, *231*, 99–105.
2. Eastman, J.A.; Choi, U.S.; Li, S.; Thompson, L.J.; Lee, S. Enhanced Thermal Conductivity through the Development of Nanofluids. *MRS Proc.* **1996**, *457*, 3–11. [CrossRef]
3. Sheikholeslami, M.; Ganji, D.D.; Javed, M.Y.; Ellahi, R. Effect of thermal radiation on magnetohydrodynamics nanofluid flow and heat transfer by means of two-phase model. *J. Magn. Magn. Mater.* **2015**, *374*, 36–43. [CrossRef]
4. Ellahi, R.; Hassan, M.; Zeeshan, A. Shape effects of nanosize particles in Cu–H_2O nanofluid on entropy generation. *Int. J. Heat Mass Transf.* **2015**, *81*, 449–456. [CrossRef]
5. Sadaf, H.; Nadeem, S. Influences of slip and Cu-blood nanofluid in a physiological study of cilia. *Comp. Meth. Prog. Biomed.* **2016**, *131*, 169–180. [CrossRef] [PubMed]
6. Noreen, S.; Rashidi, M.; Qasim, M. Blood flow analysis with considering nanofluid effects in vertical channel. *Appl. Nanosci.* **2017**, *7*, 193–199. [CrossRef]
7. Ardahaie, S.S.; Amiri, A.J.; Amouei, A.; Hosseinzadeh, K.; Ganji, D.D. Investigating the effect of adding nanoparticles to the blood flow in presence of magnetic field in a porous blood arterial. *Inform. Med. Unlocked* **2018**, *10*, 71–81. [CrossRef]
8. Elelamy, A.F.; Elgazery, N.S.; Ellahi, R. Blood flow of MHD non-Newtoniannanofluid with heat transfer and slip effects. *Int. J. Num. Meth. Heat Fluid Flow* **2020**, *30*, 4883–4908. [CrossRef]
9. Kamal, F.; Zaimi, K.; Ishak, A.; Pop, I. Stability analysis of MHD stagnation-point flow towards a permeable stretching/shrinking sheet in a nanofluid with chemical reactions effect. *Sains Malays.* **2019**, *48*, 243–250. [CrossRef]
10. Waini, I.; Ishak, A.; Groşan, T.; Pop, I. Mixed convection of a hybrid nanofluid flow along a vertical surface embedded in a porous medium. *Int. Commun. Heat Mass Transf.* **2020**, *114*, 104565. [CrossRef]
11. Waini, I.; Ishak, A.; Pop, I. Hybrid nanofluid flow past a permeable moving thin needle. *Mathematics* **2020**, *8*, 612. [CrossRef]
12. Waini, I.; Ishak, A.; Pop, I. Squeezed hybrid nanofluid flow over a permeable sensor surface. *Mathematics* **2020**, *8*, 898. [CrossRef]
13. Huang, X.; El-Sayed, M.A. Gold nanoparticles: Optical properties and implementations in cancer diagnosis and photothermal therapy. *J. Adv. Res.* **2010**, *1*, 13–28. [CrossRef]
14. Mekheimer, K.S.; Hasona, W.; Abo-Elkhair, R.; Zaher, A. Peristaltic blood flow with gold nanoparticles as a third grade nanofluid in catheter: Application of cancer therapy. *Phys. Lett. A* **2018**, *382*, 85–93. [CrossRef]
15. Lowe, G.D.; Rumley, A. The relationship between blood viscosity and blood pressure in a random sample of the population aged 55 to 74 years. *Eur. Heart J.* **1993**, *597*, 14.
16. Thurston, G.B.; Henderson, N.M. Effects of flow geometry on blood viscoelasticity. *Biorheology* **2006**, *43*, 729–746. [PubMed]
17. Khan, M.; Shahzad, A. On boundary layer flow of a Sisko fluid over a stretching sheet. *Quaest. Math.* **2013**, *36*, 137.e51. [CrossRef]
18. Munir, A.; Shahzad, A.; Khan, M. Convective flow of Sisko fluid over a bidirectional stretching surface. *PLoS ONE* **2015**, *1*, e0130342. [CrossRef] [PubMed]

19. Khan, M.; Ahmad, L.; Khan, W.A. Numerically framing the impact of radiation on magneto nanoparticles for 3D Sisko fluid flow. *J. Braz. Soc. Mech. Sci. Eng.* **2017**, *39*, 4475.e87. [CrossRef]
20. Eid, M.R.; Alsaedi, A.; Muhammad, T.; Hayat, T. Comprehensive analysis of heat transfer of gold-blood nanofluid (Sisko-model) with thermal radiation. *Results Phys.* **2017**, *7*, 4388–4393. [CrossRef]
21. Ahmad, L.; Khan, M. Numerical simulation for MHD flow of Sisko nanofluid over a moving curved surface: A revised model. *Microsyst. Technol.* **2019**, *25*, 2411–2428. [CrossRef]
22. Khan, U.; Zaib, A.; Shah, Z.; Baleanu, D.; Sherif El-Sayed, M. Impact of magnetic field on boundary-layer flow of Sisko liquid comprising nanomaterials migration through radially shrinking/stretching surface with zero mass flux. *J. Mater. Res. Technol.* **2020**, *9*, 3699–3709. [CrossRef]
23. Inoue, S.; Kobaya, M. Biological activities caused by far infrared radiation. *Int. J. Biometeorol.* **1989**, *33*, 145–150. [CrossRef]
24. Kobu, Y. Effects of infrared radiation on intraosseous blood flow and oxygen tension in rat tibia. *Kobe J. Med. Sci.* **1999**, *45*, 27–39. [PubMed]
25. Nishimoto, C.; Ishiura, Y.; Kuniasu, K.; Koga, T. Effects of ultrasonic radiation on cutaneous blood flow in the paw of decerebrated rats. *Kawasaki J. Med. Welfare* **2006**, *12*, 13–18.
26. He, Y.; Shirazaki, M.; Liu, H.; Himeno, R.; Sun, R. A numerical coupling model to analyze the blood flow, temperature, and oxygen transport in human breast tumor under laser irradiation. *Comput. Biol. Med.* **2006**, *36*, 1336–1350. [CrossRef]
27. Prakash, J.; Makinde, O.D. Radiative heat transfer to blood flow through a stenotic artery in the presence of magnetic field. *Lat. Am. Appl. Res.* **2011**, *41*, 273–277.
28. Misra, J.C.; Sinha, A. Effect of thermal radiation on MHD flow of blood and heat transfer in a permeable capillary in stretching motion. *Heat Mass Transf.* **2013**, *49*, 617–628. [CrossRef]
29. Khan, U.; Shafiq, A.; Zaib, A.; Sherif El-Sayed, M.; Baleanu, D. MHD radiative blood flow embracing gold particles via a slippery sheet through an erratic heat sink/source. *Mathematics* **2020**, *8*, 1597. [CrossRef]
30. Zaib, A.; Khan, U.; Wakif, A.; Zaydan, M. Numerical entropic analysis of mixed MHD convective flows from a non-isothermal vertical flat plate for radiative tangent hyperbolic blood biofluids conveying magnetite ferroparticles: Dual similarity solutions. *Arabian J. Sci. Eng.* **2020**, *45*, 5311–5330. [CrossRef]
31. Chen, I.H. Analysis of an intensive magnetic field on blood flow: Part 2. *J. Bioelectr.* **1985**, *4*, 55–61. [CrossRef]
32. Koriko, O.K.; Animasaun, I.; Mahanthesh, B.; Saleem, S.; Sarojamma, G.; Sivaraj, R. Heat transfer in the flow of blood-gold Carreau nanofluid induced by partial slip and buoyancy. *Heat Transf. Asian Res.* **2018**, *47*, 806–823. [CrossRef]
33. Roşca, N.C.; Pop, I. Unsteady boundary layer flow over a permeable curved stretching/shrinking surface. *Eur. J. Mech. B/Fluids* **2015**, *51*, 61–67. [CrossRef]
34. Saleh, S.H.M.; Arifin, N.M.; Nazar, R.; Pop, I. Unsteady micropolar fluid over a permeable curved stretching shrinking surface. *Math. Prob. Eng.* **2017**, *2017*, 3085249. [CrossRef]
35. Abbas, Z.; Naveed, M.; Sajid, M. Heat transfer analysis for stretching flow over a curved surface with magnetic field. *J. Eng. Thermophys.* **2013**, *22*, 337–345. [CrossRef]
36. Waini, I.; Ishak, A.; Pop, I. Flow and heat transfer along a permeable stretching/shrinking curved surface in a hybrid nanofluid. *Phys. Scr.* **2019**, *94*, 105219. [CrossRef]

Article

Double Solutions and Stability Analysis of Micropolar Hybrid Nanofluid with Thermal Radiation Impact on Unsteady Stagnation Point Flow

Nur Syazana Anuar [1] and Norfifah Bachok [1,2,*]

[1] Department of Mathematics, Faculty of Science, Universiti Putra Malaysia, Serdang 43400, Selangor, Malaysia; nursyazana931@gmail.com

[2] Institute for Mathematical Research, Universiti Putra Malaysia, Serdang 43400, Selangor, Malaysia

* Correspondence: norfifah@upm.edu.my

Abstract: The mathematical modeling of unsteady flow of micropolar $Cu-Al_2O_3$/water nanofluid driven by a deformable sheet in stagnation region with thermal radiation effect has been explored numerically. To achieve the system of nonlinear ordinary differential equations (ODEs), we have employed some appropriate transformations and solved it numerically using MATLAB software (built-in solver called bvp4c). Influences of relevant parameters on fluid flow and heat transfer characteristic are discussed and presented in graphs. The findings expose that double solutions appear in shrinking sheet case in which eventually contributes to the analysis of stability. The stability analysis therefore confirms that merely the first solution is a stable solution. Addition of nanometer-sized particle (Cu) has been found to significantly strengthen the heat transfer rate of micropolar nanofluid. When the copper nanoparticle volume fraction increased from 0 to 0.01 (1%) in micropolar nanofluid, the heat transfer rate increased roughly to an average of 17.725%. The result also revealed that an upsurge in the unsteady and radiation parameters have been noticed to enhance the local Nusselt number of micropolar hybrid nanofluid. Meanwhile, the occurrence of material parameter conclusively decreases it.

Keywords: micropolar hybrid nanofluid; dual solution; stretching/shrinking sheet; stability analysis; thermal radiation

Citation: Anuar, N.S.; Bachok, N. Double Solutions and Stability Analysis of Micropolar Hybrid Nanofluid with Thermal Radiation Impact on Unsteady Stagnation Point Flow. *Mathematics* **2021**, *9*, 276. https://doi.org/10.3390/math9030276

Received: 6 December 2020
Accepted: 22 December 2020
Published: 30 January 2021

Publisher's Note: MDPI stays neutral with regard to jurisdictional claims in published maps and institutional affiliations.

Copyright: © 2021 by the authors. Licensee MDPI, Basel, Switzerland. This article is an open access article distributed under the terms and conditions of the Creative Commons Attribution (CC BY) license (https://creativecommons.org/licenses/by/4.0/).

1. Introduction

For a number of years, the studies of micropolar fluid flow have captivated the attention of numerous scientists in understanding the fluid behavior especially in the study of rheological complex fluids, as, for example, the colloidal fluids, polymeric suspension, liquid crystals, animal blood, etc. [1]. In sight of these important applications, Eringen [2,3] was the first who originated the microfluid theory in his papers of simple microfluids and theory of micropolar fluids. This kind of fluids demonstrate the micro-rotational effect and micro-rotational inertia. Afterwards, this theory was then extended by Eringen [4] by taking into account the thermal effect and thus established the thermomicropolar fluids theory. Implementing the idea of Eringen, the micropolar fluid flow using a boundary layer approximation has been derived by many researchers in various problems such as in stagnation region [5], semi-infinite plate [6], cylinder [7], and rotating surface [8]. After some years, Nazar et al. [9] initiated the theoretical study of micropolar fluid flow when the sheet is stretch in the stagnation region, and soon after, Ishak et al. [10] and Yacob and Ishak [11] analyzed the same fluid induced by a shrinking sheet and observed the existence of nonunique solutions. Afterwards, Sandeep and Sulochana [12] undertook a numerical research of unsteady magnetohydrodynamic (MHD) micropolar fluid in both permeable shrinking and stretching sheet. The heat transfer characteristic of micropolar fluid flow driven by a shrinking sheet was discussed by Mishra et al. [13]. Soon after, Lund et al. [14]

noticed the existence of triple solutions at specific values of suction parameter in micropolar fluid when the sheet is shrunk exponentially and conducted the stability analysis. Further, a number of attempts toward this path have been made in the investigations of [15–17].

The inclusion of nanoparticles in a conventional fluid can literally change the flow and heat transfer capabilities, thereby can boost the thermal conductivity of the conventional fluid. It seems that Choi and Eastman [18] was the earlier person who conceived the idea of nanofluid, i.e., nanoparticle suspended in base fluid. Since then, nanofluids have been widely used in industrial cooling application [19], biomedical technology [20], solar thermal application [21], and many more. Numerous researchers, such as Gangadhar et al. [22], Chaudhary and Kanika [23], Naqvi et al. [24] and Anuar et al. [25,26], have scrutinized the concept of nanofluid flow and its heat transfer in their work. However, less studies are observed in micropolar nanofluid. The investigation of micropolar nanofluid driven by a stretching sheet was explored numerically by Hussain et al. [27]. Afterwards, Bourantas and Loukopoulos [28] and Noor et al. [29] scrutinized the micropolar nanofluid flow in an inclined square and vertical plate, respectively. The numerical investigation of micropolar nanofluid driven by a shrinking and stretching sheet have been made by Gangadhar et al. [30] and they pointed out that double solutions exist in certain range of parameters. Meanwhile, Dero et al. [31] point out the existence of triple solutions in their research involving micropolar nanofluid when the sheet is stretch/shrunk exponentially. The studies of micropolar nanofluid in an inclined stretching/shrinking have been scrutinized by Lund et al. [32] with consideration of convective boundary conditions. They also observed the occurrence of nonunique solutions in their work and performed the stability analysis. Recently, Abdal et al. [33], Amjad et al. [34], Rafique et al. [35] and many others have explored the micropolar nanofluid flow problem in different surfaces and aspects.

Nevertheless, a new modern kind of nanofluid which can efficiently improve the heat transfer are later being introduced in the industry are recognized as hybrid nanofluid, i.e., mixture of two types of nanoparticle dispersed into a base fluid. This new kind of fluid, however, shows a great advance in heat conductivity and it proved by the work of Madhesh and Kalaiselvam [36], Tahat and Benim [37], Devi and Devi [38], etc. Following this, mathematical investigation specifically in boundary layer flow in hybrid nanofluid has attracted a few researchers to explore it in various surfaces such as in stretching/shrinking sheet [39], curved surface [40], thin needle [41], Riga plate [42], etc. By opting the novel idea of hybrid nanofluid, Subhani and Nadeem [43] scrutinized the behavior of hybrid nanofluid (Cu-TiO_2/water) in micropolar fluid in a porous medium past an exponentially stretching sheet and point out that the heat transfer rate for micropolar hybrid nanofluid is greater than micropolar nanofluid. Afterwards, by taking into attention the simultaneous effects of MHD and slip, Nadeem and Abbas [44] examined the micropolar hybrid nanofluid flow past a circular cylinder. In another study of Abbas et al. [45] and Al-Hanaya [46], a theoretical investigation of micropolar hybrid nanofluid using carbon nanotubes (SWCNT and MWCNT) as a nanoparticle over an exponentially stretching Riga plate and curved stretching sheet have been investigated. Apparently, the research related to micropolar hybrid nanofluids are limited in number. Hence, the principal goal of this investigation is to address the behavior of micropolar hybrid nanofluid in a deformable sheet, i.e., stretching and shrinking. It is important to note that deformable sheet is not a new crucial topic among the researchers in the fluid field since their applications are well recognized in processing industries especially in polymer processing, glass fiber production, cooling, and drying of paper and many others [47].

The impact of thermal radiation is also discussed in this paper, where this effect is crucial in solar power technology, electrical power generation, astrophysical flows, and other industrial fields. In the scenario of high-temperature flow processes, thermal radiation effects are also extremely important [48]. There is a lot of comprehensive literature now available that concerns with the thermal radiation effect on the flow of the boundary layer. For instance, Sajid and Hayat [49] have been analyzing the thermal radiation effect on the viscous flow as the sheet is stretch exponentially and realized that the thermal boundary layer thickness thickens as the radiation parameter increase. Afterwards, Nadeem et al. [50] extend the investigations of [49] by considering it in Jeffrey fluid. The numerical investigation of micropolar nanofluid over the stretching sheet with the effect of thermal radiation, MHD, and heat source/sink have been examined by Pal and Mandal [51]. Again, Gireesha et al. [52] addresses the Jeffrey nanofluids problem driven by a nonlinearly permeable stretching sheet under the effect of radiation and magnetohydrodynamic. In a recent study, Yashkun et al. [53] noticed the occurrence of dual solutions in their work of MHD hybrid nanofluid past a deformable sheet with thermal radiation effect. Hence, motivated by the aforementioned work, our aim here is to scrutinize the influence of thermal radiation towards the heat transfer of micropolar hybrid nanofluid.

In brief, this research paper is an extended work of Nazar et al. [9] to the case of unsteady two-dimensional hybrid nanofluid in shrinking sheet and take into attention the effect of thermal radiation. Given the above-mentioned study, the utilization of hybrid nanofluid (Cu and Al_2O_3) as the new heat transfer fluid for the micropolar flow problem with the thermal radiation effect, has not been performed up to now. In addition, this analysis also comprises a novel era for scientists to discover the shrinking features of micropolar hybrid nanofluids. Furthermore, the novelty of this study can also be seen in the discovery of non-unique solutions and the execution of stability analysis. To the best of authors' knowledge, the results of the present work is new and still not considered and published by any researchers. Therefore, current studies are expected to bring good benefits to researchers who are experimentally working on micropolar hybrid nanofluids, and these results are also expected to reduce the cost of experimental work in the future.

2. Mathematical Framework

2.1. Basic Equations

The unsteady two-dimensional flow of micropolar Cu–Al_2O_3/water nanofluid past a deformable sheet in the stagnation region with the influence of thermal radiation impact are investigated in this work as exemplified in Figure 1. The Cartesian coordinates used are x and y, given that x−axis is considered along the sheet while y−axis normal to it, respectively, the sheet is located in the plane $y = 0$ and the fluid fill the half space at $y \geq 0$. The temperature far from the surface (inviscid flow) and at the surface are represented by T_∞ and $T_w(x,t)$. The sheet is stretch and shrunk along the x−axis with velocity $u_w(x,t)$ and the free stream velocity is denoted by $u_e(x,t)$.

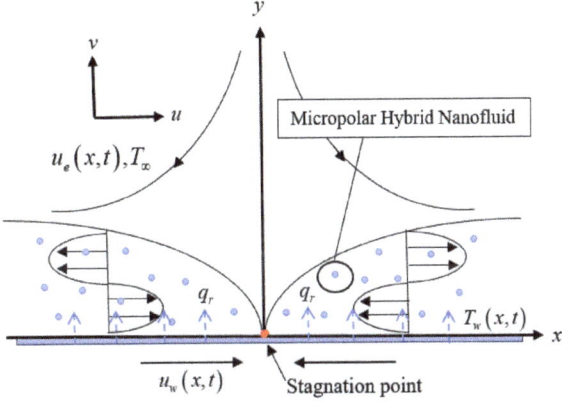

Figure 1. Schematic model of shrinking sheet.

From all of the above circumstance, the partial differential equations which govern the flow are stated as (see Nazar et al. [9], Bhattacharyya et al. [54], Roy et al. [55]):

$$\frac{\partial u}{\partial x} + \frac{\partial v}{\partial y} = 0 \tag{1}$$

$$\frac{\partial u}{\partial t} + u\frac{\partial u}{\partial x} + v\frac{\partial u}{\partial y} = \frac{\partial u_e}{\partial t} + u_e\frac{\partial u_e}{\partial x} + \frac{\mu_{hnf} + \kappa}{\rho_{hnf}}\frac{\partial^2 u}{\partial y^2} + \frac{\kappa}{\rho_{hnf}}\frac{\partial N}{\partial y} \tag{2}$$

$$\frac{\partial N}{\partial t} + u\frac{\partial N}{\partial x} + v\frac{\partial N}{\partial y} = \frac{\varsigma}{\rho_{hnf}j}\frac{\partial^2 N}{\partial y^2} - \frac{\kappa}{\rho_{hnf}j}\left(2N + \frac{\partial u}{\partial y}\right) \tag{3}$$

$$\frac{\partial T}{\partial t} + u\frac{\partial T}{\partial x} + v\frac{\partial T}{\partial y} = \frac{k_{hnf}}{(\rho C_p)_{hnf}}\frac{\partial^2 T}{\partial y^2} - \frac{1}{(\rho C_p)_{hnf}}\frac{\partial q_r}{\partial y} \tag{4}$$

Here, the velocity component in the x direction is denoted as u whereas v is the velocity component along y axis, t and T are time and temperature, N refers to the angular velocity (microrotation) in the xy–plane, q_r signifies the radiative heat flux, κ is the vortex viscosity and j is the micro inertial density. In addition, ς is the spin gradient viscosity given by (Ahmadi [6])

$$\varsigma = \left(\mu_f + \frac{\kappa}{2}\right)j \tag{5}$$

where $j = \nu_f/u_e$ is specified as the reference length. Further, k_{hnf}, ρ_{hnf}, μ_{hnf} and $(\rho C_p)_{hnf}$ are the thermal conductivity, density, dynamic viscosity, and heat capacity of Cu–Al$_2$O$_3$/water.

The accompanying conditions are

$$\begin{array}{l} u = u_w(x,t),\ v = 0,\ N = -n\frac{\partial u}{\partial y},\ T = T_w(x,t)\ \text{as } y = 0 \\ u \to u_e(x,t),\ N \to 0,\ T \to T_\infty\ \text{as } y \to \infty \end{array} \tag{6}$$

where n is the constant in the range of $[0, 1]$. It is worthwhile to note that for $n = 0$ which implies that $N = 0$ near the wall, exemplifies the microelements near the wall surface are incapable to rotate, i.e., concentrated particle flows (Jena and Mathur [56]) or also denoted as strong concentration of microelements (Guram and Smith [57]). However, for the case $n = 0.5$ which refer to a weak concentration of microelements, the disappearing of anti-symmetric part of the stress tensor is noted (Ahmadi [6]). Further, the case $n = 1$ is utilized for the modelling of turbulent boundary layer flows (Peddieson [58]). While the velocity of deformable sheet, free stream and temperature at the surface are referred from the work of Zainal et al. [59] which given as

$$u_w(x,t) = \frac{cx}{1-bt},\ u_e(x,t) = \frac{ax}{1-bt},\ T_w(x,t) = T_\infty + \frac{T_0 a x^2}{2\nu_f(1-bt)^{3/2}} \tag{7}$$

here, $a(>0)$ and $c(>0)$ are constants, b measures the unsteadiness of the problem and $T_0 > 0$ is the reference temperature.

Using the Rosseland's approximation (Brewster [60]), the q_r term can be expressed clearly as below

$$q_r = -\frac{4\sigma^*}{3k^*}\frac{\partial T^4}{\partial y} \tag{8}$$

where σ^* and k^* signify the constant of Stefan–Boltzmann and mean absorption's coefficient. Implementing the Taylor series and ignored the higher-order terms, T^4 is expanded about T_∞; hence, we have $T^4 \approx 4T_\infty^3 T - 3T_\infty^4$. Subsequently, Equation (4) become

$$\frac{\partial T}{\partial t} + u\frac{\partial T}{\partial x} + v\frac{\partial T}{\partial y} = \frac{k_{hnf}}{(\rho C_p)_{hnf}}\frac{\partial^2 T}{\partial y^2} + \frac{16\sigma^* T_\infty^3}{3k^*(\rho C_p)_{hnf}}\frac{\partial^2 T}{\partial y^2} \tag{9}$$

2.2. Thermophysical Traits of Hybrid Nanofluid

The physical traits of hybrid nanofluids are prescribed in Table 1. In Table 1, the subscript hnf, nf, f and s signify the hybrid nanofluid, nanofluid, fluid and nanoparticle, whereas $s1$ and $s2$ symbolize the first nanoparticle and second nanoparticle, respectively. Furthermore, φ_1 represents the first nanoparticle volume fraction while φ_2 denotes the second nanoparticle volume fraction. In this investigation, copper (Cu) is picked as the second nanoparticle volume fraction, alumina (Al_2O_3) is picked as the first nanoparticle volume fraction and water act as a base fluid. Table 2 displays the thermophysical traits of nanoparticles and base fluid. It is important to note that Al_2O_3 is originally disseminated into the water to achieve the appropriated hybrid nanofluid, i.e., Cu-Al_2O_3/water, and then Cu is disseminated into the Al_2O_3/water nanofluid. Additionally, the volume fraction of Al_2O_3 nanoparticle is set to 1% and Cu is fluctuated from 0 to 2%.

Table 1. Physical traits of hybrid nanofluids (Devi and Devi [38]).

Properties	Hybrid Nanofluid
Density	$\rho_{hnf} = (1-\varphi_2)\left[(1-\varphi_1)\rho_f + \varphi_1\rho_{s1}\right] + \varphi_2\rho_{s2}$
Heat capacity	$(\rho C_p)_{hnf} = \varphi_2(\rho C_p)_{s2} + (1-\varphi_2)\left[(1-\varphi_1)(\rho C_p)_f + \varphi_1(\rho C_p)_{s1}\right]$
Dynamic viscosity	$\mu_{hnf} = \frac{\mu_f}{(1-\varphi_1)^{2.5}(1-\varphi_2)^{2.5}}$
Thermal conductivity	$\frac{k_{hnf}}{k_{bf}} = \frac{k_{s2}+2k_{bf}-2\varphi_2(k_{bf}-k_{s2})}{k_{s2}+2k_{bf}+\varphi_2(k_{bf}-k_{s2})}$ where $\frac{k_{bf}}{k_f} = \frac{k_{s1}+2k_f-2\varphi_1(k_f-k_{s1})}{k_{s1}+2k_f+\varphi_1(k_f-k_{s1})}$

Table 2. Thermo physical properties (Oztop and Abu-Nada [61]).

	Physical Properties		
	$C_p\ (\mathrm{J\,kg^{-1}K^{-1}})$	$\rho\ (\mathrm{kg\,m^{-3}})$	$k\ (\mathrm{W\,m^{-1}K^{-1}})$
water	4179	997.1	0.613
Cu	385	8933	400
Al_2O_3	765	3970	40

2.3. Similarity Solutions

In this work, the subsequent similarity transformation is introduced (Roy et al. [55])

$$\eta = \left(\frac{a}{v_f(1-bt)}\right)^{1/2} y,\ \psi = \left(\frac{av_f}{1-bt}\right)^{1/2} x f(\eta),\ N = \left(\frac{a}{v_f(1-bt)}\right)^{1/2}\frac{a}{(1-bt)} x h(\eta),\ \theta(\eta) = \frac{T-T_\infty}{T_w-T_\infty} \quad (10)$$

where v_f and η are the fluid kinematic viscosity and similarity variable, while f, h and θ are the dimensionless function. Further, primes signify the differentiation with respect to η, while the stream function ψ is specified as $v = -\partial\psi/\partial x$ and $u = \partial\psi/\partial y$.

Invoking the similarity variables (10), Equation (1) is identically fulfilled and Equations (2), (3) and (9) are reduced into the following similarity equations

$$\frac{\mu_{hnf}/\mu_f}{\rho_{hnf}/\rho_f}(1+K)f''' - f'^2 + ff'' + 1 - A\left(f' - 1 + \frac{1}{2}\eta f''\right) + \frac{K}{\rho_{hnf}/\rho_f}h' = 0 \quad (11)$$

$$\frac{1}{\rho_{hnf}/\rho_f}\left(\frac{\mu_{hnf}}{\mu_f} + \frac{K}{2}\right)h'' + fh' - f'h - \frac{A}{2}(3h+\eta h') - \frac{K}{\rho_{hnf}/\rho_f}(2h+f'') = 0 \quad (12)$$

$$\frac{1}{\Pr(\rho C_p)_{hnf}/(\rho C_p)_f}\left(\frac{k_{hnf}}{k_f} + \frac{4}{3}Rd\right)\theta'' + f\theta' - 2f'\theta - \frac{A}{2}(3\theta + \eta\theta') = 0 \quad (13)$$

Here, the material parameter, unsteady parameter, Prandtl number and radiation parameter which denoted by K, A, Pr and Rd are defined by

$$K = \frac{\kappa}{\mu_f}, \quad A = \frac{b}{a}, \quad Pr = \frac{\nu_f}{\alpha_f}, \quad Rd = \frac{4\sigma^* T_\infty^3}{k_f k^*}, \qquad (14)$$

The conditions (6) become

$$\begin{array}{l} f'(0) = c/a = \lambda, \quad f(0) = 0, \quad h(0) = -nf''(0), \quad \theta(0) = 1, \\ f'(\eta) \to 1, \quad h(\eta) \to 0, \quad \theta(\eta) \to 0 \quad \text{as} \quad \eta \to \infty \end{array} \qquad (15)$$

where the stretching/shrinking parameter is denoted by λ with $\lambda > 0$ signifies the sheet is stretch, $\lambda = 0$ refers to static plate and $\lambda < 0$ denotes the sheet is shrunk.

In this investigation, the physical quantities of interest are specified as

$$C_f = \frac{1}{\rho_f u_e^2}\left[(\mu_{hnf} + \kappa)\left(\frac{\partial u}{\partial y}\right) + \kappa N\right]_{y=0}, \quad Nu_x = \frac{x}{k_f(T_w - T_\infty)}\left[-k_{hnf}\left(\frac{\partial T}{\partial y}\right)_{y=0} + q_r|_{y=0}\right] \qquad (16)$$

here, C_f is the skin friction coefficient and Nu_x is the Nusselt number. Using variables (10) and (16), the following local skin friction coefficient and local Nusselt number (heat transfer rate) are achieved

$$C_f Re_x^{1/2} = \left(\frac{\mu_{hnf}}{\mu_f} + K\right) f''(0) + Kh(0), \quad Nu_x Re_x^{-1/2} = -\left(\frac{k_{hnf}}{k_f} + \frac{4}{3} Rd\right) \theta'(0) \qquad (17)$$

where $Re_x = u_e x / \nu_f$ is the local Reynolds number.

3. Stability of the Solutions

Due to the occurrence of non-uniqueness in the present research, the stability analysis is executed by referring to the work of Merkin [62], Weidman et al. [63], and Harris et al. [64]. These analyses have been implemented by other researchers too (see for example the work of [14–16,25,26,32,39,59]). Some important steps are implemented to identify the stability of solutions, i.e., (i) introducing a new dimensionless time variables and similarity variables, (ii) implement the linear eigenvalue equations, and (iii) relax the boundary conditions.

3.1. New Similarity Transformation

A new dimensionless time variable τ need to be introduced as follows (Zainal et al. [59])

$$\tau = \frac{a}{1 - bt} t \qquad (18)$$

while the similarity variables (10) are replaced by

$$\eta = \left(\frac{a}{\nu_f(1 - bt)}\right)^{1/2} y, \quad \psi = \left(\frac{a\nu_f}{1 - bt}\right)^{1/2} x f(\eta, \tau), \quad N = \left(\frac{a}{\nu_f(1 - bt)}\right)^{1/2} \frac{a}{(1 - bt)} x h(\eta, \tau), \quad \theta(\eta, \tau) = \frac{T - T_\infty}{T_w - T_\infty} \qquad (19)$$

By applying Equations (18) and (19) in Equations (1)–(3) and (9), the new transformed differential equations are attained

$$\frac{\mu_{hnf}/\mu_f}{\rho_{hnf}/\rho_f}(1 + K)\frac{\partial^3 f}{\partial \eta^3} + f\frac{\partial^2 f}{\partial \eta^2} - \left(\frac{\partial f}{\partial \eta}\right)^2 + 1 - A\left(\frac{\partial f}{\partial \eta} + \frac{1}{2}\eta\frac{\partial^2 f}{\partial \eta^2} - 1\right) + \frac{K}{\rho_{hnf}/\rho_f}\frac{\partial h}{\partial \eta} - (A\tau + 1)\frac{\partial^2 f}{\partial \eta \partial \tau} = 0 \qquad (20)$$

$$\frac{1}{\rho_{hnf}/\rho_f}\left(\frac{\mu_{hnf}}{\mu_f} + \frac{K}{2}\right)\frac{\partial^2 h}{\partial \eta^2} + f\frac{\partial h}{\partial \eta} - \frac{\partial f}{\partial \eta}h - \frac{A}{2}\left(3h + \eta\frac{\partial h}{\partial \eta}\right) - \frac{K}{\rho_{hnf}/\rho_f}\left(2h + \frac{\partial^2 f}{\partial \eta^2}\right) - (A\tau + 1)\frac{\partial h}{\partial \tau} = 0 \qquad (21)$$

$$\frac{1}{\Pr(\rho C_p)_{hnf}/(\rho C_p)_f}\left(\frac{k_{hnf}}{k_f}+\frac{4}{3}Rd\right)\frac{\partial^2\theta}{\partial\eta^2}+f\frac{\partial\theta}{\partial\eta}-2\frac{\partial f}{\partial\eta}\theta-\frac{A}{2}\left(3\theta+\eta\frac{\partial\theta}{\partial\eta}\right)-(A\tau+1)\frac{\partial\theta}{\partial\tau}=0 \quad (22)$$

and the conditions become

$$f(0,\tau)=0,\ \frac{\partial f}{\partial\eta}(0,\tau)=\lambda,\ h(0,\tau)=-n\frac{\partial^2 f}{\partial\eta^2}(0,\tau),\ \theta(0,\tau)=1,$$
$$\frac{\partial f}{\partial\eta}(\eta,\tau)\to 1,\ h(\eta,\tau)\to 0,\ \theta(\eta,\tau)\to 0\ \text{as}\ \eta\to\infty \quad (23)$$

3.2. Introducing Linear Eigenvalue Equations

The stability of the steady flow solutions can be explored by setting $f(\eta)=f_0(\eta)$, $h(\eta)=h_0(\eta)$ and $\theta(\eta)=\theta_0(\eta)$, where it satisfied the boundary value problems (11)–(13) and (15). Thus, the following equations are introduced (Weidman et al. [63]):

$$f(\eta,\tau)=f_0(\eta)+e^{-\gamma\tau}F(\eta,\tau),\ h(\eta,\tau)=h_0(\eta)+e^{-\gamma\tau}H(\eta,\tau),\ \theta(\eta,\tau)=\theta_0(\eta)+e^{-\gamma\tau}G(\eta,\tau), \quad (24)$$

where $F(\eta,\tau)$, $H(\eta,\tau)$, $G(\eta,\tau)$ and their derivatives are small then $f_0(\eta), h_0(\eta)$ and $\theta_0(\eta)$. In addition, γ is the unknown eigenvalue which will be used to specify the stability of the solutions. Substitute Equation (24) into (20)–(22) and let $\tau\to 0$, in which $F(\eta)=F_0(\eta)$, $H(\eta)=H_0(\eta)$ and $G(\eta)=G_0(\eta)$, thereby the linearized eigenvalue equations relevant to the problem are

$$\frac{\mu_{hnf}/\mu_f}{\rho_{hnf}/\rho_f}(1+K)F_0'''+\left(f_0+\frac{A}{2}\eta\right)F_0''+F_0f_0''-(2f_0'+A-\gamma)F_0'+\frac{K}{\rho_{hnf}/\rho_f}H_0'=0 \quad (25)$$

$$\frac{1}{\rho_{hnf}/\rho_f}\left(\frac{\mu_{hnf}}{\mu_f}+\frac{K}{2}\right)H_0''+\left(f_0-\frac{A}{2}\eta\right)H_0'+F_0h_0'-F_0'h_0-\left(f_0'+\frac{3}{2}A-\gamma\right)H_0-\frac{K}{\rho_{hnf}/\rho_f}(2H_0+F_0'')=0 \quad (26)$$

$$\frac{1}{\Pr(\rho C_p)_{hnf}/(\rho C_p)_f}\left(\frac{k_{hnf}}{k_f}+\frac{4}{3}Rd\right)G_0''+\left(f_0-\frac{A}{2}\eta\right)G_0'+F_0\theta_0'-2F_0'\theta_0-\left(2f_0'+\frac{3}{2}A-\gamma\right)G_0=0 \quad (27)$$

The conditions now take the following form

$$F_0'(0)=0,\ F_0(0)=0,\ H_0(0)=-n\,F_0''(0),\ G_0(0)=0,$$
$$F_0'(\eta)\to 0,\ H_0(\eta)\to 0,\ G_0(\eta)\to 0,\ \text{as}\ \eta\to\infty \quad (28)$$

3.3. Relaxation of Boundary Conditions

To solve the stability model, we need to relax the boundary conditions as proposed by Harris et al. [64]. For that reason, the conditions $F_0'(\eta)\to 0$ as $\eta\to\infty$ can be replaced by new conditions $F_0''(0)=1$. It must be pointed out that the linearized boundary value problem (25)–(28) together with new conditions $F_0''(0)=1$ will yield the unlimited set of unknown eigenvalues $(\gamma_1<\gamma_2<\gamma_3<\ldots)$. If the smallest eigenvalues γ show a positive sign, the solutions observed an initial decay of perturbation and accordingly indicates a stable solution. On the other hand, as the smallest eigenvalues γ show a negative sign, an early growth of disruption is noticed which consequently signifies unstable solution.

4. Numerical Solutions

To solve the boundary value problems (11)–(13) with boundary conditions given by (15), we have adopted a built-in function called bvp4c from Matlab package. Further, to access the precision of current algorithm, the current results of skin friction coefficient $f''(0)$ are compared with previously reported solutions of Ishak et al. [10], who used Keller-box method in their work, Mahapatra and Nandy [65], who pursued the shooting method for their computation and Zainal et al. [59] which employed the bvp4c solver. These comparative solutions are revealed in Table 3 for selected values of shrinking parameter $(\lambda<0)$. It can be point out from these tables that there is good agreement with these methods (error is relatively small), thereby confirming the consistency of the approach used.

Furthermore, this validates the present model and proves the accuracy of the bvp4c solver in solving a boundary layer problem as the present results able to withstand the Keller-box method and shooting method which have been employed by Ishak et al. [10] and Mahapatra and Nandy [65]. In this part, the results of local skin friction $C_f \text{Re}_x^{1/2}$, Nusselt number $Nu_x\text{Re}_x^{-1/2}$, velocity profile $f'(\eta)$, microrotation profile $h(\eta)$ as well as temperature profile $\theta(\eta)$ are illustrated graphically to explore the influence of some governing parameters such as Cu nanoparticle volume fraction (φ_2), unsteady parameter (A), material parameter (K) and radiation parameter (Rd).

Table 3. Comparison values of $f''(0)$ when $A = K = n = 0$ and $\varphi_1 = \varphi_2 = 0$ for different λ values.

λ	Refs. [10] (Keller-Box Method)	Refs. [65] (Shooting Method)	Refs. [59] (bvp4c Solution)	Present Result (bvp4c Solution)
-0.25	1.402241	1.402242	1.402241	1.402241
-0.5	1.495670	1.495672	1.495670	1.495670
-0.75	1.489298	1.489296	1.489298	1.489298
-1	1.328817 [0]	1.328819 [0]	1.328817 [0]	1.328817 [0]
-1.1	1.186681 [0.049229]	1.186680 [0.049229]	1.186680 [0.049229]	1.186680 [0.049229]
-1.15	1.082231 [0.116702]	1.082232 [0.116702]	1.082231 [0.116702]	1.082231 [0.116702]
-1.2	0.932474 [0.233650]	0.932470 [0.233648]	0.932473 [0.233650]	0.932473 [0.233650]
-1.246	-	0.584374 [0.554215]	0.609826 [0.529035]	0.609826 [0.529035]
-1.2465	0.584295 [0.554283]	-	-	0.584282 [0.554296]

'[ ]' Second solution.

The effect of Cu nanoparticle volume fraction φ_2 against stretching or shrinking parameter λ on the local skin friction $C_f\text{Re}_x^{1/2}$ and Nusselt number $Nu_x\text{Re}_x^{-1/2}$ as given in Equation (17) are shown in Figure 2a,b. It is apparent from these figures that for shrinking parameter ($\lambda < 0$), the occurrence of dual solutions is noted. However, it is remarked that no solution exists when $\lambda < \lambda_c$, which indicates that the boundary layer is detach from the surface and the principle of boundary layer theory are no longer valid. Moreover, λ_c is the critical point that connected the first and second solutions. In addition, a unique solution is noticed when the sheet is stretched ($\lambda > 0$). It is clear that for $\varphi_2 = 0$, the problem reduces to the micropolar nanofluid. From these figures, it is discovered that upsurge in Cu nanoparticle volume fraction φ_2 enhances the local skin friction $C_f\text{Re}_x^{1/2}$ and local Nusselt number $Nu_x\text{Re}_x^{-1/2}$ for all domains of stretching and shrinking parameter λ in first solution but a small change is noted for the second solution. This finding proves that the increment of Cu nanoparticle volume fraction φ_2 can improve the heat transfer efficiency. This also implies that hybrid nanofluid provides a better heat performance than nanofluid. Furthermore, the enhancement of Cu nanoparticle volume fraction φ_2 on local skin friction $C_f\text{Re}_x^{1/2}$ and Nusselt number $Nu_x\text{Re}_x^{-1/2}$ fasten the detachment of boundary layer flow. Figure 3a–c exemplify the impact of Cu nanoparticle volume fraction parameter φ_2 on the velocity profile $f'(\eta)$, microrotation profile $h(\eta)$ and temperature profile $\theta(\eta)$ for shrinking sheet ($\lambda = -1.25$). It reveals that augmentation of Cu nanoparticle volume fraction φ_2 depreciates the momentum and microrotation boundary layer thickness in first and second solutions. Meanwhile, the thermal boundary layer thickness increases as Cu nanoparticle volume fraction φ_2 increases for the first solution, however a contrary observation is noted for the second solution. Furthermore, one can see that the boundary layer thickness of the first solution was slimmer than second solution. In addition, all the profiles published are asymptotically satisfied the boundary conditions (15) and eventually supported the findings obtained in Figure 2a,b.

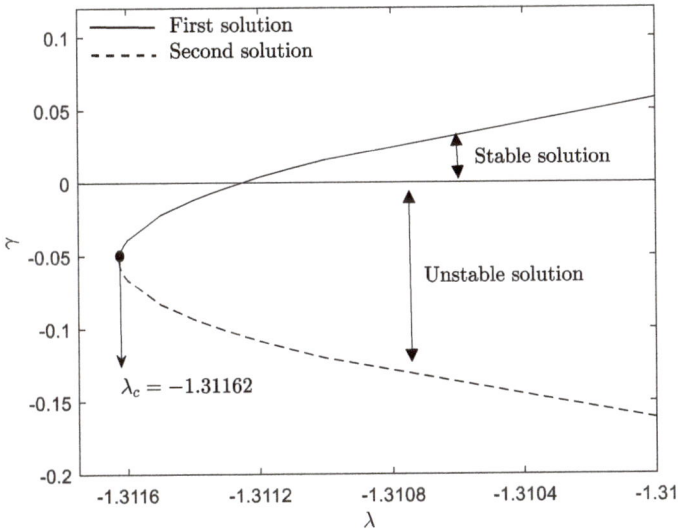

Figure 2. (a) $C_f \text{Re}_x^{1/2}$; (b) $Nu_x \text{Re}_x^{-1/2}$ with λ for various φ_2.

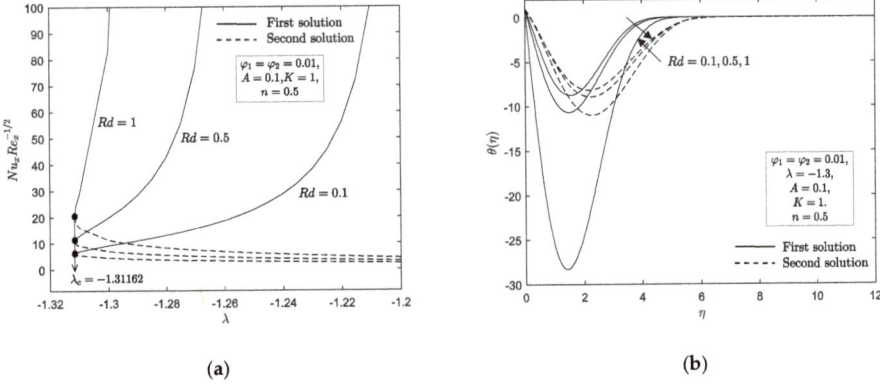

Figure 3. (a) $f'(\eta)$; (b) $h(\eta)$; (c) $\theta(\eta)$ for various φ_2.

Figure 4a,b show the impact of unsteady parameter $A = 0$, 0.1, 0.2 on the local skin friction $C_f \text{Re}_x^{1/2}$ and Nusselt number $Nu_x \text{Re}_x^{-1/2}$ towards stretching/shrinking parameter λ. The occurrence of unsteady parameter A consequently elevates the local skin friction $C_f \text{Re}_x^{1/2}$ and Nusselt number $Nu_x \text{Re}_x^{-1/2}$. It must be noted that the flow corresponds to the steady micropolar flow when $A = 0$ and it is numerically observed that dual solution exists as $-1.24641 < \lambda < -1$ and a unique solution exists when $\lambda \geq -1$. However, the physical character of flow changes when the flow becomes unsteady. For instance, as the unsteadiness parameter increase, i.e., $A = 0.1$ and $A = 0.2$, the range of similarity solutions to exist also increases where dual solutions is observed in the range of $-1.31162 < \lambda < -1$ and $-1.38029 < \lambda < -1$, respectively. The unique solution however only exists as $\lambda \geq -1$ for both case of A and concurrently no solution is noticed when $\lambda < \lambda_c$. In short, a raise in unsteady parameter A act in postponing the boundary layer detachment. On the other side, Figure 5a–c portray the discrepancy of velocity $f'(\eta)$, microrotation $h(\eta)$ and temperature $\theta(\eta)$ profiles when unsteady parameter A fluctuates from 0 to 0.15. For the first solution, the diminished of momentum boundary layer thickness is observed when the unsteady parameter A increases as shown in Figure 5a, but a reverse observation is remarked for the second solution. Furthermore, the microrotation boundary layer thickness diminishes with an increment of unsteady parameter A values in the first solution except near the sheet, while an opposing trend is remarked for the second solution. It is also interesting to observe from these figures that for both solutions, the thermal boundary layer thickness increase with an upsurge of unsteady parameter A.

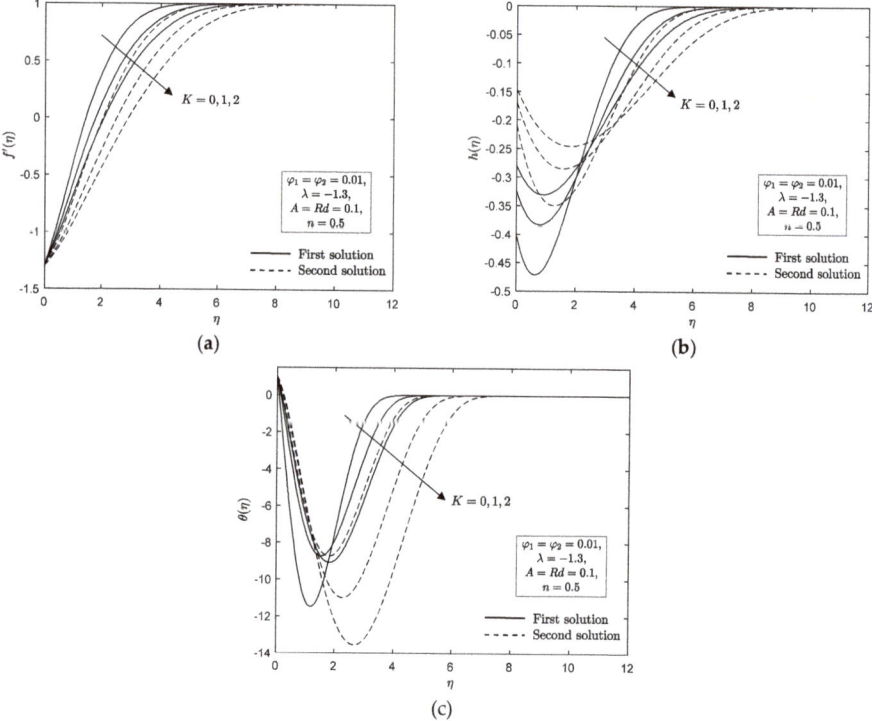

Figure 4. (a) $C_f Re_x^{1/2}$; (b) $Nu_x Re_x^{-1/2}$ with λ for various A.

Figure 6a,b and Figure 7a–c are portray to discuss the effect of material parameter K on the local skin friction $C_f Re_x^{1/2}$, local Nusselt number $Nu_x Re_x^{-1/2}$, velocity profile $f'(\eta)$, microrotation profile $h(\eta)$ and temperature profile $\theta(\eta)$ for Cu-Al$_2$O$_3$/water. It is obvious that existence of material parameter $(K = 1, 2)$ give rises to the local skin friction $C_f Re_x^{1/2}$ if compared to the absence of material parameter $(K = 0)$, i.e., no vortex viscosity. However, different results are observed for the local Nusselt number $Nu_x Re_x^{-1/2}$ where the nonexistence of micropolar fluid $(K = 0)$ cause an enhancement in comparison with the existence of material parameter $(K = 1, 2)$. This phenomenon reveals the fact that upsurge value of material parameter gives rise on the vortex viscosity in the fluid flow which consequently enhance the skin friction at the wall and decrease the rate of heat transfer at the wall. Additionally, we observed that an upsurge values of material parameter prompt the domain of similarity solutions to exist become narrow. For instance, the similarity solutions in the nonexistence of material parameter are noted in the range of $-1.31178 \leq \lambda \leq 1$, while in the existence of material parameter $(K = 1, 2)$, the range of solutions are observed to be $-1.31164 \leq \lambda \leq -1$ and $-1.31162 \leq \lambda \leq -1$. Furthermore, an upsurge values of material parameter K causing the thickness of the momentum and thermal boundary layer to increase in first and second solutions. We can see from Figure 7b that the microrotation boundary layer thickness for first and second solutions near the sheet decreases when this parameter rises while the contrary trend is observed for the large η.

(a)

(b)

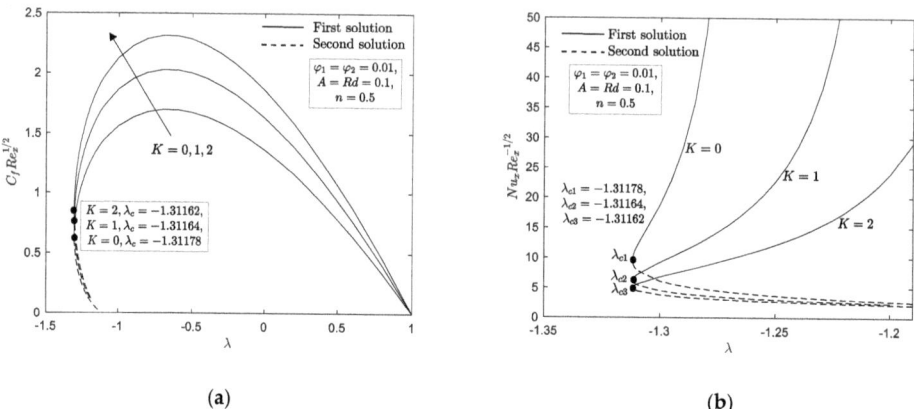

Figure 5. (**a**) $f'(\eta)$; (**b**) $h(\eta)$; (**c**) $\theta(\eta)$ for various A.

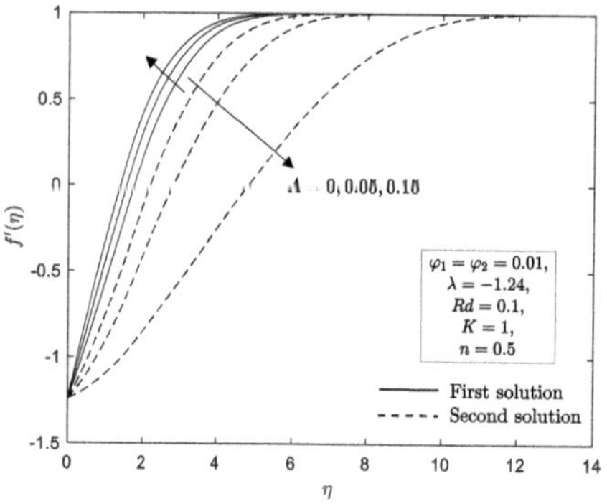

(a)

(b)

(c)

Figure 6. (**a**) $C_f \mathrm{Re}_x^{1/2}$; (**b**) $Nu_x \mathrm{Re}_x^{-1/2}$ with λ for various K.

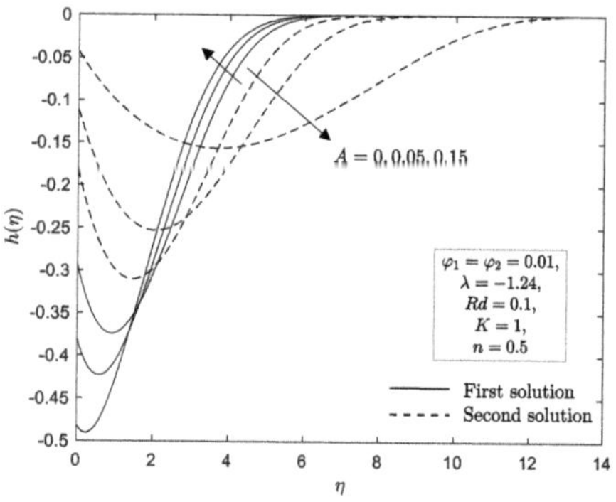

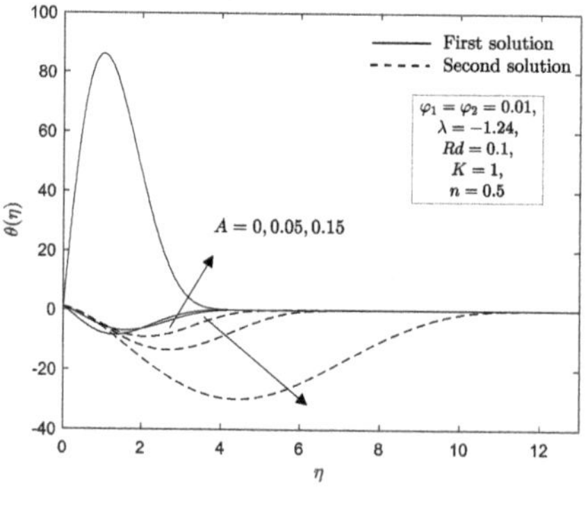

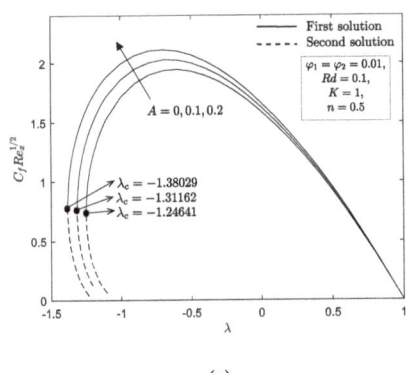
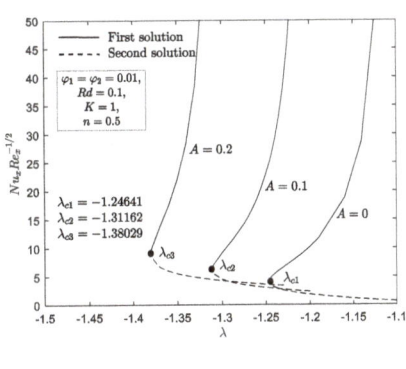

(a)   (b)

Figure 7. (a) $f'(\eta)$; (b) $h(\eta)$; (c) $\theta(\eta)$ for various K.

The influence of radiation parameter Rd on the local Nusselt number $Nu_x Re_x^{-1/2}$ and temperature profile $\theta(\eta)$ are portrayed in Figure 8a,b, respectively. We have noticed that the radiation parameter Rd has no control on the flow field, which is evident from Equations (11)–(13). The local Nusselt number is discovered to increase with an upsurge values in radiation parameter Rd. Furthermore, the thermal boundary layer thickness become thicker when this parameter raises in the first solution and it became slimmer in the second solution. As can be seen from Figure 8b, this parameter however does not give effect on the range of similarity solutions to occur, i.e., the critical value for stretching/shrinking parameter λ_c is the same for all values of radiation parameter Rd. It is noteworthy that the distribution of temperature in the fluid is significantly affected by the radiation parameter Rd. Physically, this is due to the fact that the heat is produced due to the radiation process and therefore, enhances the fluid temperature.

The boundary value problems (11)–(13) together with boundary conditions (15) observe the occurrence of non–unique solutions for some governing parameters. The phenomenon of non-unique solutions namely first and second solutions are proved and portrayed as in Figures 2–8. Accordingly, an investigation on the stability analysis has been executed in this present work so that we can identified the most stable solutions. Therefore, the linearized Equations (25)–(27) along with conditions (28) have been solved with the aid of the bvp4 function in MATLAB numerically. The smallest eigenvalues γ on the selected parameter A and λ from Figure 4a,b are listed in Table 4. This table shown that the second solution displays the negative values of γ, whereas the first solution demonstrates the positive values of γ. The smallest eigenvalues γ against λ have been plotted in Figure 9. This figure eventually supported the findings obtained from Table 4 except when the values of stretching/shrinking parameter λ approach its critical value where we observed unstable solutions for both solutions. In reference from the previous literature, we can deduce that the first solution is stable, on the contrary, the second solution is unstable. It worth noting that this analysis is important in identifying the stable solution when there is exist non-unique solutions so that the flow behavior can be predict accurately.

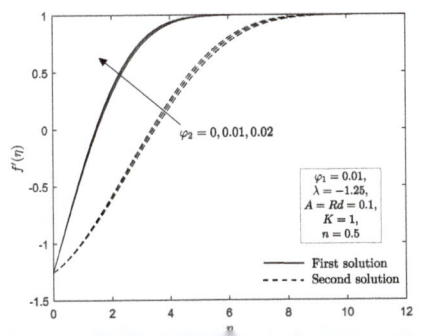
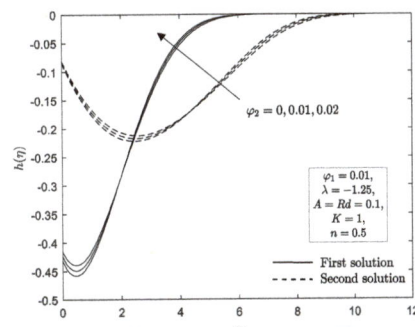

Table 4. Smallest eigenvalues γ for selected A and λ when $\varphi_1 = \varphi_2 = 0.01$, $Rd = 0.1$, $K = 1$ and $n = 0.5$.

A	λ	1st Solution	2nd Solution
0	-1.2462	0.0174	-0.0583
	-1.246	0.0323	-0.0728
	-1.24	0.1893	-0.2253
0.1	-1.3112	0.0036	-0.1086
	-1.311	0.0156	-0.1204
	-1.31	0.0577	-0.1616
0.2	-1.3785	0.0224	-0.2193
	-1.378	0.0385	-0.2348
	-1.37	0.1950	-0.3832

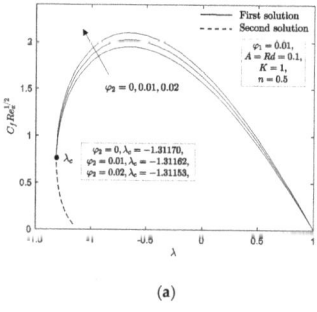
(a)

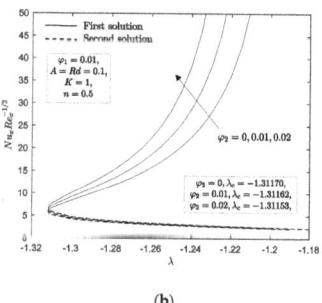
(b)

Figure 9. Smallest eigenvalues γ against λ when $\varphi_1 = \varphi_2 = 0.01$, $A = Rd = 0.1$, $K = 1$, $n = 0.5$.

5. Conclusions

Theoretical studies of unsteady micropolar Cu–Al$_2$O$_3$/water flow over a deformable sheet with thermal radiation effect has been examined numerically. The similarity solutions were produced by utilizing the bvp4c function from MATLAB software. The impact of emerging parameters has been examined and illustrated graphically. Thus, the conclusions can be outlined as follows:

- The presence of double solutions is noticeable for shrinking sheet whereas a unique solution is observed for stretching sheet.
- The stability analysis was carried out and the first solution has proven to be a stable solution, whereas the other solution is not a stable solution.
- A raise in Cu nanoparticle volume fraction φ_2 in micropolar nanofluid has tendency to improve the local Nusselt number and local skin friction for all domain of stretching/shrinking parameter λ.
- The heat transfer rate increased roughly to an average of 17.725% when the copper nanoparticle volume fraction increased from 0 to 0.01 (1%) in a micropolar nanofluid.
- The rising of unsteady parameter A and radiation parameter Rd in micropolar hybrid nanofluid increase the local Nusselt number while the reverse trend is observed with an increase of material parameter K.
- The local skin friction enhances as the value of unsteady parameter A and material parameter K increase.
- The domains of the similarity solutions decrease with a raise in Cu nanoparticle volume fraction φ_2 and material parameter K, therefore fastens the boundary layer separation. However, upsurge value of unsteady parameter A delays the boundary layer separation.

Author Contributions: Conceptualization, N.S.A. and N.B.; methodology, formal analysis and validation, N.S.A.; writing—original draft preparation, N.S.A.; writing—review and editing, supervision and funding acquisition N.B. All authors have read and agreed to the published version of the manuscript.

Funding: This research was supported by the Fundamental Research Grant Scheme (FRGS) under Ministry of Education with project number FRGS/1/2018/STG06/UPM/02/4.

Institutional Review Board Statement: Not applicable

Informed Consent Statement: Not applicable

Data Availability Statement: The data presented in this study are available on request from the corresponding author. The data are not publicly available due to its large size.

Acknowledgments: The authors wish to express their sincere thanks to the very competent reviewers for the good comments and suggestions.

Conflicts of Interest: The authors declare no conflict of interest.

Nomenclature

a, b, c	positive constants [s^{-1}]
A	unsteady parameter [-]
C_f	skin friction coefficient [-]
C_p	specific heat at constant pressure [$Jkg^{-1}K^{-1}$]
f	dimensionless stream function [-]
h	dimensionless angular velocity [-]
j	microinertia density [m^2]
k	thermal conductivity [$Wm^{-1}K^{-1}$]
k^*	mean absorption coefficient [m^{-1}]
K	dimensionless material parameter [-]
n	positive constant [s^{-1}]
N	angular velocity [ms^{-1}]
Nu_x	local Nusselt number [-]
Pr	Prandtl number [-]
q_r	radiative heat flux [Wm^{-2}]
Rd	radiation parameter [-]
Re_x	local Reynolds number [-]
t	time [s]
T	temperature [K]
u, v	velocities component in the $x-$ and $y-$ directions, respectively [ms^{-1}]
u_e	velocity of inviscid flow [ms^{-1}]
u_w	stretching/shrinking velocity [ms^{-1}]
x, y	cartesian coordinates along the surface and normal to it, respectively [m]

Greek Symbols

φ_1	nanoparticle volume fractions for Al_2O_3 (alumina) [-]
φ_2	nanoparticle volume fractions for Cu (copper) [-]
θ	dimensionless temperature [-]
γ	unknown eigenvalues [-]

λ		stretching/shrinking parameter [-]
η		similarity variable [-]
μ		dynamic viscosity [N s m^{-2}]
ν		kinematic viscosity [m^{2}s^{-1}]
ρ		density [kgm^{-3}]
τ		dimensionless time variable [-]
σ^*		Stefan–Boltzmann constant [Wm^{-2}K^{-4}]
ψ		stream function [-]
ρC_p		heat capacity [JK^{-1}m^{-3}]
ς		spin gradient viscosity [kg m s^{-1}]
κ		vortex viscosity [kg m^{-1}s^{-1}]

Subscripts

f	base fluid
hnf	hybrid nanofluid
$s1$	solid component for Al$_2$O$_3$ (alumina)
$s2$	solid component for Cu (copper)
w	condition at the surface
∞	ambient condition

Superscript

$\prime$	differentiation with respect to η

References

1. Ariman, T.; Turk, M.A.; Sylvester, N.D. Microcontinuum fluid mechanics—a review. *Int. J. Eng. Sci.* **1973**, *11*, 905–930. [CrossRef]
2. Eringen, A.C. Simple microfluids. *Int. J. Eng. Sci.* **1964**, *2*, 205–217. [CrossRef]
3. Eringen, A.C. Theory of micropolar fluids. *J. Math. Mech.* **1966**, *16*, 1–18. [CrossRef]
4. Eringen, A.C. Theory of thermomicrofluids. *J. Math. Anal. Appl.* **1972**, *38*, 480–496. [CrossRef]
5. Ebert, F. A similarity solution for the boundary layer flow of a polar fluid. *Chem. Eng. J.* **1973**, *5*, 85–92. [CrossRef]
6. Ahmadi, G. Self-Similar solution of imcompressible micropolar boundary layer flow over a semi-infinite plate. *Int. J. Eng. Sci.* **1976**, *14*, 639–646. [CrossRef]
7. Gorla, R.S.R. Buoyancy effects on the boundary layer flow of a micropolar fluid along a vertical cylinder. *Int. J. Eng. Sci.* **1988**, *26*, 883–892. [CrossRef]
8. Gorla, R.S.R.; Takhar, H.S. Boundary layer flow of micropolar fluid on rotating axisymmetric surfaces with a concentrated heat source. *Acta Mech.* **1994**, *105*, 1–10. [CrossRef]
9. Nazar, R.; Amin, N.; Filip, D.; Pop, I. Stagnation point flow of a micropolar fluid towards a stretching sheet. *Int. J. Non Linear Mech.* **2004**, *39*, 1227–1235. [CrossRef]
10. Ishak, A.; Lok, Y.Y.; Pop, I. Stagnation-Point flow over a shrinking sheet in a micropolar fluid. *Chem. Eng. Commun.* **2010**, *197*, 1417–1427. [CrossRef]
11. Yacob, N.A.; Ishak, A. Micropolar fluid flow over a shrinking sheet. *Meccanica* **2012**, *47*, 293–299. [CrossRef]
12. Sandeep, N.; Sulochana, C. Dual solutions for unsteady mixed convection flow of MHD micropolar fluid over a stretching/shrinking sheet with non-uniform heat source/sink. *Eng. Sci. Technol. Int. J.* **2015**, *18*, 738–745. [CrossRef]
13. Mishra, S.R.; Khan, I.; Al-Mdallal, Q.M.; Asifa, T. Free convective micropolar fluid flow and heat transfer over a shrinking sheet with heat source. *Case Stud. Therm. Eng.* **2018**, *11*, 113–119. [CrossRef]
14. Lund, L.A.; Omar, Z.; Khan, I.; Raza, J.; Sherif, E.S.M.; Seikh, A.H. Magnetohydrodynamic (MHD) flow of micropolar fluid with effects of viscous dissipation and joule heating over an exponential shrinking sheet: Triple solutions and stability analysis. *Symmetry* **2020**, *12*, 142. [CrossRef]
15. Khashi'ie, N.S.; Arifin, M.N.; Nazar, R.; Hafidzuddin, E.H.; Wahi, N.; Pop, I. Mixed convective flow and heat transfer of a dual stratified micropolar fluid induced by a permeable stretching/shrinking sheet. *Entropy* **2019**, *21*, 1162. [CrossRef]
16. Lund, L.A.; Omar, Z.; Khan, I.; Baleanu, D.; Sooppy Nisar, K. Triple solutions and stability analysis of micropolar fluid flow on an exponentially shrinking surface. *Crystals* **2020**, *10*, 283. [CrossRef]

17. Ahmad, F.; Almatroud, A.O.; Hussain, S.; Farooq, S.E.; Ullah, R. Numerical solution of nonlinear diff. Equations for heat transfer in micropolar fluids over a stretching domain. *Mathematics* **2020**, *8*, 854. [CrossRef]
18. Choi, S.U.S.; Eastman, J. Enhancing thermal conductivity of fluids with nanoparticles. *ASME Publ. Fed.* **1995**, *231*, 99–103.
19. Rafati, M.; Hamidi, A.A.; Niaser, M.S. Application of nanofluids in computer cooling systems (heat transfer performance of nanofluids). *Appl. Therm. Eng.* **2012**, *45*, 9–14. [CrossRef]
20. Sheikhpour, M.; Arabi, M.; Kasaeian, A.; Rabei, A.R.; Taherian, Z. Role of nanofluids in drug delivery and biomedical technology: Methods and applications. *Nanotechnol. Sci. Appl.* **2020**, *13*, 47. [CrossRef]
21. Nagarajan, P.K.; Subramani, J.; Suyambazhahan, S.; Sathyamurthy, R. Nanofluids for solar collector applications: A review. *Energy Procedia* **2014**, *61*, 2416–2434. [CrossRef]
22. Gangadhar, K.; Kannan, T.; Sakthivel, G.; DasaradhaRamaiah, K. Unsteady free convective boundary layer flow of a nanofluid past a stretching surface using a spectral relaxation method. *Int. J. Ambient Energy* **2020**, *41*, 609–616. [CrossRef]
23. Chaudhary, S.; Kanika, K.M. Viscous dissipation and Joule heating in MHD Marangoni boundary layer flow and radiation heat transfer of Cu–water nanofluid along particle shapes over an exponential temperature. *Int. J. Comput. Math.* **2020**, *97*, 943–958. [CrossRef]
24. Naqvi, S.M.R.S.; Muhammad, T.; Saleem, S.; Kim, H.M. Significance of non-uniform heat generation/absorption in hydromagnetic flow of nanofluid due to stretching/shrinking disk. *Phys. Stat. Mech. Appl.* **2020**, *553*, 123970. [CrossRef]
25. Anuar, N.S.; Bachok, N.; Arifin, N.M.; Rosali, H. Role of multiple solutions in flow of nanofluids with carbon nanotubes over a vertical permeable moving plate. *Alex. Eng. J.* **2020**, *59*, 763–773. [CrossRef]
26. Anuar, N.S.; Bachok, N.; Arifin, N.M.; Rosali, H. MHD flow past a nonlinear stretching/shrinking sheet in carbon nanotubes: Stability analysis. *Chin. J. Phys.* **2020**, *65*, 436–446. [CrossRef]
27. Hussain, S.T.; Nadeem, S.; Haq, R.U. Model-Based analysis of micropolar nanofluid flow over a stretching surface. *Eur. Phys. J. Plus* **2014**, *129*, 161. [CrossRef]
28. Bourantas, G.C.; Loukopoulos, V.C. MHD natural-convection flow in an inclined square enclosure filled with a micropolar-nanofluid. *Int. J. Heat Mass Transf.* **2014**, *79*, 930–944. [CrossRef]
29. Noor, N.F.M.; Haq, R.U.; Nadeem, S.; Hashim, I. Mixed convection stagnation flow of a micropolar nanofluid along a vertically stretching surface with slip effects. *Meccanica* **2015**, *50*, 2007–2022. [CrossRef]
30. Gangadhar, K.; Kannan, T.; Jayalakshmi, P. Magnetohydrodynamic micropolar nanofluid past a permeable stretching/shrinking sheet with Newtonian heating. *J. Braz. Soc. Mech. Sci. Eng.* **2017**, *39*, 4379–4391. [CrossRef]
31. Dero, S.; Rohni, A.M.; Saaban, A. MHD micropolar nanofluid flow over an exponentially stretching/shrinking surface: Triple solutions. *J. Adv. Res. Fluid Mech. Therm. Sci.* **2019**, *56*, 165–174.
32. Lund, L.A.; Omar, Z.; Khan, U.; Khan, I.; Baleanu, D.; Nisar, K.S. Stability analysis and dual solutions of micropolar nanofluid over the inclined stretching/shrinking surface with convective boundary condition. *Symmetry* **2020**, *12*, 74. [CrossRef]
33. Abdal, S.; Ali, B.; Younas, S.; Ali, L.; Mariam, A. Thermo-Diffusion and multislip effects on MHD mixed convection unsteady flow of micropolar nanofluid over a shrinking/stretching sheet with radiation in the presence of heat source. *Symmetry* **2020**, *12*, 49. [CrossRef]
34. Amjad, M.; Zehra, I.; Nadeem, S.; Abbas, N. Thermal analysis of Casson micropolar nanofluid flow over a permeable curved stretching surface under the stagnation region. *J. Therm. Anal. Calorim.* **2020**, 1–13. [CrossRef]
35. Rafique, K.; Anwar, M.I.; Misiran, M.; Khan, I.; Baleanu, D.; Nisar, K.S.; Seikh, A.H. Hydromagnetic flow of micropolar nanofluid. *Symmetry* **2020**, *12*, 251. [CrossRef]
36. Madhesh, D.; Kalaiselvam, S. Experimental analysis of hybrid nanofluid as a coolant. *Procedia Eng.* **2014**, *97*, 1667–1675. [CrossRef]
37. Tahat, M.S.; Benim, A.C. Experimental analysis on thermophysical properties of Al_2O_3/CuO hybrid nano fluid with its effects on flat plate solar collector. *Defect Diffus. Forum* **2017**, *374*, 148–156. [CrossRef]
38. Devi, S.U.; Devi, S.A. Heat transfer enhancement of Cu-Al_2O_3/water hybrid nanofluid flow over a stretching sheet. *J. Niger. Math. Soc.* **2017**, *36*, 419–433.
39. Anuar, N.S.; Bachok, N.; Pop, I. Cu-Al_2O_3/Water hybrid nanofluid stagnation point flow past MHD stretching/shrinking sheet in presence of homogeneous-heterogeneous and convective boundary conditions. *Mathematics* **2020**, *8*, 1237. [CrossRef]
40. Nadeem, S.; Abbas, N.; Malik, M.Y. Inspection of hybrid based nanofluid flow over a curved surface. *Comput. Methods Programs Biomed.* **2020**, *189*, 105193. [CrossRef]
41. Waini, I.; Ishak, A.; Pop, I. Hybrid nanofluid flow past a permeable moving thin needle. *Mathematics* **2020**, *8*, 612. [CrossRef]
42. Khashi'ie, N.S.; Arifin, M.N.; Pop, I. Mixed convective stagnation point flow towards a vertical riga plate in hybrid Cu-Al_2O_3/Water nanofluid. *Mathematics* **2020**, *8*, 912. [CrossRef]
43. Subhani, M.; Nadeem, S. Numerical analysis of micropolar hybrid nanofluid. *Appl. Nanosci.* **2019**, *9*, 447–459. [CrossRef]
44. Nadeem, S.; Abbas, N. On both MHD and slip effect in micropolar hybrid nanofluid past a circular cylinder under stagnation point region. *Can. J. Phys.* **2019**, *97*, 392–399. [CrossRef]
45. Abbas, N.; Nadeem, S.; Malik, M.Y. Theoretical study of micropolar hybrid nanofluid over Riga channel with slip conditions. *Phys. Stat. Mech. Appl.* **2020**, 124083. [CrossRef]
46. Al-Hanaya, A.M.; Sajid, F.; Abbas, N.; Nadeem, S. Effect of SWCNT and MWCNT on the flow of micropolar hybrid nanofluid over a curved stretching surface with induced magnetic field. *Sci. Rep.* **2020**, *10*, 1–18. [CrossRef]

47. Bhattacharyya, K. Dual solutions in boundary layer stagnation-point flow and mass transfer with chemical reaction past a stretching/shrinking sheet. *Int. Commun. Heat Mass Transf.* **2011**, *38*, 917–922. [CrossRef]
48. Bestman, A.R.; Adjepong, S.K. Unsteady hydromagnetic free-convection flow with radiative heat transfer in a rotating fluid. *Astrophys. Space Sci.* **1988**, *143*, 73–80. [CrossRef]
49. Sajid, M.; Hayat, T. Influence of thermal radiation on the boundary layer flow due to an exponentially stretching sheet. *Int. Commun. Heat Mass Transf.* **2008**, *35*, 347–356. [CrossRef]
50. Nadeem, S.; Zaheer, S.; Fang, T. Effects of thermal radiation on the boundary layer flow of a Jeffrey fluid over an exponentially stretching surface. *Numer. Algorithms* **2011**, *57*, 187–205. [CrossRef]
51. Pal, D.; Mandal, G. Thermal radiation and MHD effects on boundary layer flow of micropolar nanofluid past a stretching sheet with non-uniform heat source/sink. *Int. J. Mech. Sci.* **2017**, *126*, 308–318. [CrossRef]
52. Gireesha, B.J.; Umeshaiah, M.; Prasannakumara, B.C.; Shashikumar, N.S.; Archana, M. Impact of nonlinear thermal radiation on magnetohydrodynamic three dimensional boundary layer flow of Jeffrey nanofluid over a nonlinearly permeable stretching sheet. *Phys. Stat. Mech. Appl.* **2020**, 124051. [CrossRef]
53. Yashkun, U.; Zaimi, K.; Bakar, N.A.A.; Ishak, A.; Pop, I. MHD hybrid nanofluid flow over a permeable stretching/shrinking sheet with thermal radiation effect. *Int. J. Numer. Methods Heat Fluid Flow* **2020**. [CrossRef]
54. Bhattacharyya, K.; Mukhopadhyay, S.; Layek, G.C.; Pop, I. Effects of thermal radiation on micropolar fluid flow and heat transfer over a porous shrinking sheet. *Int. J. Heat Mass Transf.* **2012**, *55*, 2945–2952. [CrossRef]
55. Roy, N.C.; Hossain, M.; Pop, I. Analysis of dual solutions of unsteady micropolar hybrid nanofluid flow over a stretching/shrinking sheet. *J. Appl. Comput. Mech.* **2020**. [CrossRef]
56. Jena, S.K.; Mathur, M.N. Similarity solutions for laminar free convection flow of a thermomicropolar fluid past a non-isothermal vertical flat plate. *Int. J. Eng. Sci.* **1981**, *19*, 1431–1439. [CrossRef]
57. Guram, G.S.; Smith, A.C. Stagnation flows of micropolar fluids with strong and weak interactions. *Comput. Math. Appl.* **1980**, *6*, 213–233. [CrossRef]
58. Peddieson, J., Jr. An application of the micropolar fluid model to the calculation of a turbulent shear flow. *Int. J. Eng. Sci.* **1972**, *10*, 23–32. [CrossRef]
59. Zainal, N.A.; Nazar, R.; Naganthran, K.; Pop, I. Unsteady stagnation point flow of hybrid nanofluid past a convectively heated stretching/shrinking sheet with velocity slip. *Mathematics* **2020**, *8*, 1649. [CrossRef]
60. Brewster, M.Q. *Thermal Radiative Transfer and Properties*; John Wiley & Sons: Hoboken, NJ, USA, 1992.
61. Oztop, H.F.; Abu-Nada, E. Numerical study of natural convection in partially heated rectangular enclosures filled with nanofluids. *Int. J. Heat Fluid Flow* **2008**, *29*, 1326–1336. [CrossRef]
62. Merkin, J.H. On dual solutions occurring in mixed convection in a porous medium. *J. Eng. Math.* **1986**, *20*, 171–179. [CrossRef]
63. Weidman, P.D.; Kubitschek, D.G. The effect of transpiration on self-similar boundary layer flow over moving surfaces. *Int. J. Eng. Sci.* **2006**, *44*, 730–737. [CrossRef]
64. Harris, S.D.; Ingham, D.B.; Pop, I. Mixed convection boundary layer flow near the stagnation point on a vertical surface in porous medium: Brinkman model with slip. *Transp. Porous Media* **2009**, *77*, 267–285. [CrossRef]
65. Mahapatra, T.R.; Nandy, S.K. Slip effects on unsteady stagnation-point flow and heat transfer over a shrinking sheet. *Meccanica* **2013**, *48*, 1599–1606. [CrossRef]

Article

Effects of Leading-Edge Modification in Damaged Rotor Blades on Aerodynamic Characteristics of High-Pressure Gas Turbine

Thanh Dam Mai [1] and Jaiyoung Ryu [1,2,*]

[1] Department of Mechanical Engineering, Chung-Ang University, Seoul 06911, Korea; maithanhdam0610@cau.ac.kr
[2] Department of Intelligent Energy and Industry, Chung-Ang University, Seoul 06911, Korea
* Correspondence: jairyu@cau.ac.kr; Tel.: +82-2-820-5279

Received: 9 November 2020; Accepted: 5 December 2020; Published: 9 December 2020

Abstract: The flow and heat-transfer attributes of gas turbines significantly affect the output power and overall efficiency of combined-cycle power plants. However, the high-temperature and high-pressure environment can damage the turbine blade surface, potentially resulting in failure of the power plant. Because of the elevated cost of replacing turbine blades, damaged blades are usually repaired through modification of their profile around the damage location. This study compared the effects of modifying various damage locations along the leading edge of a rotor blade on the performance of the gas turbine. We simulated five rotor blades—an undamaged blade (reference) and blades damaged on the pressure and suction sides at the top and middle. The Reynolds-averaged Navier–Stokes equation was used to investigate the compressible flow in a GE-E^3 gas turbine. The results showed that the temperatures of the blade and vane surfaces with damages at the middle increased by about 0.8% and 1.2%, respectively. This causes a sudden increase in the heat transfer and thermal stress on the blade and vane surfaces, especially around the damage location. Compared with the reference case, modifications to the top-damaged blades produced a slight increase in efficiency about 2.6%, while those to the middle-damaged blades reduced the efficiency by approximately 2.2%.

Keywords: gas turbine; damaged rotor blade; leading-edge modification; aerodynamic characteristics

1. Introduction

Unlike the conventional thermal power plants using coal as the power source, combined-cycle power plants (CCPPs) utilizes natural gas to generate electricity to satisfy the industrial demand and daily consumption, reducing the CO_2 emission. Hence, for environmental concerns, CCPPs have been widely used in replacement of the thermal power plants in recent years. In CCPPs, the gas turbine is one of the most important components as it has a significant effect on power generation and overall efficiency. To enhance the power and efficiency of a CCPP, the turbine inlet temperature (TIT) is usually increased. The TIT is strongly affected by the outlet flow and temperature distribution of the combustor. This complex phenomenon in the TIT is called a hot streak (HS). The conditions of HS operation are similar to the actual operation conditions of gas turbines. Hence, it is important to examine the conditions of a HS in the analysis of gas turbine performance. Several experimental and numerical studies have been conducted to analyze the effects of a HS on the flow and heat transfer characteristics of a gas turbine [1,2]. Therefore, it is necessary to consider the HS condition instead of uniform inlet temperature conditions when performing simulations, for more accurate results and to reflect the actual operation conditions of a gas turbine.

In general, the turbine blade are fabricated with metals or alloys, and the TIT is higher than the melting temperature of metals and alloys, which usually is affected by pressure [3,4]. Hence, if the TIT is increased to enhance efficiency, the flow and heat transfer characteristics of the blade surface will be significantly affected. If a suitable cooling method cannot be provided, a sudden increase in heat loads will be produced on the blade surface, which will ultimately reduce the fatigue life of the blade or could even damage the blade after a certain time. Primary turbine blades can suffer several types of damage, such as dents, scores, and scratches, which can occur on the leading as well as trailing edge of the blade and which have a significant effect on the gas turbine performance. Replacing a damaged blade is more expensive than repairing it. Consequently, the latter is the preferred option. Kaewbumrung et al. [5] proposed a repair method for a damaged blade in the compressor, i.e., the blend method. The surface of the damaged blade becomes smoother after repair, which improves its aerodynamic performance in the compressor compared with that of unmodified blades.

The reasons behind the failure of gas turbine blades have been extensively investigated. Kumari et al. [6] examined the effects of blade surface cracks on the internal structure of the blade. They also examined the path of crack propagation within the coating barrier layer. Witek [7] conducted experiments and simulations on crack growth propagation due to vibration in compressor blades. The simulation results agreed well with experimental results. Mazur et al. [8] analyzed the effects of the failure of the first-stage nozzle of a gas turbine on the fatigue life of the blades; they concluded that the failure of the nozzle significantly reduced the fatigue life of the blade. These studies have provided valuable insight into the effects of several parameters, such as axial gap, HS, and inlet pressure conditions, on the flow and heat transfer behaviors in normal blades. Moreover, cracking and crack growth propagation in damaged blades have been successfully predicted. However, previous studies have only investigated the effects of critical damage leading to sudden failure during operation or the effects of inlet conditions on normal blades, which provide limited information regarding primary damage to blades and its effect on the flow and thermal characteristics of a turbine.

Many studies have examined the heat flow characteristics of the surface of normal turbine blades under various conditions. Choi and Ryu [9] investigated the effects of the axial gap and inlet temperature conditions on the thermal flow characteristics of a blade surface. They claimed that the thermal load on the surface of rotor blades increased when the axial gap decreased. Wang et al. [10] examined the differences between the effects of uniform and non-uniform inlet pressures and temperatures on the aerodynamic characteristics of turbine blades. Azad et al. [11] reported the effects of the tip gap and inlet turbulence intensity on the local heat transfer at the tip surface. They found that a higher tip gap, as well as a higher turbulence intensity, resulted in a higher heat transfer coefficient on the tip surface. These studies focused only on normal blades without surface damage. Therefore, it is crucial to determine the combined effects of the primary damage and inlet conditions, especially HS conditions, on the complex heat flow in a gas turbine.

Gas turbines should be simulated with multistage conditions for predicting the flow and heat transfer characteristics in the passage and on the blade surface more accurately. However, multistage gas turbine simulations are expensive; hence, previous studies have considered only one stage for the simulation [10–14]. Furthermore, the first stage of a gas turbine is significantly affected by the HS condition. In this study, we consider minor damage to a rotor blade. Therefore, 1.5 stages are sufficient to predict the combined effects of the HS condition and rotor blade damage on the flow and heat transfer characteristics in the passage and on the blade surface.

It is necessary to examine the influence of rotor blade damage on the flow and heat transfer in high-pressure gas turbines under the HS condition. This study provides a clearer understanding of the heat flow and thermal characteristics of gas turbines with blades damaged at different locations. It is important for engineers to identify the locations that require greater protection from damage, as this can reduce maintenance costs, which are considered to be the highest among the operating costs of gas turbines. Therefore, unsteady simulation was performed to analyze the combined effects of the HS condition and modification of damaged rotor blades on the aerodynamic characteristics and heat flow behaviors in a 1.5-stage high-pressure gas turbine.

2. Numerical Details

2.1. Geometry and Grid

In this study, a model of the first 1.5 stages in a GE-E^3 engine was used as the gas turbine model for the unsteady simulation. The original turbine stage has 46 stator guide vanes, 76 rotor blades in the first stage, and 48 stator guide vanes in the next half stage [15]. For accurate prediction, the pitch angles of one stator vane and two rotor blades should be the same. This assumption can be realized using a domain scaling method [16]. In this study, the number and other parameters of the rotor blades were fixed. After applying the domain scaling method, the first and second stator vanes were magnified by 46/38 times and 48/38 times, respectively. Details regarding the vanes and blades can be found in a previous numerical study [9]. In this study, we assumed that the rotor blade initially had minor damage to the middle and top sections—on both the pressure and suction sides—to investigate the influence of various damage locations on the flow, heat transfer, and aerodynamic characteristics. The damage constituted approximately 0.5% of the volume of a normal blade. The final damage used in the simulations was the smoother post-modification damage. The computational domain consisted of two stator vanes and two rotor blades, as shown in Figure 1. Figure 1b shows the designs of the undamaged reference blade and the four blades with damage at different locations.

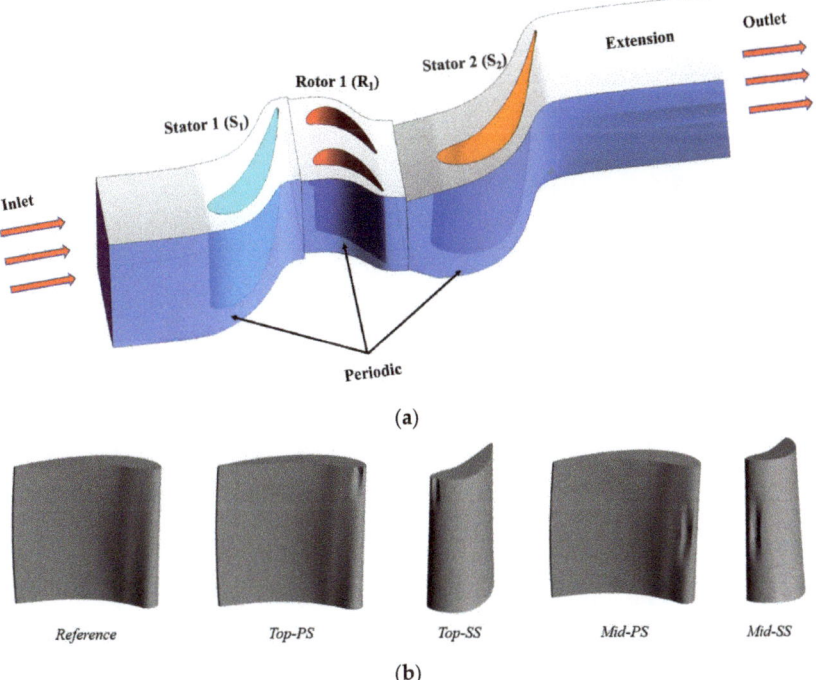

Figure 1. (a) Computational domain used for this study; (b) Damaged rotor blade after modification; Top and Mid denote the top and middle regions of the blade, respectively. PS and SS denote the pressure-side and suction-side damage, respectively.

A specialized computational fluid dynamics (CFD) tool for meshing in turbomachinery analysis, ANSYS Turbogrid [17], was adopted for the mesh generation, as shown in Figure 2. We used the blade and vane geometry of the GE-E^3 gas turbine engine. The simulation parameters and mesh generation process for the first stage were referred to from Choi and Ryu [9]. However, we used a different length for the outlet of the second stator vane (S_2). Therefore, a grid-independent test had to be conducted to find an appropriate mesh size for S_2. Consequently, we simulated a total of five mesh sizes and ultimately selected a mesh size of 2.4 million for the computation. Details of the grid-independent test for the second stator vane are shown in Table 1. After the grid-independent test, the total mesh size of the computational domain was approximately 8 million, with the y^+ value being less than 0.5 at the blade surface and less than 1 at the other walls. The variations in y^+ at different span-wise locations on the rotor blade are shown in Figure 3.

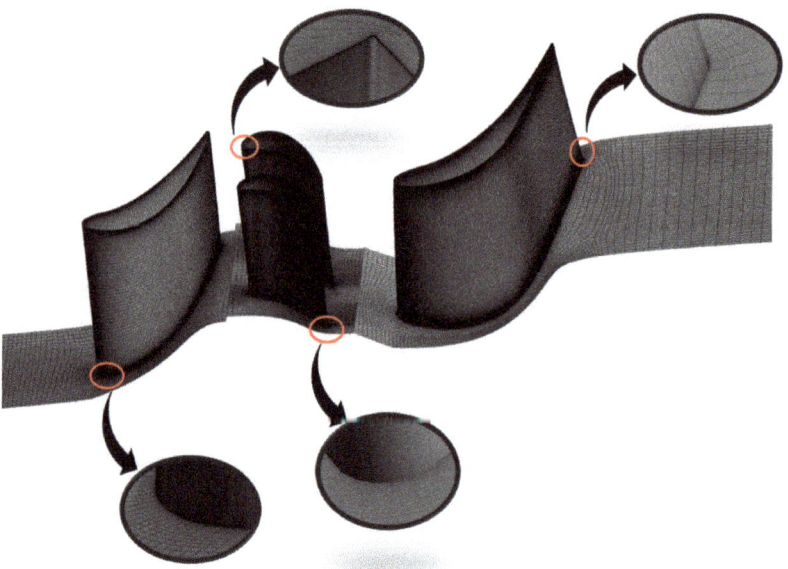

Figure 2. Detailed mesh of computational domain for the 1.5-stage GE-E3 gas turbine.

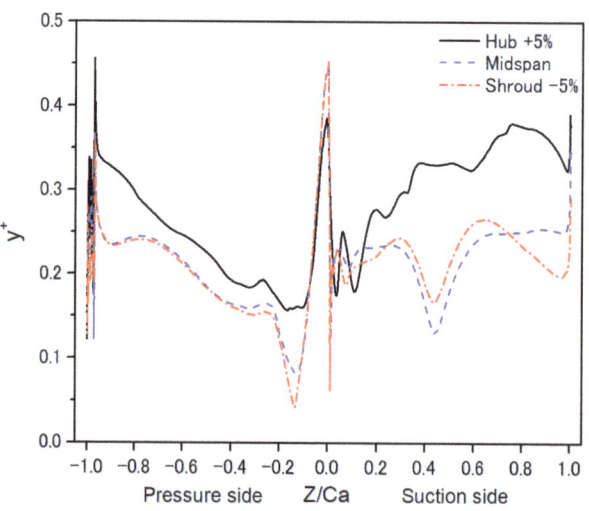

Figure 3. Variation in y^+ at different span-wise locations in the rotor blade.

Table 1. Grid-independent test.

	Domain Node Number of S_2 (×10^6)	Average Heat Flux (kW/m^2) (Relative Error)	Average Pressure (kPa) (Relative Error)
Case 1	1	274.88 (0.76%)	124.69 (1.17%)
Case 2	1.34	276.97 (1.06%)	126.15 (0.69%)
Case 3	1.8	279.91 (0.82%)	127.03 (0.58%)
Case 4	2.4	282.21 (0.43%)	127.76 (0.23%)
Case 5	3.2	283.41	128.05

2.2. Governing Equations and Turbulence Model

To analyze the three-dimensional compressible fluid flow in a gas turbine under unsteady-state conditions, we used the continuity equation, momentum equation, and energy equation, which can be expressed as shown below:

Continuity equation:

$$\frac{\partial \rho}{\partial t} + \frac{\partial}{\partial x_i}(\rho u_i) = 0. \tag{1}$$

Momentum equation:

$$\frac{\partial}{\partial t}(\rho u_i) + \frac{\partial(\rho u_i u_j)}{\partial x_i} = -\frac{\partial P}{\partial x_i} + \frac{\partial}{\partial x_j}\left[\mu\left(\frac{\partial u_i}{\partial x_j} + \frac{\partial u_j}{\partial x_i} - \frac{2}{3}\delta_{ij}\frac{\partial u_k}{\partial x_k}\right)\right] + \frac{\partial\left(-\rho\overline{u'_i u'_j}\right)}{\partial x_j}. \tag{2}$$

Energy equation:

$$\frac{\partial}{\partial t}(\rho E) + \frac{\partial}{\partial x_j}\left(u_j(\rho E + P)\right) = \frac{\partial}{\partial x_j}\left[(k_{eff})\frac{\partial T}{\partial x_j}\right] + \frac{\partial}{\partial x_j}\left[u_i\mu_{eff}\left(\frac{\partial u_i}{\partial x_j} + \frac{\partial u_j}{\partial x_i} - \frac{2}{3}\delta_{ij}\frac{\partial u_k}{\partial x_k}\right)\right] \tag{3}$$

$$i, j, k = 1, 2, \text{ and } 3,$$

where ρ is the fluid density, u is the fluid velocity, P is the fluid pressure, and μ is the fluid viscosity. In the energy equation, E is the specific internal energy, k_{eff} is the effective thermal conductivity, and μ_{eff} is the effective dynamic viscosity. A finite volume method (FVM)-based commercial CFD software, ANSYS CFX [18], was used to solve the governing equations.

Accurate prediction of the complex heat flow in a gas turbine requires an appropriate turbulence model for the simulation. Various methods, such as direct numerical simulation (DNS), large eddy simulation (LES), and the Reynolds-average Navier–Stokes (RANS) method, have been introduced for this purpose. DNS and LES can provide details of turbulence statistics but are high-cost methods [19–21]. The RANS method is usually used for CFD simulations due to its lower computational cost [22,23], especially for turbomachinery simulation. Moreover, previous studies have confirmed that the SST γ model and SST $\gamma - \theta$ model are the most suitable for analysis of transitional flows [24,25]. Furthermore, Choi and Ryu [9] concluded that results obtained using the $k - \omega$ SST γ turbulence model were in agreement with the corresponding experimental results [26]. Therefore, to accurately predict the complex fluid flow and heat transfer characteristics in a gas turbine, we used the $k - \omega$ SST γ turbulence model in this study.

The $k - \omega$ SST model was combined with free stream formulations and the $k - \omega$ formulation in the near wall using a blending function proposed by Menter [27,28]. The corresponding continuity, momentum, turbulence kinetic energy (k) equation, and eddy dissipation (ω) equations were formulated to express the $k - \omega$ baseline (BSL) model:

$$\frac{\partial}{\partial x_i}(\rho u_i) = 0, \tag{4}$$

$$\frac{\partial(\rho u_i u_j)}{\partial x_i} = -\frac{\partial P^*}{\partial x_i} + \frac{\partial}{\partial x_j}\left[\mu_{eff}\left(\frac{\partial u_i}{\partial x_j} + \frac{\partial u_j}{\partial x_i}\right)\right], \tag{5}$$

$$\frac{\partial(\rho k)}{\partial t} + \frac{\partial(\rho u_j k)}{\partial x_j} = P_K - 0.09\rho\omega k + \frac{\partial}{\partial x_j}\left[\left(\mu + \frac{\mu_t}{\sigma_k}\right)\frac{\partial k}{\partial x_j}\right], \tag{6}$$

$$\frac{\partial(\rho\omega)}{\partial t} + \frac{\partial(\rho u_j \omega)}{\partial x_j} = \frac{\gamma}{\nu_t}P_K - \beta\rho\omega^2 + \frac{\partial}{\partial x_j}\left[\left(\mu + \frac{\mu_t}{\sigma_\omega}\right)\frac{\partial k}{\partial x_j}\right] + 2\rho(1 - F_1)\frac{1}{\sigma_{\omega,2}}\frac{1}{\omega}\frac{\partial k}{\partial x_j}\frac{\partial \omega}{\partial x_j}, \tag{7}$$

where:

$$P^* = P + \frac{2}{3}\left(\rho k + (\mu + \mu_t)\frac{\partial u_k}{\partial x_k}\right), \tag{8}$$

$$P_K = \mu_t\left(\frac{\partial u_i}{\partial x_j} + \frac{\partial u_j}{\partial x_i}\right)\frac{\partial u_i}{\partial x_j} - \frac{2}{3}\frac{\partial u_k}{\partial x_k}\left(3\mu_t\frac{\partial u_k}{\partial x_k} + \rho k\right), \tag{9}$$

$\mu_t(kg/ms)$ is the turbulence viscosity calculated using the following equation:

$$\frac{1}{\rho}\mu_t = \frac{0.31k}{\max(0.31\omega, SF_2)}, \tag{10}$$

The blending functions F_1 and F_2 are defined by the following variables:

$$arg_1 = \min\left[\max\left(\frac{\sqrt{k}}{0.09\omega y}, \frac{500\nu}{y^2\omega}\right), \frac{4\rho\sigma_{\omega,2}k}{CD_{k\omega}y^2}\right], \tag{11}$$

$$arg_2 = \max\left(2\frac{\sqrt{k}}{0.09\omega y}, \frac{500\nu}{y^2\omega}\right), \tag{12}$$

as follows:

$$F_1 = \tanh(arg_1^2), \tag{13}$$

$$F_2 = \tanh(arg_2^2), \tag{14}$$

where σ_k and σ_ω are the turbulent Prandtl numbers for k and ω, respectively. The formulations for these equations are expressed below:

$$\sigma_k = \frac{1}{F_1/\sigma_{k,1} + (1 - F_1)/\sigma_{k,2}}, \tag{15}$$

$$\sigma_\omega = \frac{1}{F_1/\sigma_{\omega,1} + (1 - F_1)/\sigma_{\omega,2}}, \tag{16}$$

$$CD_{k\omega} = \max\left(2\rho\frac{1}{\sigma_{\omega,2}}\frac{1}{\omega}\frac{\partial k}{\partial x_i}\frac{\partial \omega}{\partial x_j}, 10^{-10}\right). \tag{17}$$

Moreover, it is necessary to define the transport equation for the intermittency (γ) to obtain the complete expression for the $k - \omega$ SST γ turbulence model. The transport equation of γ can be defined as:

$$\frac{\partial(\rho\gamma)}{\partial t} + \frac{\partial(\rho u_j \gamma)}{\partial x_j} = P_{\gamma 1} + P_{\gamma 2} - (E_{\gamma 1} + E_{\gamma 2}) + \frac{\partial}{\partial x_j}\left[\left(\mu + \frac{\mu_t}{\sigma_\gamma}\right)\frac{\partial k}{\partial x_j}\right]. \tag{18}$$

The transition sources are defined as follows:

$$P_{\gamma 1} = 2F_{length}\rho S(\gamma F_{onset})^{c_{\gamma 3}}, \quad (19)$$

$$E_{\gamma 1} = \gamma P_{\gamma 1}. \quad (20)$$

The destruction/relaminarization sources are defined as follows:

$$P_{\gamma 2} = 2c_{\gamma 1}\rho \Omega \gamma F_{turb}, \quad (21)$$

$$E_{\gamma 2} = \gamma P_{\gamma 2} c_{\gamma 2}. \quad (22)$$

The transition onset is controlled by the following functions:

$$R_T = \frac{\rho k}{\mu \omega}, \quad (23)$$

$$F_{onset1} = \frac{\rho y^2 S}{2.193 \mu Re_{\theta c}}, \quad (24)$$

$$F_{onset2} = \min\left(\max\left(F_{onset1}, F_{onset1}^4\right), 2\right), \quad (25)$$

$$F_{onset3} = \max\left(1 - \left(\frac{R_T}{2.5}\right)^3, 0\right), \quad (26)$$

$$F_{onset} = \max(F_{onset2} - F_{onset3}, 0), \quad (27)$$

$$F_{turb} = e^{-(0.25 R_T)^4}, \quad (28)$$

where S is the strain rate magnitude, F_{length} is an empirical correlation, Ω is the vorticity magnitude, and $Re_{\theta c}$ is the critical Reynolds number, at which the intermittency first starts to increase in the boundary layer. The other constant coefficients for the equations above are as follows [5,29]:

$$\sigma_{k,1} = 1.176, \ \sigma_{k,2} = 1.0, \ c_{\gamma 1} = 0.03, \ c_{\gamma 2} = 50,$$

$$\sigma_{\omega,1} = 2.0, \ \sigma_{\omega,2} = 1.168, \ c_{\gamma 3} = 0.5, \ \sigma_{\gamma} = 1.$$

2.3. Boundary Conditions and Unsteady Simulation

The boundary conditions used for the simulation were obtained from a report on a GE-E³ performance test [15], shown in Table 2. Ideal gas was used as the working fluid. Total pressure and temperature were set for the inlet, while static pressure was specified for the outlet. The magnitude of the total pressure used for the inlet was 344,740 Pa, with a uniform turbulence intensity of 5%. A non-uniform temperature, which was considered as the HS condition, was applied. The maximum and average temperatures of the HS were approximately 839 and 728 K, respectively. The detailed profile of the HS is shown in Figure 4. The static pressure specified for the outlet was 104,470 Pa. The speed of the rotor blade was fixed at 3600 rpm. Furthermore, to investigate the effects of rotor blade damage coupled with HS conditions, we analyzed five rotor blades cases. An undamaged rotor blade was considered as the reference case. The other cases were damage (i) at the middle of the blade on the pressure side, (ii) at the middle of the blade on the suction side, (iii) at the top of the blade on the pressure side, and (iv) at the top of the blade on the suction side. Moreover, we conducted both adiabatic and isothermal condition tests for the blade surface to calculate the heat transfer characteristics. The temperature applied to the blade surface under isothermal conditions was set as 389.95 K.

Table 2. Boundary conditions and simulation settings.

Boundary conditions	
Inlet	Total pressure: 344,740 Pa Total temperature: 839 K Turbulence intensity: 5%
Outlet	Static pressure: 104,470 Pa
Simulation settings	
Wall conditions	Adiabatic or iso-thermal
Rotor blade conditions	Undamaged or damaged at PS and SS in top and middle of blade
Vane and blade interface	Transient rotor-stator (unsteady simulation)
Rotor speed	3600 RPM

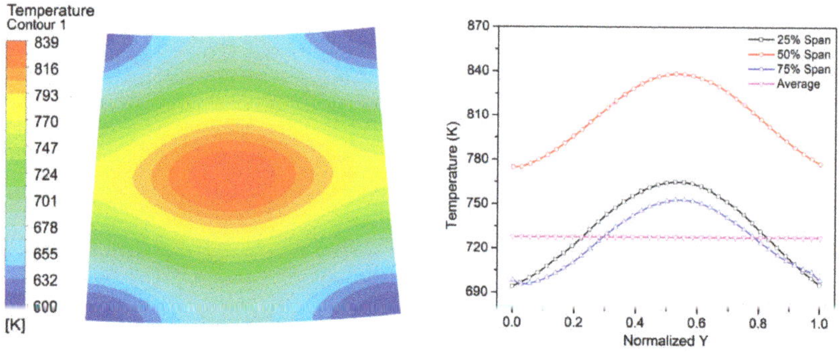

Figure 4. (a) Hot streak distribution at the inlet; (b) Radial temperature at different span-wise locations and average temperature at the inlet.

To perform an unsteady simulation, it is necessary to conduct a steady simulation, the results of which are used as an initial condition for the unsteady simulation. A frozen rotor was set as the interface between the stator and rotor in the steady simulation. The unsteady simulation was performed with a transient rotor-stator setting for the rotor–stator interface. To obtain accurate results in the unsteady simulation, the appropriate number of time steps per pitch—the number of steps when one rotor blade passes a pitch of the stator—needs to be considered. Previous studies have confirmed that 32 steps are the most suitable to guarantee convergence in such unsteady simulations [9,10].

In this study, we used a total of 20 pitches for the unsteady simulation as the results started to become periodic after 10 pitches. The first 10 pitches were considered to be the initial transient condition, and the final 10 pitches were used to analyze the results. We set several monitoring points near the pressure side of the blade and vane to check the convergence history. When the pressure and temperature of monitored points started to exhibit periodicity, the unsteady simulations were considered to have become convergent. The simulations were conducted using a 96-core workstation (4 Intel Xeon CPU E7-8890 v4 @ 2.20 GHz, RAM 512 GB), and the time needed for an unsteady simulation was approximately 72 h using 60 cores.

3. Results and Discussion

3.1. Flow Characteristics

Figure 5 shows the velocity streamlines and total pressure contours at different locations in the span-wise direction of the R_1 blade, to illustrate the effects of different blade damage locations on the flow field. It can be seen that the damage mainly affected the flow field on the suction side of the blade. The hub +5% location had a small flow-circulation zone in the reference case. However, in the damaged blades, this zone was not present, due to the effects of the altered the blade profile on the flow in span-wise direction. In the mid-span, in comparison with the reference case, the middle-damage cases exhibited more circulation, while the top-damage cases exhibited less circulation. As a result, a noticeable difference in the total pressure contours at the mid-span can be seen among the five cases, as shown in Figure 5b. The middle-damage cases exhibited lower total pressures than other cases at the mid-span since they had more flow circulation at this location. Moreover, at the shroud −5% location, a lower total pressure can be observed in the top-damage cases than in the other cases; this is because a small flow circulation was generated in the top-damage cases. It can be concluded that the damage location affected not only the circulation zones at the suction side of the blade but also the total pressure near the damage, which significantly impacted the flow and heat transfer characteristics.

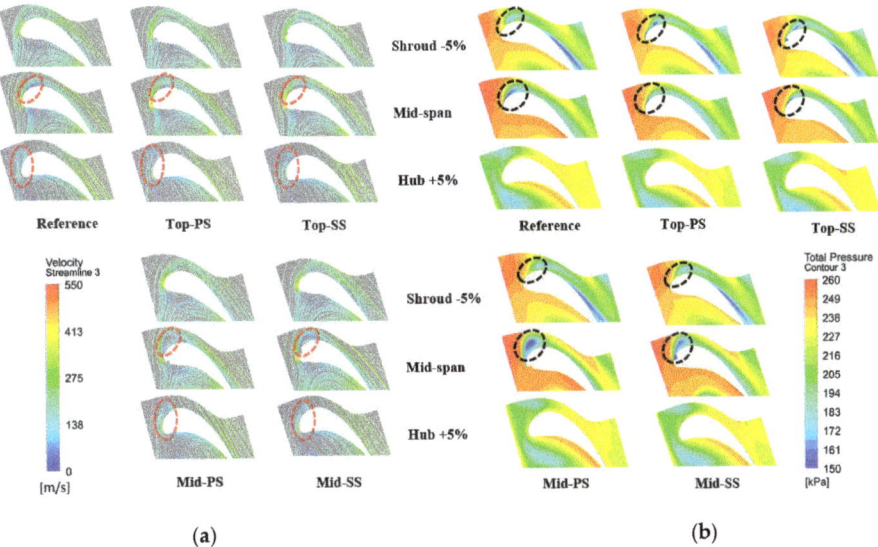

Figure 5. (a) Velocity streamlines; (b) total pressure contours at different span-wise locations of the R_1 blade.

Figure 6 shows the temperature contours on the rotor blade surface under various blade conditions. The highest temperature on the pressure side of the blade was near the mid-span of the leading edge, which exhibited more vorticity and was affected by the HS condition. The damage did not affect the temperature distribution on the pressure side, but it significantly affected the temperature distribution on the suction side. The high-temperature regions extended in the span-wise direction in the top-damage cases, and in both the span-wise and radial directions in the middle-damage cases. As shown in Figure 6, the suction-side surface of the damaged blades had a higher temperature than the corresponding surface of the reference blades, due to the effects of the damage on the passage flow and flow circulation. As a result, both the average and maximum temperatures of the blade surface increased when the blades were damaged, as shown in Figure 7. Compared with the reference case, the middle-damage case at the suction side exhibited a higher temperature—by approximately 5 (average) and 2 K (maximum). Considering the effects of the HS, the maximum temperature in the middle-damage cases was higher than that in the other cases.

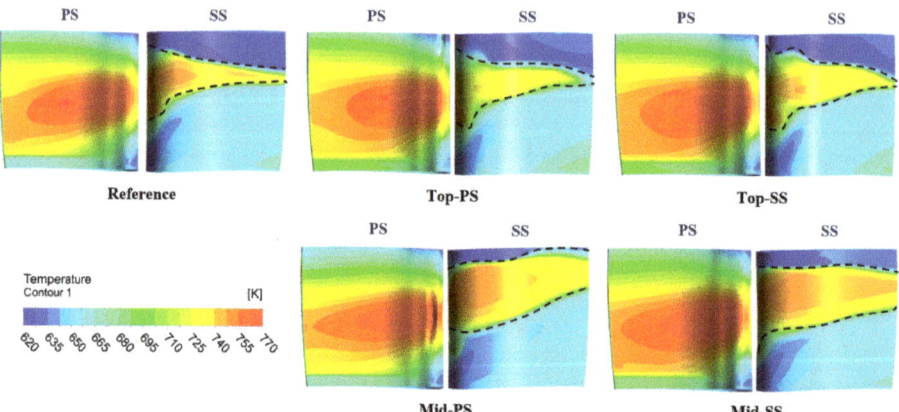

Figure 6. Temperature distribution on the R_1 blade under various blade conditions.

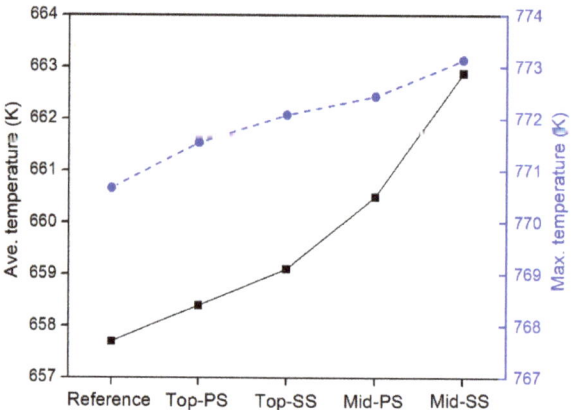

Figure 7. Average and maximum temperatures on the R_1 blade under various blade conditions.

It is necessary to examine the effects of damage locations on the flow characteristics downstream of the rotor blade. Figure 8 shows the contours of static entropy and total pressure at the R_1 outlet under various blade conditions. The static entropy was directly affected by the rotor blade conditions. Compared with the reference case, in the top-damage cases, the high-static-entropy regions extended in the radial direction, while in the middle-damage cases, these regions extended in both the span-wise and radial directions. These conditions strongly influenced the temperature distribution, which significantly affected the flow and heat transfer characteristics of the S_2 vane. Similarly, the total pressure at the R_1 outlet was significantly dependent on the blade conditions. Overall, the total pressure increased when damage occurred on the blade. The increase in total pressure resulted in an increase in the leakage flow passing through the blade tip or the main passage flow. This increment is reflected in the contours of the total pressure, shown in Figure 8b.

The attributes of heat transfer are strongly affected by the flow vortex structure [30,31]. Touil and Ghenaiet [32] investigated the effects of blade–vane interaction on the vortex structure in high-pressure gas turbines. Wei et al. [33] describe the flow structure using an iso-surface with the λ_2—criteria method. Figure 9 shows the 3-D complex vortex structure of the flow passing through R_1 under various blade conditions. The structure was expressed using the λ_2—criteria method, with the magnitudes of strength level and values of λ_2 being 10^4 and 5.14×10^6 s^{-2}, respectively. In comparison with the reference case, the top-damage cases exhibited a weaker tip leakage vortex, while the middle-damage cases exhibited a stronger tip leakage vortex. The pressure field directly affected the tip leakage flow conditions since the tip leakage flow is driven by the pressure difference between the pressure and suction sides of the rotor blade.

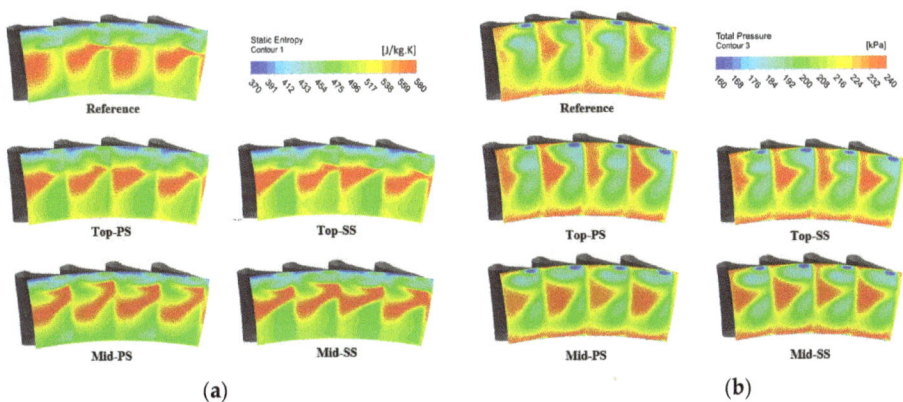

Figure 8. (**a**) Static entropy; (**b**) Total pressure at the R_1 outlet under various blade conditions.

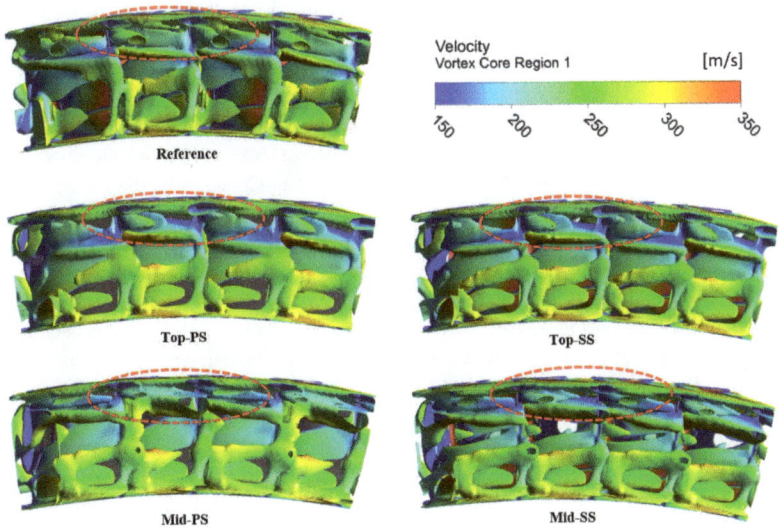

Figure 9. 3-D vortex structure at outlet of the R_1 blade.

Figure 10 shows the pressure difference between the pressure and the suction sides of the R_1 blade and the leakage flow passing through blade tip under various blade conditions. The top-damage cases had lower pressure differences, while the middle-damage cases had higher pressure differences than the reference case. As a result, the tip leakage flow in the middle-damage cases was higher than that in the reference case, whereas the opposite was true for the top-damage cases. The tip leakage flow significantly affected the heat transfer characteristics and efficiency of the gas turbine, as discussed in the following section.

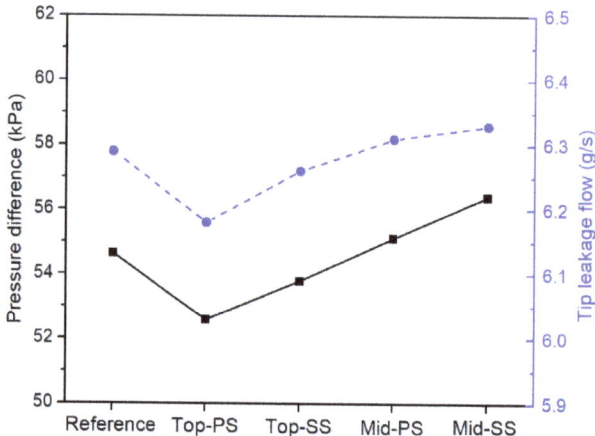

Figure 10. Pressure difference between pressure side and suction side of the R_1 blade and tip leakage flow through the R_1 blade tip.

To provide a better understanding of the effects of changes in the upstream-to-downstream flow, we first present the velocity contour at the S_2 vane entrance, as shown in Figure 11. It can be seen that the changes in the profile of the damaged blade had significant effects on the flow characteristics downstream. In the top damage cases, the flow fields arriving into S_2 were similar to those in the reference case. The flow extended from the hub—where the flow velocity was the highest—to the casing. The flow in the middle-damage cases also extended from the hub to the casing. However, the flow only developed to mid-span; the flow from 60%-span to the casing in the middle-damage cases was not significantly different from the corresponding flows in the top-damage cases and the reference case. This was due to the leakage from the various damaged rotor blade locations. In the top-damage cases, the leakage flow passing through the damage locations was not significant, resulting in flow velocity contours similar to those in the reference case. Conversely, the leakage flow in the middle-damage cases was noticeable, with a higher flow velocity around the mid-span. Moreover, the turbulence intensity of the flow increased due to changes in the blade profile resulting from the damage. The turbulence intensity also depended on the damage location—damage on the pressure side generated a higher turbulence intensity downstream than damage on the suction side did; in addition, damage at the middle created more turbulence than damage at the top did. This is because we had applied the HS for the inlet condition with the highest flow at the center, which had a more significant effect on the middle of the blade than that at the top.

From the velocity contours at the S_2 vane entrance, shown in Figure 11, the flow arriving at the suction side of the S_2 vane was more noticeable than that arriving at the pressure side. Figure 12 shows the velocity streamlines on the S_2 vane suction side, which were used for analyzing the flow characteristics. The flow formation was less near the hub of the S_2 vane in the middle-damage cases than in the reference case and top-damage cases, due to the higher tip leakage flow, which generated secondary flow and changed the flow structure in the passage and vane surface. Moreover, the flow trends near the shroud of the vane—denoted by the black rectangles in Figure 12—were strongly dependent on the damage locations. Compared with the reference case, there were fewer forming lines in the damage cases, due to the effects of damage on the secondary flow and tip leakage flow. Moreover, the difference in the non-uniform total pressure at the entrance of the S_2 vane due to the effects of various damage locations resulted in changes in the flow structure on the S_2 vane surface. Due to the effects of damage, compared with the reference case, the total pressure at the S_2 vane entrance increased by 0.25% and 0.5% in the top-damage and middle-damage cases, respectively. The changes in the rotor blade profile affected not only the passage flow but also the flow on the blade and vane surfaces due to their effects on the total pressure and consequently, on the leakage flow through the blade tip and passage.

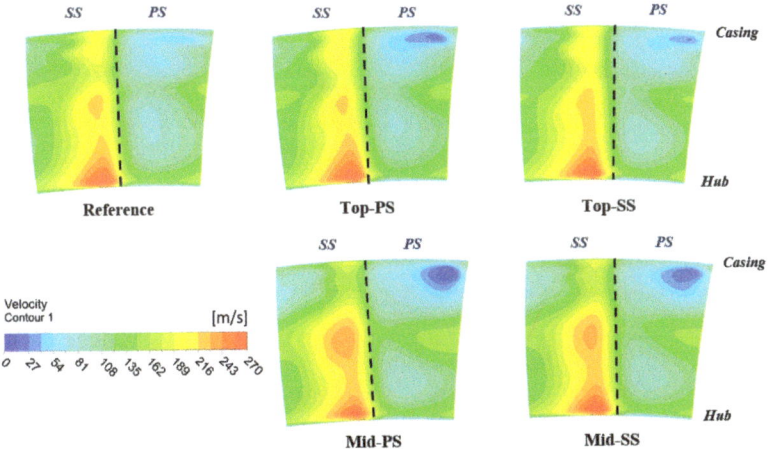

Figure 11. Velocity contours at the entrance of the S_2 vane.

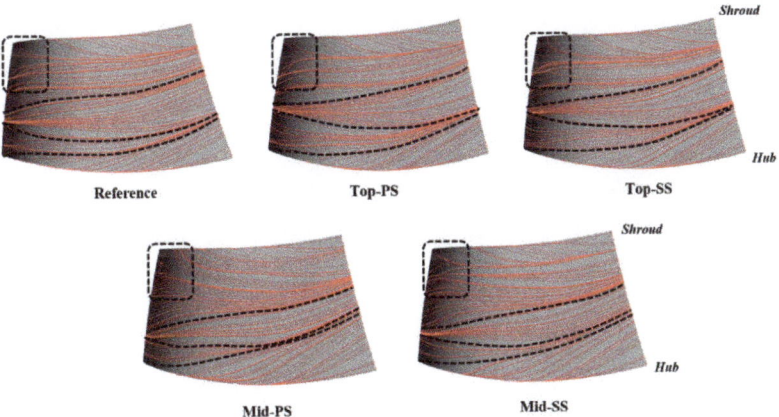

Figure 12. Velocity streamlines at the suction side of the S_2 vane surface.

The structure of the flow strongly affects its characteristics and the heat transfer properties of the vane surface. Figure 13 shows the temperature contours on the S_2 vane surface under various blade conditions. Unlike for the R_1 blade, the damage resulted in significant changes in both the pressure and suction sides for the S_2 vane. This vane received more leakage flow when the blades were damaged; hence, more flow arrived at S_2. The greater flow produced higher temperatures on both the pressure and suction sides of the S_2 vane, leading to a significant increase in both the average and maximum temperatures, as shown in Figure 14. The increments in the average and maximum temperatures were approximately 9 and 7 K, respectively. These changes are more noticeable than those for the R_1 blade surface. This increment in the temperature of the vane surface generated a higher thermal stress, which consequently reduced the fatigue life of the vane. It can be concluded that the damage on the rotor blade had more significant effects downstream than at the blade surface. Overall, the damage on the rotor blade considerably affected the flow characteristics both in the passage and on the surface of the blade and vane.

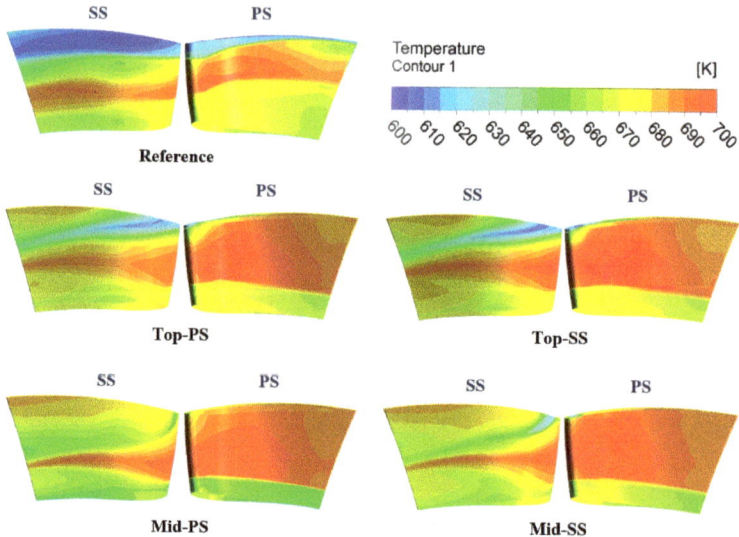

Figure 13. Temperature distribution on the S_2 vane surface.

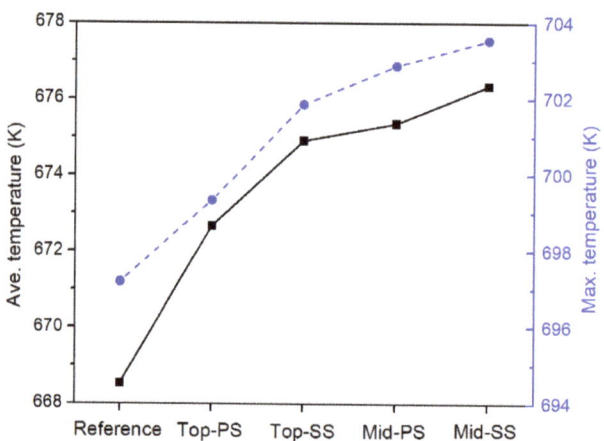

Figure 14. Average and maximum temperature of the S_2 vane surface under various blade conditions.

3.2. Heat Transfer Characteristics

The change in flow characteristics due to the effects of various damage locations on the rotor blade strongly affected the heat transfer on the blade and vane surface. Figure 15 shows the contour distribution of heat flux on the R_1 blade surface under various blade conditions. As with the temperature distribution, the heat flux on the suction side was significantly affected, while the effects on the pressure side were negligible. This occurred because of the effects of the blade profile on the flow at the leading edge of the blade. Although the profile became smoother after modification, it considerably altered the flow field, especially at the suction side of the blade. When the profile changed, it generated a stronger vortex and created a larger circulation-flow region on the suction side, as shown in Figure 5a. Therefore, the heat flux on the suction side of the blade increased noticeably when the blade was damaged. Moreover, the heat flux increased significantly at the edges of the damage locations. This caused a sudden increase in thermal stress around the damage locations, which caused the damage to become more critical and reduced the fatigue life of the blade. Another reason for the sudden increase in heat flux around the mid-span was the HS applied to the inlet flow. With the HS, the highest temperature was at the center of the flow. Coupled with the altered blade profile, it caused a significant increase in heat flux at the suction side around the mid-span location. Overall, when the rotor blade was damaged, the heat flux increased suddenly at the suction side of the blade and around the damaged region. Hence, it is necessary to provide a suitable cooling method to prevent excessive thermal stress at these locations.

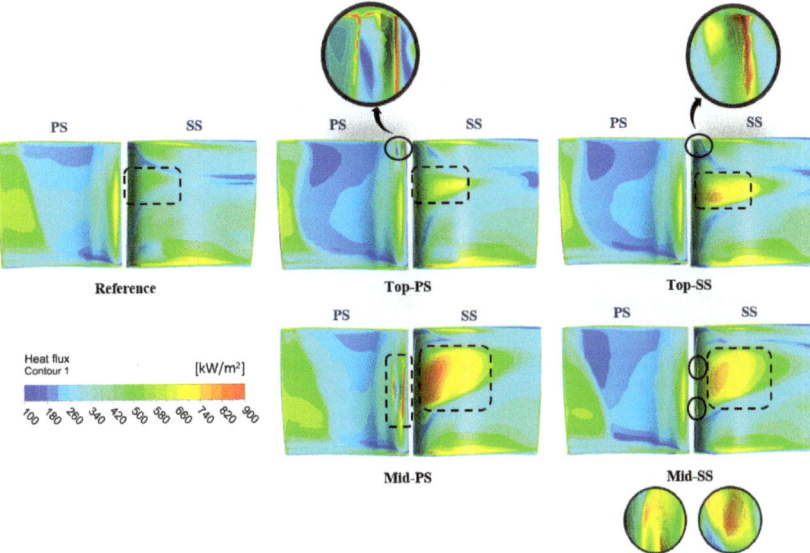

Figure 15. Heat flux distribution on the R_1 blade under various blade conditions.

Figure 16 presents the contours of heat flux distribution on the S_2 vane surface under various blade conditions. The characteristics of the flow after passing through R_1 changed significantly. This considerably affected the heat transfer behaviors on the S_2 vane surface on both the pressure and suction sides. On the pressure side, the heat flux increased noticeably when the blades were damaged. The high-heat-flux region extended in the span-wise direction in both the top-damage and middle-damage cases. This phenomenon occurred due to the increased turbulence intensity of the flow and the increased vane surface temperature in the damaged blades. The combined effects of the turbulence intensity and temperature tended to increase the heat flux on the pressure side of the S_2 vane. Similarly, the heat flux on the suction side increased due to these coupled effects. The high-heat-flux regions on the pressure side of the vane surface were located around the mid-span, while those on the suction side were located near the hub and tip. Overall, the damaged rotor blade surfaces significantly increased the heat flux on the S_2 vane surface. Hence, to protect the vane surface from sudden changes in thermal stress, efficient cooling methods need to be provided at these locations.

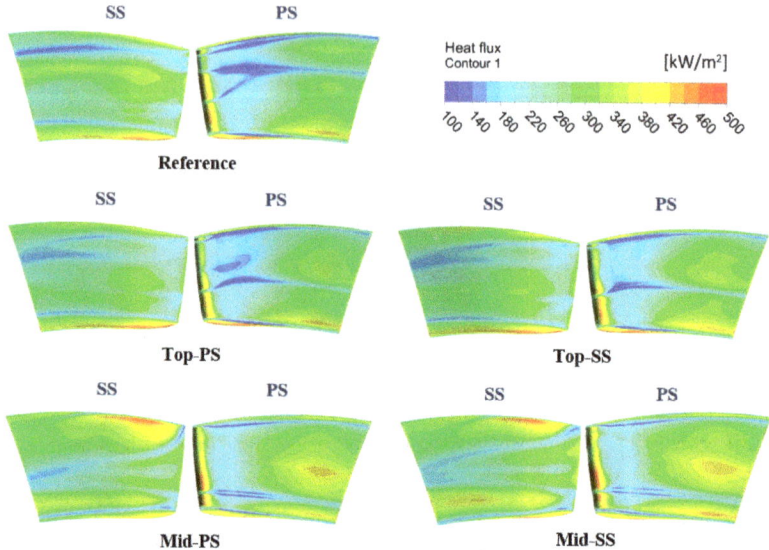

Figure 16. Heat flux distribution on the S_2 vanes surface under various blade conditions.

The tip leakage flow created by the difference in pressure between the pressure and suction sides significantly affected the heat transfer characteristics at the tip surface. To analyze the heat transfer at the blade tip under various blade conditions, contours of the Stanton number distribution on the blade tip were plotted and are shown in Figure 17. The Stanton number can be expressed as follows:

$$St = \frac{q}{(T_w - T_0)\rho_0 V_0 C_p},\qquad(29)$$

where q is the heat flux, T_w is the temperature of the wall surface, T_0 is the average total temperature of the inlet flow, C_p is the specific heat of ideal air, and ρ_0 and V_0 are the average density and average velocity of the inlet flow, respectively.

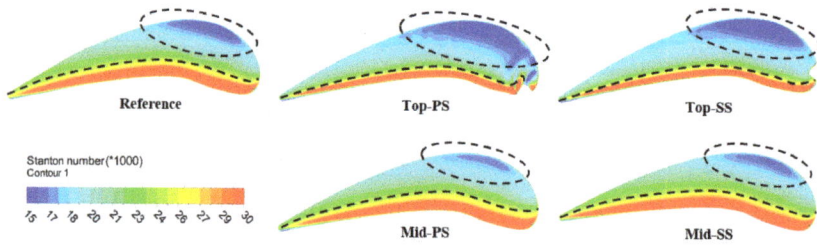

Figure 17. Stanton number distribution on blade tip under various blade conditions.

Compared with the reference case, the top-damage cases exhibited lower heat transfer, while the middle-damage cases exhibited higher heat transfer. This result is consistent with the pressure difference and tip leakage flow shown in Figure 10—where the middle-damage cases exhibit more tip leakage flow—as well as with the results of a previous study by Yang and Feng [34]. In addition, the high-heat-transfer regions were located on the pressure side of the blade, while the low-heat-transfer regions were located on the suction side. This occurred because the heat transfer at the blade tip was directly affected by the leakage flow from the pressure side to the suction side of the blade. The tip leakage flow in the top-damage cases was lower than that in the reference case, the high-heat-transfer region near the pressure side narrowed, and the low heat transfer region near the suction side widened. Conversely, the tip leakage flow in the middle-damage cases was higher than that in the reference case, the low-heat-transfer region near the suction side narrowed, and the high-heat-transfer region near the pressure side expanded. Overall, the heat transfer on the blade tip was strongly dependent on the rotor blade conditions. Especially, damage at the middle causes an increase of the heat transfer characteristics, which increased the local thermal stress at the blade tip region.

3.3. Aerodynamic and Total-to-Total Efficiencies

Herein, we calculate the aerodynamic and total-to-total efficiencies to evaluate the performance of the gas turbine under various blade conditions. These efficiencies are calculated using Equations (30) and (31):

$$\eta_a = \frac{F_L}{F_D} \tag{30}$$

$$\eta_{tot} = \frac{T\omega}{\dot{m}C_pT_0\left\{1 - (P/P_0)^{\frac{\kappa-1}{\kappa}}\right\}}, \tag{31}$$

where η_a is the aerodynamic efficiency, F_L is the lift force, F_D is the drag force, η_{tot} is the total-to-total efficiency, T is the torque, ω is the angular velocity, $\dot{m}$ is the mass flow rate, C_p is the specific heat of ideal air, κ is the ratio of specific heat, P is the outlet mass-averaged total pressure, and T_0 and P_0 are the average temperature and total pressure of the turbine inlet, respectively.

Figure 18 presents the aerodynamic and total-to-total efficiencies under various blade conditions. Both the efficiencies are strongly dependent on the damage locations and exhibit the same trends. Compared with the reference case, the top-damage cases exhibited a slight increase in efficiency. Conversely, the middle-damage cases exhibited a significant efficiency reduction. The aerodynamic efficiency is strongly affected by lift and drag forces, while the total-to-total efficiency is significantly affected by torque and outlet pressure. Figure 19 shows the drag force, lift force, torque, and outlet pressure under various blade conditions. Compared with the reference case, the middle-damage cases exhibited a significant increase in drag force, while the top-damage cases exhibited a reduction. In contrast, the middle-damage cases exhibited a decrease in lift force, while the top-damage cases exhibited a slight increase. The changes in the drag and lift forces can be explained by the velocity contour shown in Figure 5a. In the middle-damage cases, the circulation zone on the suction side was closer to the leading edge at the mid-span than in the other cases. This means that the flow separation point in the middle-damage cases was closer to the leading edge than in the other cases. According to flight theory, the closer the separation point moves to the leading edge, the higher the drag and the lower the lift that are generated. In the top-damage cases, the separation point was farther away from the leading edge than in the reference case. Therefore, these cases exhibited higher lift and lower drag forces than the reference case. As a result, the aerodynamic efficiency increased slightly in the top-damage cases but was significantly reduced in the middle-damage cases.

The torque values directly affected the total-to-total efficiency. Similar to the lift force, the torque in the top-damage cases was slightly higher than in the other cases—including the reference case—under normal blade conditions. The outlet pressure also increased in the damaged blades due to the pressure loss at the damage location. The increase in torque and outlet pressure resulted in an improvement in the total-to-total efficiency in the top-damage cases. However, in the middle-damage cases, the decrease in torque was greater than the increase in outlet pressure; for example, in comparison with the reference case, the torque decreased by approximately 2.8% while the outlet pressure only increased by approximately 1.2% in the middle-damage at the pressure side case. Hence, compared with that in the reference case, the total-to-total efficiency in the middle-damage cases were noticeably reduced. Overall, it can be concluded that the efficiency of gas turbines is strongly dependent on the blade conditions—either normal or damaged. Moreover, if the blades are damaged, the damage locations significantly affect the turbine efficiency. Specifically, if the top part of the blade is damaged, the turbine efficiency can be slightly increased after modification.

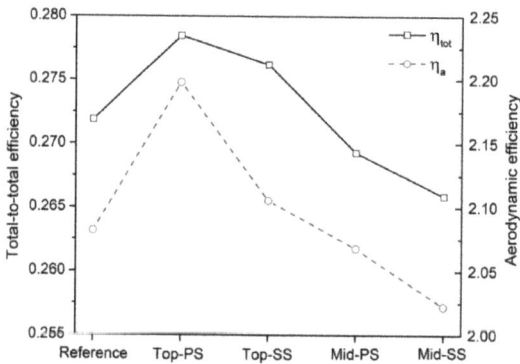

Figure 18. Total-to-total and aerodynamic efficiencies under various blade conditions.

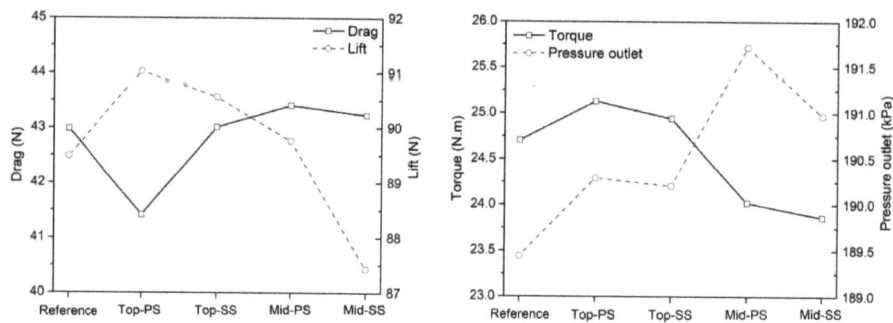

Figure 19. Drag force, lift force, torque, and outlet pressure under various blade conditions.

4. Conclusions

This paper presented a numerical investigation of the effects of modifications along the leading edge of a damaged rotor blade on the flow field and heat transfer characteristics in a 1.5-stage GE-E^3 gas turbine. This is the first study in which the effects of different damage locations of a turbine blade are examined. We analyzed five cases of rotor blades—an undamaged blade as reference and blades damaged at the top and middle on the pressure and suction sides.

The results confirmed that:

- *The average and maximum temperatures on the R_1 blade and S_2 vane surfaces of the damaged blades were higher than those in the reference case.* This was due to the effects of the altered flow field profiles on the damaged blades.
- *The tip leakage flow increased in the middle-damage cases but decreased in the top-damage cases, compared to the reference case.*
- *The heat transfer on the blade tip in the middle-damage cases was remarkably higher than the one in the other cases.*
- *The location of the damage had significant effects on the heat transfer characteristics on the blade and vane surfaces.* On the R_1 blade, the heat flux around the damage location exhibited a

sudden increase. The heat flux on the S$_2$ vane surface considerably increased around the mid-span on the pressure side and around the hub and tip on the suction side. *This led to an increase in the local thermal stress, showing a potential reduction in the fatigue life of the blade and the vane which would increase the maintenance costs.*

- Moreover, the *modifications to the top-damaged blades enhanced the aerodynamic and total-to-total efficiencies, while the same for the middle-damaged blades caused reductions in the efficiencies.*

This study investigated the effects of modification at various locations along the leading edge of the rotor blades. Hence, additional studies should be conducted on damage at other locations, such as the center and trailing edge of the blade, to provide a comprehensive overview of the effects of damage on the flow field, heat transfer, and aerodynamic performance of a gas turbine. This could provide more insight for design engineers to develop better cooling methods to enhance the fatigue life of the blades and vanes of gas turbines. Moreover, the findings of this study can facilitate damage or failure detection in gas turbines through monitoring of sudden changes in pressure and temperature fields.

Author Contributions: Conceptualization, methodology, investigation: T.D.M. and J.R.; validation, formal analysis, writing—original draft preparation, visualization: T.D.M.; writing—review and editing, supervision, project administration, funding acquisition: J.R. All authors have read and agreed to the published version of the manuscript.

Funding: This work was supported by the Korea Institute of Energy Technology Evaluation and Planning (KETEP) grant funded by the Korea government (MOTIE) (20193310100060, Evaluation of the performance for F-class or more gas turbine blade prototype). This research was supported by Korea Electric Power Corporation (Grant number: R19XO01-39). This research was supported by the MSIT (Ministry of Science and ICT), Korea, under the ITRC (Information Technology Research Center) support program (IITP-2020-2020-0-01655) supervised by the IITP (Institute of Information & Communications Technology Planning & Evaluation).

Conflicts of Interest: The authors declare no conflict of interest.

Nomenclature

ρ	Fluid density (kg/m^3)	Ω	Vorticity magnitude
u	Fluid velocity (m/s)	$Re_{\theta c}$	Critical Reynolds number
P	Fluid pressure (Pa)	St	Stanton number
μ	Fluid viscosity (Pa.s)	q	Heat flux (W/m^2)
E	Specific internal energy (J)	T_w	Temperature of the wall surface (K)
k_{eff}	Effective thermal conductivity [(W/m.K)	T_0	Average total temperature of the inlet flow (K)
μ_{eff}	Effective dynamic viscosity (Pa.s)	C_p	Specific heat of ideal air (J/kg.K)
μ_t	Turbulence viscosity (Pa.s)	ρ_0	Average density of the inlet flow (kg/m^3)
F_1, F_2	Blending functions	V_0	Average velocity of the inlet flow (m/s)
k	Turbulence kinetic energy (J/kg)	P_0	Average total pressure of inlet flow (Pa)
ω	Eddy dissipation/ Angular velocity (rad/s)	T	Torque (N.m)/Temperature (K)
σ_k	Turbulent Prandtl number for k	$\dot{m}$	Mass flow rate (kg/s)
σ_ω	Turbulent Prandtl number for ω	κ	Ratio of specific heat
γ	Intermittency	F_L	Lift force (N)
S	Strain rate magnitude (s^{-1})	F_D	Drag force (N)
F_{length}	Empirical correlation	η_a	Aerodynamic efficiency
		η_{tot}	Total-to-total efficiency

References

1. Povey, T.; Qureshi, I. Developments in hot-streak simulators for turbine testing. *J. Turbomach.* **2009**, *131*, 031009-1–031009-15. [CrossRef]
2. An, B.-T.; Liu, J.-J.; Jiang, H.-D. Numerical investigation on unsteady effects of hot streak on flow and heat transfer in a turbine stage. *J. Turbomach.* **2009**, *131*, 031015-1–031015-15. [CrossRef]
3. Anzellini, S.; Monteseguro, V.; Bandiello, E.; Dewaele, A.; Burakovsky, L.; Errandonea, D. In situ characterization of the high pressure–high temperature melting curve of platinum. *Sci. Rep.* **2019**, *9*, 1–10. [CrossRef]

4. Errandonea, D.; Burakovsky, L.; Preston, D.L.; MacLeod, S.G.; Santamaría-Perez, D.; Chen, S.; Cynn, H.; Simak, S.I.; McMahon, M.I.; Proctor, J.E. Experimental and theoretical confirmation of an orthorhombic phase transition in niobium at high pressure and temperature. *Commun. Mater.* **2020**, *1*, 1–11. [CrossRef]
5. Kaewbumrung, M.; Tangsopa, W.; Thongsri, J. Investigation of the trailing edge modification effect on compressor blade aerodynamics using SST k-ω turbulence model. *Aerospace* **2019**, *6*, 48. [CrossRef]
6. Kumari, S.; Satyanarayana, D.; Srinivas, M. Failure analysis of gas turbine rotor blades. *Eng. Fail. Anal.* **2014**, *45*, 234–244. [CrossRef]
7. Witek, L. Simulation of crack growth in the compressor blade subjected to resonant vibration using hybrid method. *Eng. Fail. Anal.* **2015**, *49*, 57–66. [CrossRef]
8. Mazur, Z.; Hernandez-Rossette, A.; Garcia-Illescas, R.; Luna-Ramirez, A. Failure analysis of a gas turbine nozzle. *Eng. Fail. Anal.* **2008**, *15*, 913–921. [CrossRef]
9. Choi, M.G.; Ryu, J. Numerical study of the axial gap and hot streak effects on thermal and flow characteristics in two-stage high pressure gas turbine. *Energies* **2018**, *11*, 2654. [CrossRef]
10. Wang, Z.; Wang, D.; Liu, Z.; Feng, Z. Numerical analysis on effects of inlet pressure and temperature non-uniformities on aero-thermal performance of a HP turbine. *Int. J. Heat Mass Transf.* **2017**, *104*, 83–97. [CrossRef]
11. Azad, G.S.; Han, J.-C.; Teng, S.; Boyle, R.J. Heat transfer and pressure distributions on a gas turbine blade tip. *J. Turbomach.* **2000**, *122*, 717–724. [CrossRef]
12. Asgarshamsi, A.; Benisi, A.H.; Assempour, A.; Pourfarzaneh, H. Multi-objective optimization of lean and sweep angles for stator and rotor blades of an axial turbine. *Proc. Inst. Mech. Eng. Part G J. Aerosp. Eng.* **2015**, *229*, 906–916. [CrossRef]
13. Pogorelov, A.; Meinke, M.; Schröder, W. Large-eddy simulation of the unsteady full 3D rim seal flow in a one-stage axial-flow turbine. *Flowturbulence Combust.* **2019**, *102*, 189–220. [CrossRef]
14. Sasao, Y.; Kato, H.; Yamamoto, S.; Satsuki, H.; Ooyama, H.; Ishizaka, K. F208 numerical and experimental investigations of unsteady 3-d flow through two-stage cascades in steam turbine model (Steam Turbine-3). In Proceedings of the International Conference on Power Engineering (ICOPE), Kobe, Japan, 16–20 November 2009.
15. Timko, L. *Energy Efficient Engine High Pressure Turbine Component Test Performance Report*; Technical Report; National Aeronautics and Space Administration (NASA): Cincinnati, OH, USA, 1 January 1984.
16. Arnone, A.; Benvenuti, E. *Three-Dimensional Navier-Stokes Analysis of a Two-Stage Gas Turbine*; American Society of Mechanical Engineers: New York, NY, USA, 1994; Volume 78835.
17. ANSYS. *ANSYS TurboGrid Tutorials, Release 15.0*; ANSYS: Canonsburg, PA, USA, 2013.
18. ANSYS. *ANSYS CFX, 12.1 Ansys Cfx-Solver Theory Guide*; ANSYS: Canonsburg, PA, USA, 2009.
19. Celik, I.; Yavuz, I.; Smirnov, A. Large eddy simulations of in-cylinder turbulence for internal combustion engines: A review. *Int. J. Engine Res.* **2001**, *2*, 119–148. [CrossRef]
20. Ryu, J.; Lele, S.; Viswanathan, K. Study of supersonic wave components in high-speed turbulent jets using an LES database. *J. Sound Vib.* **2014**, *333*, 6900–6923. [CrossRef]
21. Ryu, J.; Livescu, D. Turbulence structure behind the shock in canonical shock–vortical turbulence interaction. *J. Fluid Mech.* **2014**, *756*, R1-1–R1-13. [CrossRef]
22. Ngo, T.T.; Phu, N.M. Computational fluid dynamics analysis of the heat transfer and pressure drop of solar air heater with conic-curve profile ribs. *J. Therm. Anal. Calorim.* **2020**, *139*, 3235–3246. [CrossRef]
23. Nguyen, M.P.; Ngo, T.T.; Le, T.D. Experimental and numerical investigation of transport phenomena and kinetics for convective shrimp drying. *Case Stud. Therm. Eng.* **2019**, *14*, 100465. [CrossRef]
24. Hao, Z.-R.; Gu, C.-W.; Ren, X.-D. The application of discontinuous Galerkin methods in conjugate heat transfer simulations of gas turbines. *Energies* **2014**, *7*, 7857–7877. [CrossRef]
25. Kim, J.; Bak, J.; Kang, Y.; Cho, L.; Cho, J. A study on the numerical analysis methodology for thermal and flow characteristics of high pressure turbine in aircraft gas turbine engine. *Ksfm J. Fluid Mach.* **2014**, *17*, 46–51. [CrossRef]
26. Hylton, L.; Mihelc, M.; Turner, E.; Nealy, D.; York, R. *Analytical and Experimental Evaluation of the Heat Transfer Distribution over the Surfaces of Turbine Vanes*; Technical Report; National Aeronautics and Space Administration (NASA): Indianapolis, IN, USA, 1 May 1983.
27. Menter, F.; Rumsey, C. Assessment of two-equation turbulence models for transonic flows. In Proceedings of the Fluid Dynamics Conference, Colorado Springs, CO, USA, 20–23 June 1994; p. 2343.

28. Menter, F.R. Two-equation eddy-viscosity turbulence models for engineering applications. *Aiaa J.* **1994**, *32*, 1598–1605. [CrossRef]
29. ANSYS. *ANSYS Fluent Theory Guide*; ANSYS, Inc.: Canonsburg, PA, USA, 2013.
30. Gentry, M.; Jacobi, A.M. Heat transfer enhancement by delta-wing-generated tip vortices in flat-plate and developing channel flows. *J. Heat Transf.* **2002**, *124*, 1158–1168. [CrossRef]
31. Lee, J.; Lee, H.J.; Ryu, J.; Lee, S.H. Three-dimensional turbulent flow and heat transfer characteristics of longitudinal vortices embedded in turbulent boundary layer in bent channels. *Int. J. Heat Mass Transf.* **2018**, *117*, 958–965. [CrossRef]
32. Touil, K.; Ghenaiet, A. Simulation and analysis of vane-blade interaction in a two-stage high-pressure axial turbine. *Energy* **2019**, *172*, 1291–1311. [CrossRef]
33. Wei, Z.-J.; Qiao, W.-Y.; Chen, P.-P.; Liu, J. Formation and Transport of Secondary Flows Caused by Vortex-Blade Interaction in a High Pressure Turbine. In Proceedings of the Turbo Expo: Power for Land, Sea, and Air, Montreal, QC, Canada, 15–19 June 2015; p. V02AT38A008.
34. Yang, D.; Feng, Z. Tip leakage flow and heat transfer predictions for turbine blades. In Proceedings of the Turbo Expo: Power for Land, Sea, and Air, Montreal, QC, Canada, 14–17 May 2007; pp. 589–596.

Publisher's Note: MDPI stays neutral with regard to jurisdictional claims in published maps and institutional affiliations.

© 2020 by the authors. Licensee MDPI, Basel, Switzerland. This article is an open access article distributed under the terms and conditions of the Creative Commons Attribution (CC BY) license (http://creativecommons.org/licenses/by/4.0/).

Article

Impact of Heat Generation on Magneto-Nanofluid Free Convection Flow about Sphere in the Plume Region

Anwar Khan [1], Muhammad Ashraf [1], Ahmed M. Rashad [2] and Hossam A. Nabwey [3,4,*]

[1] Department of Mathematics, Faculty of Science, University of Sargodha, Sargodha 40100, Pakistan; anwarkhanuos@gmail.com (A.K.); muhammad.ashraf@uos.edu.pk (M.A.)
[2] Department of Mathematics, Faculty of Science, Aswan University, Aswan 81528, Egypt; am_rashad@yahoo.com
[3] Department of Mathematics, College of Science and Humanities in Al-Kharj, Prince Sattam Bin Abdulaziz University, Al-Kharj 11942, Saudi Arabia
[4] Department of Basic Engineering Science, Faculty of Engineering, Menoufia University, Shebin El Kom 32511, Egypt
* Correspondence: eng_hossam21@yahoo.com or h.mohamed@psau.edu.sa

Received: 12 October 2020; Accepted: 9 November 2020; Published: 11 November 2020

Abstract: The main aim of the current study is to analyze the physical phenomenon of free convection nanofluids heat transfer along a sphere and fluid eruption through boundary layer into a plume region above the surface of the sphere. In the current study, the effect of heat generation with the inclusion of an applied magnetic field by considering nanofluids is incorporated. The dimensioned form of formulated equations of the said phenomenon is transformed into the non-dimensional form, and then solved numerically. The developed finite difference method along with the Thomas algorithm has been utilized to approximate the given equations. The numerical simulation is carried out for the different physical parameters involved, such as magnetic field parameter, Prandtl number, thermophoresis parameter, heat generation parameter, Schmidt number, and Brownian motion parameter. Later, the quantities, such as velocity, temperature, and mass distribution, are plotted under the impacts of different values of different controlling parameters to ascertain how these quantities are affected by these pertinent parameters. Moreover, the obtained results are displayed graphically as well in tabular form. The novelty of present work is that we first secure results around different points of a sphere and then the effects of all parameters are captured above the sphere in the plume.

Keywords: nanofluids; MHD; heat generation; sphere; plume; finite difference method

1. Introduction

The conventional fluids, such as mixtures of ethylene glycol, oil, and water, have been used for the purpose of heat transportation by the research community. The heat transfer process was made very slow by the use of these fluids due to their poor thermal conductivity. The utilization of nanofluids as a cooling source increases operating and manufacturing costs. So, nanofluids are being used to speed up the heat transfer performances because of their excellent thermal conductivity. Nanofluids result from the suspension of submicron solid particles (nanoparticles) in the base fluids, such as water or any organic solvent. Nanoparticles are of growing interest as they play an effective role to strengthen the thermal conductivity of the base fluid. The inclusion of a magnetic field in the analysis of nanofluids has attracted much attention of researchers because of its growing applications in the fields of engineering, physics, and chemistry. The nanofluids which contain magnetic particles act as super-paramagnetic fluids which absorb the energy control of the flow and act as an alternating electromagnetic field. Nanofluids are employed as coolants in computer microchips and many other electronic devices which utilize micro-fluidic applications. With motivation from the above applications of magnetohydrodynamic flow, Sparrow and Cess [1] comprehensively analyzed the study of magnetohydrodynamic natural convection flow through the vertical plate by encountering both upward and downward flows with the effect of buoyancy forces. Potter and Riley [2] focused their attention on natural convection flow due to a heated sphere placed in static fluid by considering large values of the Grashof number. They discussed the characteristics of boundary layer flow into the plume numerically. Riley [3] considered the phenomenon of free convection flow along the surface of a sphere by maintaining higher temperature than the surroundings. He evaluated the model numerically for finite values of Grashof and Parndtl numbers. Andersson [4] studied the model of visco-elastic fluid over the stretching surface considering the effect of a transverse magnetic field analytically. Stephen and Eastman [5] proposed a novel type of fluid whose thermal conductivity is higher than conventional fluids and termed them as nanofluids. They concluded that such types of fluid enhance the thermal performances during the process of heat transfer. Samuel and Falade [6] investigated the stability of hydromagnetic fluid in porous media by incorporating the outcomes of variable viscosity. Their prediction for theoretical analysis was that an increase in the viscosity variation parameter creates a stability of the fluid flow. The transient form of the convective flow along the surface of a moving plate in a porous medium with uniform heat flux with the inclusion of a magnetic field has been studied by Al-Kabeir et al. [7]. Chamkha and Aly [8] presented nanofluids flow by means of free convection heat transfer over the permeable plate observing a magnetic field, transpiration parameter, heat absorption, and generation influences for main physical properties. The phenomenon of double diffusive free convection nanofluids flow over the vertical plate was examined in [9]. Rosmilaet al. [10] studied the problem of free convection magnetohydrodynamic flow of nanofluids over a linearly stretching surface by the opting shooting technique along with the Runge–Kutta method of the fourth order. Mohammad et al. [11] analyzed the flow problem of a magnetohydrodynamic boundary layer over a vertical surface for nanofluids taking into account Newtonian heating effects. Gandhar and Reddy [12] predicted heat and mass transfer mechanism for moving plate held vertically embedded in porous media due to the insertion of magnetic field. The analysis on the influences of buoyancy force, magnetic field, and a stretching and shrinking sheet on the stagnation point flow of nanofluids was performed by Makinde et al. [13]. Olanrewaju and Makinde [14] discussed the problem of natural convection flow of nanofluids over a porous surface with a stagnation point in the presence of Newtonian heating effects. Chamkha et al. [15] reviewed the available material properties of nanofluids and focused on several geometries and applications. Stagnation point flow on a vertical stretching surface by imposing the slip condition was discussed by Khairy and Ishak [16]. The analysis of the nanofluids in the presence of a chemical reaction and magnetic field has been carried by Ltu and Ochsnor [17]. Another study was conducted to assess the free convection flow of nanofluids about different circumferential points of a sphere and the fluid erupting from the boundary layer flow into the plume made above the sphere [18]. The characteristics of heat and fluid flow in the presence of nanofluids have been investigated by [19–25] along different simple and complex geometries.

With inspiration from aforesaid research attempts, we intended to elaborate the problem of natural convective flow of magnetohydrodynamic nanofluids flow at the different circumferential positions along the surface of a sphere and into the plume made above the sphere by encountering the effects of heat generation and absorption. It is necessary to highlight that no one has paid any attention towards such a problem before this attempt. In the subsequent sections, the mathematical formulation is performed and after suitable transformation of the modeled equations, a very accurate approximating technique known as the finite difference method is directly employed to get the approximate solutions of the partial differential equations. By using FORTRAN as a computing tool, asymptotic and valid solutions of the governing model satisfying the given boundary conditions are calculated. Further, in this study, the different trends/behaviors depending on various combinations of many influential parameters have been displayed graphically as well as in tabular form.

2. Statement of the Problem and Mathematical Formulation

Consider a steady, two-dimensional, viscous, incompressible, and electrically conducting boundary layer flow of nanofluid. In this analysis, water is taken as the base fluid and heat generation effects are encountered. The physical sketch and geometry of the problem are shown in Figure 1. The sphere surface is kept at constant temperature $\hat{T}_w$ and the nanoparticles volume fraction at the surface is $\hat{C}_w$. The coordinate along the surface of a sphere is $\hat{x}$ and $\hat{y}$ is taken as normal to the surface. The corresponding velocity components $\hat{u}$ and $\hat{v}$ is considered along and normal to the surface of the sphere respectively. There are three regions, namely sphere, fluid erupting from the boundary layer, and plume made above the sphere. The universal conservation equations for the current mechanism following Potter and Riley [2] take the forms given as below:

$$\frac{\partial (\hat{r}\hat{u})}{\partial \hat{x}} + \frac{\partial (\hat{r}\hat{v})}{\partial \hat{y}} = 0 \qquad (1)$$

$$\hat{u}\frac{\partial \hat{u}}{\partial \hat{x}} + \hat{v}\frac{\partial \hat{u}}{\partial \hat{y}} = \nu \frac{\partial^2 \hat{u}}{\partial \hat{y}^2} + g\beta(\hat{T}-\hat{T}_\infty)Sin\frac{\hat{x}}{a} + g\beta_c(\hat{C}-\hat{C}_\infty)Sin\frac{\hat{x}}{a} - \frac{\sigma_0 \beta_0^2}{\rho}\hat{u} \qquad (2)$$

$$\hat{u}\frac{\partial \hat{T}}{\partial \hat{x}} + \hat{v}\frac{\partial \hat{T}}{\partial \hat{y}} = \alpha \frac{\partial^2 \hat{T}}{\partial \hat{y}^2} + \tau\left\{D_B \frac{\partial \hat{C}}{\partial \hat{y}}\frac{\partial \hat{T}}{\partial \hat{y}} + \frac{D_T}{\hat{T}_\infty}\left(\frac{\partial \hat{T}}{\partial \hat{y}}\right)^2\right\} + \frac{Q_0}{\rho C_p}(\hat{T}-\hat{T}_\infty) \qquad (3)$$

$$\hat{u}\frac{\partial \hat{C}}{\partial \hat{x}} + \hat{v}\frac{\partial \hat{C}}{\partial \hat{y}} = D_B \frac{\partial^2 \hat{C}}{\partial \hat{y}^2} + \frac{D_T}{T_\infty}\frac{\partial^2 \hat{T}}{\partial \hat{y}^2} \qquad (4)$$

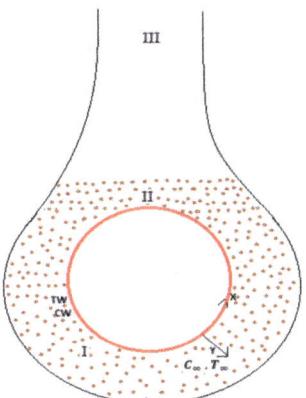

Figure 1. Coordinate System and Flow Geometry.

Subjected to the corresponding boundary conditions:

$$\begin{array}{llll} \hat{u}=0=\hat{v}, & \hat{T}=\hat{T}_w, & \hat{C}=\hat{C}_w & at & \hat{y}=0 \\ \hat{u}\to 0=\hat{v}, & \hat{T}\to\hat{T}_\infty, & \hat{C}\to\hat{C}_\infty & as & \hat{y}\to\infty \end{array} \qquad (5)$$

The symbols appeared in the above governing equations such as g, β, α, τ and β_c, are termed as gravitational acceleration, thermal expansion of temperature, thermal diffusivity, nanoparticles heat capacity to base fluid ratio and solutal thermal expansion. The Brownian diffusion coefficient, heat generation coefficient, thermophoretic diffusion coefficient, and magnetic field strength are denoted by D_B, Q_o, D_T, and B_o, respectively. To make the above proposed model dimensionless, here, non-dimensionless variables are defined as below:

$$x = \frac{\hat{x}}{a}, \quad r(\hat{x}) = a\sin\hat{x}, \quad y = G_r^{\frac{1}{4}}\frac{\hat{y}}{y}, \quad u = G_r^{-\frac{1}{2}}\frac{a}{\nu}\hat{u}, \quad v = G_r^{-\frac{1}{4}}\frac{a}{\nu}\hat{v}$$

$$G_r = \frac{g\beta\Delta\hat{T}a^3}{\nu^2}, \quad G_{rc} = \frac{g\beta_c\Delta\hat{T}a^3}{\nu^2}, \quad r = \frac{\hat{r}}{a} \tag{6}$$

where a is the sphere radius. By inserting Equation (6) into Equations (1)–(5), we obtain the following non-dimensional forms of the governing equations as given below:

$$\frac{\partial(ru)}{\partial x} + \frac{\partial(rv)}{\partial y} = 0 \tag{7}$$

$$u\frac{\partial u}{\partial x} + v\frac{\partial u}{\partial y} = \nu\frac{\partial^2 u}{\partial^2 y} + \theta\sin x + \varphi\sin x - Mu \tag{8}$$

$$u\frac{\partial \theta}{\partial x} + v\frac{\partial \theta}{\partial y} = \frac{1}{\Pr}\frac{\partial^2 \theta}{\partial^2 y} + Nb\frac{\partial \varphi}{\partial y}\frac{\partial \theta}{\partial y} + Nt\left(\frac{\partial \theta}{\partial y}\right)^2 + Q\theta \tag{9}$$

$$u\frac{\partial \varphi}{\partial x} + v\frac{\partial \varphi}{\partial y} = \frac{1}{Sc}\left(\frac{\partial^2 \varphi}{\partial^2 y} + \frac{Nt}{Nb}\frac{\partial^2 \theta}{\partial^2 y}\right) \tag{10}$$

With boundary conditions:

$$\begin{array}{cccc} u = 0 = v & \theta = 1, & \varphi = 1 & at \quad y = 0 \\ u \to 0, & \varphi \to 0, & \theta \to 0 & as \quad y \to \infty \end{array} \tag{11}$$

where

$$\Pr = \frac{\nu}{\alpha}, \quad Sc = \frac{\nu}{D_b}, \quad N_b = \frac{(\rho c)_p D_B(t5\varphi_w - \varphi_\infty)}{(\rho c)_f \nu}, \quad N_t = \frac{(\rho c)_p D_T(T_w - T_\infty)}{(\rho c)_f T_\infty \nu},$$

The parameters appearing above are the thermophoresis parameter, Schmidt number, Prandtl number, and Brownian motion parameter, which are designated Nt, Sc, Pr, and Nb. Here, $M = \frac{\sigma_0\beta_0^2}{\rho\nu}\frac{a^2}{G_r^{1/2}}$, and $Q = \frac{Q_0}{\mu C_p}\frac{a^2}{G_r^{1/2}}$ represent the magnetic field parameter and heat generation parameter, respectively.

3. Method of Solution

To adopt an ease in making the algorithm, the following primitive variables are used to make the primitive form of the above Equations (7)–(10) along with boundary conditions:

$$u = x^{\frac{1}{2}}U(X,Y), \quad v = x^{-\frac{1}{4}}V(X,Y), \quad Y = x^{-1/4}y, \quad \theta = \overline{\theta}(X,Y), \quad \varphi = \overline{\varphi}(X,Y),$$
$$x = X, \quad \theta = \overline{\theta}(X,Y), \quad \varphi = \overline{\varphi}(X,Y), \quad r(x) = a\sin X \tag{12}$$

After substitution of the variables defined in Equation (12) into Equations (7)–(11), we get the following primitive system of partial differential equations:

$$X\cos XU + \left\{X\frac{\partial U}{\partial X} - \frac{1}{4}Y\frac{\partial U}{\partial Y} + \frac{1}{2}U + \frac{\partial V}{\partial Y}\right\}\sin X = 0 \tag{13}$$

$$XU\frac{\partial U}{\partial X} + \frac{1}{2}U^2 + \left(V - \frac{1}{4}YU\right)\frac{\partial U}{\partial Y} = \frac{\partial^2 U}{\partial Y^2} - \overline{\theta}\sin X - \overline{\varphi}\sin X - X^{1/2}MU \tag{14}$$

$$XU\frac{\partial\overline{\theta}}{\partial X}+\left(V-\frac{1}{4}YU\right)\frac{\partial\overline{\theta}}{\partial Y}=\frac{1}{\text{pr}}\frac{\partial^2\overline{\theta}}{\partial Y^2}+\text{Nb}\frac{\partial\overline{\varphi}}{\partial Y}\frac{\partial\overline{\theta}}{\partial Y}+\text{Nt}\left(\frac{\partial\overline{\theta}}{\partial Y}\right)^2+Q\overline{\theta} \quad (15)$$

$$XU\frac{\partial\overline{\varphi}}{\partial X}+\left(V-\frac{1}{4}YU\right)\frac{\partial\overline{\varphi}}{\partial Y}=\frac{1}{\text{Sc}}\left(\frac{\partial^2\overline{\varphi}}{\partial^2 Y}+\frac{\text{Nt}}{\text{Nb}}\frac{\partial^2\overline{\theta}}{\partial^2 Y}\right) \quad (16)$$

The corresponding boundary conditions are:

$$\begin{array}{ll} U=0=V, \quad \overline{\theta}=1 \quad \overline{\varphi}=1 & \text{at} \quad Y=0, \\ U\to 0, \quad \overline{\varphi}\to 0, \quad \overline{\theta}\to 0 & \text{as} \quad Y\to\infty \end{array} \quad (17)$$

4. Computational Scheme

The formulated model is complex and its analytical solution cannot be found. So, we move towards the approximate solutions of the present problem with the use of very accurate approximating technique known as finite difference method. This method is directly applied to partial differential Equations (13)–(17) to convert into algebraic system of equations which is solved by coding on computing tool FORTRAN package. The backward difference is used along x-axis and central difference along y-axis. The discretization procedure is given below:

$$\frac{\partial U}{\partial X}=\frac{U_{i,j}-U_{i,j-1}}{\Delta X}. \quad (18)$$

$$\frac{\partial U}{\partial Y}=\frac{U_{i+1,j}-U_{i-1,j}}{2\Delta Y}. \quad (19)$$

$$\frac{\partial^2 U}{\partial Y^2}=\frac{U_{i+1,j}-2U_{i,j}+U_{i-1,j}}{\Delta Y^2}. \quad (20)$$

The insertion of Equations (18)–(20) into Equations (13)–(17) implies:

$$V_{i,j}=\left\{V_{i-1,j}-X_i\frac{\Delta Y}{\Delta X}(U_{i,j}-U_{i,j-1})+\frac{1}{8}Y_j(U_{i+1,j}-U_{i-1,j})-\frac{1}{2}\Delta YU_{i,j}\right\}-\frac{\cos X_i}{\sin X_i}X_iU_{i,j} \quad (21)$$

$$\begin{array}{l}\left(\left[\frac{1}{2}\Delta Y(V_{i,j}-\frac{1}{4}Y_jU_{i,j})\right]+1\right)U_{i-1,j}+\left(-\Delta Y^2\left\{\left(\frac{1}{2}+\frac{X_i}{\Delta X}\right)U_{i,j}+X_i^{\frac{1}{2}}M\right\}-2\right)U_{i,j} \\ +\left(-\left[\frac{1}{2}\Delta Y(V_{i,j}-\frac{1}{4}Y_jU_{i,j})\right]+1\right)U_{i+1,j} \\ =-X_iU_{i,j}U_{i,j-1}\frac{\Delta Y^2}{\Delta X}+\Delta Y^2(\overline{\theta}_{i,j}\sin X_i+\overline{\varphi}_{i,j}\sin X_i)\end{array} \quad (22)$$

$$\begin{array}{l}\left(\left[\frac{1}{2}\Delta Y(V_{i,j}-\frac{1}{4}Y_jU_{i,j})+\text{Nb}\frac{1}{4}(\overline{\varphi}_{i+1,j}-\overline{\varphi}_{i-1,j})+\text{Nt}\frac{1}{4}(\overline{\theta}_{i+1,j}-\overline{\theta}_{i-1,j})\right]+\frac{1}{\text{Pr}}\right)\overline{\theta}_{i-1,j} \\ +\left(-\left(U_{i,j}\frac{X_i}{\Delta X}-Q\overline{\theta}_{i,j}\right)\Delta Y^2-\frac{2}{\text{Pr}}\right)\overline{\theta}_{i,j} \\ +\left(-\left[\frac{1}{2}\Delta Y(V_{i,j}-\frac{1}{4}Y_jU_{i,j})+\text{Nb}\frac{1}{4}(\overline{\varphi}_{i+1,j}-\overline{\varphi}_{i-1,j})\right.\right. \\ \left.\left.+\text{Nt}\frac{1}{4}(\overline{\theta}_{i+1,j}-\overline{\theta}_{i-1,j})\right]+\frac{1}{\text{Pr}}\right)\overline{\theta}_{i-1,j}=-X_iU_{i,j}\frac{\Delta Y^2}{\Delta X}\overline{\theta}_{i,j-1}\end{array} \quad (23)$$

$$\begin{array}{l}\left(\left[\frac{1}{2}\Delta Y(V_{i,j}-\frac{1}{4}Y_jU_{i,j})\right]+\frac{1}{\text{Sc}}\right)\overline{\varphi}_{i-1,j}+\left(-X_iU_{i,j}\frac{\Delta Y^2}{\Delta X}-\frac{2}{\text{Sc}}\right)\overline{\varphi}_{i,j} \\ +\left(-\left[\frac{1}{2}\Delta Y(V_{i,j}-\frac{1}{4}Y_jU_{i,j})\right]+\frac{1}{\text{Sc}}\right)\overline{\varphi}_{i-1,j}=-X_iU_{i,j}\overline{\varphi}_{i,j-1}\frac{\Delta Y^2}{\Delta X}+\frac{1}{\text{Sc}}\frac{\text{Nt}}{\text{Nb}}\end{array} \quad (24)$$

With boundary conditions:

$$\begin{array}{ll} U_{i,j}=0=V_{i,j}, \quad \overline{\theta}_{i,j}=1, \quad \overline{\varphi}_{i,j}=1 & \text{at} \quad Y=0 \\ U_{i,j}\to 0, \quad \overline{\varphi}_{i,j}\to 0, \quad \overline{\theta}_{i,j}\to 0 & \text{as} \quad Y\to\infty \end{array} \quad (25)$$

5. Governing Equations for Plume Region

Considering the diagram of the geometry, we can see that nanofluid enters from the region-II to region-III. For this region, the aforesaid model is altered and a new model for the plume region is formulated by following [2]:

$$\frac{\partial(\hat{z}\hat{u})}{\partial \hat{x}} + \frac{\partial(\hat{z}\hat{w})}{\partial \hat{z}} = 0 \tag{26}$$

$$\hat{u}\frac{\partial \hat{u}}{\partial \hat{x}} + \hat{w}\frac{\partial \hat{u}}{\partial \hat{z}} = \nu \frac{1}{\hat{z}}\frac{\partial}{\partial \hat{z}}\left(\hat{z}\frac{\partial \hat{u}}{\partial \hat{z}}\right) - g\beta(\hat{T} - \hat{T}_\infty) - g\beta_c(\hat{C} - \hat{C}_\infty) - \frac{\sigma_0 \beta_0^2}{\rho}\hat{u} \tag{27}$$

$$\hat{u}\frac{\partial \hat{T}}{\partial \hat{x}} + \hat{w}\frac{\partial \hat{T}}{\partial \hat{z}} = \alpha \frac{1}{\hat{z}}\frac{\partial}{\partial \hat{z}}\left(\hat{z}\frac{\partial \hat{T}}{\partial \hat{z}}\right) + \tau\left\{D_B \frac{\partial \hat{C}}{\partial \hat{z}}\frac{\partial \hat{T}}{\partial \hat{z}} + \frac{D_T}{T_\infty}\left(\frac{\partial \hat{T}}{\partial \hat{z}}\right)^2\right\} + \frac{Q_0}{\mu C_p}\frac{a^2}{G_r^{1/2}}\theta \tag{28}$$

$$\hat{u}\frac{\partial \hat{C}}{\partial \hat{x}} + \hat{w}\frac{\partial \hat{C}}{\partial \hat{z}} = D_B \frac{1}{\hat{z}}\frac{\partial}{\partial \hat{z}}\left(\hat{z}\frac{\partial \hat{C}}{\partial \hat{z}}\right) + \frac{D_T}{T_\infty}\frac{1}{\hat{z}}\frac{\partial}{\partial \hat{z}}\left(\hat{z}\frac{\partial \hat{T}}{\partial \hat{z}}\right) \tag{29}$$

With boundary conditions:

$$\begin{aligned} \hat{w} = \frac{\partial \hat{u}}{\partial \hat{z}} = \frac{\partial \hat{T}}{\partial \hat{z}} = 0 \quad &\text{at} \quad \hat{z} = 0, \\ \hat{u} \to 0 = \hat{w}, \hat{T} \to \hat{T}_\infty, \quad &\hat{C} \to \hat{C}_\infty \quad \text{as} \quad \hat{z} \to \infty. \end{aligned} \tag{30}$$

Dimensionless variables:

$$x = \frac{\hat{x}}{a}, \quad z = G_r^{1/4}\frac{\hat{z}}{z}, \quad z = \frac{\hat{z}}{a}, \quad u = G_r^{-1/2}\frac{a}{\nu}\hat{u}, \quad w = G_r^{-1/4}\frac{a}{\nu}\hat{w}$$

$$G_r = \frac{g\beta \Delta T a^3}{\nu^2}, \quad G_{rc} = \frac{g\beta_c \Delta T a^3}{\nu^2}, \quad \tau = \frac{(\rho c)_p}{(\rho c)_f} \tag{31}$$

Dimensionless form of system of equations:

$$\frac{\partial(zu)}{\partial x} + \frac{\partial(zw)}{\partial z} = 0 \tag{32}$$

$$u\frac{\partial u}{\partial x} + w\frac{\partial u}{\partial z} = \frac{1}{z}\frac{\partial}{\partial z}\left(z\frac{\partial u}{\partial z}\right) - \theta - \varphi - M \tag{33}$$

$$u\frac{\partial \theta}{\partial x} + w\frac{\partial \theta}{\partial z} = \frac{1}{\Pr}\frac{1}{z}\frac{\partial}{\partial z}\left(z\frac{\partial \theta}{\partial z}\right) + Nb\frac{\partial \varphi}{\partial z}\frac{\partial \theta}{\partial z} + Nt\left(\frac{\partial \theta}{\partial z}\right)^2 + Q\theta \tag{34}$$

$$u\frac{\partial \varphi}{\partial x} + w\frac{\partial \varphi}{\partial z} = \frac{1}{Sc}\left(\frac{1}{z}\frac{\partial}{\partial z}\left(z\frac{\partial \varphi}{\partial z}\right) + \frac{Nt}{Nb}\frac{1}{z}\frac{\partial}{\partial z}\left(z\frac{\partial \theta}{\partial z}\right)\right) \tag{35}$$

With boundary conditions:

$$\begin{aligned} w = \frac{\partial u}{\partial z} = \frac{\partial \theta}{\partial z} = 0 \quad &\text{at} \quad z = 0 \\ u \to 0, \quad \varphi \to 0, \quad \theta \to 0 \quad &\text{as} \quad z \to \infty \end{aligned} \tag{36}$$

For the convenient form of the integration, we use the following variables for the required form:

$$\begin{aligned} u = x^{\frac{1}{2}}U(X,Z), \quad W = x^{-\frac{1}{4}}W(X,Z), \quad Z = x^{-\frac{1}{4}}z, \\ \theta = \overline{\theta}(X,Z), \quad \varphi = \overline{\varphi}(X,Z), \quad x = X. \end{aligned} \tag{37}$$

Using the above primitive variable formulation, we have the following system of equations:

$$Z\frac{\partial U}{\partial X} - \frac{Z^2}{4X}\frac{\partial U}{\partial Z} + \frac{3}{4}ZU + W + Z\frac{\partial W}{\partial Z} = 0 \tag{38}$$

$$XU\frac{\partial U}{\partial X} + \frac{1}{2}U^2 + \left(W - \frac{1}{4}ZU\right)\frac{\partial U}{\partial Z} = \frac{1}{Z}\frac{\partial}{\partial Z}\left(Z\frac{\partial U}{\partial Z}\right) - \overline{\theta} - \overline{\varphi} - X^{1/2}MU. \tag{39}$$

$$XU\frac{\partial \overline{\theta}}{\partial X} + \left(W - \frac{1}{4}ZU\right)\frac{\partial \overline{\theta}}{\partial Z} = \frac{1}{\Pr}\frac{1}{Z}\frac{\partial}{\partial Z}\left(Z\frac{\partial \overline{\theta}}{\partial Z}\right) + Nb\frac{\partial \overline{\varphi}}{\partial Z}\frac{\partial \overline{\theta}}{\partial Z} + Nt\left(\frac{\partial \overline{\theta}}{\partial Z}\right)^2 + Q\overline{\theta}. \tag{40}$$

$$XU\frac{\partial \overline{\varphi}}{\partial X} + \left(W - \frac{1}{4}ZU\right)\frac{\partial \overline{\varphi}}{\partial Z} = \frac{1}{Sc}\left(\frac{1}{Z}\frac{\partial}{\partial Z}\left(Z\frac{\partial \overline{\varphi}}{\partial Z}\right) + \frac{Nt}{Nb}\frac{1}{Z}\frac{\partial}{\partial Z}\left(Z\frac{\partial \overline{\theta}}{\partial Z}\right)\right). \tag{41}$$

With boundary conditions:

$$\begin{array}{llll}
W = \frac{\partial U}{\partial Z} = \frac{\partial \overline{\vartheta}}{\partial Z} = 0, & \overline{\varphi} = 1, & \overline{\theta} = 1 & \text{at} \quad Z = 0, \\
U \to 0, & \overline{\varphi} \to 0, & \overline{\theta} \to 0 & \text{as} \quad Z \to \infty.
\end{array} \tag{42}$$

Solution Methodology

For the numerical evaluation of the flow equations in the plume region, the finite difference scheme is implemented. The constitutive equations in discretized forms are given as below:

$$W_{i,j} = \frac{1}{(\Delta Z + Z_j)}\left\{Z_j W_{i-1,j} - Z_j(U_{i,j} - U_{i,j-1})\frac{\Delta Z}{\Delta X} + \frac{Z_j^2}{8X_i}(U_{i+1,j} - U_{i-1,j}) - \frac{3}{4}\Delta Z Z_j U_{i,j}\right\} \tag{43}$$

$$\begin{aligned}
&\left\{\left[\tfrac{1}{2}\Delta Z\left(W_{i,j} - \tfrac{1}{4}Z_j U_{i,j} - \tfrac{1}{2Z_j}\right) + 1\right]\right\}U_{i-1,j} \\
&\quad + \left\{\left[\Delta Z^2\left(-X_i U_{i,j}\tfrac{1}{\Delta X} - \tfrac{1}{2}U_{i,j} - X_i^{1/2}M\right) - 2\right]\right\}U_{i,j} \\
&\quad + \left\{\left[-\tfrac{1}{2}\Delta Z\left(W_{i,j} - \tfrac{1}{4}Z_j U_{i,j} - \tfrac{1}{2Z_j}\right) + 1\right]\right\}U_{i+1,j} \\
&\quad = \Delta Z^2\left(\overline{\theta}_{i,j} + \overline{\varphi}_{i,j} - X_i U_{i,j} U_{i,j-1}\left(\tfrac{1}{\Delta X}\right)\right)
\end{aligned} \tag{44}$$

$$\begin{aligned}
&\left[\tfrac{1}{2}\Delta Z\left(W_{i,j} - \tfrac{1}{4}Z_j U_{i,j} - \tfrac{1}{2\Pr Z_j}\right) + Nb\tfrac{1}{4}(\overline{\varphi}_{i+1,j} - \overline{\varphi}_{i-1,j}) + Nt\tfrac{1}{4}(\overline{\theta}_{i+1,j} - \overline{\theta}_{i-1,j})\right]. \\
&\quad + \tfrac{1}{\Pr}\}\overline{\theta}_{i-1,j} + \left\{-\Delta Z^2\left[\tfrac{X_i}{\Delta X}U_{i,j} - Q\right] - \tfrac{2}{\Pr}\right\}\overline{\theta}_{i,j} \\
&\quad + \left\{-\left[\tfrac{1}{2}\Delta Z\left(W_{i,j} - \tfrac{1}{4}Z_j U_{i,j} - \tfrac{1}{2\Pr Z_j}\right) + Nb\tfrac{1}{4}(\overline{\varphi}_{i+1,j} - \overline{\varphi}_{i-1,j})\right.\right. \\
&\quad \left.\left. + Nt\tfrac{1}{4}(\overline{\theta}_{i+1,j} - \overline{\theta}_{i-1,j})\right] + \tfrac{1}{\Pr}\right\}\overline{\theta}_{i+1,j} = -X_i U_{i,j}\tfrac{\Delta Z^2}{\Delta X}\overline{\theta}_{i,j-1}
\end{aligned} \tag{45}$$

$$\begin{aligned}
&\left\{\tfrac{1}{2}\Delta Z\left(W_{i,j} - \tfrac{1}{4}Z_j U_{i,j}\right) + \tfrac{1}{Sc}\left(1 - \tfrac{\Delta Z}{2Z_j}\right)\right\}\overline{\varphi}_{i-1,j} + \left\{-X_i U_{i,j}\tfrac{\Delta Z^2}{\Delta X} - \tfrac{2}{Sc}\right\}\overline{\varphi}_{i,j} \\
&\quad + \left\{-\tfrac{1}{2}\Delta Z\left(W_{i,j} - \tfrac{1}{4}Z_j U_{i,j}\right) - \tfrac{1}{Sc}\left(1 + \tfrac{\Delta Z}{2Z_j}\right)\right\}\overline{\varphi}_{i+1,j} \\
&\quad = -X_i U_{i,j}\tfrac{\Delta Z^2}{\Delta X}\overline{\varphi}_{i,j-1} \\
&\quad + \tfrac{1}{Sc}\left\{\tfrac{Nt}{Nb}\left\{\overline{\theta}_{i+1,j} + \overline{\theta}_{i-1} - 2\overline{\theta}_{i,j} + (\overline{\theta}_{i+1,j} - \overline{\theta}_{i-1,j})\tfrac{\Delta Z}{2Z_j}\right\}\right\}
\end{aligned} \tag{46}$$

With boundary conditions:

$$\begin{array}{llll}
W_{i,j} = 0, & U_{i+1,j} = U_{i-1,j}, & \overline{\theta}_{i+1,j} = \overline{\theta}_{i-1,j}, \overline{\varphi}_{i+1,j} = \overline{\varphi}_{i-1,j}, \\
& \overline{\varphi}_{i,j} = 1, & \overline{\theta}_{i,j} = 1 \quad \text{at } Z_j = 0, \\
U_{i,j} \to 0, & \overline{\varphi}_{i,j} \to 0, & \overline{\theta}_{i,j} \to 0 & \text{as} \quad Z_j \to \infty.
\end{array} \tag{47}$$

6. Analysis of the Results

This section covers the discussion and conclusion on the behaviors of velocity field U, temperature field θ, and mass field φ, along with heat transfer rate $\frac{\partial \theta}{\partial Y}$, mass transfer rate $\frac{\partial \varphi}{\partial Y}$, and skin friction $\frac{\partial U}{\partial Y}$, with the variations of different flow parameters. The effects of parameters which are taken into observation are named as magnetic field parameter, M, heat generation parameter, Q, Schmidt number, Sc, Prandtl numbers, Pr, thermophoresis parameter, Nt, and Brownian motion parameter, Nb. The obtained numerical solutions for considered governing properties are displayed in graphical form and tabulated as well. The solution detail has been split into two parts, i.e., the several locations around a sphere and in the plume region above the sphere.

6.1. Fluxes and Boundary Layers on the Sphere

In this subsection, we are going to present and discuss the obtained solutions at different circumferential stations around the surface of a sphere. The result demonstrated in Figure 2a–c are for velocity, temperature, and mass profiles with the variations of Schmidt number keeping the remaining parameters constant at different circumferential positions of a sphere. It can be viewed that, as the Schmidt number is increased at the considered positions of a sphere, that is X = 0.1, 1.0, 2.0, and 3.0, velocity and mass profiles go down, but the opposite behavior is observed in the temperature field. In addition, it is necessary to mention that maximum magnitude for velocity distribution is achieved at position X = 1.0, but for temperature and mass concentration, it is obtained at X = 3.0. In these graphs, the simultaneous momentum and mass diffusion convection processes have been highlighted very clearly. Figure 3a,b depicts the results for velocity, temperature field, and mass concentration corresponding to increasing values of heat generation parameter Q and the remaining parameters treated as fixed at several stations of a sphere. We can see that the temperature and mass distributions have decreasing behavior, but the opposite phenomenon is observed in the velocity distribution. One aspect which is necessary to highlight is that very minor variations are observed for temperature and velocity fields, but a reasonable change is found in mass distribution at the taken circumferential positions of a sphere. From these graphs, it is evident that the heat generation parameter balances the heat transfer mechanism in the fluid flow domain. Figure 4a–c represents the behavior of the aforesaid physical properties for different values of the Prandtl number Pr. The outcomes shown in Figure 4a–c imply that, owing to the enhancement of Prandtl number at different circumferential locations of a sphere, a decrease in mass and velocity distributions, but an increase in temperature distribution, are noted. It is necessary to mention that the highest magnitude for velocity is gained at circumferential points X = 1.0, while on the other hand, mass and temperature distribution secure the peak value at position X = 3.1. As the Prandtl number controls the relative thickness of the momentum and thermal boundary layer, when Pr is small, the heat diffuses quickly as compared to the velocity. The effects of Brownian motion parameters on the physical properties mentioned earlier are presented in Figure 5a–c. It is noteworthy to point out that the augmentation in the Brownian motion parameter gives birth to a rise in mass distribution, but no remarkable variations are noted in the temperature and velocity fields. Figure 6a–c highlight the outcomes of profiles of velocity, temperature, and mass concentration under the action of diverse values of magnetic field parameter. It can be noticed that fluid velocity slows down as magnetic field parameter M is increased from 0.2 to 0.8 at each contemplated point around our proposed geometry and the temperature profile and mass concentration get smaller magnitudes for the same values of the parameters and positions. It is a point of interest that top values for flow velocity are maintained at X = 1.0 and for temperature and mass distribution, the highest values are gained at position X = 3.0. Variations in fluid velocity, temperature field, and mass concentration for increasing values of the thermophoresis parameter are demonstrated in Figure 7a–c. Very profound results are determined for all proposed properties. It is worthy to mention that the velocity of the fluid and temperature are reduced for increasing values of thermophoresis parameter Nt at the proposed positions about the surface of a sphere. On the other hand, for the same parametric conditions, mass concentration is enhanced. In Figure 8a–c, heat and mass transfer rates with skin friction are plotted. Interestingly, it can be seen that the heat transfer rate grows well, but mass transfer and skin friction become weaker at every contemplated circumferential point of a sphere. Similar properties as discussed earlier are taken under discussion and displayed in Figure 9a–c. Benchmark results for velocity, temperature, and solutal gradients for different values of Prandtl number have been studied at various stations of a sphere. There is a reduction in skin friction and mass transfer, but an increment in heat transfer rate is noted. The results tabulated in Table 1 represents skin friction, heat transfer rate and mass transfer for varying values of Brownian motion parameter Nb. The outcomes in Table 1 imply that skin friction get reduced whereas heat and mass transfer rates go up as Nb is augmented at the proposed stations of a sphere. Table 2 is reflecting the influences of magnetic field parameter on aforementioned material properties. By making larger the values of magnetic field parameter all contemplated material properties get declined. Further, it is concluded that greatest magnitudes for skin friction, rate of heat transfer and mass transfer rate are assured at positions X = 2.0, X = 1.0, and X = 1.0, respectively. Heat generation effects on velocity gradient, heat transfer rate and mass transfer rate are illustrated in Table 3. We can claim from the displayed results that skin friction and mass transfer enhance, but the converse phenomenon occurred for the case of heat transfer. In Table 4, the impact of Schmidt number is shown. Tabulated results show that skin friction falls down, but the mass transfer rate and heat transfer rate are augmented.

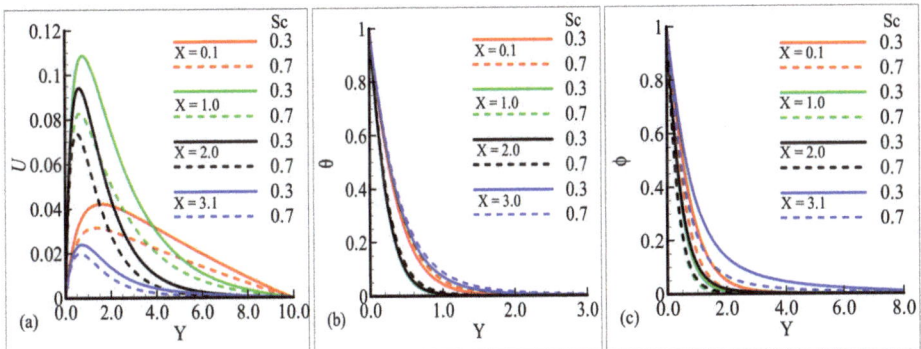

Figure 2. Physical effects on quantities (**a**) U, (**b**) θ, and (**c**) φ versus Sc when Nt = 0.02, Nb = 0.2, Pr = 0.72, M = 0.2, Q = 1.0.

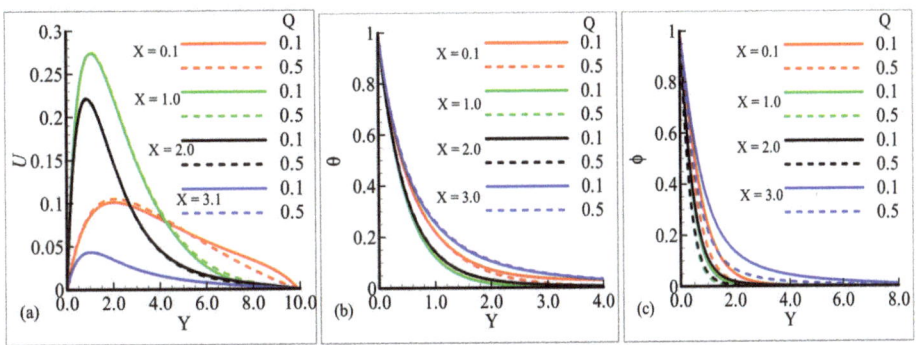

Figure 3. Physical effects on quantities (**a**) U, (**b**) θ, and (**c**) φ versus Q, when Nt = 0.02, Nb = 0.4, Pr = 0.72, M = 0.2, Sc = 1.0.

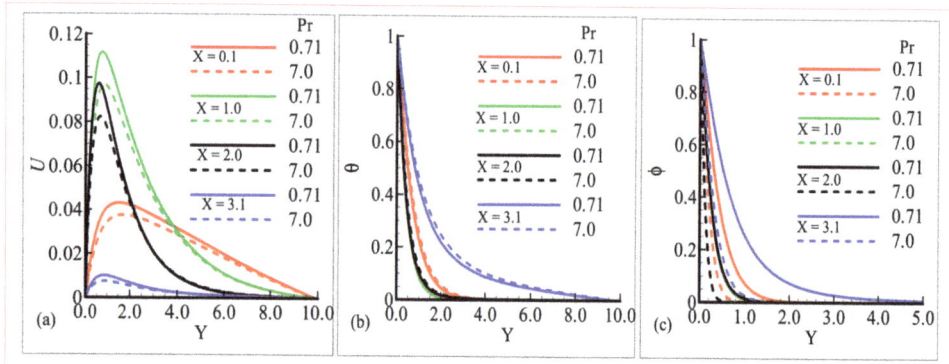

Figure 4. Physical effects on quantities (**a**) U, (**b**) θ, and (**c**) φ versus Pr, when Nt = 0.02, Nb = 0.4, SC = 1.0, M = 0.2, Q = 1.0.

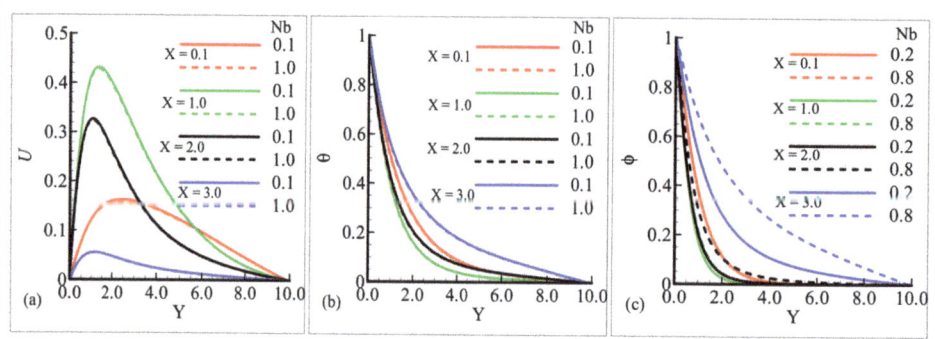

Figure 5. Physical effects on quantities (**a**) U, (**b**) θ, and (**c**) φ versus Nb, when Nt = 0.02, Sc = 1.0, Pr = 7.0, M = 0.2, Q = 1.0.

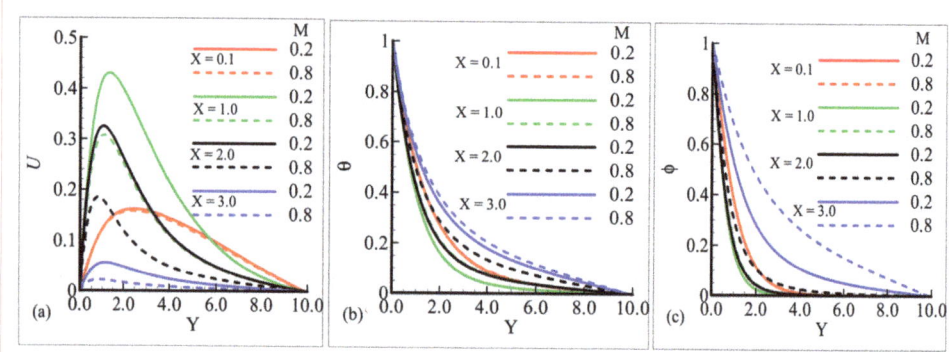

Figure 6. Physical effects on quantities (**a**) U, (**b**) θ, and (**c**) φ versus M, when Nt = 0.02, Nb = 0.4, Pr = 7.0, Sc = 10, Q = 1.0.

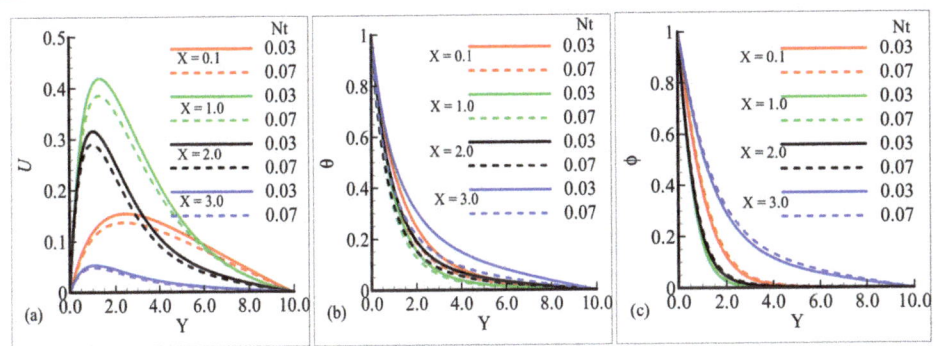

Figure 7. Physical effects on quantities (**a**) U, (**b**) θ, and (**c**) φ versus Nt, when Sc = 1, Nb = 0.4, pr = 7.0, M = 0.2, Q = 1.0.

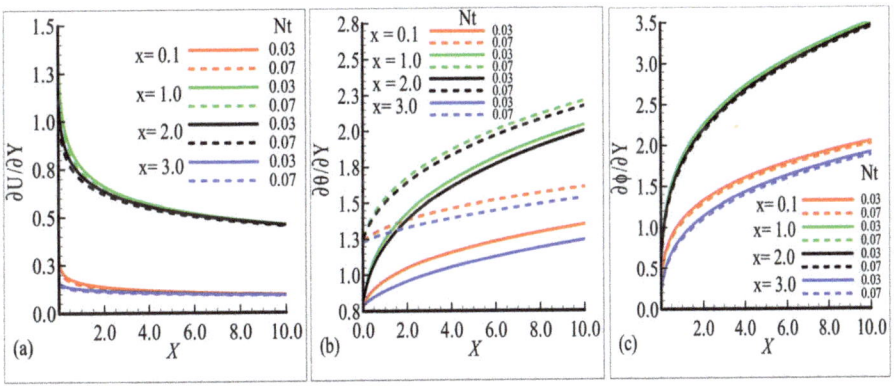

Figure 8. Physical effects on quantities (**a**) $\frac{\partial U}{\partial Y}$, (**b**) $\frac{\partial \theta}{\partial Y}$, and (**c**) $\frac{\partial \phi}{\partial Y}$ versus, Nt when Sc = 1.0, Nb = 0.4, Pr = 7.0, M = 0.2, Q = 1.0.

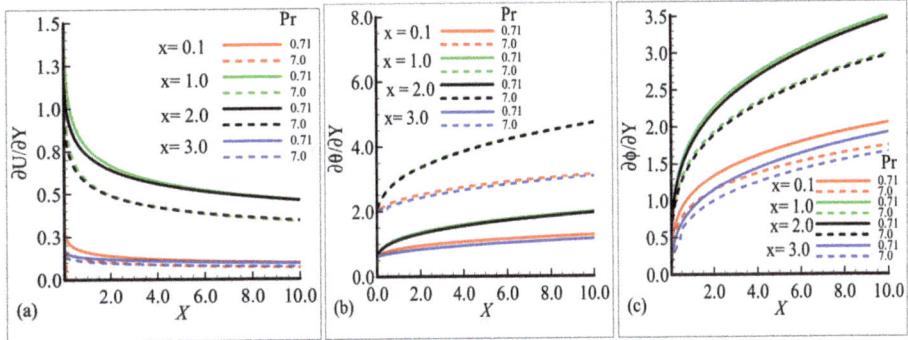

Figure 9. Physical effects on quantities (**a**) $\frac{\partial U}{\partial Y}$, (**b**) $\frac{\partial \theta}{\partial Y}$, and (**c**) $\frac{\partial \phi}{\partial Y}$ versus, Pr when Sc = 1.0, Nb = 0.4, Pr = 0.2, M = 0.2, Q = 1.0.

Table 1. Physical effects on quantities $\frac{\partial U}{\partial Y}$, $\frac{\partial \theta}{\partial Y}$, and $\frac{\partial \phi}{\partial Y}$ versus Nb, when remaining emerging parameters are constant.

	$\left(\frac{\partial U}{\partial Y}\right)$		$\left(\frac{\partial \theta}{\partial Y}\right)$		$\left(\frac{\partial \phi}{\partial Y}\right)$	
X	Nb = 0.1	Nb = 1.0	Nb = 0.1	Nb = 1.0	Nb = 0.1	Nb = 1.0
0.1	0.28019	0.27897	0.65136	0.66346	0.39999	0.40399
1.0	1.34747	1.34216	0.70666	0.72635	0.67801	0.68116
2.0	1.16031	1.15575	0.68683	0.70445	0.61784	0.62109
3.0	0.16315	0.16272	0.63540	0.64245	0.23270	0.23778

Table 2. Physical effects on quantities $\frac{\partial U}{\partial Y}$, $\frac{\partial \theta}{\partial Y}$, and $\frac{\partial \phi}{\partial Y}$ versus M, when remaining emerging parameters are constant.

	$\left(\frac{\partial U}{\partial Y}\right)$		$\left(\frac{\partial \theta}{\partial Y}\right)$		$\left(\frac{\partial \phi}{\partial Y}\right)$	
X	M = 0.2	M = 0.8	M = 0.2	M = 0.8	M = 0.2	M = 0.8
0.1	0.27951	0.27693	0.65533	0.65499	0.40432	0.40164
1.0	1.34503	1.08100	0.71309	0.68896	0.68082	0.60477
2.0	1.15823	0.57759	0.69259	0.66098	0.62072	0.45414
3.0	0.16290	0.09249	0.63772	0.63199	0.23703	0.15057

Table 3. Physical effects on quantities $\frac{\partial U}{\partial Y}$, $\frac{\partial \theta}{\partial Y}$, and $\frac{\partial \phi}{\partial Y}$ versus Q, when remaining emerging parameters are constant.

	$\left(\frac{\partial U}{\partial Y}\right)$		$\left(\frac{\partial \theta}{\partial Y}\right)$		$\left(\frac{\partial \phi}{\partial Y}\right)$	
X	Q = 0.1	Q = 0.5	Q = 0.1	Q = 0.5	Q = 0.1	Q = 0.5
0.1	0.27647	0.27777	0.69830	0.67952	0.40154	0.40235
1.0	1.33468	1.33918	0.75128	0.73451	0.67824	0.67935
2.0	1.14874	1.15285	0.73271	0.71513	0.61789	0.61911
3.0	0.16161	0.16216	0.68246	0.66294	0.23523	0.23600

Table 4. Physical effects on quantities $\frac{\partial U}{\partial Y}$, $\frac{\partial \theta}{\partial Y}$, and $\frac{\partial \phi}{\partial Y}$ versus Sc, when remaining emerging parameters are constant.

	$\left(\frac{\partial U}{\partial Y}\right)$		$\left(\frac{\partial \theta}{\partial Y}\right)$		$\left(\frac{\partial \phi}{\partial Y}\right)$	
X	Sc = 0.3	Sc = 0.7	Sc = 0.3	Sc = 0.7	Sc = 0.3	Sc = 0.7
0.1	0.29190	0.28399	0.65416	0.65502	0.24685	0.34990
1.0	1.42701	1.37282	0.71728	0.71439	0.42843	0.59370
2.0	1.23893	1.18350	0.69594	0.69366	0.37666	0.53863
3.0	0.16793	0.16482	0.63709	0.63748	0.14962	0.20333

6.2. Fluxes and Boundary Layers in the Plume Region

The present subsection deals with the analysis and demonstration of the numerical solutions of the flow model developed for the case of the plume region which occurs above the sphere. Figure 10a–c illustrate the temperature profile and nanoparticles volume fraction profile for various values of Schmidt number in the plume region. All the other parametric values are fixed. We can deduce from the figures that, as Sc is augmented from 0.3 to 0.9, velocity is lowered whereas temperature and nanoparticles volume fraction profiles curves go up. It was expected that the nanoparticles volume fraction will rise corresponding to an increase in Schmidt number Sc. The influences of thermophoresis parameter Nt, on the material properties are highlighted in Figure 11a–c. Computed results are reflecting that velocity of the flow field gets enhanced but temperature and nanoparticles volume fraction decline as Nt is increased. Graphical representations in Figure 12a–c are for the same substantial properties under the influence of several values of Brownian motion parameter Nb. It can be inferred from the displayed results that the velocity of the flow field goes down, nanoparticles volume fraction distribution goes up, but no variations are seen in the temperature field. Heat generation impacts by taking its several values on the conduct of matter properties, such as velocity, temperature, and nanoparticles volume fraction profiles, are examined in Figure 13a,b. From the sketched graphs, it is inferred that velocity and temperature field get larger magnitudes with the reduction in nanoparticles volume fraction by the augmentation of heat generation parameter Q. The results according to expectation satisfy the given boundary conditions and approach to the targets asymptotically. The graphs in Figure 14a–c are sketched for many values of magnetic field parameter M. It can be deduced from these plots that flow velocity and nanoparticles volume fraction rise, but no difference is found in the temperature field. The effects of various values of Prandtl number Pr on the already mentioned material properties are elaborated graphically in Figure 15a–c. We can see that velocity and temperature distributions decrease but nanoparticles volume fraction increases owing to increasing values of Pr. It was obvious that there is a reduction in field velocity and temperature profile.

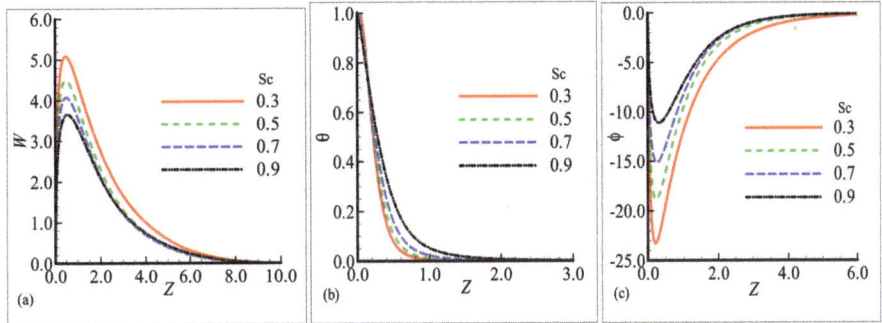

Figure 10. Physical effects on quantities (**a**) u, (**b**) θ, and (**c**) φ versus, Sc, when $Nt = 0.5$, $Nb = 0.4$, $Pr = 0.71$, $M = 0.5$, $Q = 0.4$.

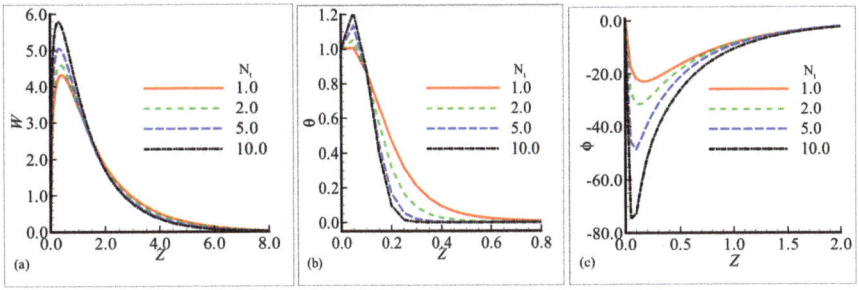

Figure 11. Physical effects on quantities (**a**) u, (**b**) θ, and (**c**) φ versus Nt, when $Sc = 0.8$, $Nb = 0.4$, $Pr = 0.71$, $M = 0.4$, $Q = 0.2$.

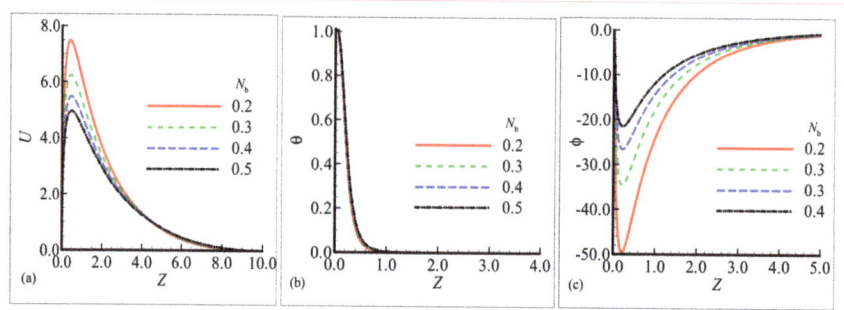

Figure 12. Physical effects on quantities (**a**) u, (**b**) θ, and (**c**) φ versus Nb, when Nt = 0.5, Sc = 0.3, Pr = 0.71, M = 0.4, Q = 0.4.

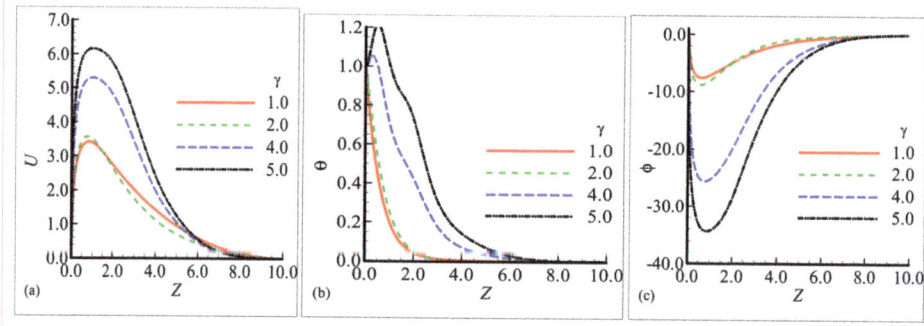

Figure 13. Physical effects on quantities (**a**) u, (**b**) θ, and (**c**) φ versus Q, when Nt = 0.1, Nb = 0.4, Pr = 0.71, M = 0.2, Sc = 0.2.

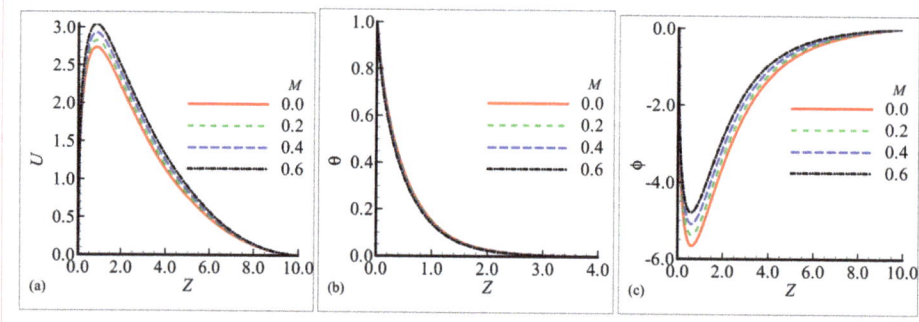

Figure 14. Physical effects on quantities (**a**) u, (**b**) θ, and (**c**) ϕ versus, when Nt = 0.1, Nb = 0.4, Pr = 0.71, Q = 0.2, Sc = 0.2.

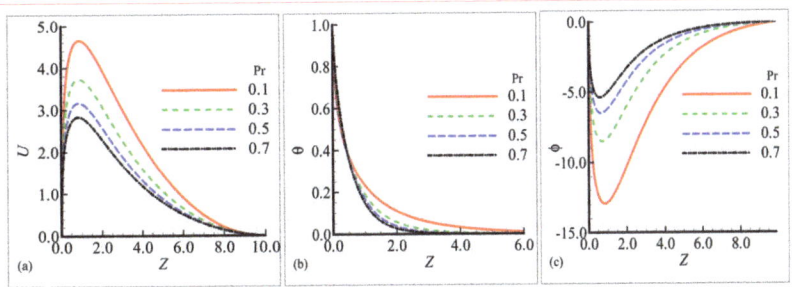

Figure 15. Physical effects on quantities (**a**) u, (**b**) θ, and (**c**) ϕ versus Pr, when Nt = 0.1, Nb = 0.4, Q = 0.2, M = 0.2, Sc = 0.2.

7. Conclusions

The phenomena of steady laminar natural convection nanofluid flow around the surface of a sphere and in the plume region are numerically examined under the impact of heat generation and an applied magnetic field. We summarize the obtained results in the following lines.

- The Schmidt number exerted a noticeable influence on the heat and fluid flow mechanism around the surface of the sphere and in the plume region in terms of velocity profile, temperature profile, and mass concentration.
- It was found that, under the action of diverse values of magnetic field parameter, the velocity of fluid slows down as the magnetic field parameter increased from 0.2 to 0.8, while on the other hand, the temperature profile and mass concentration became smaller in magnitude for the same values of the parameters and positions.
- The effect of thermophoresis parameter N_t cannot be neglected, as due to an increase in the magnitude of this parameter, the velocity profile is maximum at position $X = 1.0$, while the heat and mass transfer are reduced at the same position.
- The variation in the Brownian motion parameter N_b results in distinct changes in the thermal and flow field, depending on different positions around the surface of the sphere and in the plume region. The Brownian motion demonstrated its increasing effect for velocity at position $X = 1.0$, and dominated at the same position for temperature distribution and mass concentration.
- The graphs sketched in the plume region for many values of magnetic field parameter M show that the flow velocity and nanoparticles volume fraction rise, but no difference is found in the temperature field. The numerical solution obtained around the sphere reflects the influence of the magnetic field parameter on the aforementioned material properties, i.e., increasing the values of the magnetic field parameter.
- The Prandtl number Pr largely effects the fluid and thermal characteristics in the prescribed domain of study.
- From the obtained results of velocity profile, temperature distribution, and mass concentration around the sphere and within the plume region, it is observed that all the results are satisfied by the subjected boundary conditions.

Author Contributions: Conceptualization, A.K. and M.A.; Supervision, M.A. and H.A.N.; Investigation, A.N. and A.M.R.; Methodology, A.K., M.A. and H.A.N.; Writing—original draft, A.K., M.A. and H.A.N.; Writing—review & editing, M.A., A.M.R. and H.A.N.; Software, A.K., M.A. and A.M.R.; Formal Analysis, A.K., M.A., A.M.R. and H.A.N. All authors have read and agreed to the published version of the manuscript.

Funding: This research received no external funding.

Conflicts of Interest: The authors declare no conflict of interest.

Abbreviations

$\hat{u}$	Dimensioned velocity component in $\hat{x}$ direction
$\hat{v}$	Dimensioned velocity component in $\hat{y}$ direction
$\hat{W}$	Dimensioned velocity component in $\hat{Z}$ direction
$\hat{x}, \hat{y}$	Dimensioned axes along and normal to the surface of a sphere
$\hat{Z}$	Measured radially from the plume axis
U	Primitive variable for velocity component in X direction
V	Primitive variable for velocity component in Y direction
$g\,(ms^{-2})$	Gravitational acceleration
$T\,(K)$	Fluid temperature in boundary layer
$C(kgm^{-3})$	Mass concentration in boundary layer
$C_P(Jkg^{-1}.K^{-1})$	Specific heat at constant pressure
$a(m)$	The radius of a sphere
$\bar{r}(m)$	Dimensioned radial distance from the symmetric axis to the surface of a sphere
$D_m(m^2s^{-1})$	Mass diffusion cefficient

Greek Symbols

$\beta_t(K^{-1})$	Volumetric coefficient thermal expansion
τ	Thermophoretic diffusion coefficient
$\beta_c(K^{-1})$	Volumetric coefficient concentration expansion
$\alpha(ms^{-1})$	Thermal diffusivity
θ	Dimensionless temperature
ϕ	Dimensionless mass concentration
$\mu(Pa.s)$	Dynamic viscosity
ρ_p	Density of the particle
$\nu(m^2s^{-1})$	Kinematic viscosity
$\rho(kgm^{-3})$	Fluid density
$\kappa(Wm^{-1}.K^{-1})$	Thermal conductivity

Subscripts

∞	Ambient conditions
w	Wall conditions

References

1. Sparrow, E.M.; Cess, R.D. The effect of the magnetic field on free convection heat transfer. *Int. J. Heat Mass Transfer.* **1961**, *3*, 267–274.
2. Potter, J.M.; Riley, N. Free convection from a heated sphere at large Grashof number. *J. Fluid Mech.* **1980**, *100*, 769–783. [CrossRef]
3. Riley, N. The heat transfer from a sphere in free convective flow. *Comput. Fluids.* **1986**, *14*, 225–237.
4. Andersson, H.I. MHD flow of a viscoelastic fluid past a stretching surface. *ActaMechanica* **1992**, *95*, 227–230.
5. Choi, S.U.S.; Eastman, J.A. *Enhancing Thermal Conductivity of Fluids with Nanoparticles*; Argonne National Lab.: DuPage County, IL, USA, 1995. Available online: https://www.osti.gov/servlets/purl/196525 (accessed on 20 October 2020).
6. Adesanya, S.O.; Falade, J.A. Hydrodynamic stability analysis for variable viscous fluid flow through a porous medium. *Int. J. Differ. Equ.* **2014**, *13*, 219–230.
7. El-Kabeir, S.M.M.; Rashad, A.M.; Gorla, R.S.R. Unsteady MHD combined convection over a moving vertical sheet in a fluid saturated porous medium with uniform surface heat flux. *Math. Comp. Model* **2007**, *46*, 384–397. [CrossRef]
8. Chamkha, A.J.; Aly, A.M. MHD free convection flow of a nanofluid past a vertical plate in the presence of heat generation or absorption effects. *Chem. Eng. Commun.* **2010**, *198*, 425–441.
9. Kuznetsov, A.V.; Nield, D.A. Double-diffusive natural convective boundary-layer flow of nanofluids past a vertical plate. *Int. J. Therm. Sci.* **2011**, *50*, 712–717.

10. Rosmila, A.B.; Kandasamy, R.; Muhaimin, I. Lie symmetry group transformation for MHD natural convection flow of nanofluid over linearly porous stretching sheet in presence of thermal stratification. *Appl. Math. Mech.* **2012**, *33*, 593–604. [CrossRef]
11. Uddin, M.J.; Khan, W.A.; Ismail, A.I. MHD free convective boundary layer flow of a nanofluid past a flat vertical plate with Newtonian heating boundary condition. *PLoS ONE* **2012**, *7*, e49499.
12. Gangadhar, K.; Reddy, B. Chemically reacting MHD boundary layer flow of heat and mass transfer over a moving vertical plate in a porous medium with suction. *J. Appl. Fluid Mech.* **2013**, *6*, 107–114.
13. Makinde, O.D.; Khan, W.A.; Khan, Z.H. Buoyancy effects on MHD stagnation point flow and heat transfer of a nanofluid past a convectively heated stretching/shrinking sheet. *Int. J. Heat Mass Transfer.* **2013**, *62*, 526–533. [CrossRef]
14. Olanrewaju, A.M.; Makinde, O.D. On boundary layer stagnation point flow of a nanofluid over a permeable flat surface with Newtonian heating. *Chem. Eng. Commun.* **2013**, *200*, 836–852. [CrossRef]
15. Chamkha, A.J.; Jena, S.K.; Mahapatra, S.K. MHD convection of nanofluids: A review. *J. Nanofluids* **2015**, *4*, 271–292. [CrossRef]
16. Zaimi, K.; Ishak, A. Stagnation-point flow towards a stretching vertical sheet with slip effects. *Mathematics* **2016**, *4*, 27. [CrossRef]
17. Itu, C.; Öchsner, A.; Vlase, S.; Marin, M.I. Improved rigidity of composite circular plates through radial ribs. *Proc. Inst. Mech. Eng., Part L J. Mater. Des. Appl.* **2019**, *233*, 1585–1593. [CrossRef]
18. Ashraf, M.; Khan, A.; Gorla, R.S.R. Natural convection boundary layer flow of nanofluids around different stations of the sphere and into the plume above the sphere. *Heat Transf. Asian Res.* **2019**, *48*, 1127–1148. [CrossRef]
19. Sheikholeslami, M.; Abelman, S.; Ganji, D.D. Numerical simulation of MHD nanofluid flow and heat transfer considering viscous dissipation. *Int. J. Heat Mass Transf.* **2014**, *79*, 212–222. [CrossRef]
20. Qureshi, M.Z.A.; Rubbab, Q.; Irshad, S.; Ahmad, S.; Aqeel, M. Heat and mass transfer analysis of mhd nanofluid flow with radiative heat effects in the presence of spherical au-metallic nanoparticles. *Nanoscale Res. Lett.* **2016**, *11*, 472. [CrossRef]
21. Myers, T.G.; Ribera, H.; Cregan, V. Does mathematics contribute to the nanofluid debate? *Int. J. Heat Mass Transf.* **2017**, *111*, 279–288. [CrossRef]
22. Tlili, I.; Khan, W.A.; Ramadan, K. MHD flow of nanofluid flow across horizontal circular cylinder: Steady forced convection. *J. Nanofluids* **2019**, *8*, 179–186. [CrossRef]
23. Khan, W.A.; Rashad, A.M.; Abdou, M.M.M.; Tlili, I. Natural bioconvection flow of a nanofluid containing gyrotactic microorganisms about a truncated cone. *Eur. J. Mech.-B/Fluids* **2019**, *75*, 133–142. [CrossRef]
24. Parida, S.K.; Mishra, S.R. Heat and mass transfer of MHD stretched nanofluids in the presence of chemical reaction. *J. Nanofluids* **2019**, *8*, 143–149. [CrossRef]
25. Khan, I.; Alqahtani, A.M. MHD nanofluids in a permeable channel with porosity. *Symmetry* **2019**, *11*, 378. [CrossRef]

Publisher's Note: MDPI stays neutral with regard to jurisdictional claims in published maps and institutional affiliations.

© 2020 by the authors. Licensee MDPI, Basel, Switzerland. This article is an open access article distributed under the terms and conditions of the Creative Commons Attribution (CC BY) license (http://creativecommons.org/licenses/by/4.0/).

Article

Numerical Investigation of Aerodynamic Drag and Pressure Waves in Hyperloop Systems

Thi Thanh Giang Le [1], Kyeong Sik Jang [1], Kwan-Sup Lee [2] and Jaiyoung Ryu [1,3,*]

1. Department of Mechanical Engineering, Chung-Ang University, Seoul 06911, Korea; giangletthanh95@cau.ac.kr (T.T.G.L.); jks2620@cau.ac.kr (K.S.J.)
2. Hyper Tube Express (HTX) Research Team, Korea Railroad Research Institute, Gyeonggi-do 16105, Korea; kslee@krri.re.kr
3. School of Intelligent Energy and Industry, Chung-Ang University, Seoul 06911, Korea
* Correspondence: jairyu@cau.ac.kr; Tel.: +82-2-820-5279

Received: 7 September 2020; Accepted: 2 November 2020; Published: 6 November 2020

Abstract: Hyperloop is a new, alternative, very high-speed mode of transport wherein Hyperloop pods (or capsules) transport cargo and passengers at very high speeds in a near-vacuum tube. Such high-speed operations, however, cause a large aerodynamic drag. This study investigates the effects of pod speed, blockage ratio (BR), tube pressure, and pod length on the drag and drag coefficient of a Hyperloop. To study the compressibility of air when the pod is operating in a tube, the effect of pressure waves in terms of propagation speed and magnitude are investigated based on normal shockwave theories. To represent the pod motion and propagation of pressure waves, unsteady simulation using the moving-mesh method was applied under the sheer stress transport $k-\omega$ turbulence model. Numerical simulations were performed for different pod speeds from 100 to 350 m/s. The results indicate that the drag coefficient increases with increase in BR, pod speed, and pod length. In the Hyperloop system, the compression wave propagation speed is much higher than the speed of sound and the expansion wave propagation speed that experiences values around the speed of sound.

Keywords: Hyperloop system; transonic speed; aerodynamic drag; drag coefficient; pressure wave; shockwave

1. Introduction

Hyperloop is an innovative transportation system first outlined in a 2013 white paper by a joint team from Tesla Inc. and SpaceX Corp. As described in the Hyperloop Alpha document, it is a new, alternative, very high-speed mode of transport with benefits in terms of comfort, convenience, time, and cost [1,2]. The Hyperloop system consists of Hyperloop pods (or capsules) transporting cargo and passengers at very high speeds in a near-vacuum tube. The basic concept of the Hyperloop is similar to the evacuated tube transportation system proposed by Oster [3] in 1977. Estimates suggest that the trip between Los Angeles and San Francisco of roughly 350 miles, which usually takes 2.5 h by a high-speed train, would take only 35 min by the Hyperloop system (estimated time at average speeds of approximately 1000 km/h and 240 km/h for Hyperloop pod and high-speed train, respectively) [1,4]. This high-speed pod–tube configuration could be faster and more energy-efficient than trains or cars, as well as cheaper and less polluting than aircraft. SpaceX founder Elon Musk and his team called it the fifth form of public transportation after planes, trains, cars, and boats [1].

There are two big differences between Hyperloop and traditional rail. First, the pods are designed to float on air bearings or by magnetic levitation to reduce friction, unlike trains or cars. Second, the pods (or capsules) transport passengers through tubes in which most of the air is evacuated to reduce air resistance. This should allow the pods to travel at approximately 1250 km/h (or nearly 350 m/s), which is 3–4 times higher than the fastest high-speed train at present (350–380 km/h). Such high-speed operation, however, would cause a larger aerodynamic drag. Addressing this challenge, the closed partial-vacuum tube could drastically lower the aerodynamic drag. Hence, the Hyperloop white paper proposed maintaining a constant pressure of 1/1000 atm (101.325 Pa) inside the tube. In addition, when the pod is operating at transonic speeds, the restricted air owing to air compressibility could lead to large variations in pressure at the front and rear of the pod. Hence, variation of pressure wave propagation should also be considered.

Oh et al. [5] conducted a large parametric study in steady state by applying a two-dimensional axisymmetric model to a compressible flow to investigate the effects of blockage ratio (BR), pod speed/length, and tube pressure/temperature on the aerodynamic drag in a Hyperloop system. The study suggested that these parameters strongly affected drag, except the pod length and tube temperature, which have negligible influences. However, the simulation in their study could not analyze pressure wave propagation. In steady-state simulation, the pressure waves produced by the nose and tail of the pod affect the inlet and outlet boundaries. This limitation can influence the accuracy of the results. Yang et al. [6] concluded that aerodynamic drag increased proportionally with the internal tube pressure and the square of the operating speed. Gillani et al. [7], Singh et al. (2019) [8], and Choi et al. [9] studied the relationship between the pod shape and the aerodynamic drag of the pod. They proposed that the elliptical train shape efficiently reduced the aerodynamic drag at the tube pressure of 1013.25 Pa. Zhang [10] carried out steady computational fluid dynamic (CFD) simulations and indicated that the increase of BR increased the aerodynamic drag. Similar conclusions were presented by Kang et al. [11], who performed parametric simulations of transonic vehicles in an evacuated tube. The authors implied that the BR and internal tube pressure strongly affected aerodynamic drag of transonic trains; the maximized drag coefficient was obtained at Mach 0.7. Kim et al. [12] concluded that the occurrence of shockwaves greatly increased the aerodynamic drag in the tube–train system.

The compressibility of air should be considered in cases where a pod travels through a tube at high speed. The motion of the pod generates a series of compression waves in the front. These pressure waves propagate forward and backward with the direction of the operating pod. Meanwhile, expansion waves propagate behind the pod. The propagation of compression waves intensifies the pressure ahead of the pod, whereas the expansion waves reduce the pressure behind the pod. The difference between the pressures at the nose and the rear of the pod increases as the pod speed increases, causing a sharp rise in pressure drag. A study by Oh et al. [5] observed choked flow in the Hyperloop system at a pod speed of 180 m/s and a BR of 0.36, which significantly increased the drag. The normal shockwave is similar to a step-change of the compression wave. The presence of shockwaves produces discontinuous changes in the flow parameters. The normal and oblique shockwaves in the rear of the pod strongly interfere with the aerodynamic characteristics of the pod–tube system. Hence, in this study, the pressure wave propagation speed is also evaluated.

The analysis of pressure wave propagation requires time-dependent CFD simulation. In addition, a moving overset mesh method was employed to represent the pod motion and determine the influence of a moving pod on the compression waves and expansion waves inside the tube. The overset mesh is a renowned method for moving-mesh simulations and has high accuracy and computational speed by reducing the re-meshing effects during the simulation.

Oh et al. showed that symmetric models of a semicircular nose and tail make the difference between three- and two-dimensional models insignificant (just 4%) [5]. Therefore, the two-dimensional axisymmetric model is sufficient for simulating the Hyperloop with an idealized geometrical shape. Therefore, in this study, a two-dimensional axisymmetric model was constructed. Moreover, the main objective of this study is to analyze the aerodynamic drag and pressure wave propagation, which can be fully described by two-dimensional simulations. The low computational cost of the two-dimensional model allowed us to consider a long tube and carry out a comprehensive parametric study.

2. Numerical Method

The simulation was conducted on ANSYS Fluent 18.1 (Ansys Inc., Canonsburg, PA, USA) with the shear stress transport (SST) $k - \omega$ viscous model. The compressible ideal gas condition was applied by using the density-based implicit solver. The viscosity was assumed as the Sutherland model, in which the viscosity varies only with temperature.

2.1. Computational Domain

The pressure waves propagate in both forward and backward directions through the tube. If an exceedingly short tube was used in the numerical simulations, the pressure waves would reflect off the boundaries and cause unpredictable pressure variations [9,13]. Hence, a sufficiently long tube was chosen to fully analyze the pressure wave propagation. The simulation was conducted in a two-dimensional axisymmetric model. The nose and tail of the Hyperloop pod were assumed to have idealized semicircular geometries.

The BR is determined by the following equation:

$$BR = \frac{\text{Cross} - \text{sectional area of pod}}{\text{Cross} - \text{sectional area of tube}} = \frac{d_{pod}^2}{d_{tube}^2} \quad (1)$$

where d_{pod} is the diameter of the pod and d_{tube} is the diameter of the tube.

The pod's dimensions were ø 3 m × 43 m. The BR of 0.36 gave a tube diameter of 5 m and the BR of 0.25 gave a tube diameter of 6 m. The length of the tube was 1200 m. The computational geometry and boundary conditions used in the simulation are shown in Figure 1. Figure A1 (Appendix A) informs that the designed domain used in this simulation is long enough for the pressure wave to fully develop without losses (reflections).

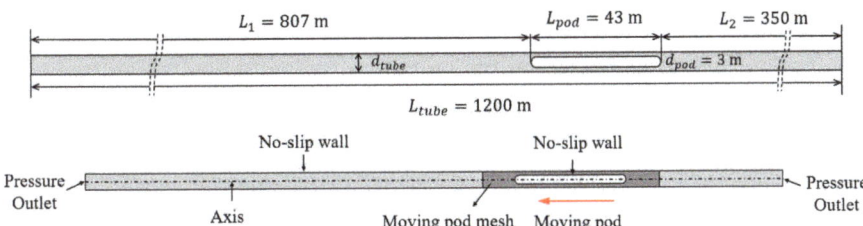

Figure 1. Geometry and boundary conditions of the simulation. $d_{tube} = 5$ m for BR = 0.36 and $d_{tube} = 6$ m for BR = 0.25.

2.2. Overset Meshing and Computational Grid

For conducting the moving-mesh simulation, the overset and dynamic mesh options were applied simultaneously. The overset mesh method allows multiple disconnected meshes to sufficiently overlap each other. Using the overset mesh method to perform a moving mesh helps achieve more efficient computation time and prevents re-meshing that otherwise reduces the accuracy by causing the generation of poor cells. Generally, the mesh used for the overset method consists of two parts: a background zone and a separate component zone. Background zones are the mesh of the off-body or fluid space. The component zones are meshes that contain the objects of analysis; they require overset boundaries. These meshes must be of high quality and should cover the solution domain [14]. The component zones overlay the background zones near the overset boundaries; near these regions, the background and other component zones unify into one zone. An advantage of overset meshing is that individual parts of the overset mesh are created independently, and hence any zone can be easily replaced without having to recreate the whole geometry [15].

In this study, two mesh zones were independently generated. The background zone is the tube mesh and the component zone is the pod mesh. Figure 2 presents the mesh and the generation of the overset mesh used in this study. The grids were generated on ANSYS ICEM 18.1. Both zones were treated with a hexahedral mesh. A finer mesh is created near the pod and the wall to ensure that the maximum y^+ is maintained around 0.5, except at the nose of the pod, where the highest y^+ values reach 1.5. Figure 3 shows the variation of the y^+ values around the pod surface and the tube wall for the highest pod speed case, in other words, 350 m/s. In this study, the maximum and minimum y^+ values were observed at the nose and tail of the pod, respectively.

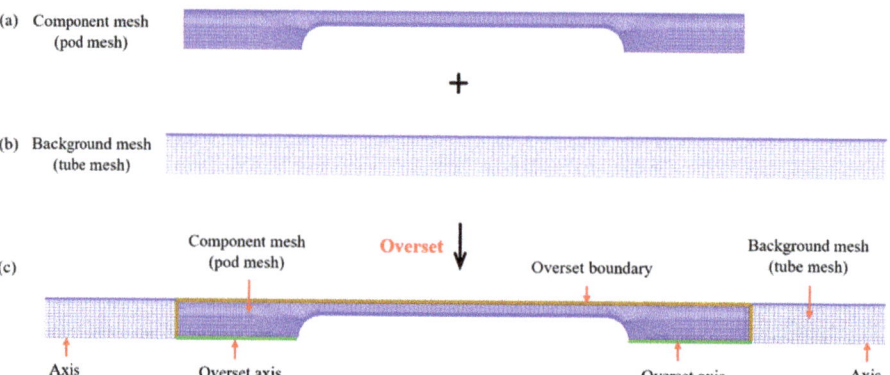

Figure 2. Schematic of the overset mesh generation. The number of meshes in this figure is only 1/10 of the number of final meshes. Yellow lines represent the overset boundaries where the pod mesh (**a**) and tube mesh connect (**b**). When two meshes overlap, the redundant mesh of the tube mesh vanishes as shown in (**c**).

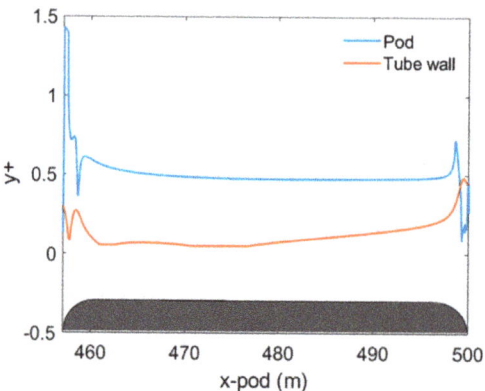

Figure 3. Variation of y^+ around the pod surface and tube wall at a pod speed of 350 m/s. The maximum y^+ was obtained at the nose of the pod.

The number of elements in the tube mesh was fixed at 979,951 cells. For the grid independence test, only the pod mesh was demonstrated by comparing the results of two main objectives of this study, in other words, drag force and pressure obtained with several different grids. The pod speed of 350 m/s and tube pressure of 101.325 Pa (1/1000 atm) were used to evaluate the mesh. Table 1 lists the independence test results of the grid computed from three meshes, in other words, coarse mesh (mesh 1), medium mesh (mesh 2), and fine mesh (mesh 3). The difference between the medium mesh and the fine mesh is 0.02% in total drag and 0.004% in pressure; hence, a fine mesh was unnecessary and mesh 2 was chosen. All simulations in this study were conducted with a mesh composed of 1,684,782 cells.

Table 1. Grid independent test: Only the pod zone is demonstrated. Drag force and maximum pressure are used to estimate the grid. D_p—pressure drag, D_f—friction drag, D_t—total drag, and P_{max}—maximum pressure. Mesh 2 is applied in all simulations. Total number of elements in this simulation is 1,684,782.

Case	Total Number of Cells (Pod Zone)	D_p (N)	D_f (N)	D_t (N)	P_{max} (Pa)
Mesh 1	638,067	1108.33 (0.13%)	134.62 (0.72%)	1242.95 (0.19%)	238.09 (1.34%)
Mesh 2	704,331	1106.89 (0.005%)	133.60 (0.26%)	1240.49 (0.02%)	241.28 (0.004%)
Mesh 3	792,607	1106.83	133.31	1240.20	241.27

2.3. Boundary Conditions

Figure 1 describes the boundary conditions of the simulation. For the tube, the pressure-outlet boundary condition was applied at the two tube exits with a constant value of 101.325 Pa (1/1000 atm). The pod and the tube walls are stationary walls with no-slip and adiabatic conditions. The pod was placed at a fixed position and instantly starts to move from right to left at a specified speed. Eight pod speeds were considered, from 100 to 350 m/s. Meanwhile, the pod-mesh zone also moved with the same speed and in the same direction as the pod.

The Reynolds number (Re) was calculated using the formula $Re = \rho v_P d_h / \mu$, where ρ, v_P, d_h, and μ are the reference air density, pod speed, hydraulic diameter, and viscosity, respectively. The hydraulic diameter $d_h = d_{tube} - d_{pod}$ was determined as 2 m for BR = 0.36 and 3 m for BR = 0.25. Hence, for the range of pod speed considered in this study, the Reynolds number Re ranged from 13×10^3 to 45×10^3 for BR = 0.36, which indicates turbulent flow. Table 2 shows the variation of the Reynolds number Re with respect to the increase in pod speed. As the pod speed increases, the Reynolds number Re also increases. As BR increases, the Reynolds number Re decreases.

Table 2. Variation of Reynolds number (Re) with respect to pod speed and BR.

Pod Speed (m/s)		100	150	200	225	250	275	300	350
Re	BR = 0.36	12,752	9128	25,504	28,692	31,880	35,068	38,256	44,632
	BR = 0.25	19,128	28,692	38,256	43,037	47,820	52,602	57,383	66,947

The simulations in this study were carried out under unsteady conditions. The variation of drag and the propagation of compression waves in Figure 4 show that the drag and compression waves tend to stabilize after 0.2 s. Owing to this stationary condition, one second of simulation time was enough to fully develop the flow field and investigate the behavior of the pressure waves. Therefore, in this study, we took a total simulation time of 1 s. The values of drag were estimated at 1 s, and results of pressure waves were exported after initial transients. In the steady-state condition, the pressure wave propagation cannot be examined. Furthermore, the pressure waves that arrive at the boundaries may alter the specified boundary conditions and affect drag evaluation. Hence, the unsteady-state simulation was used to explain the pressure wave phenomenon, as well as estimate the drag variation more accurately.

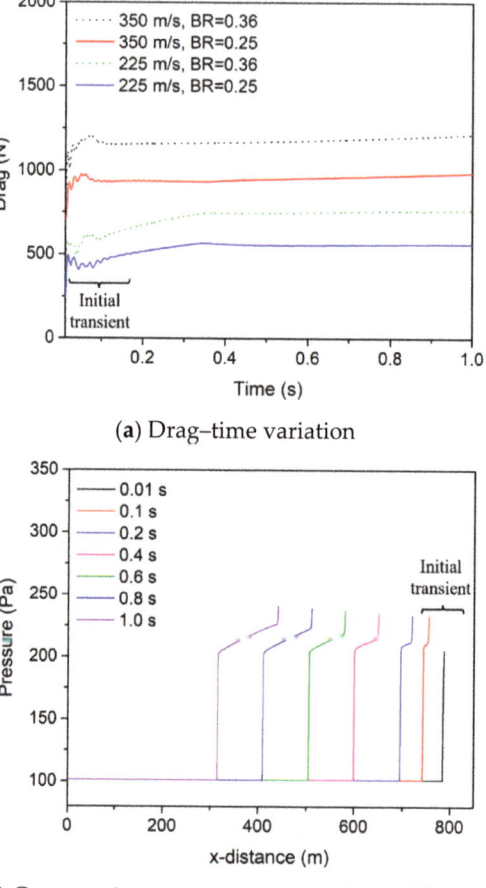

Figure 4. Variation of (**a**) drag and (**b**) compression wave propagation with respect to time in unsteady conditions. All results after the initial transients are analyzed.

2.4. Mathematical Model

The governing equations are the conservation of mass, momentum (Navier–Stokes), and energy equations, which are described as follows in that order:

$$\frac{\partial \rho}{\partial t} + \frac{\partial}{\partial x_i}(\rho u_i) = 0 \qquad (2)$$

$$\frac{\partial}{\partial t}(\rho u_i) + \frac{\partial}{\partial x_j}(\rho u_i u_j) = -\frac{\partial P}{\partial x_i} + \frac{\partial}{\partial x_j}\left[\mu\left(\frac{\partial u_i}{\partial x_j} + \frac{\partial u_j}{\partial x_i} - \frac{2}{3}\delta_{ij}\frac{\partial u_k}{\partial x_k}\right)\right] + \frac{\partial}{\partial x_j}\left(-\rho\{u'_i u'_j\}\right) \qquad (3)$$

$$\frac{\partial}{\partial t}(\rho E) + \frac{\partial}{\partial x_j}\left(u_j(\rho E + P)\right) = \frac{\partial}{\partial x_j}\left[(k_{eff})\frac{\partial T}{\partial x_j}\right] + \frac{\partial}{\partial x_j}\left[u_i \mu_{eff}\left(\frac{\partial u_i}{\partial x_j} + \frac{\partial u_j}{\partial x_i} - \frac{2}{3}\delta_{ij}\frac{\partial u_k}{\partial x_k}\right)\right] \qquad (4)$$

$$i, j, k = 1, 2, 3$$

Here, ρ is the fluid density, u is the fluid velocity, P is the fluid pressure, and μ is the fluid viscosity; E is the specific internal energy, k_{eff} is the effective thermal conductivity, and μ_{eff} is the effective dynamic viscosity.

To solve the governing equations, a density-based solver built using the finite volume method was employed. The implicit Roe's flux difference scheme was used as the spatial discretization scheme. The least-squares cell-based method was selected for gradient calculation. Furthermore, the flow was discretized using the second-order upwind scheme. The turbulent kinetic energy and specific dissipation rate were set according to the first-order upwind scheme. Although the first-order upwind scheme may not yield results of greater accuracy compared with the second-order upwind scheme, the first-order upwind scheme easily converges and minimizes the computational cost.

For the time integration, the first-order implicit scheme was used to process the unsteady simulation. The unsteady constant-time stepping with a time step size of 8×10^{-5} was adopted, and the maximum number of iterations per time step was selected as 20 to reach a residual convergence of 10^{-4}. Additionally, the Courant number was set at 2 in all the simulations.

A suitable turbulence model can enhance the precision and reliability of the CFD simulation. Direct numerical simulation (DNS) and large eddy simulation (LES) can effectively analyze the complex fluid-flow model, especially for the turbulence structure [16–20]. However, these turbulence models require a high computational cost. The Reynolds-averaged Navier–Stokes (RANS) model can balance the computational cost and performance efficiency of the numerical analysis. A few RANS models such as $k-\varepsilon$, $k-\omega$, SST γ, transition SST, et cetera. are widely used for compressible flow simulations. In this study, the pod operates in the transonic to supersonic speed regimes. Therefore, to specify the turbulent flow, the SST $k-\omega$ model was applied. This is a hybrid model combining the advantages of the Wilcox (standard) $k-\omega$ and the $k-\varepsilon$ models. The two variables, the turbulence kinetic energy $k\,(\mathrm{m}^2/\mathrm{s}^2)$ and the specific dissipation rate $\omega\,(\mathrm{s}^{-1})$, were respectively determined by the following two equations [15,21,22]:

$$\frac{\partial(\rho k)}{\partial t} + \frac{\partial(\rho u_j k)}{\partial x_j} = P_K - \beta^* \rho \omega k + \frac{\partial}{\partial x_j}\left[\left(\mu + \frac{\mu_t}{\sigma_k}\right)\frac{\partial k}{\partial x_j}\right] \tag{5}$$

$$\frac{\partial(\rho \omega)}{\partial t} + \frac{\partial(\rho u_j \omega)}{\partial x_j} = \frac{\gamma}{v_t} P_K - \beta \rho \omega^2 + \frac{\partial}{\partial x_j}\left[\left(\mu + \frac{\mu_t}{\sigma_\omega}\right)\frac{\partial k}{\partial x_j}\right] + 2\rho(1-F_1)\frac{1}{\sigma_{\omega,2}}\frac{1}{\omega}\frac{\partial k}{\partial x_j}\frac{\partial \omega}{\partial x_j} \tag{6}$$

where

$$P_K = \tau_{ij}\frac{\partial u_i}{\partial x_j}, \tag{7}$$

$$\tau_{ij} = 2\mu_t S_{ij} - \frac{2}{3}\rho k \delta_{ij}. \tag{8}$$

where τ_{ij} denotes the Reynolds stresses ($\mathrm{kg\,m^{-1}\,s^{-2}}$), S_{ij} denotes the mean rate of deformation component (s^{-1}), and δ_{ij} denotes the Kronecker delta function.

$$F_1 = \tanh\left\{\left\{\min\left[\max\left(\frac{\sqrt{k}}{\beta^*\omega y}, \frac{500v}{y^2\omega}\right), \frac{4\rho\sigma_{\omega,2}k}{CD_{k\omega}y^2}\right]\right\}^4\right\}, \tag{9}$$

$$CD_{k\omega} = \max\left(2\rho\frac{1}{\sigma_{\omega,2}}\frac{1}{\omega}\frac{\partial k}{\partial x_i}\frac{\partial \omega}{\partial x_j}, 10^{-10}\right), \tag{10}$$

$$\sigma_k = \frac{1}{F_1/\sigma_{k,1} + (1-F_1)/\sigma_{k,2}}, \tag{11}$$

$$\sigma_\omega = \frac{1}{F_1/\sigma_{\omega,1} + (1-F_1)/\sigma_{\omega,2}}. \tag{12}$$

σ_k and σ_ω are the turbulent Prandtl numbers for k and ω, respectively. μ_t (kg/ms) is the turbulence viscosity, which is calculated by the following equations:

$$\frac{1}{\rho}\mu_t = \frac{a_1 k}{\max(a_1\omega, SF_2)}, \tag{13}$$

$$F_2 = \tanh\left(\max\left(2\frac{\sqrt{k}}{\beta^*\omega y}, \frac{500\nu}{y^2\omega}\right)^2\right). \tag{14}$$

In Equation (13), the term $S = (2S_{ij}S_{ij})^{1/2}$ is the invariant measure of the strain rate, and a_1 is a constant equal to 0.31. The other constant values in the above equations are given as follows [15]:

$$\beta^* = 0.09,$$

$$\sigma_{k,1} = 2.0, \ \sigma_{k,2} = 1.0,$$

$$\sigma_{\omega,1} = 2.0, \ \sigma_{\omega,2} = 1.168.$$

This turbulence model has been widely used to simulate the aerodynamic characteristics of high-speed trains [6,8,10,23–25]; it has improved the accuracy and reliability of free shear flows and transonic shockwaves predictions [15]. For more details of this model, readers are directed to [6,21,22].

2.5. Validation

To ensure the stability of the numerical solution, the sphere tests of Charters and Thomas [26] were chosen for validation. Figure 5 shows a comparison of the drag coefficient obtained from our simulations and the work of Charters and Thomas, in which experiments were conducted with small spheres at various velocities. A 2D axisymmetric sphere was simulated with an unsteady moving mesh. The numerical method used for the validation was the same as those mentioned in Section 2.4. Figure 5 shows a similar trend in the drag coefficient between the two studies, indicating a good agreement. Therefore, our selected numerical method is appropriate to simulate the moving pod–tube configuration.

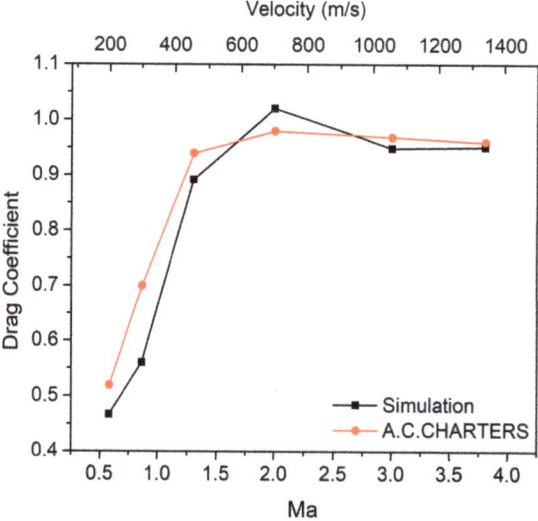

Figure 5. Comparison of drag coefficients between the 2D axisymmetric simulation and the experimental data for a 14.28-mm sphere reported by Charters and Thomas [26].

3. Results

Any object moving through a fluid will experience an aerodynamic drag that is produced by both pressure and shear forces acting on its surface.

$$D_t = D_p + D_f, \tag{15}$$

where D_t is the total drag, D_p is the pressure drag, and D_f is the friction drag. Pressure drag is strongly dependent on the shape or form of the object; friction drag is a function of the wall shear stress, which is affected by surface roughness and the Reynolds number [27]. The drag coefficient is defined by the following equations:

$$C_D = \frac{D_t}{\frac{1}{2}\rho v_P^2 A} \quad \text{for calculating total drag coefficient} \tag{16}$$

$$C_{D_p} = \frac{D_p}{\frac{1}{2}\rho v_P^2 A} \quad \text{for calculating pressure drag coefficient} \tag{17}$$

$$C_{D_f} = \frac{D_f}{\frac{1}{2}\rho v_P^2 A} \quad \text{for calculating friction drag coefficient} \tag{18}$$

where A is the frontal area, which is the cross-sectional area of the pod calculated by $\pi d_{pod}^2/4$. Thus, in this study, the reference area in Equations (16)–(18) is approximately 7.065 m². ρ is the reference density of air in the tube pressure and temperature of 300 K. v_P is the pod speed.

This study mainly focuses on the influence of various parameters on the aerodynamic drag. The amplitude and speed of pressure waves are also presented. The main variables considered here include BR, operating pod speed, internal tube pressure, and pod length. Owing to a fixed cross-sectional area of the pod, the variation in BR is obtained by changing the tube diameter to 6 m for BR = 0.25 (5 m for BR = 0.36). The pod speed was varied from 100 to 350 m/s in intervals of 50 m/s, in other words, 100, 150, 200, 250, 300, 350 m/s. For investigating the flows near the critical Mach number, two more pod speeds were chosen, namely, 225 and 275 m/s. With changing tube pressure, the pod speed of 300 m/s was selected to evaluate four different tube pressures, namely, 101.325, 500, 750, and 1013.25 Pa under a BR of 0.36 and a pod length of 43 m. The Hyperloop Alpha documents used a pod length of 43 m to carry 28 passengers per trip. To investigate the effect of pod length on drag coefficient, the case of L_{ref} = 43 m and two reference cases $L_{ref}/2$ and $2L_{ref}$ were selected.

3.1. Effect of Blockage Ratio and Pod Speed

As indicated in Figure 6, the changes of friction and pressure drag coefficient vary by two values of BR from v_P of 100 to 350 m/s. The variation of friction and pressure drag is given in insets. Besides, Figures A2 and A3 (Appendix B) also show the variation of friction and pressure drag coefficients on Re and Mach number. In Figure 6, at the same v_P, the friction drag coefficient C_{D_f} and the pressure drag coefficient C_{D_p} increase as the BR increases. When the BR is higher, the area where the flow passes through decreases. Thus, the flow becomes harder to bypass, increasing the friction drag generated along the pod surface and resulting in C_{D_f} increases as BR increases (Figure 6a). It should be noted that from the v_P of 200 m/s, the effect of BR on C_{D_f} reduces due to severe choking.

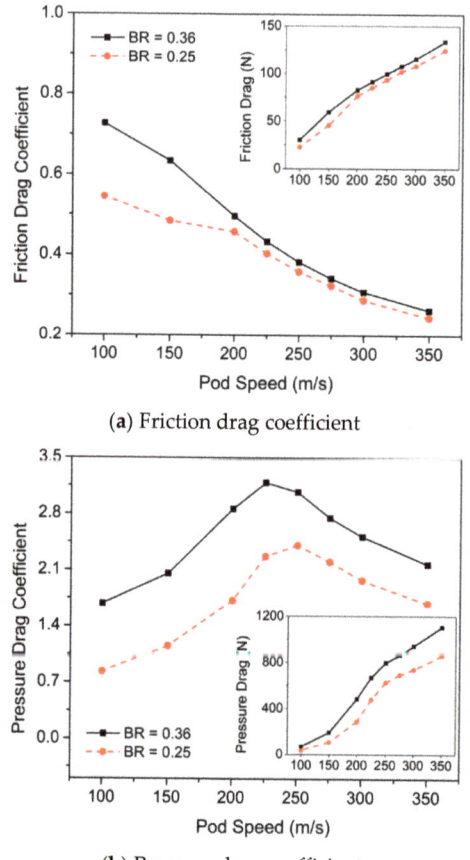

Figure 6. Variations of (**a**) friction and (**b**) pressure drag coefficients with pod speed and blockage ratio. Insets illustrate the variation of drag.

The change in pressure drag depends on the pressure difference between the nose and tail of the pod. With increasing BR, the air becomes more compressible, which increases the pressure magnitude along the pod surface as well as the pressure difference between the nose and the tail, resulting in an increase in pressure drag; hence C_{D_p} increases. With increasing v_P, pressure and friction drag increase continuously and significantly. Unlike drag, the drag coefficients witness different tendencies. As shown in Figure 6b, at BR = 0.36, C_{D_p} increases and reaches the maximum at a v_P of 225 m/s. Subsequently, it exhibits a continuous drop. By contrast, C_{D_f} presents an incessant decrease with the increment of v_P. The results show tendencies similar with those in previous literature [5,11,23,28]. Note that, at lower values of BR, the maximum value of C_{D_p} changes. In this study, at BR = 0.25, C_{D_p} reaches a maximum at a v_P of 250 m/s, which will be compared later with BR = 0.36.

Generally, aerodynamic drag is composed of pressure and friction drags. Unlike in the open air, the Hyperloop pod moves in a tunnel, experiencing more aerodynamic drag owing to the increase in pressure generated by its interaction with the tunnel walls [29]. At higher speed, there will be higher pressure. When the pod passes through the tube at a high speed, a high-pressure region is formed in the front of the pod nose. Meanwhile, the pod tail experiences an increase in the velocity of flow and reduces the pressure behind the pod. This phenomenon is similar to the behavior of flow through a convergent–divergent nozzle and results in a greater pressure difference between the nose and the tail of the pod; this leads to an increase in the pressure drag. Besides, the low pressure in the tube reduces the friction drag in the Hyperloop system. Hence, in the pod–tube system, pressure drag is more dominant than friction drag.

To evaluate the portions of pressure and friction drag in the total drag, the ratios of the two drag components to the total drag are shown in Figure 7. At BR = 0.36 and v_P = 100 m/s, pressure drag dominates the total drag by 70%; the remaining percentage belongs to friction drag. However, pressure drag becomes more dominant as v_P increases owing to the increase of the difference in pressure at the nose and tail; the friction drag does not change much. This results in the pressure drag becoming more dominant at v_P of 350 m/s, making up 89.2% of the total drag, whereas friction drag only makes up 10.8%. The portion of D_p/D_t decreases with reducing BR, whereas the portion of D_f/D_t increases. This is because the effect of BR on the pressure drag is much higher than on friction drag. As BR reduces from 0.36 to 0.25, and at v_P = 100 m/s, pressure drag decreases by more than 50%, whereas the friction decreases by approximately 25%. This gap is greater at v_P = 350 m/s, where the margins of decrease for pressure drag and friction drag are 22.4% and 6.9%, respectively. Note that the ratios of the pressure and friction drags to the total drag begin to converge from a v_P of 250 m/s.

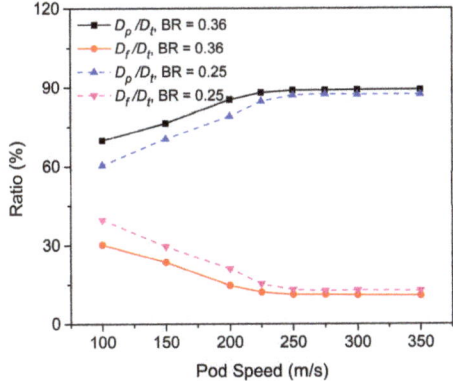

Figure 7. Ratios of the pressure and friction drags to the total drag. The results are similar to those of Oh et al. (2019) [5], who conducted steady-state simulations.

The pressure contours in Figure 8 show that the pressure in front of the pod increases with increase of BR. Figure 8d,e,j,k reveals shockwaves at the rear end of the pod; these shockwaves are made of normal and oblique shockwaves generated by the interaction and reflection of pressure waves between the tube walls and pod surface. When the local flow velocity exceeds the speed of sound, the interaction and connection of the airflow and pod surface create oblique shockwaves [28]. With increasing v_P, the shockwaves become more distinct, expand backward, are reflected by the upper and lower walls of the tube [16], and are then weakened downstream owing to the friction of the airflow. This oblique shockwave intensifies the airflow pressure at the rear end of the pod, resulting in an abrupt pressure increase at the tail (Figure 9a). Note that at lower BR, the formation of oblique shockwaves is delayed. The existence of shockwaves affects the variation of C_{D_p}. In this study, shockwaves are noticed before the v_P where C_{D_p} is maximized. At BR = 0.25, the shockwave structure can be observed from a v_P of 225 m/s (Figure 8c), and from 200 m/s at BR = 0.36 (Figure 8h). The previous study conducted by Kim et al. (2011) [12] also indicated that the impact of shockwaves was reduced when BR was reduced. Hence, to limit the effect of shockwaves, we considered the influence of BR [8,12].

With increasing BR, the percentage of pressure drag to total drag increases. To explain this phenomenon, Figure 9a illustrates the pressure difference between the nose and the tail of the pod, which is one of the factors affecting pressure drag. As the v_P increases, the pressure at the nose increases significantly, which is in contrast with the slight decrease in pressure at the tail. This difference causes a substantial increase in the pressure difference between the nose and the tail, resulting in a sharp increase in the pressure drag. This pressure difference in turn increases with the increase of v_P and BR. At the same v_P, the pressure at the nose with BR = 0.36 is higher than in the case when BR = 0.25 because of the generation of stronger compression waves. Otherwise, the tail pressure experiences a smaller pressure drop when the BR is smaller, because the air in the larger tube expands easily [11,12,30].

Figure 9b illustrates the pressure variation across the pod surface. There is a sudden increase in the pressure at the tail from a v_P of 250 m/s owing to the formation of strong oblique shockwaves. A noticeable increase in the surface pressure occurs from 200 to 250 m/s, where the choking flow becomes severe. Note that, although an increment in the surface pressure was observed from 100 to 250 m/s, the pressure magnitude at the tail exhibited a considerable drop owing to the stronger flow expansion with increasing v_P. As oblique shockwaves appear at the tail from a v_P of 250 m/s, the tail pressure slightly increases.

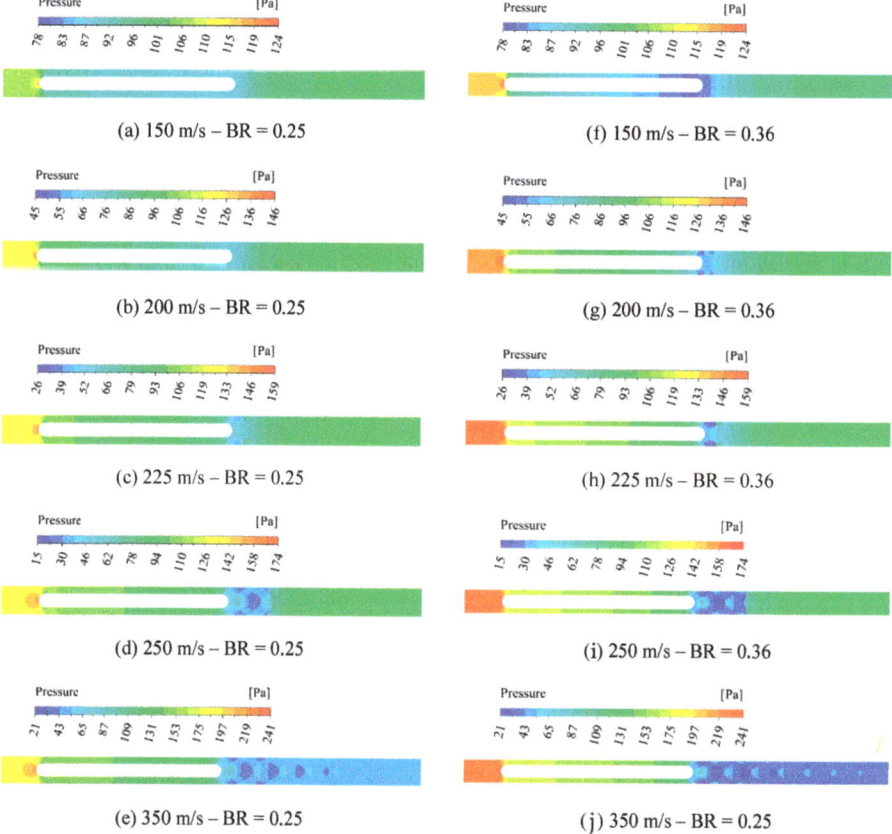

Figure 8. Pressure contours for selected pod speeds with P_{tube} = 101.325 Pa, L_{pod} = 43 m, and (**a**–**e**) for BR = 0.25, (**f**–**j**) for BR = 0.36. Contour levels are fixed for each pod speed.

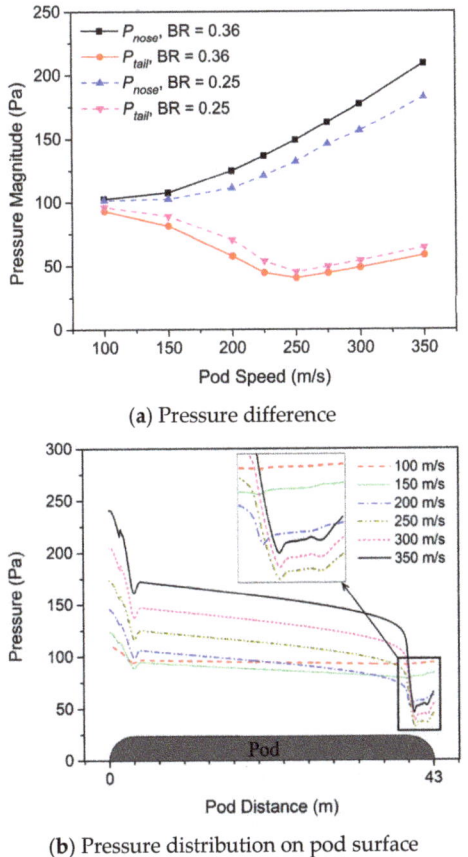

Figure 9. (a) Pressure difference between the nose and the tail. (b) Pressure distribution across the pod surface $\left(\text{BR} = 0.36,\ P_{tube} = 101.325\ \text{Pa},\ L_{pod} = 43\ \text{m}\right)$. Inset shows the magnified view of the tail.

3.2. Effect of Tube Pressure

The Hyperloop Alpha documents recommended an ideal tube pressure of 1/1000 atm for operating the Hyperloop systems. In this section, to verify this factor, different tube pressures were applied for the pod speed of 300 m/s with a BR of 0.36 and a pod length of 43 m. Figure 10a illustrates the changes in drag and drag coefficient with the tube pressure. As shown in the sub-figure, aerodynamic drag linearly increases with tube pressure. With increasing tube pressure, pressure drag becomes more dominant, whereas the variation of friction drag is minor. The tenfold increase of tube pressure has increased the total drag by more than nine times. Therefore, it is better to maintain a lower tube pressure to have a smaller drag. The variation of drag coefficient is inversely proportional to the flow density, reference area, and the square of operating speed. As mentioned earlier, the cross-sectional area and operating speed were fixed to investigate the effect of tube pressure. Hence, in this study, the change of drag coefficient with respect to tube pressure is mostly affected by the density of air, which is calculated by $\rho = P_t/RT$, where P_t is the tube pressure, R is the gas constant, and T is the temperature. The higher tube pressure causes higher density, resulting in a larger Re and a marginal decrease in the drag coefficient. Figure 10b shows the proportion of pressure and friction drag in the total drag. Note that tube pressure has only a small influence on the proportion of the drag components to the total drag.

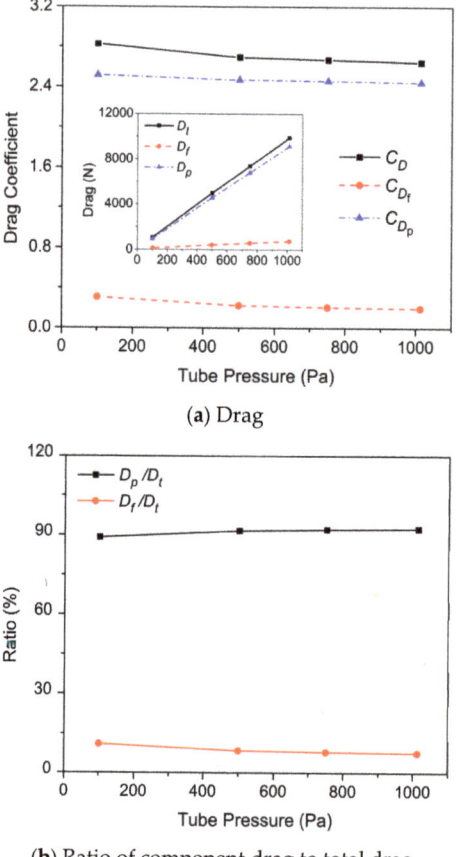

(a) Drag

(b) Ratio of component drag to total drag

Figure 10. Effect of tube pressure: (**a**) Variation of drag and drag coefficient with respect to the tube pressure. (**b**) Ratios of the pressure and friction drags to the total drag.

3.3. Effect of Pod Length

This section investigates the effect of pod length on the drag coefficient. Three cases were examined based on the reference pod length from the Hyperloop Alpha documents. The pod length of L_{ref} = 43 m was used as a reference case to analyze the two cases of $L_{ref}/2$ and $2L_{ref}$. Figure 11 examines the effect of pod length on the drag coefficient and the ratios of pressure and friction drags to the total drag.

As indicated in Figure 11a, the pod length does not have a significant effect on C_{D_p}, whereas it leads to a considerable increase in C_{D_f}. The variation of drag coefficient with respect to pod length follows a similar trend as the variation of drag force. The variation of drag force with pod length is presented in the sub-figure. Friction drag is generally generated along the pod surface; hence, when increasing the pod length, friction drag increases nearly proportionally. The drag acting on a shorter pod mostly depends on the pressure drag. The increase of pressure drag is mostly affected by the pressure distribution in the front and rear of the pod; hence, the extension of pod length has little effect on the pressure drag. Nevertheless, friction drag is significantly altered because it is strongly dependent on the surface area. C_{D_f} is nearly doubled when doubling the pod length, whereas the increase of C_{D_p} is negligible. Consequently, with the increase of pod length, the increase of total drag is mostly because of the increase in friction drag. To further explain the statement, the ratios of the pressure and friction drags to the total drag are presented in Figure 11b. With $L = L_{ref}/2$, the ratio of pressure drag to friction drag is approximately 12:1. However, with $L = 2L_{ref}$, this ratio reduces to 5:1.

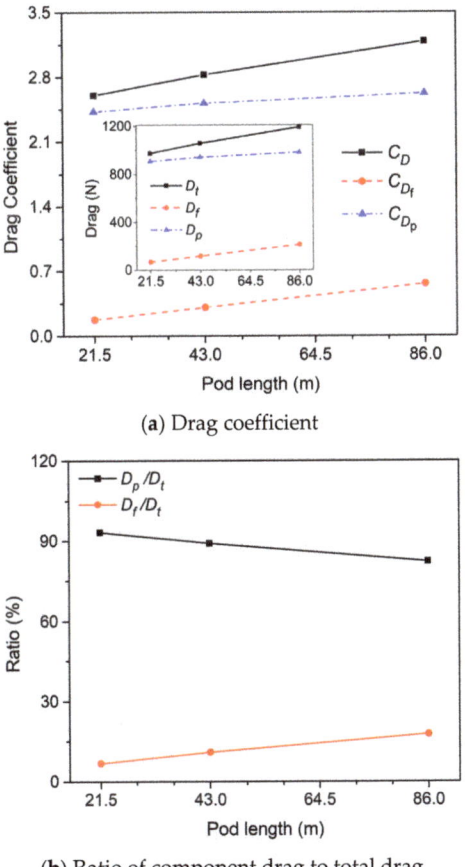

(a) Drag coefficient

(b) Ratio of component drag to total drag

Figure 11. Effect of pod length: (**a**) Variation of drag and drag coefficient with respect to pod length. The results are similar to those of the steady state study by Oh et al. (2019) [5] (the difference is below 17%). (**b**) Ratio of pressure and friction drag to total drag.

3.4. Pressure Wave Propagation

The speed of sound in gases is dependent on the temperature. Under the assumption of isentropic flow, the speed of sound is estimated by $c = \sqrt{\gamma R T}$. γ and R are the ratios of specific heats and the individual gas constant with the assigned values of 1.4 and 287.058 J/kg.K, respectively; at an air temperature of 300 K, c = 347.1 m/s. Under atmosphere pressure and linear wave assumption, the pressure waves propagate at the speed of sound [31]. However, in the Hyperloop system, the reference pressure is reduced to 1/1000 atm and the pressure waves have high amplitude, causing the nonlinear wave phenomenon. Therefore, the pressure wave propagation speeds are greater than the speed of sound. Figure 12 compares the pressure wave propagation speeds induced at the front and rear end of the pod with the speed of sound. A higher BR creates higher pressure wave propagation speed at the front end of the pod $(v_{s,\,front})$, but it barely affects the pressure wave propagation speed at the end of the pod $(v_{s,\,rear})$.

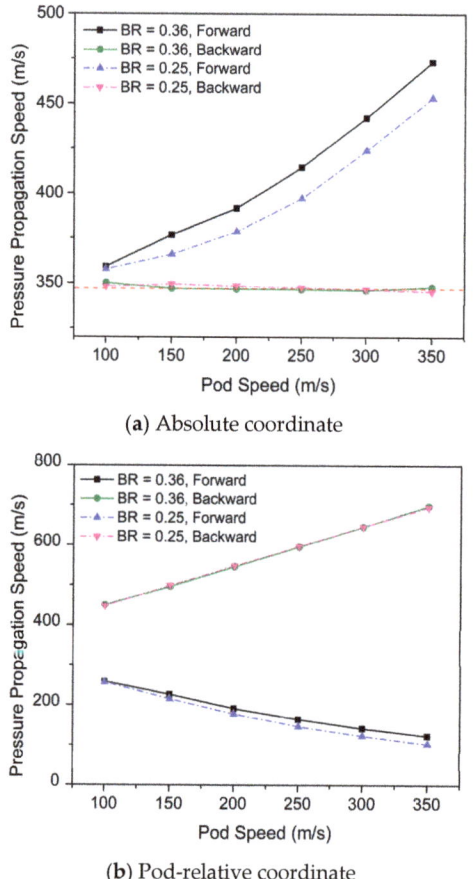

Figure 12. Forward and backward pressure wave propagation speeds observed in (**a**) absolute coordinates (the reference frame is fixed) and (**b**) pod-relative coordinates (the reference frame moves with the pod). The red dashed line represents the speed of sound (347.1 m/s).

Figure 12a shows the values of pressure wave propagation speed (v_s) in absolute coordinates. The range of $v_{s,front}$ is from 355 to 473 m/s, which is much higher than the normal speed of sound of 347.1 m/s. As v_P increases, $v_{s,front}$ increases steeply. In the rear end of the pod, $v_{s,rear}$ did not change much and maintained a speed around the speed of sound. At v_P of 350 m/s, the differences between speed of sound and v_s for forward and backward flows are 26.6% and 0.6%, respectively. Note that in this study, the analysis of v_s is conducted under ideal operating conditions, where the tube walls are smooth and straight and without the application of a vacuum pump.

As shown in Figure 12b, the pod-relative coordinates and the absolute coordinates are vastly different. $v_{s,rear}$ is much higher than $v_{s,front}$. This is because the directions of v_P and $v_{s,front}$ are the same, in contrast to the direction of $v_{s,rear}$, which is the opposite of v_P. That means as v_P increases, the difference between v_P and the compression wave speed decreases, whereas the difference between v_P and the expansion wave speed increases.

Figure 13 illustrates the pressure variation along the tube axis and the pod surface. As v_P increases, the pressure magnitude at the front of the pod significantly increases, whereas the pressure behind the pod slightly decreases. Compression waves are generated in front of the pod and propagate mainly in the forward direction faster than the speed of sound. These compression waves cause the pressure variation in front of the pod to fluctuate abruptly and generate a high-pressure region. Meanwhile, expansion waves are generated behind the pod, and propagate both in the forward and backward directions of the pod at around the speed of sound. The forward expansion waves pass the pod surface and merge with the incident compression waves, decreasing the pressure of the front waves. This phenomenon is probably observed at lower speeds, namely, 100 m/s and 150 m/s. This pressure wave distribution is similar to the variation in the high-speed train–tunnel system presented by Zonglin et al. [13]. When v_P increases, the compressibility of air and the compression wave propagation speed increase, and consequently alter the forms of the high-pressure region in front of the pod. Figure A4 (Appendix C) describes this high-pressure region in more detail. By contrast, the position of the low-pressure region shows only a minor change. The oblique shockwaves caused by the reflection of waves between the tube walls and the pod tail are illustrated in Figure 13 for $v_P \geq 250$ m/s.

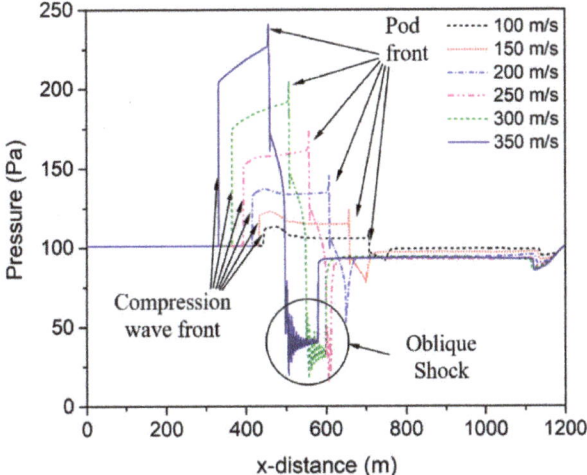

Figure 13. Pressure wave distribution along the centerline of the tube and pod surface at $t = 1.0$ s. Four cases of selected pod speeds under $P_{tube} = 101.325$ Pa, $L = 43$ m, and BR = 0.36 are presented. Pod moves from right to left.

Figure 14 shows the pressure variation along the tube axis in the front and rear of the pod (depicted in Figure 15) for $t = 0.2$–1.0 s. Compression waves are generated ahead of the pod and propagate in the same direction, whereas expansion waves propagate in the opposite direction. The following equation presents the pressure ratio for flows across a normal shock [27,31]:

$$\frac{p_2}{p_1} = \frac{2\gamma M_s^2 - (\gamma - 1)}{\gamma + 1} \tag{19}$$

with

$$M_s = \frac{v_s}{\sqrt{\gamma R T_1}} \tag{20}$$

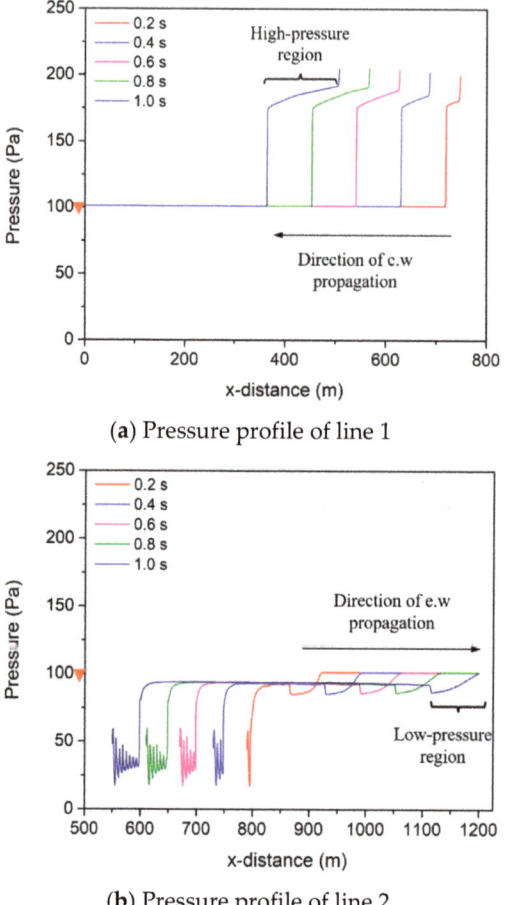

(a) Pressure profile of line 1

(b) Pressure profile of line 2

Figure 14. Pressure distributions of line 1 (the axis line in front of the pod) and line 2 (the axis line at the rear of the pod) for $t = 0.2$–1.0 ($v_P = 300$ m/s, BR $= 0.36$, $P_{tube} = 101.325$ Pa, $L_{pod} = 43$ m). Pod moves from right to left. Red triangles represent the tube pressure of 101.325 Pa.

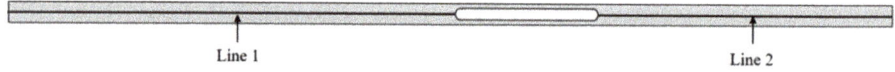

Figure 15. Descriptions of line 1 and line 2 from Figure 12.

The term p_2 in Equation (19) is the pressure term of the front shockwave shown in Figure 16a. Figure 16a compares the pressure magnitude of the front shockwave between BRs of 0.25 and 0.36. The results indicate that a higher BR produces a higher front shockwave pressure. In Figures 12–14, stronger compression waves form in front of the pod with increase of v_P. This behavior of front waves is similar to the phenomenon in the normal shockwave theory. Hence, Figure 16b compares the compression wave propagation speed generated in front of the pod from the simulation with the ones calculated by Equations (19) and (20). There is a good agreement between the two results. Therefore, the compression wave generated in front of the pod in this study conforms to the normal shockwave theory.

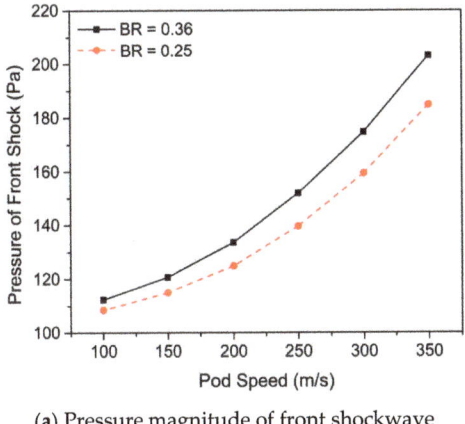

(a) Pressure magnitude of front shockwave

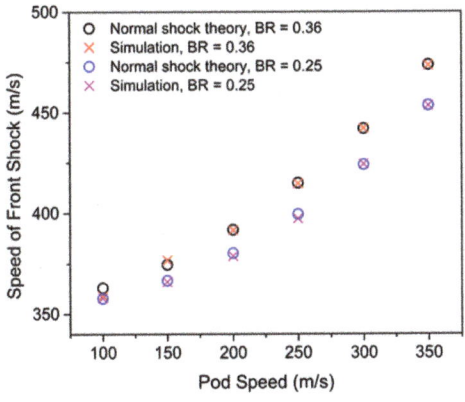

(b) Speed of front shockwave

Figure 16. Pressure magnitude and speed of front shockwave. The compression wave is well-matched with the normal shockwave theory. (**a**) Pressure magnitude of compression wave traveling in the forward direction of the pod. (**b**) Comparison of compression wave propagation speeds calculated using the simulation and normal shockwave equations.

4. Conclusions

This study simulated the unsteady conditions of a Hyperloop system using the overset moving mesh to investigate the influences of pod speed, BR, tube pressure, and pod length on the aerodynamic drag and pressure waves induced in the system.

The results provide a clear picture of the variation of the drag coefficient. The drag coefficient is maximized at lower pod speeds with a higher BR, in other words, 225 m/s for BR = 0.36 and 250 m/s for BR = 0.25. The drag coefficient increases with the increase of pod speed, BR, and pod length. In the Hyperloop system, pressure drag is regarded as a considerable component of total drag, whereas the influence of friction drag is minor. In addition, the pressure difference between the nose and the tail significantly impacts the pressure drag, and consequently the total drag. The drag increases proportionally with the tube pressure, whereas the drag coefficient decreases slightly. The increase of total drag with the increase of pod length is mostly dependent on the increment of the friction drag.

The presence of compression waves and expansion waves generates the opposite tendency of pressure in the front and rear of the pod. Once the local flow speed exceeds the supersonic speed, oblique shockwaves occur, vastly influencing the tail pressure of the pod. In the smaller tube (higher BR), the compressed air pushes the pod nose and significantly increases the pressure at the front while decreasing the pressure behind the pod.

The compression waves and expansion waves, along with their speeds were investigated. In the Hyperloop system, these waves become faster than the local speed of sound at lower internal tube pressures and higher operating speeds. As BR increases, the speed of compression wave propagation is largely affected, while the expansion waves propagate at a similar speed. The study also suggested that the normal shockwave theory can be used to predict the variation of compression wave propagation speed.

Author Contributions: Conceptualization, methodology, and investigation: T.T.G.L. and J.R.; validation and writing—original draft preparation: T.T.G.L.; formal analysis, data curation, and visualization: T.T.G.L. and K.S.J.; writing—review and editing, supervision, project administration, and funding acquisition: K.-S.L. and J.R. All authors have read and agreed to the published version of the manuscript.

Funding: This research was supported by the MSIT (Ministry of Science and ICT), Korea, under the ITRC (Information Technology Research Center) support program (IITP-2020-2020-0-01655) supervised by the IITP (Institute of Information and Communications Technology Planning and Evaluation), and by the National Research Foundation of Korea (NRF) grant funded by the Korean government (MEST) (No. 2019R1A2C1087763). This research was supported by "Core Technology Development of Subsonic Capsule Train" of the Korea Railroad Research Institute under Grant PK2001A1A, Korea.

Conflicts of Interest: The authors declare no conflict of interest.

Appendix A. Pressure Wave Propagation with Respect to Simulation Time

In this study, 1 s was sufficient for the flow to fully develop. Therefore, the tube was created long enough and the pod was placed at the given position to prevent the pressure waves from being reflected by the boundaries in 1 s. The following figures show that at $t = 1$ s, the compression waves and expansion waves do not reflect off the boundaries. Therefore, the designed geometry is reasonable.

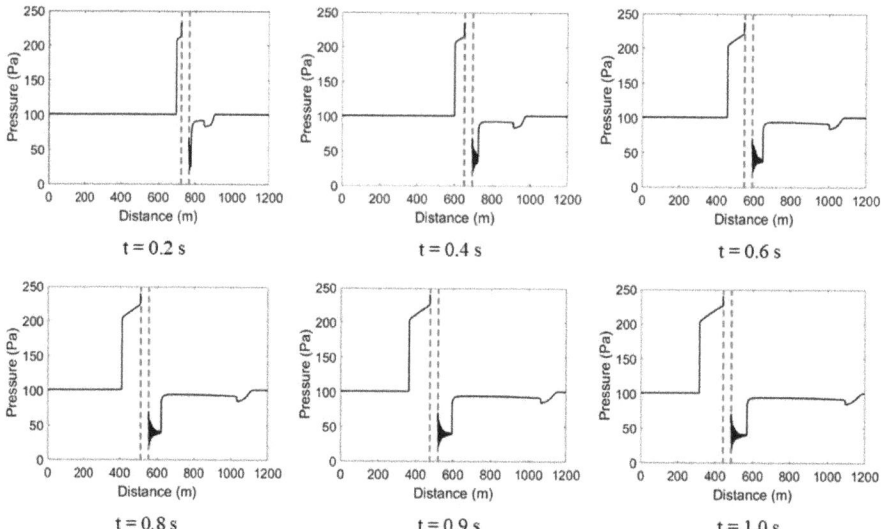

Figure A1. Pressure line plot along the centerline in the front and rear of the pod varied with simulation time ($v_P = 350$ m/s). The simulation was terminated before the expansion wave could reflect off from the outlet boundary.

Appendix B

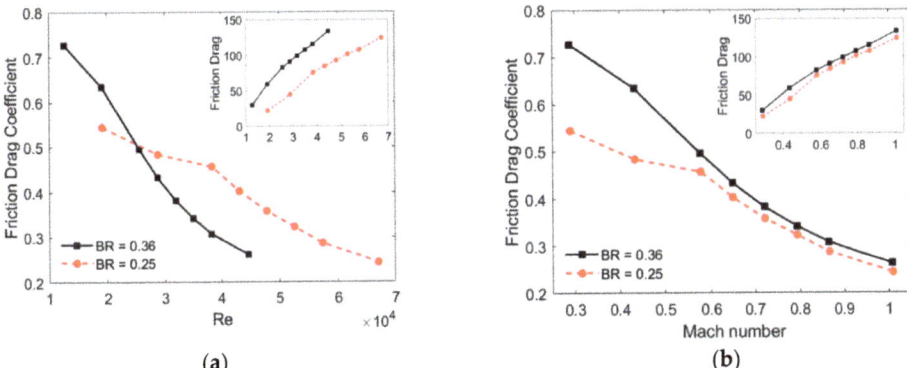

Figure A2. Variation of friction drag coefficients on blockage ratio and (**a**) Re, (**b**) Mach number. Insets represent the variation of drag.

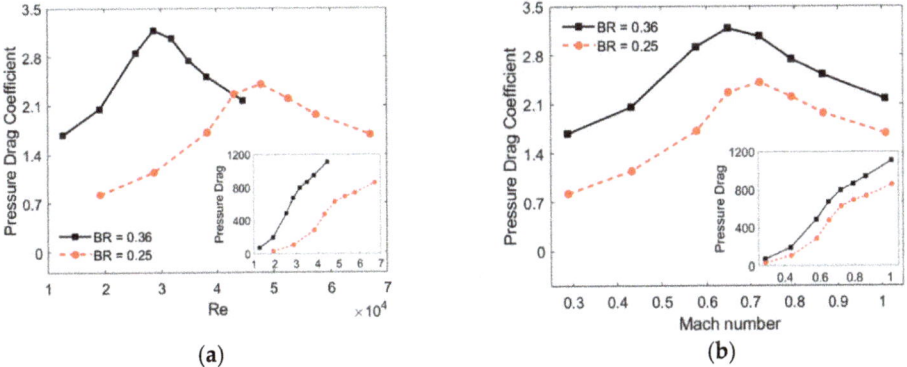

Figure A3. Variation of pressure drag coefficients on blockage ratio and (**a**) Re, (**b**) Mach number. Insets represent the variation of drag.

Appendix C. High-Pressure Region

The pressure contours at the front and rear of the pod are shown in Figure A4 to describe the high-pressure region and the form of oblique shockwaves coexist with normal shockwaves. Figure A4 indicates that because the front wave propagation speed is much higher than the operating pod speed (150 m/s), a high-pressure region is created in places far from the nose. This is contrary to the phenomenon that occurs at a v_P of 350 m/s, in which the high-pressure region occurs right in front of the nose. This means that when the local flow exceeds supersonic speed, the compression wave gets increasingly closer. This high-pressure region separates the pressure in front of the pod into two distinct regions: a disturbed region (orange-red color) and an undisturbed region (green color) that is at the initial tube pressure. A severe shock phenomenon occurs at a v_P of 350 m/s, which did not appear at a v_P of 150 m/s. The low-pressure region is not significant compared to other pressure regions behind the pod. Hence, it was not observed in the pressure contour.

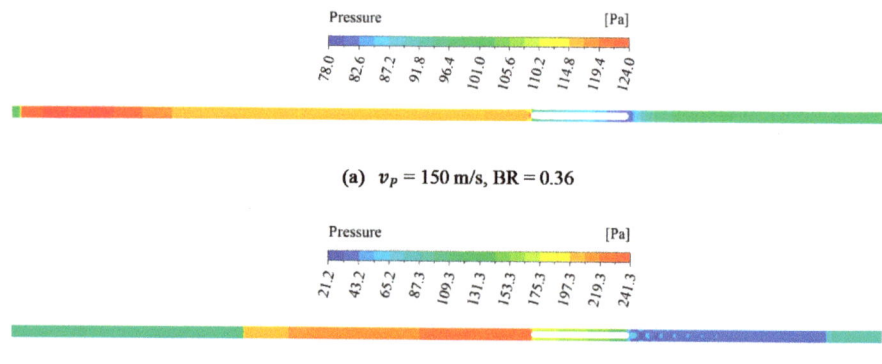

Figure A4. Pressure contour of pressure waves generated at the front and rear of the pod. The largest red regime describes the high-pressure regime.

References

1. Musk, E. *Hyperloop Alpha Documents*; SapceX: Hawthorne, CA, USA, 2013.
2. Opgenoord, M.M.J.; Merian, C.; Mayo, J.; Kirschen, P.; O'Rourke, C.; Izatt, G. *MIT Hyperloop Final Report*; Massachusetts Institute of Technology: Cambridge, MA, USA, 2017.
3. Oster, D. Evacuated Tube Transport. U.S. Patent 5950543A, 14 September 1999.
4. Taylor, C.L.; Hyde, D.J.; Barr, L.C. *Hyperloop Commercial Feasibility Analysis: High Level Overview*; John, A., Ed.; Volpe National Transportation Systems Center: Cambridge, MA, USA, 2016.
5. Oh, J.-S.; Kang, T.; Ham, S.; Lee, K.-S.; Jang, Y.-J.; Ryou, H.-S.; Ryu, J. Numerical analysis of aerodynamic characteristics of hyperloop system. *Energies* **2019**, *12*, 518. [CrossRef]
6. Yang, Y.; Wang, H.; Benedict, M.; Coleman, D. Aerodynamic Simulation of High-Speed Capsule in the Hyperloop System. In Proceedings of the 35th AIAA Applied Aerodynamics Conference, Denver, CO, USA, 5–9 June 2017.
7. Gillani, S.A.; Panikulam, V.P.; Sadasivan, S.; Yaoping, Z. CFD Analysis of aerodynamic drag effects on vacuum tube trains. *J. Appl. Fluid Mech.* **2019**, *12*, 303–309. [CrossRef]
8. Singh, Y.K.; Kamyar, M. Numerical analysis for aerodynamic behaviour of hyperloop pods. *Preprints* **2019**. [CrossRef]
9. Yoon, T.S.; Lee, S.; Hwang, J.H.; Lee, D.H. Prediction and validation on the sonic boom by a high-speed train entering a tunnel. *J. Sound Vibrat.* **2001**, *247*, 195–211. [CrossRef]
10. Zhang, Y. Numerical simulation and analysis of aerodynamic drag on a subsonic train in evacuated tube transportation. *J. Modern Transp.* **2012**, *20*, 44–48. [CrossRef]
11. Kang, H.; Jin, Y.; Kwon, H.; Kim, K. A Study on the aerodynamic drag of transonic vehicle in evacuated tube using computational fluid dynamics. *Int. J. Aeronaut. Space Sci.* **2017**, *18*, 614–622. [CrossRef]
12. Kim, T.-K.; Kim, K.-Y.; Kwon, H.Y. Aerodynamic characteristics of a tube train. *J. Wind Eng. Ind. Aerodyn.* **2011**, *99*, 1187–1196. [CrossRef]
13. Zonglin, J.; Matsuoka, K.; Sasoh, A.; Takayama, K. Numerical and experimental investigation of wave dynamic processes in high-speed train/tunnels. *Acta Mech. Sin.* **2002**, *18*, 209–226. [CrossRef]
14. Steinbrenner, J.P. Automatic Structured and Unstructured Grid Cell Remediation for Overset Meshes. In Proceedings of the 52nd Aerospace Sciences Meeting, Harbor, MD, USA, 13–17 January 2014.
15. ANSYS. *ANSYS Fluent Version, 18.1 User's Guide*; ANSYS: Canonsburg, PA, USA, 2015.
16. Cao, D.; He, G.; Qin, F.; Wei, X.; Shi, L.; Liu, B.; Huang, Z. LES Study on flow features of the supersonic mixing layer affected by shock waves. *Int. Commun. Heat Mass Transf.* **2017**, *85*, 114–123. [CrossRef]
17. Choi, M.G.; Ryu, J. Numerical study of the axial gap and hot streak effects on thermal and flow characteristics in two-stage high pressure gas turbine. *Energies* **2018**, *11*, 2654. [CrossRef]

18. Moin, P.; Mahesh, K. Direct numerical simulation: A tool in turbulence research. *Ann. Rev. Fluid Mech.* **1998**, *30*, 539–578. [CrossRef]
19. Ryu, J.; Lele, S.K.; Viswanathan, K. Study of supersonic wave components in high-speed turbulent jets using an les database. *J. Sound Vib.* **2014**, *333*, 6900–6923. [CrossRef]
20. Ryu, J.; Livescu, D. Turbulence structure behind the shock in canonical shock–vortical turbulence interaction. *J. Fluid Mech.* **2014**, *756*, 143–195. [CrossRef]
21. Menter, F. Zonal Two Equation k-ω Turbulence Models for Aerodynamic Flows. In Proceedings of the 23rd Fluid Dynamics, Plasmadynamics, and Lasers Conference, Orlando, FL, USA, 6–9 July 1993.
22. Rocha, P.A.; Costa, H.H.; Barbosa, R.; Moura Carneiro, F.O.; Vieira da Silva, M.E.; Valente Bueno, A. k-ω SST (Shear Stress Transport) turbulence model calibration: A case study on a small scale horizontal axis wind turbine. *Energy* **2014**, *65*, 412–418. [CrossRef]
23. Hruschka, R.; Klatt, D. In-pipe aerodynamic characteristics of a projectile in comparison with free flight for transonic mach numbers. *Shock Waves* **2019**, *29*, 297–306. [CrossRef]
24. Zhou, P.; Zhang, J.; Li, T. Effects of blocking ratio and mach number on aerodynamic characteristics of the evacuated tube train. *Int. J. Rail Transp.* **2020**, *8*, 27–44. [CrossRef]
25. Zhou, P.; Zhang, J.; Li, T.; Zhang, W. Numerical study on wave phenomena produced by the super high-speed evacuated tube maglev train. *J Wind Eng. Ind. Aerodyn.* **2019**, *190*, 61–70. [CrossRef]
26. Charters, A.C.; Thomas, R.N. The aerodynamic performance of small spheres from subsonic to high supersonic velocities. *J. Aeronaut. Sci.* **1945**, *12*, 468–476. [CrossRef]
27. Munson, B.R.; Okiishi, T.H.; Huebsch, W.W.; Rothmayer, A.P. *Fluid Mechanics*; Wiley: Singapore, 2013.
28. Sui, Y.; Niu, J.; Yuan, Y.; Yu, Q.; Cao, X.; Wu, D.; Yang, X. An aerothermal study of influence of blockage ratio on a supersonic tube train system. *J. Therm. Sci.* **2020**, *29*. [CrossRef]
29. Raghunathan, R.S.; Kim, H.-D.; Setoguchi, T. Aerodynamics of high-speed railway train. *Progress Aerospace Sci.* **2002**, *38*, 469–514. [CrossRef]
30. Baron, A.; Mossi, M.; Sibilla, S. The alleviation of the aerodynamic drag and wave effects of high-speed trains in very long tunnels. *J. Wind Eng. Ind. Aerodyn.* **2001**, *89*, 365–401. [CrossRef]
31. Anderson, J.D. *Modern Compressible Flow: With Historical Perspective*; McGraw-Hill: New York, NY, USA, 1990; Volume 12.

Publisher's Note: MDPI stays neutral with regard to jurisdictional claims in published maps and institutional affiliations.

 © 2020 by the authors. Licensee MDPI, Basel, Switzerland. This article is an open access article distributed under the terms and conditions of the Creative Commons Attribution (CC BY) license (http://creativecommons.org/licenses/by/4.0/).

Article

Unsteady Stagnation Point Flow of Hybrid Nanofluid Past a Convectively Heated Stretching/Shrinking Sheet with Velocity Slip

Nurul Amira Zainal [1,2], Roslinda Nazar [1,*], Kohilavani Naganthran [1] and Ioan Pop [3]

1. Department of Mathematical Sciences, Faculty of Science and Technology, Universiti Kebangsaan Malaysia, Bangi 43600, Malaysia; nurulamira@utem.edu.my (N.A.Z.); kohi@ukm.edu.my (K.N.)
2. Fakulti Teknologi Kejuruteraan Mekanikal dan Pembuatan, Universiti Teknikal Malaysia Melaka, Hang Tuah Jaya, Melaka 76100, Malaysia
3. Department of Mathematics, Babeş-Bolyai University, R-400084 Cluj-Napoca, Romania; popm.ioan@yahoo.co.uk or ipop@math.ubbcluj.ro
* Correspondence: rmn@ukm.edu.my

Received: 27 July 2020; Accepted: 22 September 2020; Published: 24 September 2020

Abstract: Unsteady stagnation point flow in hybrid nanofluid (Al_2O_3-Cu/H_2O) past a convectively heated stretching/shrinking sheet is examined. Apart from the conventional surface of the no-slip condition, the velocity slip condition is considered in this study. By incorporating verified similarity transformations, the differential equations together with their partial derivatives are changed into ordinary differential equations. Throughout the MATLAB operating system, the simplified mathematical model is clarified by employing the bvp4c procedure. The above-proposed approach is capable of producing non-uniqueness solutions when adequate initial assumptions are provided. The findings revealed that the skin friction coefficient intensifies in conjunction with the local Nusselt number by adding up the nanoparticles volume fraction. The occurrence of velocity slip at the boundary reduces the coefficient of skin friction; however, an upward trend is exemplified in the rate of heat transfer. The results also signified that, unlike the parameter of velocity slip, the increment in the unsteady parameter conclusively increases the coefficient of skin friction, and an upsurge attribution in the heat transfer rate is observed resulting from the increment of Biot number. The findings are evidenced to have dual solutions, which inevitably contribute to stability analysis, hence validating the feasibility of the first solution.

Keywords: hybrid nanofluid; unsteady stagnation point; velocity slip; convective boundary condition; stability analysis

1. Introduction

The most common problem in boundary layer flow that had been treated so far is much focused on those for steady flows. Even though it is the steady flows that seem to have the utmost significance in real-world demands, some cases of time-varying in the boundary layer which is unsteady indicate an important role in several engineering problems. Some of the examples are start-up processes where the motions in rest are transits from a steady flow to another, and periodic motions of the working fluid [1]. The behavior of unsteady boundary layer flow describes an unusual pattern compared to the steady flow owing to the additional time-dependent terms in the governing equations, which exaggerated the separation of boundary layer and the fluid motion arrangement [2,3]. The thermal and mechanical properties of such an unsteady mechanism in the boundary layer approximation have been studied both analytically and numerically. Elbashbeshy and Bazid [4] presented the numerical investigation towards the unsteady stretching surface with heat transfer analysis. At the same time, Bhattacharya [5] managed to prove the existence of dual solutions in unsteady stagnation point flow towards a shrinking sheet by employing the shooting method approach coupled with a Runge–Kutta integration scheme. Bachok et al. [6] concluded that inclusion of the unsteadiness parameter offers a significant impact towards the boundary layer flow in nanofluid and Fan et al. [7] presented analytical solutions using the homotopy analysis method (HAM) and managed to advertise a highly precise analytical estimation which is in excellent agreement with the numerical results offered by the Keller box scheme. It is worth mentioning that a considerable amount of reviews on the unsteady stagnation point flow due to a stretching/shrinking surface have been accomplished by numerous researchers, including [8–11].

The stagnation point flow is one of the important topics in mechanics of fluid, in the way that stagnation point generally occurs in both engineering and science flow fields. The stagnation point flow could be identified in the extrusion process, polymer industry, and plane counter jet [12,13]. The ground-breaking research in this topic was first initiated by Hiemenz [14] who exposed an analytical explanation of two-dimensional stagnation point flow, and soon after, Homann [15] conducted a classical study of stagnation point in three-dimensional flow with regard to an axisymmetric case; whereas Howarth [16] tackled the problem of non-axisymmetric flow close to the stagnation area in three-dimensional analysis. Recently, Khashi'ie et al. [17,18], Fang and Wang [19], Waini et al. [20], and Zainal et al. [21] have scrutinized the stagnation point flow problems in diverse aspects with no-slip boundary conditions. Nevertheless, in numerous engineering occasions, the slip effect should be comprised, such as flow over lubricated or coated surfaces, rough or striated surfaces [22] and internal rare field gas flow [23]. Examples of industrial applications involving the slip boundary conditions are fluid flow on multiple interfaces, rare field fluid problems, and also the reacting flow in reactors [24,25]. Navier [26] and Maxwell [27] were the primary researchers who pioneered the study of linear slip boundary conditions, while Wang [28] has well reflected a comprehensive theoretical analysis considering the no-slip boundaries concentrating on the stagnation point flow. Rao and Rajagopal [29] have conducted an extensive evaluation and argument between the slip and no-slip condition, and Jusoh et al. [30] deliberated a modified nanofluid model towards a stretching/shrinking surface by considering a velocity slip parameter in three-dimensional flow. The study revealed that an increase in the velocity slip magnitude contributed to the intensification of skin friction coefficients.

In certain cases, the velocity slip or the non-adherence of the fluid to a solid boundary phenomenon was witnessed, for instance, in the micro-scale devices [31]. Fluids promoting slip are critical in technical applications such as polishing artificial heart valves and internal cavities [32]. The flow behavior and the shear stress in the fluid are rather distinctive with a slip at the wall boundary compared to those with no-slip condition. Besides that, the velocity slip effect does influence the heat transfer rate and was confirmed by Mukhopadhyay [33], who had investigated the slip impact of the unsteady mixed convective flow towards a porous stretching surface with heat transfer. The analysis found that the heat transfer rate declines with the velocity slip parameter, while it upsurges with the unsteadiness parameter. Mahapatra and Nandy [34] conducted a numerical study of the unsteady stagnation point flow past a shrinking sheet and heat transfer with the presence of slip effects in a viscous fluid. The results conveyed that with the increase of the velocity slip and unsteady parameter, the heat transfer rate is reported to escalate. Meanwhile, in nanofluid flow, Majumder et al. [35] specified that exertion of the partial velocity slip against the sheet surface is common. By relying on the finding in [35], Noghrehabadi et al. [36] then examined the impact of partial velocity slip on the nanofluid boundary layer flow and heat transfer past a stretching sheet. The work in [36] reported that an increment in the velocity slip effect decreases the momentum boundary layer thickness. Van Gorder et al. [37], who examined the nanofluid boundary layer flow over a stretching surface, also conveyed a similar result as [36] and further explained that no-slip condition is not applicable for fluid flows at nanoscales. Besides that, Dinarvand and Rostami [38] studied the rotating nanofluid flow and heat transfer with the presence of internal heating, velocity slip, and different shapes of nanoparticles. They showed that an increment in the velocity slip effect reduces the skin friction coefficient significantly. Researchers also tend to analyze the effect of velocity slip in the unsteady nanofluid flow as unsteady flow problems are more relatable to real-world applications. For instance, Seth et al. [39] studied the unsteady hydromagnetic nanofluid flow and heat transfer past a non-linearly stretching surface with Navier's velocity slip and presented the analysis of entropy generation. Other valuable references regarding unsteady nanofluid flow with the velocity slip effect can be found in [40–42].

Ever since the outstanding inventions achieved by Choi and Eastman [43], who originated the brilliant idea of demonstrating the nanoparticle suspension in a base fluid and came out with the nanofluid term, a better type of working fluid is still being pursued. Acknowledging the sufficient improvement in the thermal conductivity of the conventional fluid is crucial, an advanced nanofluid form known as hybrid nanofluid is introduced, which intends to have highly developed heat conductivity. This modern type of fluid agent has fascinated numerous researchers owing to its reputation in the emergence and improvement of thermal characteristics in realistic applications, including micro-channel, heat pipes, heat exchangers, air conditioning systems, and mini-channel heat sink [44,45]. Gupta et al. [46] and Xian et al. [47] have reviewed the preparation method of hybrid nanoparticles along with the stabilization and its significance in industrial sectors. One of the critical elements in establishing a sustainable hybrid nanofluid suspension is selecting an appropriate combination of nanoparticles. The most widely used nanoparticles for the formation of hybrid nanofluid suspension are carbon materials (graphite, MWCNTs, CNTs), metals (Cu, Ag), metal oxides (Al_2O_3, CuO, Fe_2O_3), metal carbide, and a metal nitride. Madhesh and Kalaiselvam [48] conducted an experimental study to examine the features of hybrid nanofluid as a coolant agent, Tahat and Benim [49] examined the efficiency of hybrid nanofluid on flat plate solar collector, and they verified that the viscosity, thermal conductivity, and density of the working fluid had increased together with the concentration of Al_2O_3/CuO concentration, thus enhancing the solar collector proficiency. Some early research on hybrid nanofluid that employed the numerical method was done by Labib et al. [50], who investigated the impact of base fluids and hybrid nanofluid using a two-phase mixture model in forced convective heat transfer. Moghadassi et al. [51] revealed that the heat transfer performance is enhanced by adding the nanoparticles of Al_2O_3-Cu hybrid nanofluid while creating a small pressure drop in the system regime, Devi and Devi [52] focused on the mathematical inspection towards a stretching sheet. In contrast, the evaluation of heat transfers in the natural convection of Al_2O_3/water nanofluid and Al_2O_3-Cu/water hybrid nanofluid with a discrete heat source was explored by Takabi and Salehi [53]. Additional details on this topic are well described in the literatures [54–56].

To the best of the authors' knowledge, the existing literature does not consider the unsteady stagnation point flow of hybrid nanofluids with the presence of velocity slip parameter and stability analysis in their models. Thus, the addressed issues above have inspired the authors to perform a numerical study in unsteady stagnation point flow towards a convectively heated stretching/shrinking sheet in alumina–copper/water (Al_2O_3-Cu/H_2O) with the impact of velocity slip on heat transfer. The hybrid nanofluid is recognized by dispersing Al_2O_3 nanoparticles into H_2O, followed by Cu with different volume fractions and the thermophysical properties of the hybrid nanofluid are adopted from Ghalambaz et al. [57] and Takabi and Salehi [53], which were based on the feasible physical assumptions and are in agreement with the conservation of mass and energy. The present work also utilized the bvp4c approach, which can be accessed in the MATLAB programming system towards solving the formulated problem. The existence of more than one solution is predictable; thus, an analysis of solution stability is completed to confirm the steadiness of the solutions which has an actual physical interpretation. The explanation of the results and the convergence of the obtained solutions are deliberated on in detail. Particular cases of current findings are evaluated in accordance with those of Mahapatra and Nandy [34] and Wang [58]. Furthermore, the consensus between previous and current findings is outstanding, and the agreement is excellent.

2. Mathematical Model

The unsteady two-dimensional stagnation-point flow of a hybrid Al_2O_3-Cu/H_2O nanofluid over a convectively heated stretching/shrinking sheet with the influence of velocity slip is considered in this research work, as illustrated in Figure 1 (see Dzulkifli et al. [59]). The stretching/shrinking velocity is denoted by $u_w(x,t) = bx/(1-ct)$, where b denotes a constant corresponds to stretching ($b > 0$) and shrinking ($b < 0$) cases while c signifies the unsteadiness problem and $u_e(x,t) = ax/(1-ct)$ is the velocity of the free stream where $a > 0$ represents the strength of the stagnation flow. The ambient temperature and the reference temperature are T_∞ and T_0, respectively. Now, we let the bottom of the sheet be heated by convection from a hot fluid at a specific temperature $T_f(x,t) = T_\infty + T_0 \frac{ax^2}{2\nu}(1-ct)^{-3/2}$ which supplies a coefficient of heat transfer, expressed by h_f. From all of the assumptions above; the governing boundary layer equations can be acknowledged as [34].

$$\frac{\partial u}{\partial x} + \frac{\partial v}{\partial y} = 0, \tag{1}$$

$$\frac{\partial u}{\partial t} + u\frac{\partial u}{\partial x} + v\frac{\partial u}{\partial y} = \frac{\partial u_e}{\partial t} + u_e\frac{\partial u_e}{\partial x} + \frac{\mu_{hnf}}{\rho_{hnf}}\frac{\partial^2 u}{\partial y^2}, \tag{2}$$

$$\frac{\partial T}{\partial t} + u\frac{\partial T}{\partial x} + v\frac{\partial T}{\partial y} = \frac{k_{hnf}}{(\rho C_p)_{hnf}}\frac{\partial^2 T}{\partial y^2}, \tag{3}$$

where u denotes the component of velocity in $x-$ axis, v is the velocity component in $y-$ axis, μ_{hnf} is the Al$_2$O$_3$-Cu/H$_2$O dynamic viscosity, ρ_{hnf} the density of Al$_2$O$_3$-Cu/H$_2$O, T is the Al$_2$O$_3$-Cu/H$_2$O temperature, k_{hnf} is the thermal/heat conductivity of Al$_2$O$_3$-Cu/H$_2$O and $(\rho C_p)_{hnf}$ is the Al$_2$O$_3$-Cu/H$_2$O heat capacity. The boundary conditions, together with the partial slip for velocity, are set to

$$\begin{array}{c} u = u_w(x,t) + H_1\nu\frac{\partial u}{\partial y}, \quad v = 0, \quad -k_{hnf}\frac{\partial T}{\partial y} = h_f + (T_f - T) \text{ at } y = 0, \\ u \to u_e(x,t), \quad T \to T_\infty \text{ as } y \to \infty, \end{array} \tag{4}$$

where $H_1 = H(1-ct)^{1/2}$ is the velocity slip factor, in which H refers to the initial value of the velocity slip factor. The copper (Cu) thermophysical properties, along with aluminum oxide (Al$_2$O$_3$) and water (H$_2$O) nanoparticles, are provided in Table 1, as demonstrated by [60]. In the meantime, Table 2 issued the thermophysical properties hybrid nanofluid as established by [53,57]. The nanoparticles solid volume fraction is represented by ϕ, ρ_f indicates the H$_2$O density, and ρ_s is the density of the hybrid nanoparticle, C_p is the constant pressure of heat capacity, while k_f denotes the thermal conductivity of H$_2$O and k_s is the hybrid nanoparticles thermal conductivity.

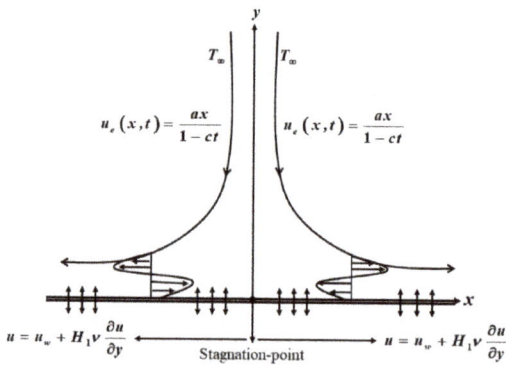

Figure 1. The schematic of problem flow (Dzulkifli et al. [59]).

Table 1. Cu thermophysical properties along with Al$_2$O$_3$ and H$_2$O (Oztop and Abu Nada [60]).

Properties	k(W/mK)	ρ (kg/m^3)	C$_p$(J/kgK)	β × 10^{-5}(mK)
Cu	400	8933	385	1.67
Al$_2$O$_3$	40	3970	765	0.85
H$_2$O	0.613	997.1	4179	21

Table 2. Hybrid Al$_2$O$_3$-Cu/H$_2$O nanofluids thermophysical properties (Takabi and Salehi [53], Ghalambaz et al. [57]).

Properties	Hybrid Nanofluid
Dynamic viscosity	$\mu_{hnf} = \dfrac{1}{(1-\phi_{hnf})^{2.5}}$
Density	$\rho_{hnf} = (1-\phi_{hnf})\rho_f + \phi_1\rho_{s1} + \phi_2\rho_{s2}$
Thermal capacity	$(\rho C_p)_{hnf} = (1-\phi_{hnf})(\rho C_p)_f + \phi_1(\rho C_p)_{s1} + \phi_2(\rho C_p)_{s2}$
Thermal conductivity	$\dfrac{k_{hnf}}{k_f} = \dfrac{\left[\left(\frac{\phi_1 k_{s1}+\phi_2 k_{s2}}{\phi_{hnf}}\right)+2k_f+2(\phi_1 k_{s1}+\phi_2 k_{s2})-2\phi_{hnf}k_f\right]}{\left[\left(\frac{\phi_1 k_{s1}+\phi_2 k_{s2}}{\phi_{hnf}}\right)+2k_f-(\phi_1 k_{s1}+\phi_2 k_{s2})+\phi_{hnf}k_f\right]}$

In order to express the governing Equations (1)–(3) concerning the boundary conditions (4) in a much simpler form, the subsequent similarity transformations are presented [34]

$$\psi = \left(\frac{av}{1-ct}\right)^{1/2} x f(\eta), \theta(\eta) = \frac{T-T_\infty}{T_f-T_\infty}, \eta = \left(\frac{a}{v(1-ct)}\right)^{1/2} y, \quad (5)$$

where ψ is the stream function that can be specified as $u = \partial\psi/\partial y, v = -\partial\psi/\partial y$ and η is the similarity variable. Thus, we attain

$$u = \frac{ax}{(1-ct)} f'(\eta), v = -\left(\frac{av}{1-ct}\right)^{1/2} f(\eta). \quad (6)$$

In view of the above relations, by employing the similarity variables (5) and (6), Equations (2) and (3) reduce to the following set of nonlinear similarity differential equations

$$\frac{\mu_{hnf}/\mu_f}{\rho_{hnf}/\rho_f} f''' + f f'' - f'^2 + 1 - \varepsilon\left(f' + \frac{1}{2}\eta f'' - 1\right) = 0, \quad (7)$$

$$\frac{1}{\Pr} \frac{k_{hnf}/k_f}{(\rho C_p)_{hnf}/(\rho C_p)_f} \theta'' + f\theta' - 2f'\theta + \frac{\varepsilon}{2}(\eta\theta' + 3\theta) = 0. \quad (8)$$

Here, ε measures the unsteadiness parameter with $\varepsilon = c/a$, Pr represents the Prandtl number where $\Pr = v_f/\alpha_f$. Next, the initial and boundary conditions (4) now transform into

$$f(0) = 0, f'(0) = \lambda + \gamma f''(0), -\frac{k_{hnf}}{k_f}\theta'(0) = \text{Bi}[1-\theta(0)], \quad (9)$$
$$f'(\eta) \to 1, \theta(\eta) \to 0, \text{ while } \eta \to \infty.$$

From Equation (9), λ symbolises as the ratio of velocity parameter, γ and Bi are the dimensionless velocity slip parameter and Biot number, respectively, which are described as

$$\lambda = \frac{b}{a}, \gamma = H(av)^{1/2}, \text{Bi} = \frac{h_f}{k_f}\sqrt{\frac{v(1-ct)}{a}}. \quad (10)$$

Next, we define the skin friction coefficient (C_f) and the local Nusselt number (Nu_x) as

$$C_f = \frac{\tau_w}{\rho_f u_e^2}, \quad Nu_x = \frac{x q_w}{k_f (T_f - T_\infty)}. \tag{11}$$

The shear stress along the $x-$ direction is represented by τ_w, while q_w signifies the surface heat flux that accentuated by

$$\tau_w = \mu_{hnf} \left(\frac{\partial u}{\partial y}\right)_{y=0}, \quad q_w = -k_{hnf} \left(\frac{\partial T}{\partial y}\right)_{y=0}. \tag{12}$$

By applying (5) and (12) into (11), we acquire

$$\sqrt{Re_x} C_f = \frac{\mu_{hnf}}{\mu_f} f''(0), \quad \frac{1}{\sqrt{Re_x}} Nu_x = -\frac{k_{hnf}}{k_f} \theta'(0), \tag{13}$$

provided that $Re_x = \frac{u_e x}{\nu_f}$ is the local Reynolds number in $x-$ axis.

3. Analysis of Solution Stability

By obeying the efforts of Merkin [61] and Merill et al. [62] from their outstanding discoveries of stability analysis scheme, the unsteady equations need to be deliberated in order to ultimately identify the reliable and stable solution since we notice the appearance of non-uniqueness solutions in the boundary value problem (7)–(9). Now, in accordance with the unsteady-state problem, a new similarity conversion is proposed

$$u = \frac{ax}{1-ct} \frac{\partial f}{\partial \eta}(\eta, \tau), v = -\left(\frac{av}{1-ct}\right)^{1/2} f(\eta, \tau), \theta(\eta, \tau) = \frac{T - T_\infty}{T_f - T_\infty},$$
$$\eta = \sqrt{\frac{a}{\nu(1-ct)}} y, \tau = \frac{a}{1-ct} t. \tag{14}$$

Employing the similarity variables of Equation (14) to Equations (7) and (8), we now obtain the following converted differential equations

$$\frac{\mu_{hnf}/\mu_f}{\rho_{hnf}/\rho_f} \frac{\partial^3 f}{\partial \eta^3} + \left(f + \frac{\varepsilon}{2}\eta\right) \frac{\partial^2 f}{\partial \eta^2} - \left(\frac{\partial f}{\partial \eta}\right)^2 - \varepsilon \frac{\partial f}{\partial \eta} - (1+\varepsilon\tau) \frac{\partial^2 f}{\partial \eta \partial \tau} + \varepsilon + 1 = 0, \tag{15}$$

$$\frac{1}{Pr} \frac{k_{hnf}/k_f}{(\rho C_p)_{hnf}/(\rho C_p)_f} \frac{\partial^2 \theta}{\partial \eta^2} + f \frac{\partial \theta}{\partial \eta} - 2\theta \frac{\partial f}{\partial \eta} - \frac{\varepsilon}{2} \eta \frac{\partial \theta}{\partial \eta} - \frac{\varepsilon}{2} 3\theta - (1+\varepsilon\tau) \frac{\partial \theta}{\partial \tau} = 0, \tag{16}$$

with respect to

$$f(0, \tau) = 0, \frac{\partial f}{\partial \eta}(0, \tau) = \lambda + \gamma \frac{\partial^2 f}{\partial \eta^2}(0, \tau), -\frac{k_{hnf}}{k_f} \frac{\partial \theta}{\partial \eta}(0, \tau) = Bi[1 - \theta(0, \tau)],$$
$$\frac{\partial f}{\partial \eta}(\eta, \tau) \to 1, \theta(\eta, \tau) \to 0, \text{ as } \eta \to \infty. \tag{17}$$

In accordance with Weidman et al. [63], to test the stability of the steady flow $f(\eta) = f_0(\eta)$ and $\theta(\eta) = \theta_0(\eta)$ which fulfil the boundary value problem and boundary conditions (refer to (7)–(9)), we write

$$f(\eta, \tau) = f_0(\eta) + e^{-\omega\tau} F(\eta), \quad \theta(\eta, \tau) = \theta_0(\eta) + e^{-\omega\tau} G(\eta), \tag{18}$$

by which ω is the eigenvalue of unidentified variables, while functions $F(\eta)$ and $G(\eta)$ are relatively small to $f_0(\eta)$ and $\theta_0(\eta)$. The eigenvalue problems (15) and (16) result in an infinite group of eigenvalues $\omega_1 < \omega_2 < \omega_3$...... that detect an early decay when ω_1 is positive, while an early growth of disruptions is observed when ω_1 is negative, which exposes the unstable flow. Substituting (18) into (15)–(17), we develop

$$\frac{\mu_{hnf}/\mu_f}{\rho_{hnf}/\rho_f}\frac{\partial^3 F}{\partial \eta^3} + \left(f_0 + \frac{\varepsilon}{2}\eta\right)\frac{\partial^2 F}{\partial \eta^2} + F\frac{\partial^2 f_0}{\partial \eta^2} - 2\frac{\partial f_0}{\partial \eta}\frac{\partial F}{\partial \eta} + (\omega - \varepsilon)\frac{\partial F}{\partial \eta} = 0, \tag{19}$$

$$\frac{1}{\Pr}\frac{k_{hnf}/k_f}{(\rho C_p)_{hnf}/(\rho C_p)_f}\frac{\partial^2 G}{\partial \eta^2} + \left(f_0 - \frac{\varepsilon}{2}\eta\right)\frac{\partial G}{\partial \eta} - 2\left(\theta_0\frac{\partial F}{\partial \eta} + G\frac{\partial f_0}{\partial \eta}\right) + F\frac{\partial \theta_0}{\partial \eta} + \left(\omega - \frac{3}{2}\varepsilon\right)G = 0, \tag{20}$$

and the boundary conditions are

$$F(0,\tau) = 0, \frac{\partial F}{\partial \eta}(0,\tau) - \gamma\frac{\partial^2 F}{\partial \eta^2}(0,\tau) = 0, -\frac{k_{hnf}}{k_f}\frac{\partial G}{\partial \eta}(0,\tau) - \text{Bi}G(0,\tau) = 0,$$
$$\frac{\partial F}{\partial \eta}(\eta,\tau) \to 0, G(\eta,\tau) \to 0, \text{ as } \eta \to \infty. \tag{21}$$

The heat transfer stability and steady-state flow solutions $f_0(\eta)$ and $\theta_0(\eta)$ was implemented via $\tau \to 0$, therefore $F = F_0(\eta)$ and $G = G_0(\eta)$ in (19)–(21). As a consequence, an early growth of Equation (18) is detected, and the subsequent generalized eigenvalue problem is recognized

$$\frac{\mu_{hnf}/\mu_f}{\rho_{hnf}/\rho_f}F_0''' + \left(f_0 + \frac{\varepsilon}{2}\eta\right)F_0'' + F_0 f_0'' - \left(2f_0' - \omega + \varepsilon\right)F_0' = 0, \tag{22}$$

$$\frac{1}{\Pr}\frac{k_{hnf}/k_f}{(\rho C_p)_{hnf}/(\rho C_p)_f}G_0'' + \left(f_0 - \frac{\varepsilon}{2}\eta\right)G_0' + F_0\theta_0' - 2\left(\theta_0 F_0' + G_0 f_0'\right) + \left(\omega - \frac{3}{2}\varepsilon\right)G_0 = 0, \tag{23}$$

subject to

$$F_0(0) = 0, F_0'(0) - \gamma F_0''(0) = 0, -\frac{k_{hnf}}{k_f}G_0'(0) - \text{Bi}G_0(0) = 0,$$
$$F_0'(\eta) \to 0, G_0(\eta) \to 0, \text{ as } \eta \to \infty. \tag{24}$$

The range of possible eigenvalues can be calculated by resting a boundary condition [64] of the present problem. In this study, we choose to repose $F'_0(\eta) \to 0$, and the linear eigenvalue problems (22)–(24) are disclosed as $F''_0(0) = 1$ for a fixed value of ω_1. It is worth mentioning that the values of ω_1 are proficient in measuring the stability of the corresponding solutions $f_0(\eta)$ and $\theta_0(\eta)$.

4. Results and Discussion

The mathematical computations of this research work were achieved by employing the bvp4c function in the MATLAB programming system subject to the governing ordinary differential Equations (7) and (8) together with the boundary conditions (9). A bvp4c method is a notable tool for solving the boundary value problem that has been extensively established by various researchers to clarify the boundary value concern. In order to ensure the obtainment of desired solutions, early estimation of the primary mesh point and variations step size is crucial. Also, a reasonable assumption of the thickness of the boundary layer along with effective preliminary approximation is essential for defining the non-uniqueness solutions. The consistency of the results generated in the present study is evaluated with those in [34,58], as accessible in Table 3, which is in excellent agreement.

In this study, we recognized the hybrid Al_2O_3-Cu/H_2O nanofluid by dispersing the first nanoparticle ϕ_1 (alumina) into the base fluid (water) followed by the second nanoparticle ϕ_2 (copper) with various amounts of volume fractions. It is important to declare that the dispersion of discrete nanoparticle Al_2O_3/Cu is capable of developing Al_2O_3-Cu/H_2O and Cu-H_2O nanofluids. Apart from that, the Prandtl (Pr) number is set to be fixed at Pr = 6.2 corresponds to water as the reference-based fluid, while as for the hybrid Al_2O_3-Cu/H_2O nanofluid, the size of the nanoparticles is assumed to be standardized, and the thermophysical properties effect of nanoparticles agglomeration is ignored. The main purpose of the present study is to examine the influence of the control parameter such as the nanoparticles volume fraction (ϕ_1, ϕ_2), the unsteadiness parameter (ε), the velocity slip parameter

(γ), and the Biot number (Bi) towards the coefficients of skin friction variations ($f''(0)$) and the heat transfer rate ($\theta'(0)$). The alumina and copper nanoparticles volume fraction in the present work is chosen within the range of $0.00 \leq \phi_2 \leq 0.04$ which motivated by the experimental work done by Suresh et al. [65] who conducted the synthesis, characterization of $Al_2O_3 - Cu/H_2O$ nanocomposite powder for different volume concentrations 0.1%, 0.33%, 0.75%, 1%, and 2%. In their valuable study, the stability of the prepared nanofluids was determined by measuring the pH of nanofluids, and the nanofluid stability was found to diminish with increasing volume concentration. On another note, various values of the controlling parameter that has been used, are set within the following extent; $0.1 \leq \varepsilon \leq 0.2$, $0.1 \leq \gamma \leq 0.4$, and $0.2 \leq Bi \leq 0.7$ to ensure the certainty of the obtained solutions. The present study also interested in witnessing the non-uniqueness solutions that appear in the ordinary governing differential of the ensuing problem, hence confirm the real and valid solutions through the stability analysis by utilizing the bvp4c approach in MATLAB operating system (MATLAB R2019b, MathWorks, Natick, MA, USA).

Table 3. Evaluation values of $f''(0)$ when $\varepsilon = \gamma = Bi = 0$, by certain values of λ.

λ	Present Result		Mahapatra and Nandy [34]		Wang [58]	
	First Solution	Second Solution	First Solution	Second Solution	First Solution	Second Solution
−0.25	1.402241	-	1.402242	-	1.4022404	-
−0.50	1.495670	-	1.495672	-	1.4956704	-
−0.75	1.489298	-	1.489296	-	1.4893004	-
−1.00	1.328817	0.000000	1.328819	0.000000	1.3288204	0.000000
−1.10	1.186680	0.049229	1.186680	0.049229	-	-
−1.15	1.082231	0.116702	1.082232	0.116702	1.082230	0.116702
−1.20	0.932473	0.233650	0.932470	0.233648	-	-
−1.246	0.609826	0.529035	0.584374	0.554215	0.5543004	-

The generated results of Equations (7) and (8) perceive the non-uniqueness (dual) solutions together with the boundary conditions (9) to a specific scope of λ_c where λ_c manifests the non-uniqueness solutions meeting point, and this too is regarded as a critical point. There is a single solution at a critical point, and that is the distinct line. The flow separation happens to occur after the critical point, and the flow is no longer laminar and does not obey the principle of boundary layer theory. According to the numerical results attained in this research work, it is proven that a non-uniqueness solutions appear, namely first and second solutions, as depicted in Figures 2–15. The existence of the non-uniqueness solutions contributes to the analysis of solution stability so that one may be able to verify the theoretically relevant solution. The first solution, however, is predicted to be reliable and fundamentally exist in practice. The smallest eigenvalues ω_1 for some values of λ when $\varepsilon = 0.1, \phi_1 = \phi_2 = 0.02, Bi = 0.2$, and $Pr = 6.2$ with $\gamma = 0.1, 0.2, 0.4$, are tabulated in Table 4. The values of ω_1 is noted approaching zero in the first and second solutions for $\lambda \rightarrow \lambda_c$ in selected cases of γ. Hence, the authors may assure that the stability formulation and practices of the current problem are accurate and reliable. The flow is measured unstable if ω_1 is negative because it implies an early development of disruptions that leads to flow separation. The smallest eigenvalue ω_1 with positive value implies the flow is practically attainable and sustainable, which elucidates the solution stabilizing property to overwhelm the allowing disruptions. Also, it denotes an initial deterioration of disruptions that appear. The positive value of ω_1 describes the stable mode of the flow as in the first solution. The significance of dual solutions that contribute to stability analysis is in revealing the other possibilities of flow behavior, which may be useful for future references in the extrusion process. For example, in the present study, we have presented that the second solutions are unstable solutions with negative eigenvalues. These unstable solutions with the negative eigenvalues infer the growth of the disturbance in the solutions, especially at some range of the stretching/shrinking parameter ($\lambda < −1.0$). When the shrinking rate increases, it limits other external forces' effect at the sheet and eventually showed the opposite behavior of the transport phenomena than the first solutions in the same range of $\lambda < -1.0$, as the respective parameter varies. Moreover, any boundary value problem can generate more than one solution because of the nonlinearity in the boundary value problem; for example, see the mathematical model (7)–(9). Also, changes in the governing parameter values cause bifurcations in solutions that yield the existence of dual solutions [1]. Therefore, the mathematical model (7)–(9), which obeyed the boundary layer assumptions, managed to exhibit the variety in the fluid flow and heat transfer behavior through the uniqueness and existence of the solutions.

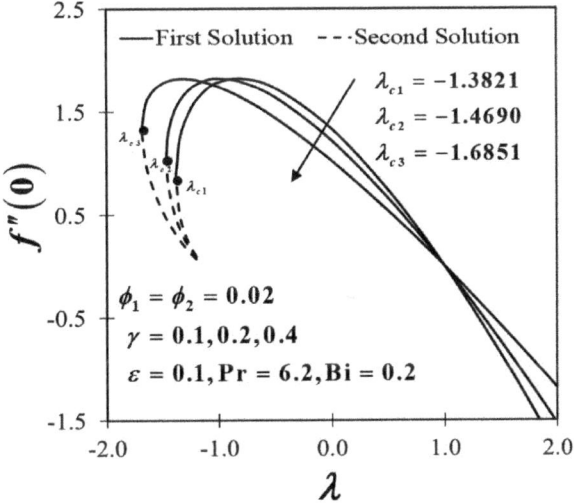

Figure 2. Variants of $f''(0)$ towards λ with $\gamma = 0.1, 0.2, 0.4$.

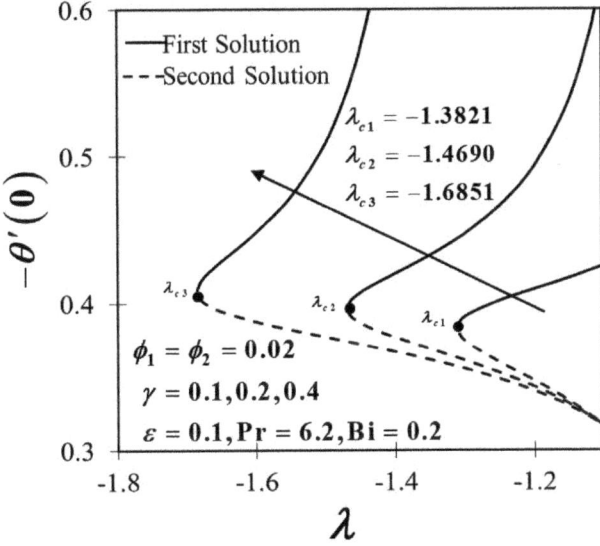

Figure 3. Variants of $-\theta'(0)$ towards λ with $\gamma = 0.1, 0.2, 0.4$.

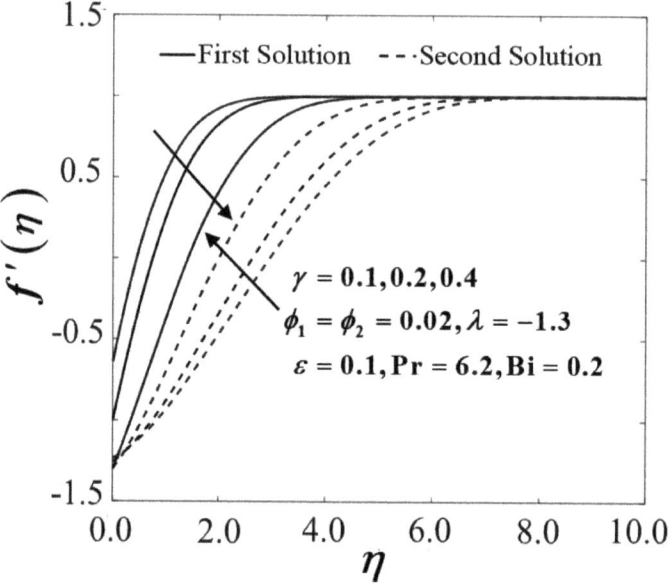

Figure 4. Velocity profiles of $f'(\eta)$ with $\lambda = -1.3$ (shrinking case).

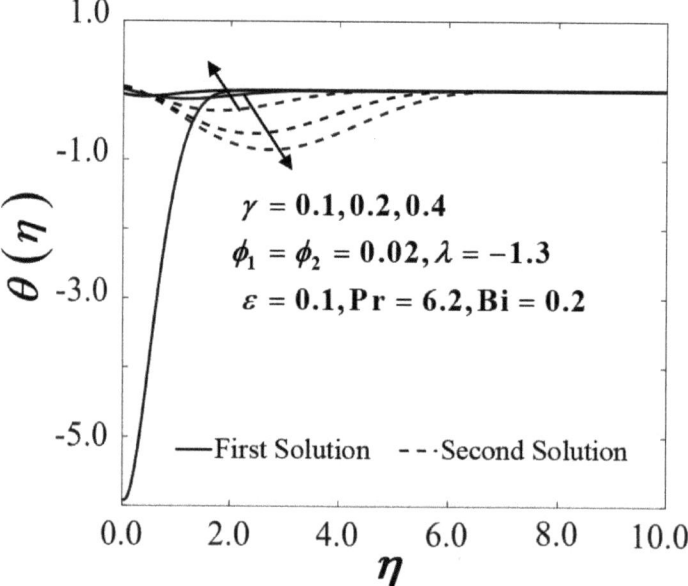

Figure 5. Temperature profiles of $\theta(\eta)$ with $\lambda = -1.3$ (shrinking case).

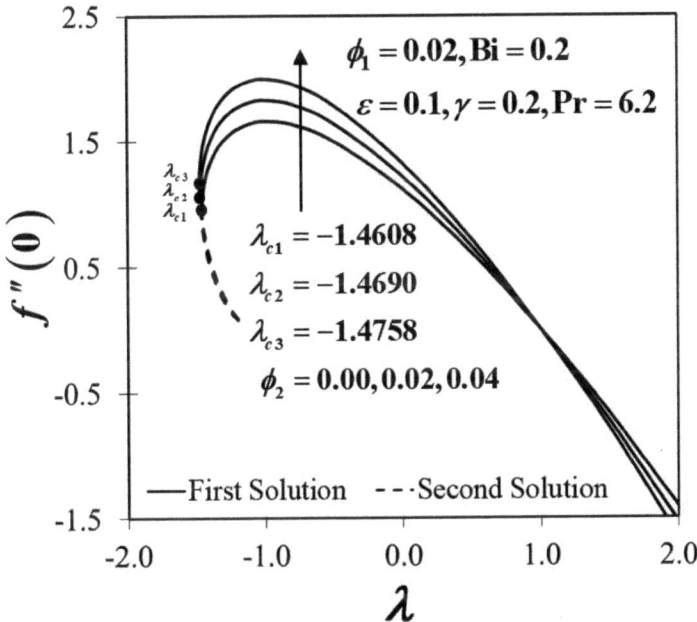

Figure 6. Variants of $f''(0)$ towards λ with $\phi_2 = 0.00, 0.02, 0.04$.

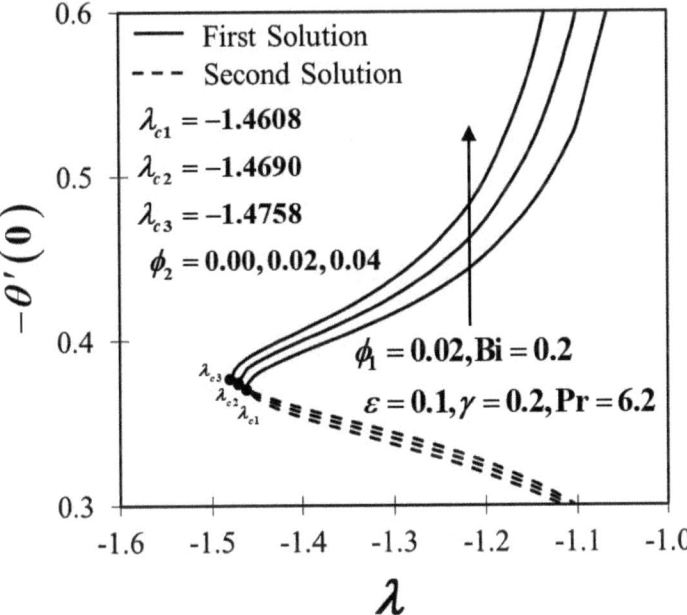

Figure 7. Variants of $-\theta'(0)$ towards λ with $\phi_2 = 0.00, 0.02, 0.04$.

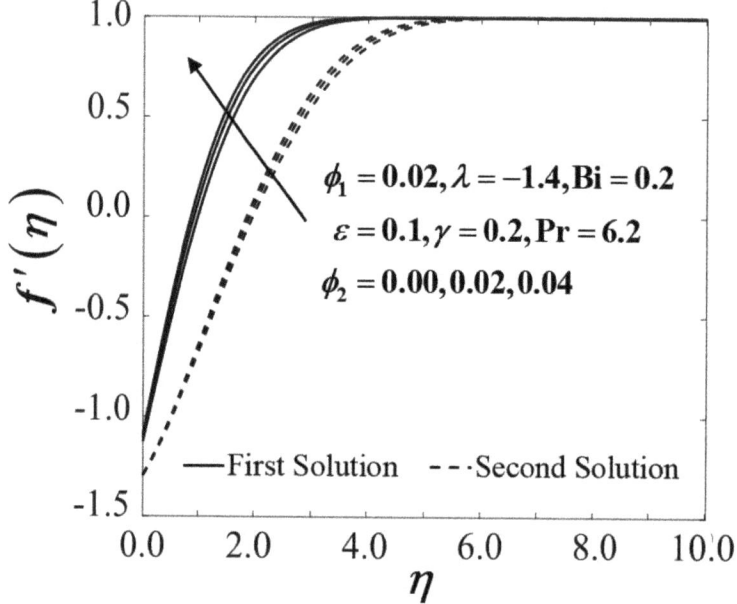

Figure 8. Velocity profiles of $f'(\eta)$ with $\lambda = -1.4$ (shrinking case).

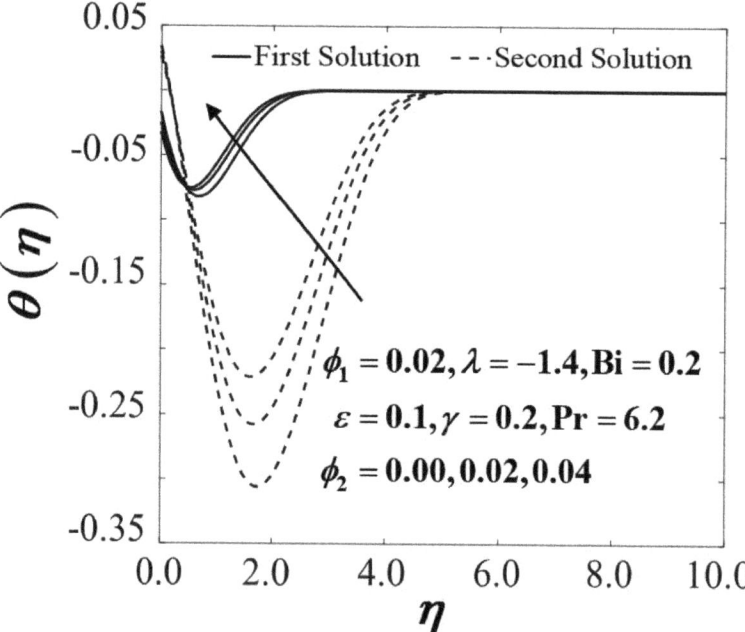

Figure 9. Temperature profiles of $\theta(\eta)$ with $\lambda = -1.4$ (shrinking case).

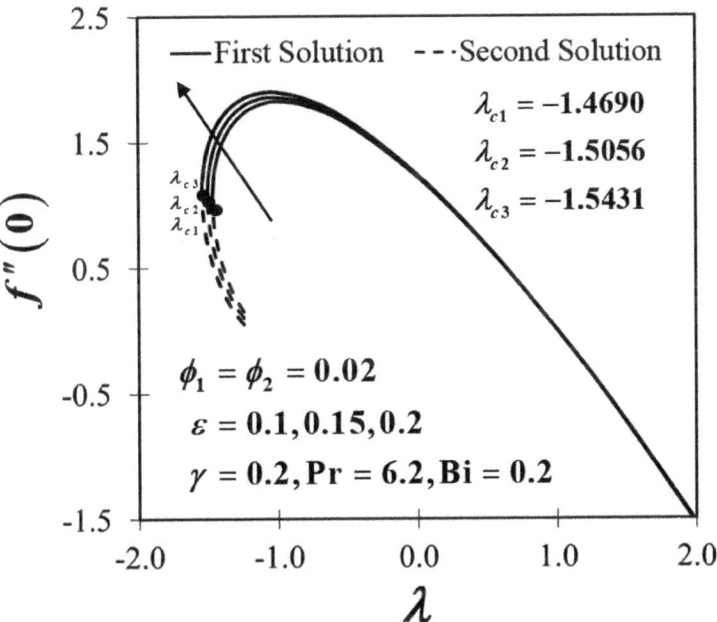

Figure 10. Variants of $f''(0)$ towards λ with $\varepsilon = 0.10, 0.15, 0.20$.

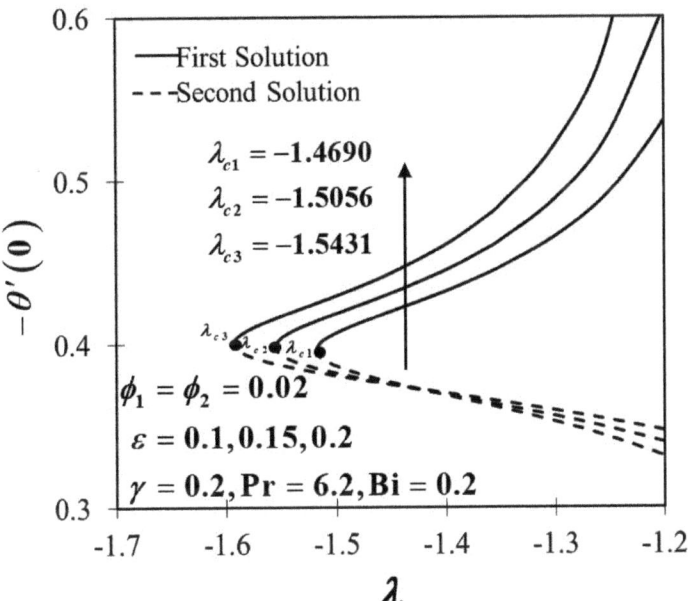

Figure 11. Variants of $-\theta'(0)$ towards λ with $\varepsilon = 0.10, 0.15, 0.20$.

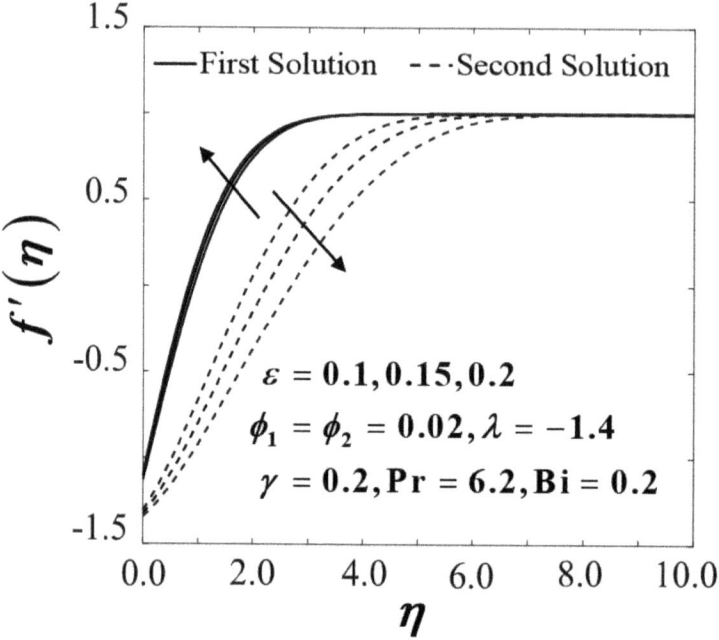

Figure 12. Velocity profiles of $f'(\eta)$ with $\varepsilon = 0.10, 0.15, 0.20$.

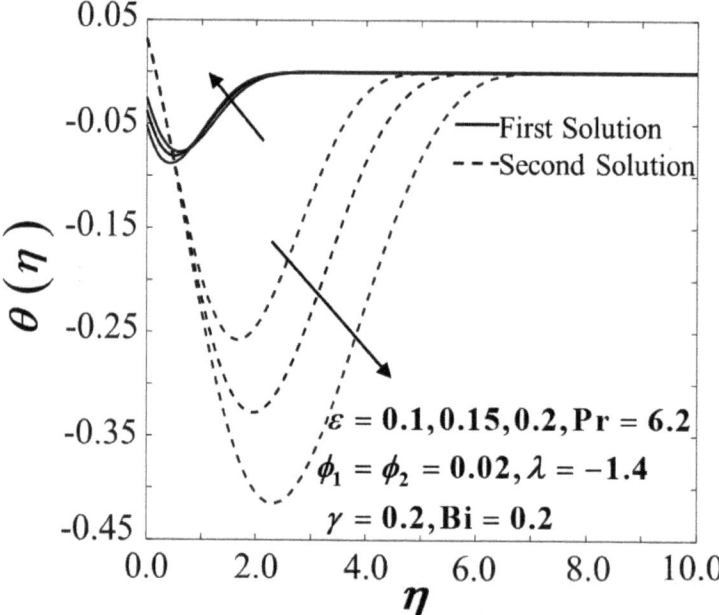

Figure 13. Temperature profiles of $\theta(\eta)$ with $\varepsilon = 0.10, 0.15, 0.20$.

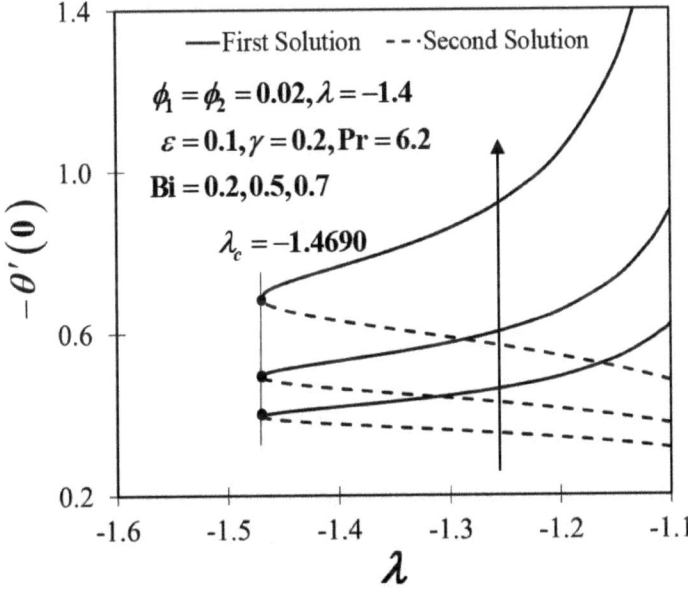

Figure 14. Variants of $-\theta'(0)$ towards λ with Bi = 0.2, 0.5, 0.7.

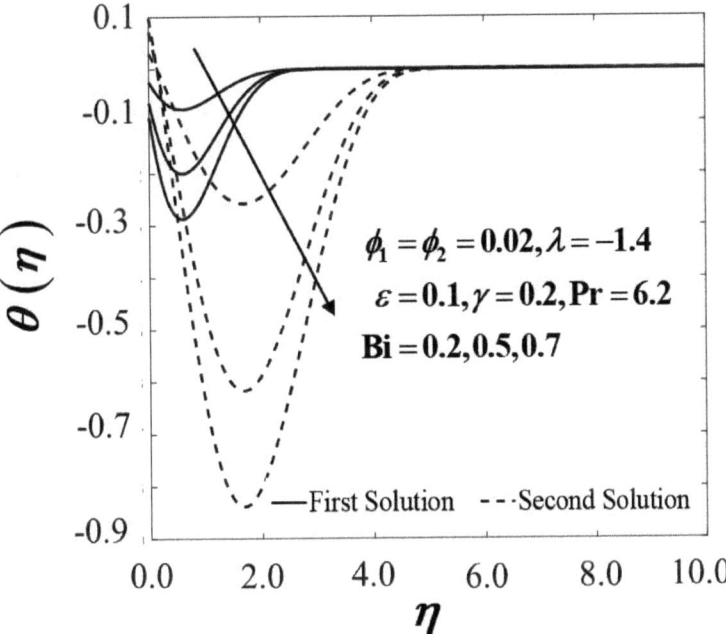

Figure 15. Temperature profiles of $\theta(\eta)$ with Bi = 0.2, 0.5, 0.7.

Table 4. Smallest eigenvalues ω_1 for some values of λ when $\gamma = 0.1, 0.2, 0.4$.

γ	λ	ω_1 First Solution	ω_1 Second Solution
0.1	−1.30	0.7646	−0.7093
	−1.380	0.1015	−0.1398
	−1.3820	0.0095	−0.0500
0.2	−1.40	0.6870	−0.6316
	−1.460	0.2387	−0.2435
	−1.4690	0.0088	−0.0225
0.4	−1.68	0.1822	−0.1576
	−1.684	0.0912	−0.0696
	−1.6851	0.0276	−0.0067

The influence of velocity slip on the skin friction coefficient and the local Nusselt number past a convectively heated stretching/shrinking sheet of hybrid Al_2O_3-Cu/H_2O nanofluid are displayed in Figures 2–5. Figure 2 presents the coefficient of skin friction $(f''(0))$ towards the stretching/shrinking parameter λ, which revealed a decline of $f''(0)$ with the occurrence of velocity slip effect ($\gamma = 0.1, 0.2, 0.4$) at the boundary. The same results are obtained in the previous literature, as reported by Dzulkifli et al. [59]. Based on a physical perspective, an improvement in the slip parameter reflects the fact that the vorticity produced by the stretching/shrinking velocity is gradually decreased, thus the vorticity stays restricted within the boundary layer for greater stretching/shrinking velocity with the same straining velocity of the stagnation flow, and subsequently, the steady solution is achievable for some broad values of λ, as stated by Mahapatra and Nandy [34]. Figure 2 emphasizes that the solution for a certain value of γ persist prior to a critical value $\lambda = (\lambda_c < 0)$ at which the boundary layer splits from the convectively heated stretching/shrinking sheet and the solution on the basis of the boundary layer approximations is not feasible. In short, no solution exists for $\lambda_c < -1.6851$. Moreover, the expansion of γ results in the increment of $|\lambda_c|$ suggesting that the velocity slip parameter is efficiently influenced by raising the range of dual solutions. Thus, it is proven that the existence of a slip velocity impact may prolong the separation of the boundary layer. Figure 2 also highlights as the sheet is stretching at the rate of $\lambda = 1$, the value of $f''(0) = 0$ which describes no frictional drag exerted at the convectively heated stretching/shrinking sheet.

Figure 3 exposes an upward trend in $-\theta'(0)$ when the velocity slip parameter arises on the convectively heated stretching/shrinking sheet, which is proportionate to the heat transfer rate. Those findings are in line with the remarkable work reviewed by Mahapatra and Nandy [34] and Khashi'ie et al. [66]. Evidently, in the case of the first solution, the rate of heat transfer presents an ascendant trend with an increase of γ while the second solution permits a top-down direction when the velocity slip γ enlarges. From the existing and current findings, the authors can conclude that the velocity slip contributes to the improvement of the heat transfer rate significantly, prior to this case study. However, the authors would also like to declare that such results may vary if different control parameters are taken into consideration. The distribution of velocity $(f'(\eta))$ and temperature $(\theta(\eta))$ profiles over several values of γ in the case of a convectively heated shrinking sheet is certified in Figures 4 and 5. As demonstrated in Figure 4, it is confirmed that as γ increases, $f'(\eta)$ is subsequently decreased in the first solution and revealed an upward trend in the velocity profile in the second solution with the presence of velocity slip condition. As the slip occurs, the velocity flow near the sheet in no longer equal to stretching velocity. This is due to the slip condition where the pulling force of the stretching sheet is partly shifted to the fluid, resulting in the decrement of the velocity profiles in the first solution. In short, the slip condition reduces the momentum transfer from the sheet to the fluid. A contradict results are obtained in $\theta(\eta)$ as the velocity slip effect is exaggerated, as shown in Figure 5. The first solution in the temperature profile distribution increases as γ intensifies, while it diminishes in the second solution. This may occur due to the appearance of slip boundary conditions on the wall, which triggers the hybrid nanofluid to retain the velocity on the walls, and such slip may prevent from total heat exchange of the hybrid nanofluid. It is therefore noticed when the slip coefficient is applied, the difference in temperature rate is increased. Additionally, the effect of convective heating progress that has been reflected in the present study, which controls the temperature of the stretching/shrinking surface also might contribute to this phenomenon. As a matter of fact, greater convection leads to increased surface temperatures, which permit the thermal impact to penetrate deeper within the hybrid Al_2O_3-Cu/H_2O nanofluid.

Figures 6 and 7 expose the coefficient of skin friction $(f''(0))$ and the rate of heat transfer $(-\theta'(0))$ of a conventional nanofluid $(\phi_1 = 0.02, \phi_2 = 0)$ and hybrid nanofluid $(\phi_1 = 0.02, \phi_2 = 0.02, 0.04)$ past a convectively heated stretching/shrinking sheet when ϕ_2 varies from 0.00 to 0.02 where $\varepsilon = 0.1, \gamma = 0.2, Bi = 0.2$, and $Pr = 6.2$. Figure 6 manifests that addition in ϕ_2 which indicates the transformation of the conventional Al_2O_3-H_2O fluid to the hybrid Al_2O_3-Cu/H_2O nanofluid, upsurges the values of $f''(0)$ once the sheet is shrinking. The viscosity of hybrid Al_2O_3-Cu/H_2O nanofluid rises when ϕ_2 increases, which eventually improves the fluid velocity over the convectively heated shrinking sheet, as proven in Figure 8. The velocity profile in Figure 8 clarifies that the momentum boundary layer thickness was diminished in response to the rise of ϕ_2, thereby raising the velocity of the fluid and boosting the gradient of velocity. In fact, the thinner momentum boundary layer continues to evolve the wall shear stress as well as the convectively heated shrinking sheet, leading to an enhancement of $f''(0)$. As $f''(0)$ increases, the result implies the increase of the frictional drag exerted on the convectively heated shrinking surface, which may delay the boundary layer flow separation. Besides, Figure 6 also highlights when the sheet is stretching at the rate of $\lambda = 1$, the value of $f''(0) = 0$ which explains no frictional drag exerted at the sheet surface. Meanwhile, Figure 7 illustrates an increasing trend of the heat transfer characteristic or $-\theta'(0)$ when the values of ϕ_2 past a convectively heated shrinking sheet, and this trend holds true to the first solution but is in dispute with the results in the second solution. In essence, as the conventional Al_2O_3-H_2O nanofluid becomes the hybrid Al_2O_3-Cu/H_2O nanofluid, the heat transfer efficiency improves. This finding upholds the assumption of the convective heat transfer system can be improved by optimizing the nanoparticle concentration when ϕ_2 increased. The results obtained in Figures 6 and 7 are consistent with Waini et al. [20] and Zainal et al. [21], whereby adding the concentrations of hybrid nanoparticles may contribute to the improvement of the heat transfer rate, accordingly. The temperature profile in Figure 9 describes the temperature variations when the conventional Al_2O_3-H_2O nanofluid becomes the hybrid Al_2O_3-Cu/H_2O nanofluid in both first and second solutions. The incline in the temperature of hybrid nanofluid proliferates the thermal conductivity, which may be triggered by the extra energy dispersed through the increment of the nanoparticles volume fraction over the state of convectively heated stretching/shrinking sheet.

Figures 10–13 show the impact of the unsteadiness parameter (ε) towards a convectively heated stretching/shrinking sheet when ε shifts from 0.1 to 0.2. The hybrid Al_2O_3-Cu/H_2O nanofluid characteristic with regard to the coefficient of skin friction $(f''(0))$ is depicted as in Figure 10. Figure 10 captures that when the sheet shrinks, the increment in ε conclusively increases the trend of $f''(0)$ in the first solution. An increment in the unsteadiness parameter results in the reduction of the boundary layer thickness, as depicted in Figure 12 and consequently upsurge the velocity gradient on the convectively heated stretching/shrinking sheet, thus $f''(0)$ improves. Furthermore, the existence of nanoparticle volume fraction in the working fluid (Al_2O_3-Cu/H_2O) might also trigger the increment of $f''(0)$ owing to an uplift of the hybrid nanofluid viscosity. This result is aligned with the study done by Ismail et al. [67]. Figure 11 presents the unsteadiness parameter effect towards the rate of heat transfer $-\theta'(0)$. According to the generated results, the heat transfer rate increases when ε increases when the convectively heated sheet is shrinking. The forced convective heat transfer is indeed proportional to the effectiveness of nanofluid heat conductivity, hence raising the reduced local Nusselt number notably. The dimensionless velocity profiles $f'(\eta)$ with a different value of ε are depicted in Figure 12, where the presence of dual velocity profiles is also observed. As illustrated in Figure 12, the first solution increases proportionally to the increment of ε values while the second solution displayed contradict results of the first solution, possibly because of the enhancement in the unsteadiness of the flow. Meanwhile, the same trend of the solution in Figure 12 also reflected the graph of temperature distribution $\theta(\eta)$ in the convectively heated stretching/shrinking sheet with the existence of unsteadiness parameter, as portrayed in Figure 13. Apart from that, it is proven that the second solution in both profiles, i.e., velocity and temperature distribution showed larger boundary layer thickness in each solution of the unsteady cases than those of the first solution.

Figures 14 and 15 depict the variants of heat transfer rate $-\theta'(0)$ and temperature distribution profile $\theta(\eta)$ with a different value of the Biot number (Bi = 0.2, 0.5, 0.7) towards the convectively heated stretching/shrinking sheet. In Figure 14, the heat transfer rate shows an upsurge trend in the dual solutions along with the augmentation of Bi values. The Biot number signifies the conduction resistance ratio within the sheet to convection resistance at the sheet. The critical values of the different usage of Biot numbers suggest no substantial impact on the magnitude of $-\theta'(0)$, as highlighted by Jusoh et al. [68]. An increment in the Biot number related to the improvement of convective heating is observed to decrease the fluid temperature proficiently in the first and second solutions, as displayed in Figure 15. This result is in contrast with the idea of a large Biot number representing larger internal thermal resistance of the sheet compare to the boundary layer thermal resistance since the temperature distribution profile spike as the value of Bi increases. However, note that the Biot number is specifically correlated to the coefficient of heat transfer h_f; therefore, it is conversely related to the thermal resistance of the current problem. Consequently, the heat resistance decreases as the Biot number increases, thereby increasing the heat transfer rate at the stretching/shrinking sheet and decreasing the temperature distribution (see Figure 15).

5. Conclusions

An analysis of the unsteady stagnation point flow of hybrid nanofluid over a convectively heated stretching/shrinking sheet incorporating the velocity slip impact on heat transfer was verified in this study. The results were generated by employing the bvp4c features in the MATLAB programming platform. The effect of diverse controlling parameters—namely, the nanoparticle volume fraction, the velocity slip, the Biot number, and the unsteadiness parameter—were reviewed. Our discoveries happen to determine that the presence of non-uniqueness solutions (first and second solutions) is verifiable within the hybrid Al_2O_3-Cu/H_2O nanofluid for a specific range of control parameters, and the stability analysis authorizes the reliability of our first solution. The nanoparticle volume fraction increment improved both the skin friction coefficient and local Nusselt number in the hybrid nanofluid (Al_2O_3-Cu/H_2O). From this, it is proven that the heat transfer rate improves when the ordinary Al_2O_3-H_2O nanofluid becomes the hybrid nanofluid (Al_2O_3-Cu/H_2O) by expanding the nanoparticles concentration. The recent addition of the velocity slip parameter at the boundary had encouraged a reduction in the skin friction coefficient and velocity profile. However, it increased the rate of heat transfer significantly. The temperature profiles escalate as the magnitude of velocity slip upsurge because of such slip may prevent from exchange of total heat in the hybrid nanofluid. An increase in the unsteadiness parameter consequently raises the velocity gradient on the convectively heated stretching/shrinking sheet, thus improving the skin friction coefficient. Besides, an upsurge in the Biot number intensity boosts the heat transfer rate since the Biot number is directly associated with the heat transfer rate coefficient. Thus, it is conversely interrelated to the thermal resistance of the designated problem. Apart from that, the critical values of the different practices of the Biot number imply no significant outcome towards the magnitude of the heat transfer coefficient.

Author Contributions: Research design, N.A.Z., R.N., K.N., and I.P.; Formulation and methodology, N.A.Z.; Result analysis, N.A.Z.; Validation, R.N. and K.N.; Article preparation, N.A.Z.; Review and editing, N.A.Z., R.N., K.N., and I.P. All authors have read and agreed to the published version of the manuscript.

Funding: The present work is endorsed by the research award (DIP-2017-009) from UKM.

Acknowledgments: All authors value the productive feedbacks by the competent reviewers.

Conflicts of Interest: The authors declare no conflict of interest.

Nomenclature

Roman letters

a, b, c	constant (−)
Bi	Biot number (−)
C_f	skin friction coefficient (−)
C_p	specific heat at constant pressure $(Jkg^{-1}K^{-1})$
H_1	velocity slip factor (−)
h_f	heat transfer coefficient $(Wm^{-2}K^{-2})$
$f(\eta)$	dimensionless stream function (−)
k	thermal conductivity of the fluid $(Wm^{-1}K^{-1})$
Nu_x	local Nusselt number (−)
(pC_p)	heat capacitance of the fluid $(JK^{-1}m^{-3})$
Pr	Prandtl number (−)
Re_x	local Reynolds number in $x-$ axis (−)
t	time (s)
T	fluid temperature (K)
T_0	reference temperature (K)
T_∞	ambient temperature (K)
u, v	velocities component in the $x-$ and $y-$ directions, respectively (ms^{-1})
u_e	velocities of the free stream in (ms^{-1})
u_w	velocities of the stretching/shrinking surface (ms^{-1})
x, y	rtesian coordinates (m)

Greek symbols

ψ	stream function (−)
η	similarity variable (−)
θ	dimensionless temperature (−)
ε	unsteadiness parameter (−)
λ	ratio of the velocity parameter (−)
γ	velocity slip parameter (−)
μ	dynamic viscosity of the fluid $(kgm^{-1}s^{-1})$
ν	kinematic viscosity of the fluid (m^2s^{-1})
ρ	density of the fluid (kgm^{-3})
τ	dimensionless time variable (−)
τ_w	wall shear stress $(kgm^{-1}s^{-2})$
ϕ_1	nanoparticle volume fractions for Al_2O_3 (alumina) (−)
ϕ_2	nanoparticle volume fractions for Cu (copper) (−)
ω	eigenvalue (−)
ω_1	smallest eigenvalue (−)

Subscripts

f	base fluid (−)
nf	nanofluid (−)
hnf	hybrid nanofluid (−)
$s1$	solid component for Al_2O_3 (alumina) (−)
$s2$	solid component for Cu (copper) (−)

Superscript

$'$	differentiation with respect to η (−)

References

1. Schlichting, H.; Gersten, K. *Boundary Layer Theory*; Springer: Berlin/Heidelberg, Germany, 2016.

2. Smith, F.T. Steady and unsteady boundary layer separation. *Annu. Rev. Fluid Mech.* **1986**, *18*, 197–220. [CrossRef]
3. White, F.M. *Viscous Fluid Flow*; McGraw-Hill: New York, NY, USA, 1991.
4. Elbashbeshy, E.M.A.; Bazid, M.A.A. Heat transfer over an unsteady stretching surface. *Heat Mass Transf.* **2004**, *41*, 1–4. [CrossRef]
5. Bhattacharyya, K. Dual solutions in unsteady stagnation-point flow over a shrinking sheet. *Chin. Phys. Lett.* **2011**, *28*, 084702. [CrossRef]
6. Bachok, N.; Ishak, A.; Pop, I. The boundary layers of an unsteady stagnation-point flow in a nanofluid. *Int. J. Heat Mass Transf.* **2012**, *55*, 6499–6505. [CrossRef]
7. Fan, T.; Xu, H.; Pop, I. Unsteady stagnation flow and heat transfer towards a shrinking sheet. *Int. Commun. Heat Mass Transf.* **2010**, *37*, 1440–1446. [CrossRef]
8. Zainal, N.A.; Nazar, R.; Naganthran, K.; Pop, I. Unsteady three-dimensional MHD non-axisymmetric Homann stagnation point flow of a hybrid nanofluid with stability analysis. *Mathematics* **2020**, *8*, 784. [CrossRef]
9. Kamal, F.; Zaimi, K.; Ishak, A.; Pop, I. Stability analysis on the stagnation-point flow and heat transfer over a permeable stretching/shrinking sheet with heat source effect. *Int. J. Numer. Methods Heat Fluid Flow* **2018**, *28*, 2650–2663. [CrossRef]
10. Basir, F.; Hafidzuddin, E.H.; Naganthran, K.; Chaharborj, S.S.; Kasihmuddin, M.S.M.; Nazar, R. Stability analysis of unsteady stagnation-point gyrotactic bioconvection flow and heat transfer towards the moving sheet in a nanofluid. *Chin. J. Phys.* **2020**, *65*, 538–553. [CrossRef]
11. Jusoh, R.; Nazar, R.; Pop, I. Impact of heat generation/absorption on the unsteady magnetohydrodynamic stagnation point flow and heat transfer of nanofluids. *Int. J. Numer. Methods Heat Fluid Flow* **2019**, *30*, 557–574. [CrossRef]
12. Zheng, Y.; Ahmed, N.A.; Zhang, W. Heat dissipation using minimum counter-flow jet ejection during spacecraft re-entry. *Procedia Eng.* **2012**, *49*, 271–279. [CrossRef]
13. Fisher, E.G. *Extrusion of Plastics*; Wiley: New York, NY, USA, 1976.
14. Hiemenz, K. Die Grenzschicht an einem in den gleichförmigen Flüssigkeitsstrom eingetauchten geraden Kreiszylinder. *Dinglers Polytech. J.* **1911**, *326*, 321–324.
15. Homann, F. Der Einfluss grosser Zähigkeit bei der Strömung um den Zylinder und um die Kugel. *Z. Angew. Math. Mech.* **1936**, *16*, 153–164.
16. Howarth, L. CXLIV. The boundary layer in three-dimensional flow—Part II. The flow near a stagnation point. *Lond. Edinb. Dublin Philos. Mag. J. Sci.* **1951**, *42*, 1433–1440.
17. Khashi'ie, N.S.; Arifin, N.M.; Pop, I. Mixed convective stagnation point flow towards a vertical Riga plate in hybrid Cu–Al$_2$O$_3$/water nanofluid. *Mathematics* **2020**, *8*, 912.
18. Khashi'ie, N.S.; Hafidzuddin, E.H.; Ariffin, N.M.; Wahi, N. Stagnation point flow of hybrid nanofluid over a permeable vertical stretching/shrinking cylinder with thermal stratification effect. *CFD Lett.* **2020**, *12*, 80–94.
19. Fang, T.G.; Wang, F.J. Momentum and heat transfer of a special case of the unsteady stagnation-point flow. *Appl. Math. Mech. Engl.* **2020**, *41*, 51–82.
20. Waini, I.; Ishak, A.; Pop, I. Hybrid nanofluid flow towards a stagnation point on a stretching/shrinking cylinder. *Sci. Rep.* **2020**, *10*, 1–12.
21. Zainal, N.A.; Nazar, R.; Naganthran, K.; Pop, I. MHD mixed convection stagnation point flow of a hybrid nanofluid past a vertical flat plate with convective boundary condition. *Chin. J. Phys.* **2020**, *66*, 630–644.
22. Wang, C.Y. Flow over a surface with parallel grooves. *Phys. Fluids.* **2003**, *15*, 1114–1121.
23. Sharipov, F.; Seleznev, V. Data on internal rarefied gas flows. *J. Phys. Chem. Ref. Data* **1998**, *27*, 657–706.
24. Hafidzuddin, E.H.; Nazar, R.; Arifin, N.M.; Pop, I. Effects of anisotropic slip-on three-dimensional stagnation-point flow past a permeable moving surface. *Eur. J. Mech. B Fluids.* **2017**, *65*, 515–521. [CrossRef]
25. Pavlišič, A.; Huš, M.; Prašnikar, A.; Likozar, B. Multiscale modelling of CO_2 reduction to methanol over industrial Cu/ZnO/Al$_2$O$_3$ heterogeneous catalyst: Linking ab initio surface reaction kinetics with reactor fluid dynamics. *J. Clean. Prod.* **2020**, *275*, 122958. [CrossRef]
26. Navier, C.L. Memorie sur les lois du lois du mouvement des fluides. *Mem. Acad. Sci. Inst. France* **1827**, *6*, 298–440.

27. Maxwell, J. On stresses in rarefied gases arising from inequalities of temperature. *Philos. Trans. R. Soc. Lond.* **1879**, *27*, 231–256.
28. Wang, C.Y. Stagnation flows with slip: Exact solutions of the Navier-Stokes equations. *Z. Fur Angew. Math. Und Phys.* **2003**, *54*, 184–189. [CrossRef]
29. Rao, I.J.; Rajagopal, K.R. Effect of the slip boundary condition on the flow of fluids in a channel. *Acta Mech.* **1999**, *135*, 113–126. [CrossRef]
30. Jusoh, R.; Nazar, R.; Pop, I. Three-dimensional flow of a nanofluid over a permeable stretching/shrinking surface with velocity slip: A revised model. *Phys. Fluids* **2018**, *30*, 033604. [CrossRef]
31. Yoshimura, A.; Prud'homme, R.K. Wall slip corrections for couette and parallel disk viscometers. *J. Rheol.* **1988**, *32*, 53–67. [CrossRef]
32. Vajravelu, K.; Mukhopadhyay, S. *Fluid Flow, Heat and Mass Transfer at Bodies of Different Shapes: Numerical Solutions*; Academic Press: Cambridge, MA, USA, 2015.
33. Mukhopadhyay, S. Effects of slip-on unsteady mixed convective flow and heat transfer past a porous stretching surface. *Nucl. Eng. Des.* **2011**, *241*, 2660–2665. [CrossRef]
34. Mahapatra, R.T.; Nandy, S.K. Slip effects on unsteady stagnation-point flow and heat transfer over a shrinking sheet. *Meccanica* **2013**, *48*, 1599–1606. [CrossRef]
35. Majumder, M.; Chopra, N.; Andrews, R.; Hinds, B.J. Nanoscale hydrodynamics: Enhanced flow in carbon nanotubes. *Nature* **2005**, *438*, 930. [CrossRef]
36. Noghrehabadi, A.; Pourrajab, R.; Ghalambaz, M. Effect of partial slip boundary condition on the flow and heat transfer of nanofluids past stretching sheet prescribed constant wall temperature. *Int. J. Therm. Sci.* **2012**, *54*, 253–261. [CrossRef]
37. Van Gorder, R.A.; Sweet, E.; Vajravelu, K. Nano boundary layers over stretching surfaces. *Commun. Nonlinear Sci. Numer. Simulat.* **2010**, *15*, 1494–1500. [CrossRef]
38. Dinarvand, S.; Rostami, M.N. Rotating Al_2O_3-H_2O nanofluid flow and heat transfer with internal heating, velocity slip and different shapes of nanoparticles. *Multidiscip. Model. Mater. Struct.* **2020**. Available online: https://doi.org/10.1108/MMMS-01-2020-0017 (accessed on 17 August 2020). [CrossRef]
39. Seth, G.S.; Bhattacharyya, A.; Kumar, R.; Chamkha, A.J. Entropy generation in hydromagnetic nanofluid flow over a non-linear stretching sheet with Navier's velocity slip and convective heat transfer. *Phys. Fluids* **2018**, *30*, 122003. [CrossRef]
40. Rahman, J.U.; Khan, U.; Ahmad, S.; Ramzan, M.; Suleman, M.; Lu, D.; Inam, S. Numerical Simulation of Darcy–Forchheimer 3D Unsteady Nanofluid Flow Comprising Carbon Nanotubes with Cattaneo–Christov Heat Flux and Velocity and Thermal Slip Conditions. *Processes* **2019**, *7*, 687. [CrossRef]
41. Reddy, R.C.S.; Reddy, P.S. A comparative analysis of unsteady and steady Buongiorno's Williamson nanoliquid flow over a wedge with slip effects. *Chin. J. Chem. Eng.* **2020**, *28*, 1767–1777. [CrossRef]
42. Reddy, J.V.R.; Sugunamma, V.; Sandeep, N. Thermophoresis and Brownian motion effects on unsteady MHD nanofluid flow over a slendering stretching surface with slip effects. *Alex. Eng. J.* **2018**, *57*, 2465–2473. [CrossRef]
43. Choi, S.U.; Eastman, J. Enhancing thermal conductivity of fluids with nanoparticles. *ASME Publ. Fed.* **1995**, *231*, 99–103.
44. Shah, T.R.; Ali, H.M. Applications of hybrid nanofluids in solar energy, practical limitations and challenges: A critical review. *Sol. Energy* **2019**, *183*, 173–203. [CrossRef]
45. Huminic, G.; Huminic, A. Hybrid nanofluids for heat transfer applications—A state-of-the-art review. *Int. J. Heat Mass Transf.* **2018**, *125*, 82–103.
46. Gupta, M.; Singh, V.; Kumar, S.; Kumar, S.; Dilbaghi, N. Up to date review on the synthesis and thermophysical properties of hybrid nanofluids. *J. Clean. Prod.* **2018**, *190*, 169–192.
47. Xian, H.W.; Azwadi, N.; Sidik, C.; Aid, S.R.; Ken, T.L.; Asako, Y. Review on preparation techniques, properties and performance of hybrid nanofluid in recent engineering applications. *J. Adv. Res. Fluid Mech. Therm. Sci.* **2018**, *45*, 1–13.
48. Madhesh, D.; Kalaiselvam, S. Experimental analysis of hybrid nanofluid as a coolant. *Procedia Eng.* **2014**, *97*, 1667–1675.
49. Tahat, M.S.; Benim, A.C. Experimental analysis on thermophysical properties of Al_2O_3/CuO hybrid nanofluid with its effects on flat plate solar collector. *Defect Diffus. Forum* **2017**, *374*, 148–156.

50. Labib, M.N.; Nine, M.J.; Afrianto, H.; Chung, H.; Jeong, H. Numerical investigation on effect of base fluids and hybrid nanofluid in forced convective heat transfer. *Int. J. Therm. Sci.* **2013**, *71*, 163–171.
51. Moghadassi, A.; Ghomi, E.; Parvizian, F. A numerical study of water-based Al_2O_3 and Al_2O_3–Cu hybrid nanofluid effect on forced convective heat transfer. *Int. J. Therm. Sci.* **2015**, *92*, 50–57.
52. Devi, S.U.; Devi, S.P.A. Heat transfer enhancement of Cu–Al_2O_3/water hybrid nanofluid flow over a stretching sheet. *J. Niger. Math. Soc.* **2017**, *36*, 419–433.
53. Takabi, B.; Salehi, S. Augmentation of the heat transfer performance of a sinusoidal corrugated enclosure by employing hybrid nanofluid. *Adv. Mech. Eng.* **2014**, *6*, 147059.
54. Aladdin, N.A.L.; Bachok, N.; Pop, I. Cu–Al_2O_3/water hybrid nanofluid flow over a permeable moving surface in presence of hydromagnetic and suction effects. *Alex. Eng. J.* **2020**, *59*, 657–666.
55. Plant, R.D.; Hodgson, G.K.; Impellizzeri, S.; Saghir, M.Z. Experimental and numerical investigation of heat enhancement using a hybrid nanofluid of copper oxide/alumina nanoparticles in water. *J. Therm. Anal. Calorim.* **2020**, *141*, 1951–1968. [CrossRef]
56. Lund, L.A.; Omar, Z.; Khan, I.; Sherif, E.S.M. Dual solutions and stability analysis of a hybrid nanofluid over a stretching/shrinking sheet executing MHD flow. *Symmetry* **2020**, *12*, 276. [CrossRef]
57. Ghalambaz, M.; Roşca, N.C.; Roşca, A.V.; Pop, I. Mixed convection and stability analysis of stagnation-point boundary layer flow and heat transfer of hybrid nanofluids over a vertical plate. *Int. J. Numer. Methods Heat Fluid Flow* **2019**, *30*, 3737–3754. [CrossRef]
58. Wang, C.Y. Stagnation flow towards a shrinking sheet. *Int. J. Non-Linear Mech.* **2008**, *43*, 377–382. [CrossRef]
59. Dzulkifli, N.F.; Bachok, N.; Yacob, N.A.; Arifin, N.M.; Rosali, H. Unsteady stagnation-point flow and heat transfer over a permeable exponential stretching/shrinking sheet in nanofluid with slip velocity effect: A stability analysis. *Appl. Sci.* **2018**, *8*, 2172. [CrossRef]
60. Oztop, H.F.; Abu-Nada, E. Numerical study of natural convection in partially heated rectangular enclosures filled with nanofluids. *Int. J. Heat Fluid Flow* **2008**, *29*, 1326–1336.
61. Merkin, J.H. Natural-convection boundary-layer flow on a vertical surface with Newtonian heating. *Int. J. Heat Fluid Flow* **1994**, *15*, 392–398.
62. Merrill, K.; Beauchesne, M.; Previte, J.; Paullet, J.; Weidman, P. Final steady flow near a stagnation point on a vertical surface in a porous medium. *Int. J. Heat Mass Transf.* **2006**, *49*, 4681–4686.
63. Weidman, P.D.; Kubitschek, D.G.; Davis, A.M.J. The effect of transpiration on self-similar boundary layer flow over moving surfaces. *Int. J. Eng. Sci.* **2006**, *44*, 730–737.
64. Harris, S.D.; Ingham, D.B.; Pop, I. Mixed convection boundary-layer flow near the stagnation point on a vertical surface in a porous medium: Brinkman model with slip. *Transp. Porous Media* **2009**, *77*, 267–285. [CrossRef]
65. Suresh, S.; Venkitaraj, K.P.; Selvakumar, P. Synthesis, characterisation of Al_2O_3–Cu nanocomposite powder and water-based nanofluids. *Adv. Mater. Res.* **2011**, *328*, 1560–1567. [CrossRef]
66. Khashi'ie, N.S.; Arifin, N.M.; Pop, I.; Nazar, R.; Hafidzuddin, E.H.; Wahi, N. Three-dimensional hybrid nanofluid flow and heat transfer past a permeable stretching/shrinking sheet with velocity slip and convective condition. *Chin. J. Phys.* **2020**, *66*, 157–171. [CrossRef]
67. Ismail, N.S.; Arifin, N.M.; Nazar, R.; Bachok, N. Stability analysis of unsteady MHD stagnation point flow and heat transfer over a shrinking sheet in the presence of viscous dissipation. *Chin. J. Phys.* **2019**, *57*, 116–126. [CrossRef]
68. Jusoh, R.; Nazar, R.; Pop, I. Flow and heat transfer of magnetohydrodynamic three-dimensional Maxwell nanofluid over a permeable stretching/shrinking surface with convective boundary conditions. *Int. J. Mech. Sci.* **2017**, *124*, 166–173. [CrossRef]

© 2020 by the authors. Licensee MDPI, Basel, Switzerland. This article is an open access article distributed under the terms and conditions of the Creative Commons Attribution (CC BY) license (http://creativecommons.org/licenses/by/4.0/).

Article

Effects of Entropy Generation, Thermal Radiation and Moving-Wall Direction on Mixed Convective Flow of Nanofluid in an Enclosure

Sivasankaran Sivanandam [1,*], Ali J. Chamkha [2,3], Fouad O. M. Mallawi [1], Metib S. Alghamdi [4] and Aisha M. Alqahtani [5]

1 Department of Mathematics, King Abdulaziz University, Jeddah 21589, Saudi Arabia; fmallawi@hotmail.com
2 Institute of Research and Development, Duy Tan University, Da Nang 550000, Vietnam; alichamkha@duytan.edu.vn
3 Institute of Theoretical and Applied Research (ITAR), Duy Tan University, Hanoi 100000, Vietnam
4 Mathematics Department, King Khalid University, Abha 61421, Saudi Arabia; dr.matabalghamdi@gmail.com
5 Mathematics Department, Princess Nourah bint Abdulrahman University, Riyadh 84428, Saudi Arabia; aalqahtani@gmail.com
* Correspondence: smsivanandam@kau.edu.sa or sd.siva@yahoo.com

Received: 16 July 2020; Accepted: 27 August 2020; Published: 1 September 2020

Abstract: A numeric investigation is executed to understand the impact of moving-wall direction, thermal radiation, entropy generation and nanofluid volume fraction on combined convection and energy transfer of nanoliquids in a differential heated box. The top wall of the enclosed box is assumed to move either to the left or the right direction which affects the stream inside the box. The horizontal barriers are engaged to be adiabatic. The derived mathematical model is solved by the control volume technique. The results are presented graphically to know the impact of the dissimilar ways of moving wall, Richardson number, Bejan number, thermal radiation, cup mixing and average temperatures. It is concluded that the stream and the thermal distribution are intensely affected by the moving-wall direction. It is established that the thermal radiation enhances the convection energy transport inside the enclosure.

Keywords: mixed convection; thermal radiation; entropy; nanoliquid; moving wall

1. Introduction

The combined convective movement and thermal energy transfer have been examined in a huge number of studies for decades because of its applications in numerous fields of technological sciences. Since the communal interaction among the viscous, buoyancy, and inertia forces on the stream has been a vital matter for joint convection in a lid-driven enclosed box, the moving wall's direction of the cavity becomes significant in these studies [1–4]. Therefore, the current work keenly involves the influence of moving-wall direction on convective stream in lid-driven cavities. Combined convection together with heat transfer have been examined under several conditions in enormous studies [5–9]. Sivasankaran et al. [10] numerically explored the mixed convective stream and the energy transport in an inclined enclosed space with discrete heating. Sivasankaran and Pan [11] discovered the influence of discrete heaters and coolers on convection in a closed box. Mekroussi et al. [12] explored the combined convection in a top-driven inclined wavy walled box. Combined convection flow due to nonuniform heating in an enclosed box is discovered in some studies [13–15].

Nanofluids are pioneering fluids in the field of thermal science and it has been used actively to analyze the energy transport in thermal systems [16–22]. Sheremet et al. [21] discovered the buoyant flow and entropy generation of nanoliquid in a closed box with variable border temperature.

Alsabery et al. [22] numerically explored the entropy generation and convection of nanoliquid in a wavy walled box. Santra et al. [23] deliberated the energy transfer augmentation of a water–copper nanoliquid in a differentially heated box. Abu-Nada and Oztop [24] discovered the outcome of inclination of the box on convection of a Cu–water nanofluid. Ghasemi and Aminossadati [25] explored the buoyant convection of a CuO nanoliquid in an inclined box numerically. Bhuvaneswari et al. [26] completed a numeric work to get the impact of variable liquid properties on convective stream of a nanoliquid in a square box. Sivasankaran et al. [27] inspected the partial slip influence on magneto-convection in a 2-sided wall-driven porous enclosed space filled with a Cu–water nanoliquid. Rashad et al. [28] discovered the magneto-convection of heat generating nanoliquids in a trapezoidal box with discrete heating.

The interaction connecting natural/mixed convection and thermal radiation has gained significant consideration due to its uses in various arenas. Very few studies on the interaction of thermal radiation and convective stream have been reported in the literature [29–34]. Mansour et al. [29] discovered the outcome of radiation on buoyant convection in a porous wavy enclosed space using the non-equilibrium thermal model. They found that average heat transport decreased by increasing the surface waviness of the wall. The doubly diffusive convection with radiation in an enclosed box was explored by Moufekkir et al. [30]. Mahapatra et al. [31] explored the influence of heat generation and thermal radiation on magneto-convective stream in an inclined enclosed space with one hot side and chilled from the adjacent side. They concluded that the direction of the magnetic field influenced much on the stream pattern. Saleem et al. [32] scrutinized the impact of radiation on buoyant convection in an open box. They demonstrated that radiative heat transport increased as the optical thickness of the liquid increased. Zhang et al. [33] explored the effects of thermal radiation on magneto-convection in a cavity.

Since no study on combined convection of a nanoliquid in a wall-driven box with thermal radiation and entropy generation is reported in the literature, the current investigation is interested to investigate numerically the effect of entropy, thermal radiation and the direction of wall movement of an enclosed box on the convective stream and energy transfer of a nanoliquid.

2. Mathematical Modeling

The physical model displayed in Figure 1 is a 2-dimensional square enclosed box of size L packed with a water-based Al_2O_3-nanofluid. The stream is unsteady, incompressible and laminar. The velocity components (u, v) in Cartesian coordinates (x, y) are pointed to in Figure 1. The vertical walls of the enclosed domain have uniform temperature distributions. The horizontal barriers are thermally insulated. The gravity performances in the opposite of y-direction. The nanoliquid in the enclosed box is considered as a dilute liquid–solid mixture with a constant volume fraction of nanosized particles (Al_2O_3) distributed within the water. The nanoparticles and water are in thermal-equilibrium. The nanoliquid properties are presumed to be constant, except the density. The linear variation of density (with temperature) is given as $\rho = \rho_0[1 - \beta(\theta - \theta_0)]$, where β being the quantity of thermal expansion (Boussinesq approximation), θ is temperature and ρ_0 is density at reference. The viscous dissipation is discounted here. The mathematical model for conservation of quantities is:

$$\frac{\partial u}{\partial x} + \frac{\partial v}{\partial y} = 0 \qquad (1)$$

$$\frac{\partial u}{\partial t} + u\frac{\partial u}{\partial x} + v\frac{\partial u}{\partial y} = -\frac{1}{\rho_{nf}}\frac{\partial p}{\partial x} + \frac{\mu_{nf}}{\rho_{nf}}\left(\frac{\partial^2 u}{\partial x^2} + \frac{\partial^2 u}{\partial y^2}\right) \qquad (2)$$

$$\frac{\partial v}{\partial t} + u\frac{\partial v}{\partial x} + v\frac{\partial v}{\partial y} = -\frac{1}{\rho_{nf}}\frac{\partial p}{\partial y} + \frac{\mu_{nf}}{\rho_{nf}}\left(\frac{\partial^2 v}{\partial x^2} + \frac{\partial^2 v}{\partial y^2}\right) + \frac{(\rho\beta)_{nf}}{\rho_{nf}}g(\theta - \theta_0) \qquad (3)$$

$$\frac{\partial \theta}{\partial t} + u\frac{\partial \theta}{\partial x} + v\frac{\partial \theta}{\partial y} = \alpha_{nf}\left(\frac{\partial^2 \theta}{\partial x^2} + \frac{\partial^2 \theta}{\partial y^2}\right) + \frac{1}{(\rho c_p)_{nf}}\left(\frac{\partial q_r}{\partial x} + \frac{\partial q_r}{\partial y}\right) \qquad (4)$$

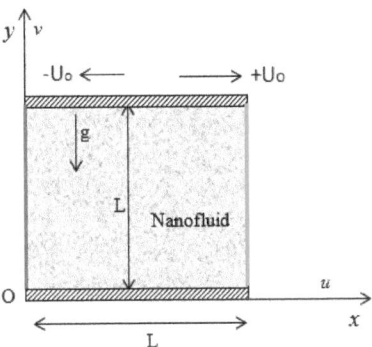

Figure 1. Physical model.

The subscript "nf" and "0" denote the nanofluid and reference state, respectively. The parameters $c_p, g, p, t, \alpha, \mu$ are specific heat, acceleration due to gravity, pressure, time, thermal diffusivity and the dynamic viscosity, respectively. The heat flux due to radiation along the x and y directions are set by $qr_x = \frac{-4\sigma^*}{3K'}\frac{\partial \theta^4}{\partial x}$ and $qr_y = \frac{-4\sigma^*}{3K'}\frac{\partial \theta^4}{\partial y}$, where σ^* is Stefan-Boltzmann constant and K' is mean absorption coefficient. By Rosseland estimate for radiation (medium is optically thick), the thermal variances within the stream are reflected to be too small. Expanding θ^4 about θ_0 through Taylor series and neglecting the higher order terms obtained from Taylor series, θ^4 is expressed as a function of temperature θ. That is,

$$\theta^4 = \theta_0^4 + 4\theta_0^3(\theta - \theta_0) + \ldots$$

Then, by approximating we get,

$$\theta^4 \cong 4\theta_0^3 \theta - 3\theta_0^4$$

Therefore, the radiative heat flux reduces to

$$qr_x = \frac{-16\sigma^* \theta_0^3}{3K'}\frac{\partial \theta}{\partial x} \text{ and } qr_y = \frac{-16\sigma^* \theta_0^3}{3K'}\frac{\partial \theta}{\partial y} \qquad (5)$$

Substituting Equation (5) into Equation (4), we get

$$\frac{\partial \theta}{\partial t} + u\frac{\partial \theta}{\partial x} + v\frac{\partial \theta}{\partial y} = \alpha_{nf}\left(\frac{\partial^2 \theta}{\partial x^2} + \frac{\partial^2 \theta}{\partial y^2}\right) + \frac{1}{(\rho c_p)_{nf}}\frac{-16\sigma^* \theta_0^3}{3K'}\left(\frac{\partial^2 \theta}{\partial x^2} + \frac{\partial^2 \theta}{\partial y^2}\right) \qquad (6)$$

Initially, the velocity and temperature are zero. When $t > 0$, $u = v = 0$ except at top wall and $u = +U_0$ (Case -1), $u = -U_0$ (Case-2), $v = 0$ on the top wall. For temperature, $\frac{\partial \theta}{\partial y} = 0$ on the top and the bottom portions. The right and left walls are lower ($\theta = \theta_c$) and higher ($\theta = \theta_h$) temperature.

The properties of the nanoliquid in the current model are defined below.

Density:

$$\rho_{nf} = \rho_f(1 - \phi) + \phi \rho_p \qquad (7)$$

Thermal expansion coefficient:

$$(\rho\beta)_{nf} = (\rho\beta)_f(1 - \phi) + \phi(\rho\beta)_p \qquad (8)$$

Specific heat:

$$(\rho c_p)_{nf} = (\rho c_p)_f(1 - \phi) + \phi(\rho c_p)_p \qquad (9)$$

The Maxwell formula is used for thermal conductivity:

$$k_{nf} = k_f \left[\frac{2 + k^*_{pf} + 2\phi(k^*_{pf} - 1)}{2 + k^*_{pf} - \phi(k^*_{pf} - 1)} \right], \quad k^*_{pf} = \frac{k_p}{k_f} \tag{10}$$

The dynamic viscosity of nanoliquid (Ho et al. [35]) is calculated as:

$$\mu_{nf} = \mu_f (1 - \phi)^{-2.5} \tag{11}$$

where the subscript "f" and "p" denote base–fluid and nanoparticle, respectively. The physical constants of the water and nanoparticles (Al$_2$O$_3$) are available in Ref [35].

The leading equations are nondimensionalized by using the subsequent variables:

$$(U, V) = \frac{(u, v)}{U_0}, \quad T = \frac{\theta - \theta_0}{\Delta \theta}, \quad (X, Y) = \frac{(x, y)}{L}, \quad \tau = \frac{tU_0}{L}, \cdots \text{and } P = \frac{p}{\rho_{nf} U_0^2} \tag{12}$$

The consequent nondimensional model equations are

$$\frac{\partial U}{\partial X} + \frac{\partial V}{\partial Y} = 0 \tag{13}$$

$$\frac{\partial U}{\partial \tau} + U \frac{\partial U}{\partial X} + V \frac{\partial U}{\partial Y} = -\frac{\partial P}{\partial X} + \left(\frac{\mu_{nf}}{\rho_{nf} v_f} \right) \left(\frac{1}{\text{Re}} \right) \left(\frac{\partial^2 U}{\partial X^2} + \frac{\partial^2 U}{\partial Y^2} \right) \tag{14}$$

$$\frac{\partial V}{\partial \tau} + U \frac{\partial V}{\partial X} + V \frac{\partial V}{\partial Y} = -\frac{\partial P}{\partial Y} + \left(\frac{\mu_{nf}}{\rho_{nf} \alpha_f} \right) \left(\frac{1}{\text{Re}} \right) \left(\frac{\partial^2 V}{\partial X^2} + \frac{\partial^2 V}{\partial Y^2} \right) + \frac{(\rho \beta)_{nf}}{\rho_{nf} \beta_f} Ri\, T \tag{15}$$

$$\frac{\partial T}{\partial \tau} + U \frac{\partial T}{\partial X} + V \frac{\partial T}{\partial Y} = \left(\frac{\alpha_{nf}}{\alpha_f} \frac{1}{\text{RePr}} \right) \left(1 + \frac{4k_f}{3k_{nf}} Rd \right) \left(\frac{\partial^2 T}{\partial X^2} + \frac{\partial^2 T}{\partial Y^2} \right) \tag{16}$$

The nondimensional quantities appearing above are the Grashof number $Gr = (g\beta_f \Delta \theta L^3)/(v_f^2)$, Radiation parameter $Rd = (4\sigma^* \theta_0^3)/(k_f K')$, Richardson number $Ri = Gr/\text{Re}^2$, Reynolds number $\text{Re} = (U_0 L)/(v_f)$ and the Prandtl number $Pr = v_f/\alpha_f$. The boundary settings are

$$U = V = 0, \quad X = 0, 1 \,\&\, Y = 0$$

$$U = +1 \text{ (Case 1)}, \,\&\, U = -1 \text{ (Case 2)}, \, V = 0, \, Y = 1$$

$$\frac{\partial T}{\partial Y} = 0 \, Y = 0 \,\&\, 1 \tag{17}$$

$$T = 1 \, X = 0 \,\&\, T = 0 \, X = 1$$

when $U = +1$ indicates that the wall moves to the right-side and $U = -1$ implies that the wall moves to the left-side in its axis, respectively.

The drag coefficient estimates the total frictional drag exerted on the wall. The drag coefficient along the moving top wall is calculated as $Cf_x = \left(\frac{\partial U}{\partial Y} \right)_{Y=1}$, respectively. The averaged drag coefficient is calculated as

$$\overline{Cf_x} = \int_0^1 Cf_x \, dX, \text{ respectively.} \tag{18}$$

The energy transport rate across the enclosed box is a vital parameter in thermal industrial applications. The local Nusselt number alongside the hot barrier of the enclosed box is defined as

$Nu = \left(-\frac{k_{nf}}{k_f}\left(1 + \frac{4k_f}{3k_{nf}}Rd\right)\frac{\partial T}{\partial X}\right)_{X=0}$. The averaged Nusselt number alongside the heated barrier is expressed as follows:

$$\overline{Nu} = \int_0^1 Nu\, dY \tag{19}$$

3. Cup Mixing Temperature and RMSD

The temperature of cup mixing is defined to discover the thermal mixing inside the chamber. The velocity weighted average temperature is most appropriate for convection flow than space averaged temperature. The temperature of cup mixing, and averaged temperature based on area are given as [34]

$$T_{Cup} = \frac{\iint \hat{V}(X,Y)\, T(X,Y) dXdY}{\iint \hat{V}(X,Y) dXdY} \tag{20}$$

where $\hat{V}(X,Y) = \sqrt{U^2 + V^2}$ and

$$T_{avg} = \frac{\iint T(X,Y) dXdY}{\iint dXdY} \tag{21}$$

The root-mean square deviation (RMSD) is deduced to calculate the degree of temperature uniformity in all considered cases. They are deduced based on temperature of cup mixing and average temperature based on area as follows:

$$RMSD_{T_{cup}} = \sqrt{\frac{\sum_{i=1}^{N}(T_i - T_{Cup})^2}{N}} \tag{22}$$

$$RMSD_{T_{avg}} = \sqrt{\frac{\sum_{i=1}^{N}(T_i - T_{avg})^2}{N}} \tag{23}$$

The greater values of RMSD point out poorer temperature regularity in the chamber and vice versa. Moreover, RMSD cannot exceed one because the dimensionless temperature differs between zero and one. These parameters are estimated by the gained values of flow and thermal fields in the same computational code.

4. Entropy Generation

The buoyance induced convection in a closed chamber discovers significant awareness in thermal engineering applications. However, the practice of entropy generation supports to spot the ideal conditions for many applications. Since the generation of entropy is as a result of the irreversible procedure of transfer of heat and viscosity, generation of entropy can be estimated from the well-known thermal and velocity fields.

The entropy generation is expressed by two quantities, i.e., heat transfer (first term in below equation) and liquid friction (last term in below equation) [18,21,22,34].

$$S_{Gen} = \frac{k_{nf}}{T_c^2}\left[\left(\frac{\partial \theta}{\partial x}\right)^2 + \left(\frac{\partial \theta}{\partial y}\right)^2 + \left(\frac{16\sigma^*\theta_0^3}{3K'}\right)\left(\left(\frac{\partial \theta}{\partial x}\right)^2 + \left(\frac{\partial \theta}{\partial y}\right)^2\right)\right] + \left(\frac{\mu_{nf}}{T_c}\right)\left\{2\left[\left(\frac{\partial u}{\partial x}\right)^2 + \left(\frac{\partial v}{\partial y}\right)^2\right] + \left(\frac{\partial u}{\partial y} + \frac{\partial v}{\partial x}\right)^2\right\} \tag{24}$$

The dimensionless entropy generation is acquired by using (10)

$$S_{total} = S^*_{heat} + S^*_{fluid}$$

$$S^*_{heat} = \left(\frac{k_{nf}}{k_f}\right)\left(1 + \frac{4Rd}{3}\right)\left[\left(\frac{\partial T}{\partial X}\right)^2 + \left(\frac{\partial T}{\partial Y}\right)^2\right] \tag{25}$$

$$S^*_{fluid} = \phi_2 \left(\frac{\mu_{nf}}{\mu_f}\right) \left\{ 2\left[\left(\frac{\partial U}{\partial X}\right)^2 + \left(\frac{\partial V}{\partial Y}\right)^2\right] + \left(\frac{\partial U}{\partial Y} + \frac{\partial V}{\partial X}\right)^2 \right\} \quad (26)$$

where $\phi_2 = \frac{U_0}{\theta_0 L^2}$. The global entropy generation attains by integrating the local entropy production inside the chamber.

$$SG_{total} = \int_V S_{total}(X,Y) dA \quad (27)$$

The local Bejan number states the strength of generation of entropy owing to thermal transference irreversibility. It is derived as

$$Be_{loc} = \frac{S^*_{heat}}{S^*_{total}} \quad (28)$$

For any point in the chamber, when $Be_{loc} > \frac{1}{2}$, the heat transfer irreversibility is dominating. When $Be_{loc} < \frac{1}{2}$, the liquid friction irreversibility dominates. If $Be_{loc} = \frac{1}{2}$, the thermal and viscous irreversibilities are equal. The average value of Bejan number demonstrates the relative importance of the thermal energy transfer irreversibility for the entire chamber.

$$Be = \frac{\int_A Be_{loc}(X,Y) dA}{\int_A dA} \quad (29)$$

5. Numeric Technique

The nondimensional Equations (12)–(15) with boundary conditions (16) are solved by the control volume technique with the "SIMPLE algorithm". A nonuniform grid of 122 × 122 is taken to investigate the problem. The justification of the numeric code is very essential in the simulation. An internal code is tested against the available results for free convection of nanoliquid in a box [35] and it is shown in Table 1. Second, the problem of combined convection stream in a lid-driven box [36,37] is employed to compare the results of the current code (See Table 2). A good agreement among these results is obtained. Hence, the results offer a guarantee in the accuracy of the current computational code to inspect the problem.

Table 1. Comparison of $\overline{Nu}$ results for free convection of nanoliquids in a square box.

Ra (Rayleigh Number)	Volume Fraction	$\overline{Nu}$ Ho et al. [35]	Present
10^3	0.01	1.129	1.137
	0.04	1.199	1.205
10^4	0.01	2.264	2.229
	0.04	2.305	2.335
10^5	0.01	4.699	4.683
	0.04	4.810	4.791
10^6	0.01	9.165	9.170
	0.04	9.428	9.513

Table 2. Comparison of average Nusselt numbers for mixed convection in a lid-driven box.

Gr	Re = 400			Re = 1000		
	Present Work	Iwatsu et al. [37]	Sharif [36]	Present Work	Iwatsu et al. [37]	Sharif [36]
10^2	4.09	3.84	4.05	6.48	6.33	6.55
10^4	3.85	3.62	3.82	6.47	6.29	6.50
10^6	1.10	1.22	1.17	1.66	1.77	1.81

6. Results and Discussion

Numeric simulations are executed to examine the mixed convective stream and energy transfer of nanoliquids in a wall-driven enclosed box with thermal radiation and entropy generation. The average and cup mixing temperature and its RMSD values are also calculated. The calculations are carried out for a Richardson number (Ri) ranging from 0.01 to 10^2, a volume fraction (ϕ) of nanoparticles from 0–4 and a radiation parameter from 0 to 10. The Grashof number is used as 10^4 and the Reynolds number varies from 10 to 10^3. The Prandtl number is taken as $Pr = 6.7$. The influence of convective stream and energy transport are assessed for several values of the volume fraction of nanoparticles, Richardson number, radiation parameter and the moving-wall directions. The results are depicted graphically for various combinations of parameters and the discussions are given below.

Figure 2 depicts the stream arrangement for several values of the pertinent parameters Rd and Ri for Case 1 ($U_0 = +1$) with $\Phi = 0.02$. In Case 1, wall is moving towards the right side, whereas the lid moves from the right-side to left-side in Case 2. The moving-wall direction is very important and produces the shear force with the adjoining fluid along the upper portion of the box. Since the convective flow is driven by both the buoyant force and the shear stress due to the moving lid, the Richardson number clearly demonstrates the three regimes of convection (free, mixed and forced). The single clockwise rotating eddy appears in the forced convective regime ($Ri < 1$) for all given values of the radiation parameter. Due to the strong shear force, the core area of the eddy travels towards the right–top corner of the enclosed box. When $Ri = 1$, that is, in the combined convective regime, the magnitude of both forces (shear and buoyancy) are comparable, the core region moving the center part of the portion of the enclosed box. In the buoyant convective regime, that is, $Ri = 100$, the variation on the flow pattern is clearly visible here. There is no change on the stream pattern in the forced convection regime when changing the radiation parameter. However, the evidence on the effect of the radiation parameter is clearly seen in the buoyancy convection regime upon raising the values of the radiation parameter for Case 1.

Figure 3 exhibits the convective stream for several values of Ri and Rd for Case 2, with $\Phi = 0.02$. The flow pattern is completely different from Case 1. The dual cell gets for all values of Ri and Rd as it occupies the entire box. Since the shear force is dominant at $Ri = 0.01$, the core section of the eddy moves towards the left–top corner of the enclosed box. The counter acting eddy could not occupy the whole space as in Case 1, where the movement of liquid particles is aiding with the buoyant force. The buoyant force by the hot liquid along the hot wall produces the clockwise-rotating eddy along the hot wall. However, the shear force dominates here, the eddy by the moving-lid occupies most the box. When rising the Richardson number values to $Ri = 1$, the mixed convection exists, where both the shear and buoyancy forces are comparable, and the eddy produced by these forces occupies about half of the enclosure in the situation. The natural convection mode at $Ri = 100$ depicts different phenomena on the stream pattern compared to the other two modes. The eddy by the buoyancy force dominates and occupies most the enclosed box. It is also detected that the eddy by the shear force is weakened on raising the values of the radiation parameter.

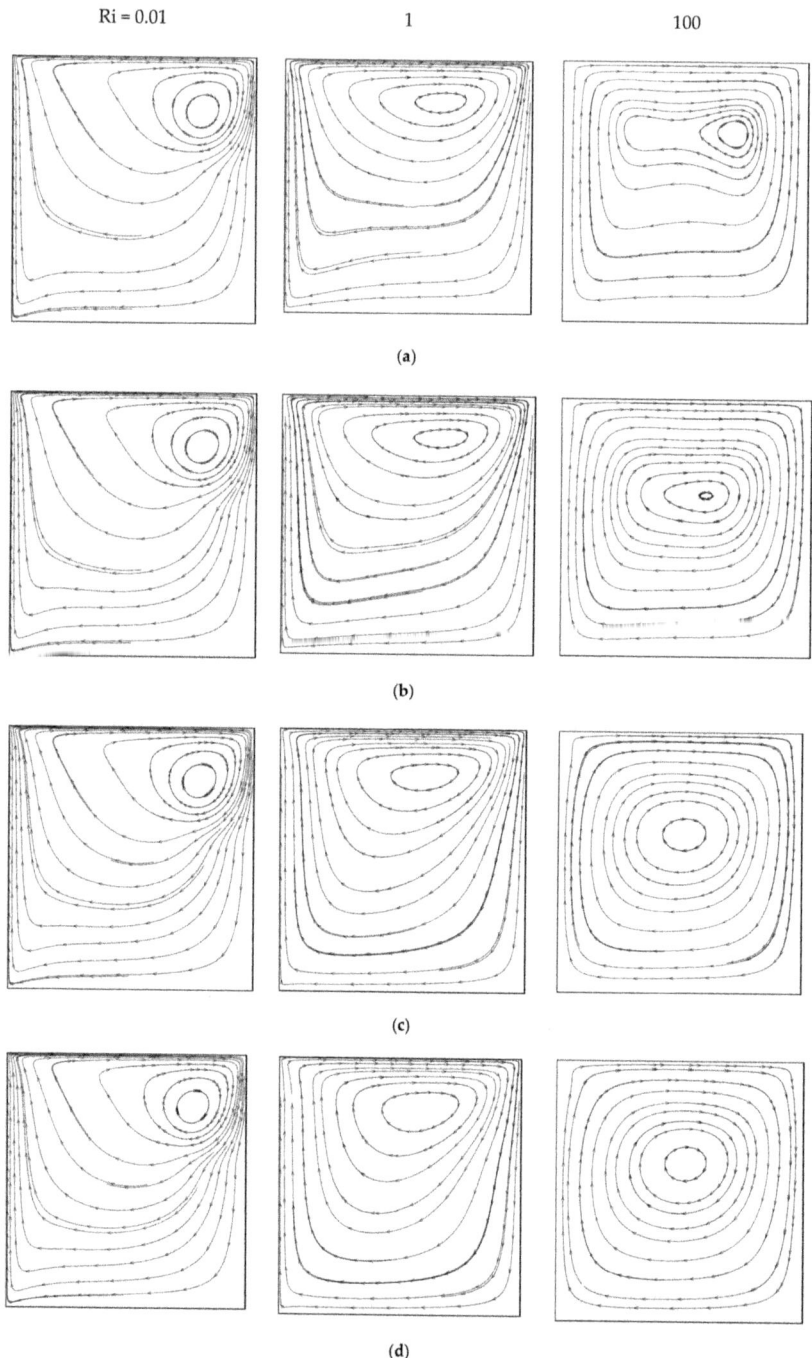

Figure 2. Streamlines for different Rd and Ri values with $U_0 = +1$ (Case 1), and $\Phi = 0.02$. (**a**) Rd = 0; (**b**) Rd = 1; (**c**) Rd = 5; (**d**) Rd = 10.

Figure 3. Streamlines for different Rd and Ri values with $U_0 = -1$ (Case 2) and $\Phi = 0.02$. (**a**) Rd = 0; (**b**) Rd = 1; (**c**) Rd = 5; (**d**) Rd = 10.

Figure 4 depicts the thermal distribution for several values of the radiation parameter and the Richardson number for Case 1 with $\Phi = 0.02$. The thermal boundary layers are shaped along the hot wall for all assumed values of the radiation parameter in the forced convection regime. The temperature boundary layers is weakened for higher values of the radiation (Rd = 10) in the combined convection regime. The horizontal thermal stratification appears in the central region of the enclosed box in the absence of radiation or lower values of the radiation parameter for the natural convection regime. The temperature gradients near wall(s) disappear on rising the value of the radiation parameter. Figure 5 exhibits the isotherms for an opposite moving lid (Case 2) with the same parameters in Figure 4. The thermal layers at the boundary do not appear along the hot wall in forced convection regime as in Case 1. Due to the dual cell structure in the flow field, the thermal layers at the boundary are collapsed along the hot wall in Case 2.

Figure 6 depicts the drag coefficient for several values of Rd and Ri for both cases of the moving lid directions. In Case 1, the skin friction declines upon raising the values of Ri. However, in Case 2, the skin friction behaves nonlinearly, that is, the skin friction grows up to Ri = 1 and then it declines upon raising the values of Ri. It is detected that there is no change on the averaged skin friction for numerous values of Rd when Ri = 0.01 and Ri = 0.1, that is, in the forced convective regime. The skin friction declines upon rising the values of the radiation parameter in the combined and natural convective regimes.

Since the energy transport rate is a key factor in the thermal systems, the (average) energy transfer rate is depicted via the Nusselt number to explore the effect on various pertinent parameters. The local energy transport along the heated wall is computed by the local Nusselt number and it is depicted in Figure 7 for both cases of moving-wall directions. It is clearly exhibited from Figure 7a,c,e that the energy transport is diminished upon raising the values of the Richardson number for Case 1. That is, the local energy transport along the hot wall is enhanced in the forced convective regime. It is almost thrice the value of local Nusselt number for free-convection regime. Case 2 also provides a similar trend on the energy transport upon raising the values of Ri number. It is detected that the local heat transport rises upon raising the radiation parameter for all convection regimes. The highest local energy transfer is observed at the bottom of the heated wall for Case 1 and then it decreases along the wall height. However, the highest local energy transfer is detected at the top of the heated wall for Ri = 0.01 and Ri = 1 in Case 2. However, the opposite trend is found for the free-convection regime in Case 2. The moving lid direction supports the fluid motion with the aiding of the buoyancy force. However, in Case 2, the moving lid direction suppresses the buoyancy force at the top section of the heated wall, and it results the dual cellular motion inside the enclosure. In the dual cell structure, the two cells hit at the top–left corner and provides the highest local heat energy transfer at this point, which is clearly seen from Figure 7b. The high amount of shear force has driven the heated fluid particles vigorously at this situation. Hence, the local heat energy transfer gives a similar trend in both cases for the natural convection regime.

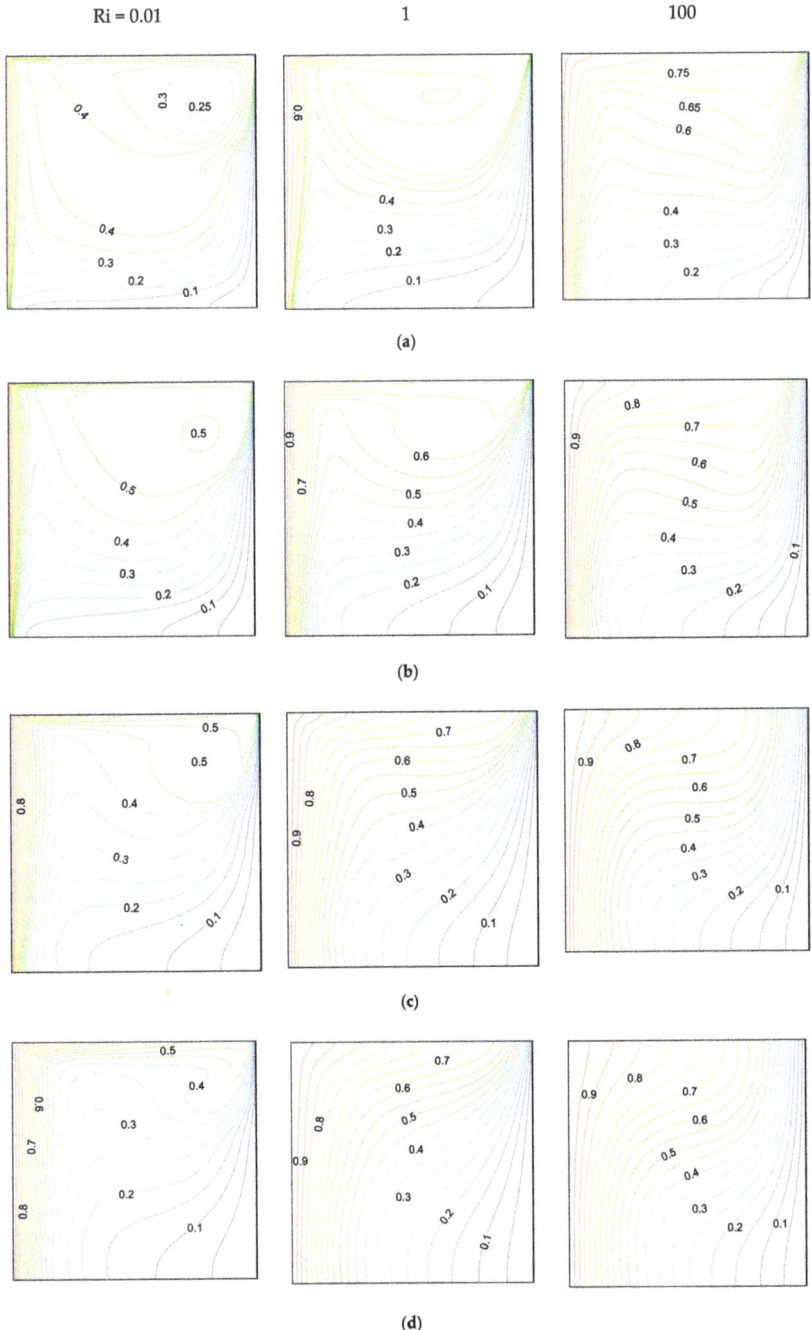

Figure 4. Isotherms for diverse Rd and Ri values with $U_0 = +1$ (Case 1) and $\Phi = 0.02$. (**a**) Rd = 0; (**b**) Rd = 1; (**c**) Rd = 5; (**d**) Rd = 10.

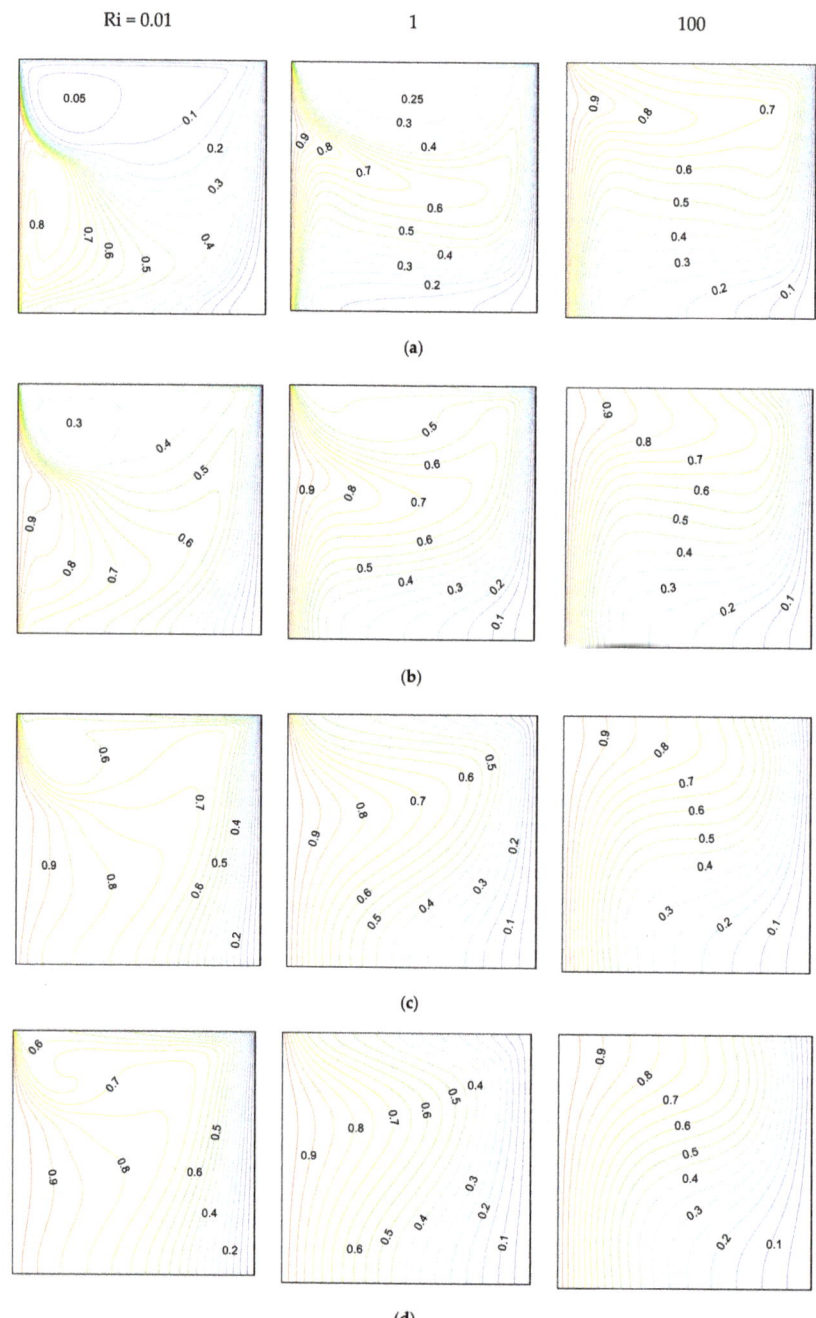

Figure 5. Isotherms for different Rd and Ri values with $U_0 = -1$ (Case 2) and $\Phi = 0.02$. (**a**) Rd = 0; (**b**) Rd = 1; (**c**) Rd = 5; (**d**) Rd = 10.

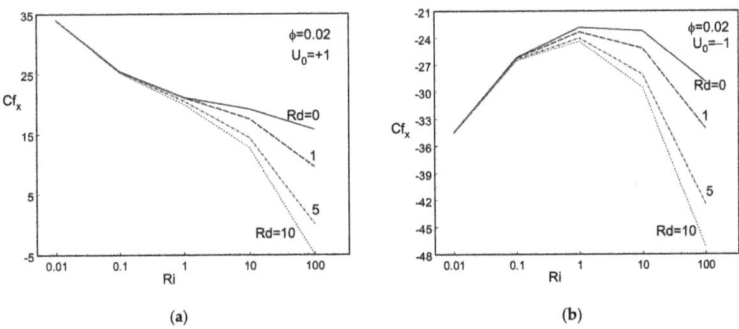

Figure 6. Drag coefficient versus Ri for different Rd. (**a**) Case 1; (**b**) Case 2.

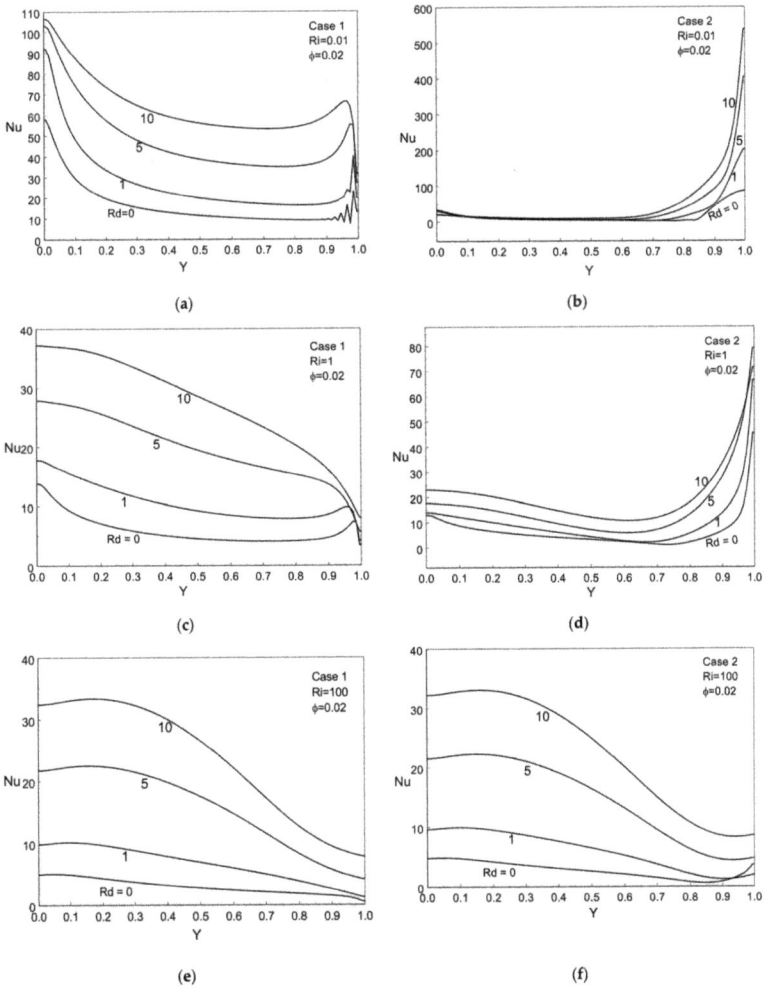

Figure 7. Local Nusselt number for diverse Ri and Rd. (**a,c,e**) Case 1; (**b,d,f**) Case 2.

Figure 8 demonstrates the averaged Nusselt number for several values of Rd and Ri for Case 1 ($U_0 = +1$) and Case 2 ($U_0 = -1$). The averaged heat transport rate is enhanced upon raising the values of the radiation parameter for both cases of the moving-wall directions. It is detected that the averaged heat transfer declines upon raising the values of Ri. Further, scrutinizing these figures, it is found that the moving-wall direction affects the thermal energy transfer rate evidently. When the wall moves from the right-side to left-side (Case 2), the heat energy transfer rate is less due to the dual-eddy structure. The effect of nanometer sized particle volume fraction on the averaged energy transport is examined and it is portrayed in Figure 9a,b for several values of the Richardson number and two cases of moving-wall directions in the presence of radiation with Rd = 5. The averaged heat transport rate decreases upon raising the values of the nanoparticle volume fraction from 0%~4% in mixed and free convective regimes for both moving-wall cases. But, the averaged heat transport rate rises with the nanoparticles volume fraction in Case 1 at Ri = 0.01. In Case 2 at Ri = 0.01, the averaged heat transfer increases first up to Φ = 2% and then it decreases upon raising the value of Φ. Comparing these two cases in Figure 9a,b, it is detected that the averaged Nusselt number is always high for Case 1 than that of Case 2. This is because of the dual eddy structure in Case 2. The energy transfer from the hotter region to the colder region taken by a single cell is faster than the energy transport by the two cells inside the enclosed box. Since the energy exchange between the two cells takes some time which slows down the overall energy transport within the enclosed box.

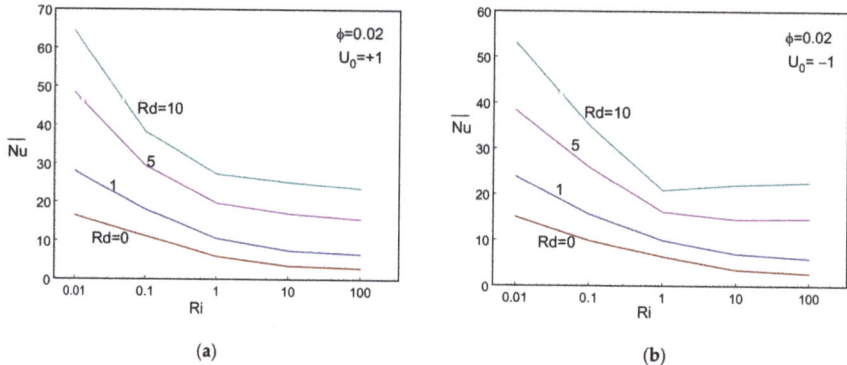

Figure 8. Averaged Nusselt number versus Ri for different Rd. (**a**) Case 1; (**b**) Case 2.

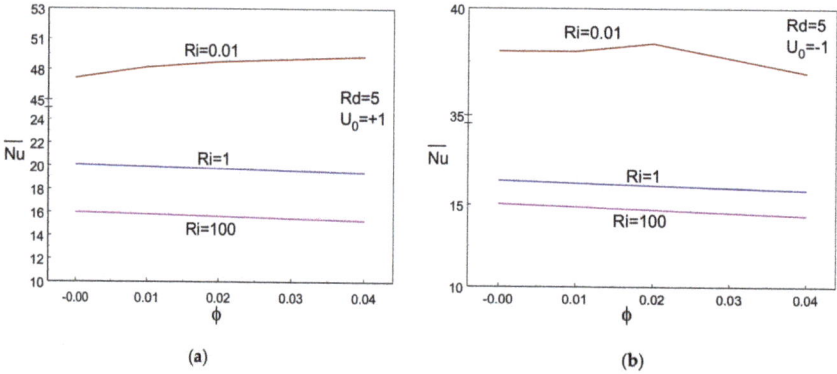

Figure 9. Averaged Nusselt number versus Φ for different Ri with Rd = 5. (**a**) Case 1; (**b**) Case 2.

Figure 10 shows the increment level of the averaged energy transport for different radiation values compared with the absence of radiation parameter. The data clearly show the increasing level of averaged energy transport while raising the values of Rd in both cases of moving wall. The increment level is very high in the natural convection regime in both cases. Figure 11 demonstrates the cup-mixing temperature for various values of Ri and Rd parameters. The behavior of cup-mixing temperature is nonlinear fashion for Case 1, however, Case 2 shows almost a linear fashion. The deviation in cup-mixing temperature with Rd is high at forced convection regime for Case 2. However, it is almost same in free-convection case. The T_{cup} values are almost constant when changing the values of Rd in free-convection flow for Case 2. Figure 12 demonstrates the average temperature for different Ri and Rd values. The higher T_{cup} values indicates the well mixing of fluid with higher temperature. It is obviously seen from Figure 12 that the T_{avg} is almost constant for all values of Rd in free-convection regime. The maxima of T_{avg} attains at Ri = 100 for all Rd values in Case 1, see Figure 12a. From Figure 12b, we observe that the deviation of T_{avg} is high at Ri = 0.01 in Case 2.

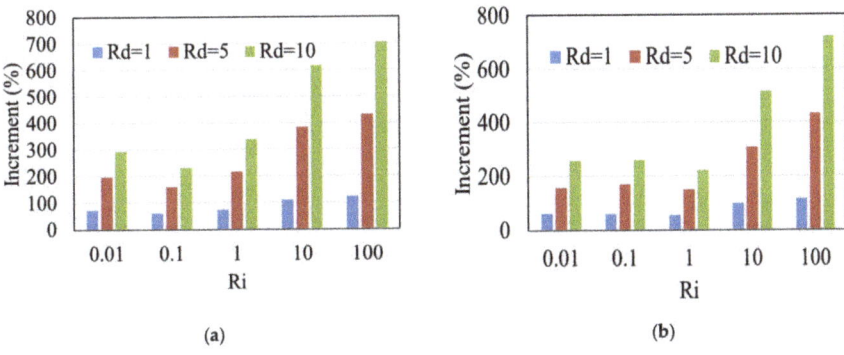

Figure 10. Increment of averaged Nusselt number. (a) Case 1; (b) Case 2.

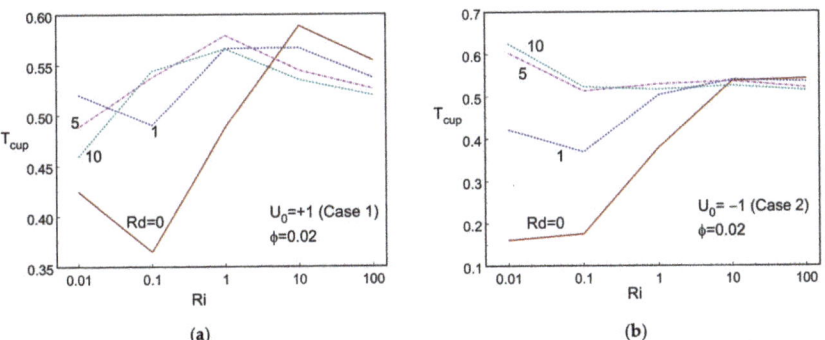

Figure 11. Cup-mixing temperature for different Ri and Rd values. (a) Case 1; (b) Case 2.

Figures 13 and 14 portray the $RMSD_{Tcup}$ and $RMSD_{Tavg}$ for both cases with different Ri and Rd values. Since the nondimensional temperature varies between 0 and 1, the RMSD values are below 1 in the present examination. It is noticed from Figure 13a that the $RMSD_{Tcup}$ increases first and decreases on raising the Ri values for Case 1. The opposite trend is observed for Case 2 in the absence of thermal radiation. However, $RMSD_{Tcup}$ increases linearly with Ri for Rd ≥ 5 for Case 2. It is observed from Figure 14, $RMSD_{Tavg}$ rises linearly with Richardson number in Case 1 for all values of Rd. It is also detected from Figure 14 that the $RMSD_{Tavg}$ rises when growing the Rd values. However, in Case 2, it behaves nonlinearly for either absence of Rd or low values of Rd. However, it acts as same as Case 1 for higher values of Rd (≥5). $RMSD_{Tavg}$ attains its maxima at strong free-convection region in the presence of thermal radiation. Since the RMSD values are lower in all cases, we get higher temperature uniformity inside the box.

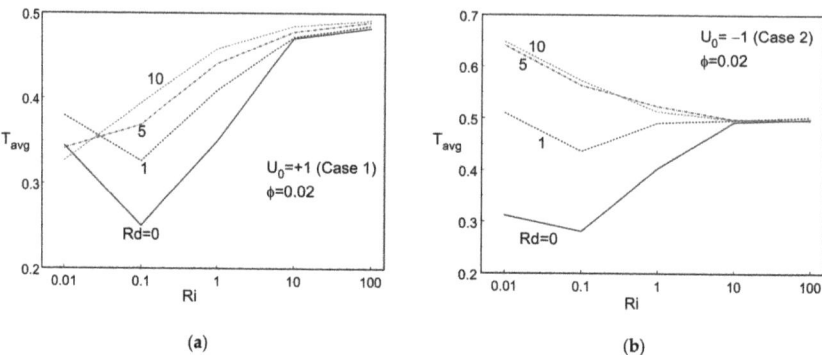

Figure 12. Averaged temperature for different Ri and Rd values. (**a**) Case 1; (**b**) Case 2.

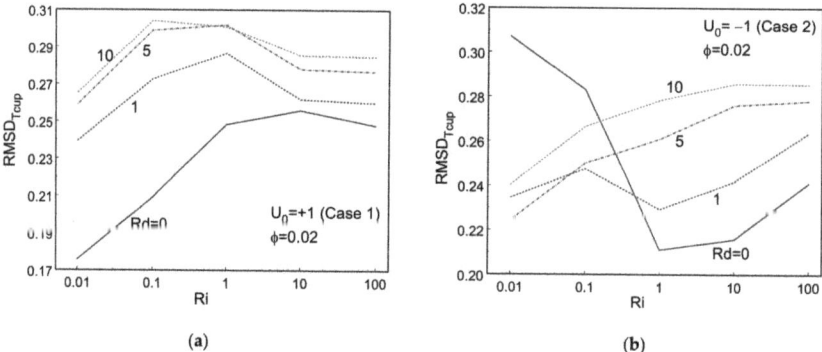

Figure 13. RMSD$_{Tcup}$ for different Ri and Rd values. (**a**) Case 1; (**b**) Case 2.

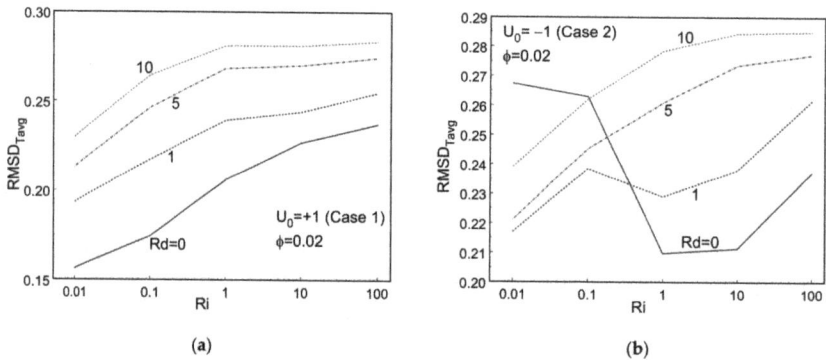

Figure 14. RMSD$_{Tavg}$ for different Ri and Rd values. (**a**) Case 1; (**b**) Case 2.

Figure 15 portrays the influence of Bejan number for both cases with different Ri and Rd values. The values of Be are almost constant on raising the Ri values until Ri = 10, but, after this, it suddenly fall down at Ri = 100 for both direction of moving-wall. When raising the Rd values, the Bejan number is increased. It results that the radiation parameter boosted up the entropy generation inside the box. It is clear that Be lies between 0 and 1. If Be tends to 0 then the irreversibility due to fluid friction controls. If Be tends to 1, the irreversibility due to thermal transfer is leading. In all cases, the values of Be is tends to 1, it results that the irreversibility due to thermal transfer is dominant here.

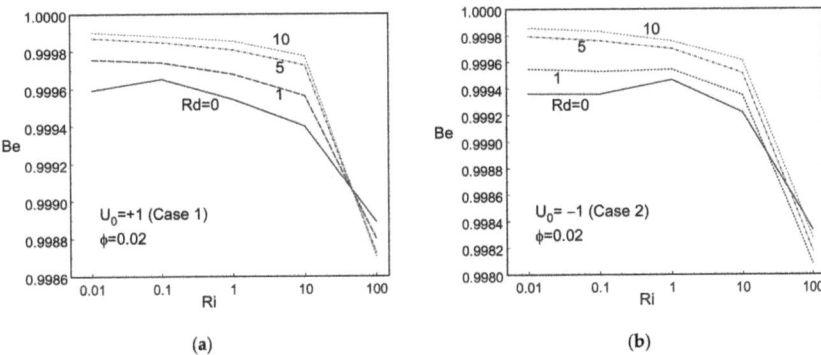

Figure 15. Bejan number for different Ri and Rd values. (**a**) Case 1; (**b**) Case 2.

7. Conclusions

The impacts of the direction of a moving wall, thermal radiation and entropy on combined convective stream and energy transfer of nanoliquids in a lid-driven enclosed box is numerically explored. The leading mathematical model is solved by the control volume technique. The following remarkable discoveries are detected from the study:

- The moving-wall direction drastically affects the stream field inside the enclosure. Single and dual cell structures are formed in Case 1 and Case 2, respectively for all values of Ri, radiation parameter and all nanoliquids;
- The skin friction declines upon raising the values of the Richardson number for Case 1. It increases up to Ri = 1 and then decreases upon raising the Richardson number in Case 2;
- The higher local energy transport is attained at bottom of the heat wall for Case 1 and at top of the hot wall for Case 2 in the forced and mixed convective flow regimes. The free-convection mode provides a similar trend on both cases, that is, the highest heat transfer attains near the bottom of the barrier;
- The thermal radiation parameter enhances the energy transport across the enclosure for all given values of Ri and ϕ in both directions of the moving wall;
- The moving-wall direction greatly influences the energy transfer rate. The Case 1 (moving-wall from left to right) provides a higher heat transfer rate than that of Case 2 for all values of Ri and the radiation parameter;
- The averaged heat transport declines upon rising the volume fraction of nanoparticle in free and mixed convection regimes for both moving-wall directions. The averaged heat transport increases with the nanoparticles volume fraction in Case 1. It rises first and then declines upon raising the values of nanoparticles volume fraction in Case 2;
- The Bejan number enhances on raising the Rd values. Entropy generation dominates by thermal transfer;
- The lower values of RMSD in all cases illustrates the higher temperature uniformity inside the box;
- The T_{cup} and T_{avg} values are almost constant when changing the values of Rd in free-convection flow for Case 2. The cup-mixing temperature behaves non-linear fashion for Case 1 and almost a linear fashion for Case 2.

Author Contributions: Conceptualization, S.S.; methodology, S.S.; software, S.S.; validation, S.S.; formal analysis, S.S.; investigation, S.S.; writing—original draft preparation, S.S.; writing—review and editing, S.S. and A.J.C.; visualization, S.S.; supervision, S.S.; project administration, S.S.; funding acquisition, S.S.; F.O.M.M.; M.S.A. and A.M.A. All authors have read and agreed to the published version of the manuscript.

Funding: This project was funded by the research and development office (RDO), at the Ministry of Education, Kingdom of Saudi Arabia, under Grant No. (H1Q1-40-2019).

Acknowledgments: This project was funded by the research and development office (RDO), at the Ministry of Education, Kingdom of Saudi Arabia, under Grant No. (H1Q1-40-2019). The authors also acknowledge with thanks research and development office (RDO-KAU) at King Abdulaziz University for technical support.

Conflicts of Interest: The authors declare no conflict of interest.

References

1. Alsabery, A.I.; Tayebi, T.; Roslan, R.; Chamkha, A.J.; Hashim, I. Entropy Generation and Mixed Convection Flow Inside a Wavy-Walled Enclosure Containing a Rotating Solid Cylinder and a Heat Source. *Entropy* **2020**, *22*, 606. [CrossRef]
2. Chamkha, A.J.; Selimefendigil, F.; Oztop, H.F. MHD mixed convection and entropy generation in a lid-driven triangular cavity for various electrical conductivity models. *Entropy* **2018**, *20*, 903. [CrossRef]
3. Bhuvaneswari, M.; Sivasankaran, S.; Kim, Y.J. Numerical Study on double diffusive mixed convection in a two-sided lid driven cavity with Soret effect. *Numer. Heat Transf. Part A* **2011**, *59*, 543–560. [CrossRef]
4. Sivasankaran, S.; Cheong, H.T.; Bhuvaneswari, M.; Ganesan, P. Effect of moving wall direction on mixed convection in an inclined lid-driven square cavity with sinusoidal heating. *Numer. Heat Transf. Part A* **2016**, *69*, 630–642. [CrossRef]
5. Mahmoudinezhad, S.; Rezania, A.; Yousefi, T. Adiabatic partition effect on natural convection heat transfer inside a square cavity: Experimental and numerical studies. *Heat Mass Transf.* **2018**, *54*, 291–304. [CrossRef]
6. Selimefendigil, F.; Öztop, H.F.; Chamkha, A.J. Analysis of mixed convection of nanofluid in a 3D lid-driven trapezoidal cavity with flexible side surfaces and inner cylinder. *Int. Commun. Heat Mass Transf.* **2017**, *87*, 40–51. [CrossRef]
7. Oztop, H.F.; Almeshaal, M.A.; Kolsi, L.; Rashidi, M.M.; Ali, M.E. Natural convection and irreversibility evaluation in a cubic cavity with partial opening in both top and bottom sides. *Entropy* **2019**, *21*, 116. [CrossRef]
8. Safaei, M.R.; Shadloo, M.S.; Goodarzi, M.S.; Hadjadj, A.; Goshayeshi, H.R.; Afrand, M.; Kazi, S.N. A survey on experimental and numerical studies of convection heat transfer of nanofluids inside closed conduits. *Adv. Mech. Eng.* **2016**, *8*, 1687814016673569. [CrossRef]
9. Chamkha, A.J.; Selimefendigil, F. MHD free convection and entropy generation in a corrugated cavity filled with a porous medium saturated with nanofluids. *Entropy* **2018**, *20*, 846. [CrossRef]
10. Sivasankaran, S.; Sivakumar, V.; Hussein, A.K. Numerical study on mixed convection in an inclined lid-driven cavity with discrete heating. *Int. Commun. Heat Mass Transf.* **2013**, *46*, 112–125. [CrossRef]
11. Sivasankaran, S.; Pan, K.L. Lattice Boltzmann simulation for a lid-driven cavity with discrete heating/cooling sources. *AIAA J. Thermophys. Heat Transf.* **2016**, *30*, 573–586. [CrossRef]
12. Mekroussi, S.; Nehari, D.; Bouzit, M.; Chemloul, N.E.S. Analysis of mixed convection in an inclined lid-driven cavity with a wavy wall. *J. Mech. Sci. Technol.* **2013**, *27*, 2181–2190. [CrossRef]
13. Sivakumar, V.; Sivasankaran, S. Mixed convection in an inclined lid-driven cavity with non-uniform heating on both sidewalls. *J. Appl. Mech. Tech. Phys.* **2014**, *55*, 634–649. [CrossRef]
14. Sivasankaran, S.; Ananthan, S.S.; Abdul Hakeem, A.K. Mixed convection in a lid-driven cavity with sinusoidal boundary temperature at the bottom wall in the presence of magnetic field. *Sci. Iran. Trans. B Mech. Eng.* **2016**, *23*, 1027–1036. [CrossRef]
15. Sivasankaran, S.; Ananthan, S.S.; Bhuvaneswari, M.; Abdul Hakeem, A.K. Double-diffusive mixed convection in a lid-driven cavity with non-uniform heating on sidewalls. *Sadhana* **2017**, *42*, 1929–1941. [CrossRef]
16. Kasmani, R.M.D.; Sivasankaran, S.; Bhuvaneswari, M.; Hussein, A.K. Analytical and numerical study on convection of nanofluid past a moving wedge with Soret and Dufour effects. *Int. J. Numer. Methods Heat Fluid Flow* **2017**, *27*, 2333–2354. [CrossRef]
17. Öztop, H.F.; Sakhrieh, A.; Abu-Nada, E.; Al-Salem, K. Mixed convection of MHD flow in nanofluid filled and partially heated wavy walled lid-driven enclosure. *Int. Commun. Heat Mass Transf.* **2017**, *86*, 42–51. [CrossRef]
18. Rashidi, M.M.; Nasiri, M.; Shadloo, M.S.; Yang, Z. Entropy Generation in a Circular Tube Heat Exchanger Using Nanofluids: Effects of Different Modeling Approaches. *Heat Transf. Eng.* **2017**, *38*, 853–866. [CrossRef]

19. Sivasankaran, S.; Narrein, K. Numerical investigation of two-phase laminar pulsating nanofluid flow in helical microchannel filled with a porous medium. *Int. Commun. Heat Mass Transf.* **2016**, *75*, 86–91. [CrossRef]
20. Khashi'ie, N.S.; Md Arifin, N.; Pop, I. Mixed convective stagnation point flow towards a vertical riga plate in hybrid Cu-Al$_2$O$_3$/water nanofluid. *Mathematics* **2020**, *8*, 912. [CrossRef]
21. Sheremet, M.A.; Grosan, T.; Pop, I. Natural convection and entropy generation in a square cavity with variable temperature side walls filled with a nanofluid: Buongiorno's mathematical model. *Entropy* **2017**, *19*, 337. [CrossRef]
22. Alsabery, A.I.; Ismael, M.A.; Chamkha, A.J.; Hashim, I. Numerical investigation of mixed convection and entropy generation in a wavy-walled cavity filled with nanofluid and involving a rotating cylinder. *Entropy* **2018**, *20*, 664. [CrossRef]
23. Santra, A.K.; Sen, S.; Chakraborty, N. Study of heat transfer augmentation in a differentially heated square cavity using copper-water nanofluid. *Int. J. Therm. Sci.* **2008**, *47*, 1113–1122. [CrossRef]
24. Abu-Nada, E.; Oztop, H.F. Effects of inclination angle on natural convection in enclosures filled with Cu-water nanofluid. *Int. J. Heat Fluid Flow* **2009**, *30*, 669–678. [CrossRef]
25. Ghasemi, B.; Aminossadati, S.M. Natural convection heat transfer in an inclined enclosure filled with a water CuO nanofluid. *Numer. Heat Transf. Part A* **2009**, *55*, 807–823. [CrossRef]
26. Bhuvaneswari, M.; Ganesan, P.; Sivasankaran, S.; Viswanathan, K.K. Effect of variable fluid properties on natural convection of nanofluids in a cavity with linearly varying wall temperature. *Math. Probl. Eng.* **2015**, *2015*, 391786. [CrossRef]
27. Sivasankaran, S.; Mansour, M.A.; Rashad, A.M.; Bhuvaneswari, M. MHD mixed convection of Cu–water nanofluid in a two-sided lid-driven porous cavity with a partial slip. *Numer. Heat Transf. Part A* **2016**, *70*, 1356–1370. [CrossRef]
28. Rashad, A.M.; Sivasankaran, S.; Mansour, M.A.; Bhuvaneswari, M. Magneto-convection of nanofluids in a lid-driven trapezoidal cavity with internal heat generation and discrete heating. *Numer. Heat Transf. Part A* **2017**, *71*, 1223–1234. [CrossRef]
29. Mansour, M.A.; Abd El-Aziz, M.M.; Mohamed, R.A.; Ahmed, S.E. Numerical simulation of natural convection in wavy porous cavities under the influence of thermal radiation using a thermal non-equilibrium model. *Transp. Porous Media* **2011**, *86*, 585–600. [CrossRef]
30. Moufekkir, F.; Moussaoui, M.A.; Mezrhab, A.; Bouzidi, M.; Lemonnier, D. Combined double diffusive convection and radiation in a square enclosure filled with semitransparent fluid. *Comput. Fluids* **2012**, *69*, 172–178. [CrossRef]
31. Mahapatra, T.R.; Pal, D.; Mondal, S. Mixed convection flow in an inclined enclosure under magnetic field with thermal radiation and heat generation. *Int. Commun. Heat Mass. Transf.* **2013**, *41*, 47–56. [CrossRef]
32. Saleem, M.; Hossain, M.A.; Saha, S.C.; Gu, Y.T. Heat transfer analysis of viscous incompressible fluid by combined natural convection and radiation in an open cavity. *Math. Probl. Eng.* **2014**, *2014*, 412480. [CrossRef]
33. Zhang, J.K.; Li, B.W.; Dong, H.; Luo, X.H.; Lin, H. Analysis of magnetohydrodynamics (MHD) natural convection in 2D cavity and 3D cavity with thermal radiation effects. *Int. J. Heat Mass. Transf.* **2017**, *112*, 216–223. [CrossRef]
34. Alzahrani, A.K.; Sivasankaran, S.; Bhuvaneswari, M. Numerical Simulation on Convection and Thermal Radiation of Casson Fluid in an Enclosure with Entropy Generation. *Entropy* **2020**, *22*, 229. [CrossRef]
35. Ho, C.J.; Chen, M.W.; Li, Z.W. Numerical simulation of natural convection of nanofluid in a square enclosure: Effects due to uncertainties of viscosity and thermal conductivity. *Int. J. Heat Mass. Transf.* **2008**, *51*, 4506–4516. [CrossRef]
36. Sharif, M.A.R. Laminar mixed convection in shallow inclined driven cavities with hot moving lid on top and cooled from bottom. *Appl. Therm. Eng.* **2007**, *27*, 1036–1042. [CrossRef]
37. Iwatsu, R.; Hyun, J.M.; Kuwahara, K. Mixed convection in a driven cavity with a stable vertical temperature gradient. *Int. J. Heat Mass. Transf.* **1993**, *36*, 1601–1608. [CrossRef]

© 2020 by the authors. Licensee MDPI, Basel, Switzerland. This article is an open access article distributed under the terms and conditions of the Creative Commons Attribution (CC BY) license (http://creativecommons.org/licenses/by/4.0/).

Article

Radiative MHD Sutterby Nanofluid Flow Past a Moving Sheet: Scaling Group Analysis

Mohammed M. Fayyadh [1], Kohilavani Naganthran [2], Md Faisal Md Basir [3], Ishak Hashim [2,*] and Rozaini Roslan [1]

1. Department of Mathematics and Statistics, Faculty of Applied Science & Technology, Universiti Tun Hussein Onn Malaysia, Pagoh 86400, Malaysia; abuzeen@gmail.com (M.M.F.); rozaini@uthm.edu.my (R.R.)
2. Department of Mathematical Sciences, Faculty of Science & Technology, Universiti Kebangsaan Malaysia, UKM Bangi 43600, Malaysia; kohi@ukm.edu.my
3. Department of Mathematical Sciences, Faculty of Science, Universiti Teknologi Malaysia, UTM Johor Bahru 81310, Malaysia; mfaisalmbasir@utm.my
* Correspondence: ishak_h@ukm.edu.my

Received: 20 July 2020; Accepted: 24 August 2020; Published: 26 August 2020

Abstract: The present theoretical work endeavors to solve the Sutterby nanofluid flow and heat transfer problem over a permeable moving sheet, together with the presence of thermal radiation and magnetohydrodynamics (MHD). The fluid flow and heat transfer features near the stagnation region are considered. A new form of similarity transformations is introduced through scaling group analysis to simplify the governing boundary layer equations, which then eases the computational process in the MATLAB bvp4c function. The variation in the values of the governing parameters yields two different numerical solutions. One of the solutions is stable and physically reliable, while the other solution is unstable and is associated with flow separation. An increased effect of the thermal radiation improves the rate of convective heat transfer past the permeable shrinking sheet.

Keywords: scaling group analysis; Sutterby fluid; nanofluid; magnetohydrodynamics (MHD); stability analysis

1. Introduction

The viscoelastic fluid is a type of non-Newtonian fluid that manifests the viscous and elasticity features under deformation. Sutterby fluid is an example of the viscoelastic fluid, and it well portrays the dilute polymer solutions [1,2]. Specifically, the Sutterby model fluid resembles the shear thinning and shear thickening aspects in high polymer aqueous solutions such as carboxymethyl cellulose (CMC), hydroxyethyl cellulose (HEC) and methyl cellulose (MC) [3]. The dilute polymer solutions have a wide range of functions in industrial practice, for instance, spray applications of agricultural chemicals [4], drag reducers in pipe flows [5], and production of domestic cleaning products [6]. The work of Fujii et al. [7] is one of the earliest studies to address the natural convection boundary layer flow in a Sutterby fluid past a vertical motionless isothermal plane and achieved an excellent comparison with the experimental results. Fujii et al. [8] revisited their work in [7] to investigate the impact of uniform heat flux under the same settings. However, the Sutterby model fluid received less attention from the boundary layer researchers at that time. Later, a new type of heat conductive fluid was introduced by Choi [9] named nanofluid. Nanofluid was also claimed to be a brilliant fluid due to its excellent heat transfer performance in engineering applications such as cooling of electronic appliances, and systems of solar water heating [10]. Nanofluid has now attracted significant interest from researchers, and boundary layer models have been studied under various settings [11–13]. After a long discontinuity in the theoretical works of the Sutterby boundary layer fluid flow, some numerical investigations of the Sutterby fluid under the Cattaneo–Christov heat flux [14,15],

Soret and Dufour effect [16], peristaltic flow [17–19], squeezed flow [20], Joule heating effect [21], homogeneous–heterogeneous reactions [22], and hybrid nanoparticles [23] have been reported recently.

Magnetohydrodynamics (MHD) is another technological conception that is widespread in engineering practice. Electromagnetic casting [24], plasma confinement, and MHD power generation [25] are examples of notable applications. Thermal radiation is a type of energy that works in conjunction with the MHD effect. Thermal radiation emits and absorbs energy in the form of waves or molecules through a non-scattering medium. The successful combination of thermal radiation and MHD in an electrically conducting fluid has significant applications in solar power technology and electrical power generation [26]. Acknowledging these applications, researchers began to examine thermal radiation and MHD effects in the boundary layer flow past a stretching/shrinking surface, and many theoretical works have been reported. Recently, Sabir et al. [27] explored the stagnation-point flow of a Sutterby fluid with the effects of an inclined magnetic field and thermal radiation past a stretching surface, and observed the declining trend of the convective heat transfer with the stronger influence of thermal radiation. Bilal et al. [28] examined the ohmically dissipated Darcy–Forchheimer slip flow of an MHD Sutterby fluid past a radiating stretching sheet and found a decrement in the convective heat transfer with increasing slip effects.

By comparison, the boundary layer equations, which were proposed by Prandtl [29], disclosed many invariant closed-form solutions. Prandtl's boundary layer equations can be reduced to a less complicated form that is in a system of ordinary differential equations. These boundary layer equations also allow many different types of symmetry groups, of which the Lie group analysis is prominent. Lie group analysis helps to identify the transformation point that represents the given boundary layer equations [30]. In Lie group analysis, the group-invariant solutions are the similarity solutions, and these similarity solutions are used to reduce the independent variables in a fluid flow problem [31]. A special form of the Lie group analysis exists, namely, the scaling group of transformation, and this has been employed by researchers in valuable contributions, for instance, see [32,33].

Regarding studies of stagnation-point flow in a Sutterby fluid, Azhar et al. [34] investigated the effect of entropy generation on the stagnation-point flow of a Sutterby nanofluid past a stretching sheet. Azhar et al. [35] reconsidered the work of [34] by incorporating the Cattaneo–Christov heat flux model and omitting the nanoparticles. Both of the studies of [34,35] solved the flow problem numerically and presented unique solutions. A number of considerable research gaps were found in the theoretical works available in the stagnation-point flow and heat transfer in a Sutterby nanofluid, for instance: inspecting fluid flow behavior and heat transfer characteristics past a shrinking sheet together with the suction effect; conducting scaling group analysis; obtaining dual solutions; and performing stability analysis. Thus, the present work is devoted to numerically solving the problem of boundary layer Sutterby nanofluid flow and heat transfer near the stagnation region over a permeable moving (stretching/shrinking) sheet. The fluid flow and heat transfer characteristics under the magnetic and thermal radiation effects are observed. Scaling group analysis is employed to obtain the apt similarity transformations so that the complex governing boundary layer equations can be brought to a soluble form. The simplified form of the mathematical model is then solved numerically in the boundary value problem solver or bvp4c function in MATLAB. Two different numerical solutions are identified with the governing parameters' variation. Further, stability analysis is undertaken in the present work to justify the presence of dual solutions. These contributions are essentially original, and all numerical results are presented and discussed in detail.

2. Problem Formulation

Contemplate an incompressible two-dimensional stagnation-point flow of a Sutterby nanofluid across a stretching/shrinking sheet as shown in Figure 1, where x and y are the Cartesian coordinates with the x-axis positioned in the horizontal direction, and the y-coordinate is normal to the x-coordinate. The free stream velocity is denoted by $\bar{u}_e$, and $\bar{u}_w$ signifies the velocity of the moving sheet, where $\bar{u}_w > 0$ infers the state of the stretching sheet, $\bar{u}_w < 0$ the embodies shrinking sheet, and $\bar{u}_w = 0$ typifies the stationary sheet. The moving (stretching or shrinking) sheet is penetrable and there is a uniform surface mass flux, of velocity $\bar{v}_w$, with $\bar{v}_w > 0$ to imply the injection situation and $\bar{v}_w < 0$ for the suction state. The free stream temperature and the wall temperature are denoted by T_∞ and T_w, respectively.

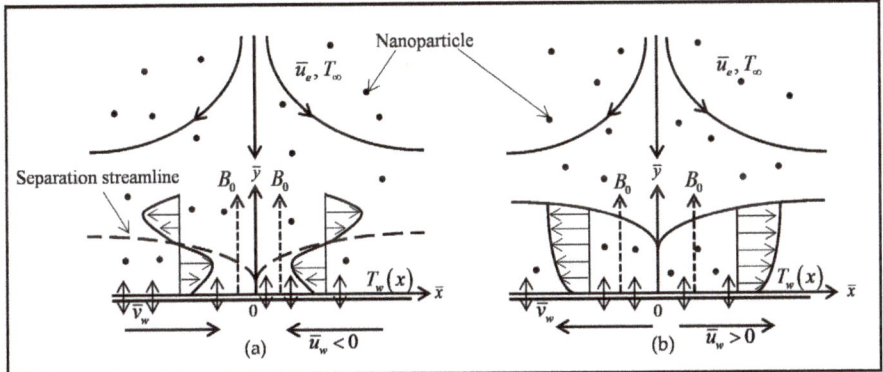

Figure 1. Schematic diagram of the present problem: (**a**) shrinking sheet ($\bar{u}_w < 0$); (**b**) stretching sheet ($\bar{u}_w > 0$).

Sutterby [1,2] introduced the constitutive law for the Sutterby fluid by expressing the Cauchy stress tensor (**T**) as:

$$\mathbf{T} = -p\mathbf{I} + \mathbf{S}, \tag{1}$$

where p is the pressure, **I** is the identity vector, and **S** is the extra stress tensor which can be defined as follows [21]:

$$\mathbf{S} = \mu_0 \left[\frac{\sinh^{-1}(E\dot{\gamma})}{(E\dot{\gamma})} \right]^m \mathbf{A}_1. \tag{2}$$

Here, μ_0 is the viscosity at low shear rates, E is the material time constant, $\dot{\gamma} = \sqrt{\mathrm{tr}(\mathbf{A}_1)^2/2}$ is the second invariant strain tensor, $\mathbf{A}_1$ is the first order Rivlin–Erickson tensor or deformation rate tensor which is defined as $\mathbf{A}_1 = (\nabla \mathbf{V}) + (\nabla \mathbf{V})^T$, and m is the power-law index. The Sutterby model in Equation (2) is a versatile model when the value of m changes. For instance, when $m = 0$, the Sutterby model imitates the Newtonian fluid behavior, when $m = 1$, the model is reduced to the Eyring model, and this model also predicts specifically the pseudo-plastic (shear thinning) and dilatant (shear thickening) fluid properties when $m > 0$ and $m < 0$, respectively. By taking the velocity field as $\mathbf{V} = [u(x,y), v(x,y)]$, and under the assumptions mentioned earlier, the governing boundary layer equations in the dimensional form can be formed as follows [36]:

$$\frac{\partial \bar{u}}{\partial \bar{x}} + \frac{\partial \bar{v}}{\partial \bar{y}} = 0, \tag{3}$$

$$\bar{u}\frac{\partial \bar{u}}{\partial \bar{x}} + \bar{v}\frac{\partial \bar{u}}{\partial \bar{y}} = \frac{\mu_{nf}}{\rho_{nf}}\frac{\partial^2 \bar{u}}{\partial \bar{y}^2} + \frac{\mu_{nf}}{\rho_{nf}}\frac{mE^2}{2}\left(\frac{\partial \bar{u}}{\partial \bar{y}}\right)^2 \frac{\partial^2 \bar{u}}{\partial \bar{y}^2} + \bar{u}_e \frac{d\bar{u}_e}{d\bar{x}} - \frac{\sigma B_0^2 (\bar{u}_e - \bar{u})}{\rho_{nf}}, \tag{4}$$

$$\bar{u}\frac{\partial T}{\partial \bar{x}} + \bar{v}\frac{\partial T}{\partial \bar{y}} = \frac{k_{nf}}{(\rho C_p)_{nf}}\frac{\partial^2 T}{\partial \bar{y}^2} + \frac{16\sigma_1 T_\infty^3}{3(\rho C_p)_{nf} k_1}\frac{\partial^2 T}{\partial \bar{y}^2}, \tag{5}$$

along with the respective boundary conditions:

$$\bar{u} = \bar{u}_w, \quad \bar{v} = \bar{v}_w, \quad T = T_w(x) \quad \text{at} \quad \bar{y} = 0.$$
$$\bar{u} = \bar{u}_e, \quad \frac{d\bar{u}}{d\bar{y}} \to 0, \quad T \to T_\infty \quad \text{as} \quad \bar{y} \to \infty, \tag{6}$$

where $\bar{u}$ and $\bar{v}$ denote velocity components in the $\bar{x}$ and $\bar{y}$ directions, respectively, μ_{nf} is the dynamic viscosity of the nanofluid, σ is the electrical conductivity, B_0 is the magnetic field strength, ρ_{nf} is the density of the nanofluid, σ_1 is the Stefan Boltzmann constant, k_1 is the Rosseland mean absorption coefficient, k_{nf} is the thermal conductivity of the nanofluid, and $(C_p)_{nf}$ is the specific heat capacity of the nanofluid. The detailed definitions of the nanofluid parameters are given by the following expressions, which are valid when the nanoparticles are of spherical shape or similar to a spherical shape [37]:

$$\mu_{nf} = \frac{\mu_{bf}}{(1-\phi)^{2.5}}, \quad \alpha_{nf} = \frac{k_{nf}}{(\rho C_p)_{nf}}, \quad (\rho C_p)_{nf} = (1-\phi)(\rho C_p)_{bf} + \phi(\rho C_p)_s,$$
$$\rho_{nf} = (1-\phi)\rho_{bf} + \phi\rho_s, \quad \frac{k_{nf}}{k_{bf}} = \frac{(k_s + 2k_{bf}) - 2\phi(k_{bf} - k_s)}{(k_s + 2k_{bf}) + \phi(k_{bf} - k_s)}, \tag{7}$$

where ϕ denotes the nanoparticle volume fraction, μ_{bf} denotes the dynamic viscosity of the base fluid, α_{nf} denotes the thermal diffusivity of the nanofluid, k_{bf} is the thermal conductivity of the base fluid, k_s is the thermal conductivity of the solid fractions, C_p is the specific heat capacity, and ρ_{bf} and ρ_s are the densities of the base fluid and solid fractions, respectively. The Sutterby model reflects the dilute polymer solution where the polymer is diluted in the appropriate solvent. Hence, for the present study, n-Hexane is chosen as the base fluid (solvent). Table 1 displays the specific values for the respective thermophysical features of n-Hexane and magnetite nanofluid [38].

Table 1. The thermo physical characteristics of the essential fluid and nanoparticles.

Physical Properties	Fluid Phase (n-Hexane /$CH_3(CH_2)_4CH_3$)	Solid Phase Magnetite (Fe_3O_4)
C_p(kJ/kg·K)	2.78	670
ρ(kg/m^3)	551	5180
k(W/mK)	82	9.7
Pr	4.36	–

3. Non-Dimensionalization of the Governing Equations

Considering the following the non-dimensional variables:

$$x = \frac{a\bar{x}}{u_0}, \quad y = \sqrt{\frac{a}{v_{bf}}}\bar{y}, \quad u = \frac{\bar{u}}{u_0}, \quad u_w = \frac{\bar{u}_w}{u_0}, \quad v = \frac{\bar{v}}{\sqrt{av_{bf}}},$$
$$v_w = \frac{\bar{v}_w}{\sqrt{av_{bf}}}, \quad u_e = \frac{\bar{u}_e}{u_0}, \quad \theta = \frac{T - T_\infty}{T_0}, \tag{8}$$

where u_0 is the characteristic velocity and introducing the stream function ψ, which can be defined by $u = \frac{\partial \psi}{\partial y}$ and $v = -\frac{\partial \psi}{\partial x}$, Equations (4) and (5) become:

$$\frac{\partial \psi}{\partial y}\frac{\partial^2 \psi}{\partial x \partial y} - \frac{\partial \psi}{\partial x}\frac{\partial^2 \psi}{\partial y^2} = \frac{A_1}{A_2}\frac{\partial^3 \psi}{\partial y^3} + \frac{A_1}{A_2}\frac{mDe}{2}\left(\frac{\partial^2 \psi}{\partial y^2}\right)^2 \frac{\partial^3 \psi}{\partial y^3} + u_e\frac{du_e}{dx} - \frac{\sigma B_0^2}{\rho_{bf}a}\frac{1}{A_2}\left(u_e - \frac{\partial \psi}{\partial y}\right), \tag{9}$$

$$\frac{\partial \psi}{\partial y}\frac{\partial \theta}{\partial x} - \frac{\partial \psi}{\partial x}\frac{\partial \theta}{\partial y} = \frac{A_4}{A_3}\frac{1}{Pr}\frac{\partial^2 \theta}{\partial y^2} + \frac{1}{A_3}\frac{4}{3}\frac{Rd}{Pr}\frac{\partial^2 \theta}{\partial y^2}, \tag{10}$$

with the corresponding boundary conditions:

$$\frac{\partial \psi}{\partial y} = u_w, \quad \frac{\partial \psi}{\partial x} = -v_w, \quad \theta T_0 = T_w(x) \quad \text{at} \quad y = 0,$$
$$\frac{\partial \psi}{\partial y} \to u_e, \quad \frac{\partial^2 \psi}{\partial y^2} \to 0, \quad \theta \to 0 \quad \text{as} \quad y \to \infty, \tag{11}$$

while satisfying the continuity equation of Equation (3). In Equations (9) and (11), $M = \frac{\sigma B_0^2}{\rho_{bf} a}$ is the magnetic parameter, $Rd = \frac{4\sigma_1 T_\infty^3}{k_1 k_{bf}}$ is the radiation parameter, $Pr = \frac{\mu_{bf}(c_p)_{bf}}{k_{bf}}$ is the Prandtl number, $De = \frac{u_0^2 a E^2}{\nu_{bf}}$ is the Deborah number, ϕ is the nanoparticle volume fraction, and terms A_1, A_2, A_3, and A_4 are expressed as:

$$A_1 = \frac{1}{(1-\phi)^{2.5}}, \quad A_2 = 1 - \phi + \phi \frac{\rho_s}{\rho_{bf}}, \quad A_3 = 1 - \phi + \phi \frac{(\rho c_p)_s}{(\rho c_p)_{bf}},$$
$$A_4 = \frac{k_{nf}}{k_f} = \frac{k_s + 2k_{bf} - 2\phi(k_{bf} - k_s)}{k_{bf} + 2k_{bf} + \phi(k_{bf} - k_s)}. \tag{12}$$

The functions u_w, v_w and $T_w(x)$ are assumed to be in the following form to ensure that similarity solution exists:

$$u_w = \frac{u_1}{u_0} x^{\frac{2}{5}}, \quad v_w = \frac{v_1}{\sqrt{a \nu_{bf}}} x^{-\frac{2}{5}}, \quad T_w(x) = T_0 x^{\frac{2}{5}}, \tag{13}$$

where u_1 is the reference velocity, v_1 is the normal reference velocity, and T_0 is the reference temperature.

4. Scaling Group Analysis

The governing boundary layer flow and heat transfer problem in the form of partial differential equations (PDEs) is complex and hard to solve by means of mathematical software. Therefore, it needs to be reduced to a simpler form so that it can be solved. Suitable similarity variables can facilitate the transformation and, at this point, scaling group analysis is required to form the specified similarity transformations for the present problem. The newly formed similarity variable will then transform the PDEs to a system of ordinary differential equations (ODEs), and the model can be solved by the desired mathematical software. Therefore, the following scaling group of transformations G is introduced:

$$G : x^* = xG^{\omega_1}, \quad y^* = yG^{\omega_2}, \quad \psi^* = \psi G^{\omega_3}, \quad \sigma^* = \sigma G^{\omega_4},$$
$$\theta^* = \theta G^{\omega_5}, \quad u_e^* = u_e G^{\omega_6}, \quad u_1^* = u_1 G^{\omega_7}, \quad m^* = m G^{\omega_8}, \tag{14}$$

where ω_i are constants to be determined in which $i = 1, \ldots 8$. The transformation G is the transformation point which transforms the $(x, y, \psi, \sigma, \theta, u_e, u_1, m,)$ coordinates to the new coordinates $(x^*, y^*, \psi^*, \sigma^*, \theta^*, u_e^*, u_1^*, m^*)$.

Next, the substitution of (14) into Equations (9)–(11) yields the following expressions:

$$\frac{A_1}{A_2} G^{[2\omega_3 - 2\omega_2 - \omega_1]} \left(\frac{\partial \psi^*}{\partial y^*} \frac{\partial^2 \psi^*}{\partial x^* \partial y^*} - \frac{\partial \psi^*}{\partial x^*} \frac{\partial^2 \psi^*}{\partial y^{*2}} \right) = G^{[\omega_3 - 3\omega_2]} \left(\frac{\partial^3 \psi^*}{\partial y^{*3}} \right)$$
$$+ \frac{A_1}{A_2} \frac{De}{2} G^{[\omega_8 + 3\omega_3 - 7\omega_2]} \left[m^* \left(\frac{\partial^2 \psi^*}{\partial y^{*2}} \right)^2 \frac{\partial^3 \psi^*}{\partial y^{*3}} \right] + G^{[2\omega_6 - \omega_1]} u_e^* \frac{du_e^*}{dx^*} \tag{15}$$
$$- \frac{B_0^2}{\rho_{bf} a} \frac{1}{A_2} \left[G^{[\omega_3 + \omega_4 - \omega_2]} \left(\sigma^* \frac{\partial \psi^*}{\partial y^*} \right) - G^{[\omega_4 + \omega_6]} (\sigma^* u_e^*) \right],$$

$$G^{[\omega_3 + \omega_5 - \omega_1 - \omega_2]} \left(\frac{\partial \psi^*}{\partial y^*} \frac{\partial \theta^*}{\partial x^*} - \frac{\partial \psi^*}{\partial x^*} \frac{\partial \theta^*}{\partial y^*} \right) = \frac{1}{Pr} G^{[\omega_5 - 2\omega_2]} \left(\frac{\partial^2 \theta^*}{\partial y^{*2}} \right) + \frac{4}{3} \frac{Rd}{Pr} G^{[\omega_5 - 2\omega_2]} \left(\frac{\partial^2 \theta^*}{\partial y^{*2}} \right), \tag{16}$$

209

along with the boundary conditions:

$$G^{[\omega_3-\omega_2]}\left(\frac{\partial \psi^*}{\partial y^*}\right) = G^{[\omega_7+\frac{2}{5}\omega_1]}\left(\frac{u_1^*}{u_0}x^{*\frac{2}{5}}\right),$$
$$G^{[\omega_3-\omega_1]}\left(\frac{\partial \psi^*}{\partial x^*}\right) = -\frac{v_1}{\sqrt{av_{bf}}}G^{[-\frac{2}{5}\omega_1]}x^{*-\frac{2}{5}}, \quad G^{[\omega_5]}\theta^* = G^{[\frac{2}{5}\omega_1]}\left(x^{*\frac{2}{5}}\right) \quad \text{at} \quad y = 0 \quad (17)$$
$$G^{[\omega_3-\omega_2]}\left(\frac{\partial \psi^*}{\partial y^*}\right) \to G^{[\omega_6]}(u_e^*) \quad \text{as} \quad y \to \infty.$$

To retain the invariance of the system under G, the parameters defined in Equation (14), the following relations must hold:

$$2\omega_3 - 2\omega_2 - \omega_1 = \omega_3 - 3\omega_2 = 3\omega_3 - 7\omega_2 + \omega_8 = 2\omega_6 - \omega_1 = \omega_3 - \omega_2 + \omega_4 = \omega_4 + \omega_6$$
$$= \omega_3 + \omega_5 - \omega_1 - \omega_2 = \omega_5 - 2\omega_2. \quad (18)$$

From the boundary conditions of Equations (17), we also obtain the following relations among the parameters:

$$\omega_3 - \omega_2 = \omega_7 + \frac{2}{5}\omega_1, \quad \omega_3 - \omega_1 = -\frac{2}{5}\omega_1, \quad \omega_5 = \frac{2}{5}\omega_1, \quad \omega_3 - \omega_2 = \omega_6. \quad (19)$$

The absolute invariant can be determined by eliminating the parameter G of the group and hence Equations (18) and (19) provide the following expressions:

$$\omega_2 = \frac{2}{5}\omega_1, \quad \omega_3 = \frac{3}{5}\omega_1, \quad \omega_4 = -\frac{4}{5}\omega_1, \quad \omega_5 = \frac{2}{5}\omega_1,$$
$$\omega_6 = \frac{1}{5}\omega_1, \quad \omega_7 = -\frac{1}{5}\omega_1, \quad \omega_8 = \frac{2}{5}\omega_1 \quad (20)$$

From Equations (13), (14), and (20), we achieve the absolute invariants under the group G similarity transformations as follows:

$$\eta = \frac{y}{x^{\frac{1}{5}}}, \quad \psi = x^{\frac{3}{5}}f(\eta), \quad \sigma = \sigma_0 x^{-\frac{4}{5}}, \quad \theta = \theta_0(\eta) x^{\frac{2}{5}},$$
$$u_e = (u_e)_0 x^{\frac{1}{5}}, \quad u_1 = (u_1)_0 x^{-\frac{1}{5}}, \quad m = m_0 x^{\frac{2}{5}}. \quad (21)$$

The similarity transformations in Equation (21) are new and, by employing them in the governing boundary layer equations of Equations (9) and (11), the reduced version of the model in the form of ordinary differential equations can be attained as follows while satisfying Equation (9):

$$\frac{A_1}{A_2}f'''\left[1 + \frac{m_0 De}{2}(f'')^2\right] - \frac{1}{5}(f')^2 + \frac{3}{5}ff'' + \frac{1}{5} - \frac{M}{A_2}(1 - f') = 0, \quad (22)$$

$$\left(A_4 + \frac{4}{3}Rd\right)\theta'' - \frac{2}{5}A_3\text{Pr}\,f'\,\theta + \frac{3}{5}A_3\text{Pr}\,f\,\theta' = 0, \quad (23)$$

with the associated boundary conditions:

$$f(0) = \frac{5}{3}f_w, \quad f'(0) = \varepsilon, \quad \theta(0) = 1, \quad f'(\infty) = 1, \quad \theta(\infty) = 0. \quad (24)$$

Here $\varepsilon = (u_1)_0/u_0$ is the stretching/shrinking parameter, where $\varepsilon > 0$ indicates the stretching sheet, $\varepsilon = 0$ specifies the stationary sheet, and $\varepsilon < 0$ represents the state of shrinking sheet. Furthermore, $f_w = -v_1/\sqrt{v_{bf}\,a}$ is the constant mass transfer parameter, and $f_w > 0$ typifies the suction effect at the surface of the moving sheet and $f_w < 0$ epitomizes the injection state. For simplicity, we choose $(u_e)_0 = 1$. The power-law index is denoted by m_0; when $m_0 = 0$, the reduced model imitates the Newtonian fluid behavior. Moreover, when $m_0 = 1$, the model is reduced to the Eyring model, whereas the present model in Equations (22)–(24) also predicts specifically the pseudo-plastic (shear thinning) and dilatant (shear thickening) fluid properties when $m_0 < 0$ and $m_0 > 0$, respectively.

The physical quantities of interest in the present work are the local skin friction coefficient $(C_{f\bar{x}})$ and the local Nusselt number $(Nu_{\bar{x}})$ which are defined as follows:

$$C_{f\bar{x}} = \frac{\tau_w}{\rho_{bf} \bar{u}_e^2}, \quad Nu_{\bar{x}} = \frac{\bar{x} q_w}{k_{bf}(T_w - T_\infty)}, \tag{25}$$

where τ_w is the wall shear stress and q_w is the heat flux at the surface of the sheet, and can be further defined as [34]:

$$\tau_w = \mu_{nf}\left\{1 + \frac{mE^2}{6}\left[2\left(\frac{\partial \bar{u}}{\partial \bar{x}}\right)_{\bar{y}=0}^2 + \left(\frac{\partial \bar{u}}{\partial \bar{y}}\right)_{\bar{y}=0}^2\right]\right\}\left(\frac{\partial \bar{u}}{\partial \bar{y}}\right)_{\bar{y}=0}, \quad q_w = -k_{nf}\left(\frac{\partial T}{\partial \bar{y}}\right)_{\bar{y}=0}. \tag{26}$$

The reduced skin friction coefficient $\mathrm{Re}_{\bar{x}}^{1/2} C_{f\bar{x}} x^{1/5}$ and the local Nusselt number $\mathrm{Re}_{\bar{x}}^{-1/2} Nu_{\bar{x}} x^{-1/5}$ can be obtained using the similarity transformations of Equation (21) and the expressions in Equations (25) and (26) as follows:

$$\begin{aligned} C_{f\bar{x}}\mathrm{Re}_{\bar{x}}^{1/2} &= \sqrt{A_1 A_2} f''(0) + \frac{m_0}{6\sqrt{A_1 A_2}} De[f''(0)]^3, \\ \mathrm{Re}_{\bar{x}}^{-1/2} Nu_{\bar{x}} x^{-2/5} &= -\frac{1}{\sqrt{A_1 A_2}}\left(1 + \frac{1}{A_4}\frac{4}{3}Rd\right)\frac{\theta'(0)}{\theta(0)}, \end{aligned} \tag{27}$$

where $\mathrm{Re}_{\bar{x}} = \bar{u}_e \bar{x}/\nu_{bf}$ denotes the local Reynolds number.

5. Stability Analysis

Merkin [39,40] established an improved version of the stability analysis, which is prominent among researchers for the examination of the stability of numerical solutions. Because we observed dual solutions in the present work, we assess the solution's stability to determine the flow behavior. To initiate the linear stability analysis, the model equations in Equations (3)–(5) need to be considered in the unsteady form as follows:

$$\frac{\partial \bar{u}}{\partial \bar{x}} + \frac{\partial \bar{v}}{\partial \bar{y}} = 0, \tag{28}$$

$$\frac{\partial \bar{u}}{\partial \bar{t}} + \bar{u}\frac{\partial \bar{u}}{\partial \bar{x}} + \bar{v}\frac{\partial \bar{u}}{\partial \bar{y}} = \frac{\mu_{nf}}{\rho_{nf}}\frac{\partial^2 \bar{u}}{\partial \bar{y}^2} + \frac{\mu_{nf}}{\rho_{nf}}\frac{mE^2}{2}\left(\frac{\partial \bar{u}}{\partial \bar{y}}\right)^2\frac{\partial^2 \bar{u}}{\partial \bar{y}^2} + \bar{u}_e\frac{d\bar{u}_e}{d\bar{x}} - \frac{\sigma B_0^2(\bar{u}_e - \bar{u})}{\rho_{nf}}, \tag{29}$$

$$\frac{\partial T}{\partial \bar{t}} + \bar{u}\frac{\partial T}{\partial \bar{x}} + \bar{v}\frac{\partial T}{\partial \bar{y}} = \frac{k_{nf}}{(\rho C_p)_{nf}}\frac{\partial^2 T}{\partial \bar{y}^2} + \frac{16 \sigma_1 T_\infty^3}{3(\rho C_p)_{nf} k_1}\frac{\partial^2 T}{\partial \bar{y}^2}, \tag{30}$$

with the boundary conditions of Equation (6). Then, we introduce the dimensional time variable, $\bar{t}$ with a new similarity variable $(\tau = \tau_0/x^{4/5})$ through scaling group analysis. The new similarity transformation is given as:

$$\begin{aligned} \eta &= \frac{y}{x^{2/5}}, \quad \psi = x^{\frac{3}{5}} f(\eta, \tau), \quad \sigma = \sigma_0 x^{-\frac{4}{5}}, \quad \theta = \theta_0(\eta, \tau) x^{\frac{2}{5}}, \\ u_e &= (u_e)_0 x^{\frac{1}{5}}, \quad u_1 = (u_1)_0 x^{-\frac{1}{5}}, \quad m = m_0 x^{\frac{2}{5}}, \quad \tau = \frac{\tau_0}{x^{\frac{4}{5}}}. \end{aligned} \tag{31}$$

Employing (31) in the dimensionless form of Equations (28)–(30) and (6) gives the following system of equations:

$$\frac{A_1}{A_2}\frac{\partial^3 f}{\partial \eta^3}\left[1 + \frac{m_0 De}{2}\left(\frac{\partial^2 f}{\partial \eta^2}\right)^2\right] - \frac{1}{5}\left(\frac{\partial f}{\partial \eta}\right)^2 + \frac{3}{5}f(\eta,\tau)\frac{\partial^2 f}{\partial \eta^2} + \frac{1}{5} - \frac{M}{A_2}\left(1 - \frac{\partial f}{\partial \eta}\right) - \frac{\partial^2 f}{\partial \eta \partial \tau} = 0, \tag{32}$$

$$\left(A_4 + \frac{4}{3}Rd\right)\frac{\partial^2 \theta}{\partial \eta^2} - \frac{2}{5}A_3 \Pr \frac{\partial f}{\partial \eta}\theta(\eta,\tau) + \frac{3}{5}A_3 \Pr f(\eta,\tau)\frac{\partial \theta}{\partial \eta} - \frac{\partial \theta}{\partial \tau} = 0, \tag{33}$$

with the boundary conditions:

$$f(0,\tau) = \frac{5}{3}f_w, \quad \frac{\partial f}{\partial \eta}(0,\tau) = \varepsilon, \quad \theta(0,\tau) = 1, \quad \frac{\partial f}{\partial \eta}(\infty,\tau) = 1, \quad \theta(\infty,\tau) = 0. \tag{34}$$

It is assumed that the solutions of (32)–(34) are expressed by the formulas of Equation (35):

$$f(\eta,\tau) = f_0(\eta) + e^{-\gamma\tau}F(\eta,\tau), \quad \theta(\eta,\tau) = \theta_0(\eta) + e^{-\gamma\tau}G(\eta,\tau), \tag{35}$$

where $f(\eta) = f_0(\eta)$ and $\theta(\eta) = \theta_0(\eta)$ are the solutions found in the previous section, in which the disturbance is superimposed to determine their stability. Here, the unknown eigenvalue parameter is denoted by γ, and $F(\eta,\tau)$ and $G(\eta,\tau)$ are relatively small compared to the steady state solutions ($f_0(\eta)$ and $\theta_0(\eta)$). The substitution of Equation (35) into Equations (32)–(34) gives the following system:

$$\frac{\partial^3 F}{\partial \eta^3}\left(1 + \frac{A_1}{A_2}\frac{m_0 De}{2}\left(f_0''\right)^2\right) + \frac{\partial^2 F}{\partial \eta^2}\left(\frac{3}{5}f_0 - \frac{2}{5}f_0'\right) + \frac{3}{5}f_0''F + \left(\frac{M}{A_2} + \gamma\right)\frac{\partial F}{\partial \eta} - \frac{\partial^2 F}{\partial \eta \partial \tau} = 0, \tag{36}$$

$$\left(A_4 + \frac{4}{3}Rd\right)\frac{\partial^2 G}{\partial \eta^2} + \frac{3}{5}A_3 Pr f_0 \frac{\partial G}{\partial \eta} + \frac{3}{5}A_3 Pr F \theta_0 - \frac{2}{5}A_3 Pr f_0' G - \frac{2}{5}A_3 Pr \theta_0 \frac{\partial F}{\partial \eta} + \gamma G - \frac{\partial G}{\partial \tau} = 0, \tag{37}$$

subject to the boundary conditions:

$$F(0,\tau) = 0, \quad \frac{\partial F}{\partial \eta}(0,\tau) = 0, \quad G(0,\tau) = 0, \quad \frac{\partial F}{\partial \eta}(\infty,\tau) = 0, \quad G(\infty,\tau) = 0. \tag{38}$$

Referring to Merkin [39,40], $\tau \to 0$ is fixed to examine the stability of the steady state boundary layer flow. Thus, $F = F_0(\eta)$ and $G = G_0(\eta)$ in Equations (37)–(39), yielding the following linearized eigenvalue problem:

$$F_0'''\left(1 + \frac{A_1}{A_2}\frac{m_0 De}{2}\left(f_0''\right)^2\right) + F_0''\left(\frac{3}{5}f_0 - \frac{2}{5}f_0'\right) + \frac{3}{5}f_0'' F_0 + \left(\frac{M}{A_2} + \gamma\right)F_0' = 0, \tag{39}$$

$$\left(A_4 + \frac{4}{3}Rd\right)G_0'' + \frac{3}{5}A_3 Pr f_0 G_0' + \frac{3}{5}A_3 Pr F_0 \theta_0 - \frac{2}{5}A_3 Pr f_0' G_0 - \frac{2}{5}A_3 Pr \theta_0 F_0' + \gamma G_0 = 0, \tag{40}$$

with the boundary conditions:

$$\begin{aligned} F_0(\eta) = 0, \quad F_0'(\eta) = 0, \quad G_0(\eta) = 0, \quad &\text{at } \eta = 0, \\ F_0'(\eta) = 0, \quad G_0(\eta) = 0 \quad &\text{as } \eta \to \infty. \end{aligned} \tag{41}$$

It is necessary to replace one of the outer boundary conditions with a normalizing boundary condition to obtain the eigenvalues. Therefore, the boundary condition $F_0'(\infty) = 0$ is substituted with $F_0''(0) = 1$. The system of equations in Equations (38)–(40) with the new boundary condition is solved by the MATLAB boundary value problem solver (bvp4c) to obtain the lowest eigenvalues as the governing parameter varies. These lowest eigenvalues are classified according to their sign. If the lowest eigenvalue falls in the positive range of values, then the respective numerical solution is accepted as a stable solution. Meanwhile, the negative lowest eigenvalue suggests the numerical solution is unstable. Further explanation about the stable and unstable solutions is provided in the next section.

6. Results and Discussion

The mathematical model in Equations (23)–(25) was solved numerically by means of the boundary value problem solver function bvp4c in the MATLAB software. The numerical results were derived while limiting the relative tolerance to 1×10^{-10}. Some of the governing parameter values were fixed throughout the computation process to align with the motivation of this study. For example, the Prandtl number (Pr) value was fixed at 4.36 because it represents the base fluid, n-hexane. The power-law index (m_0) was fixed at 1.5 to investigate the dilatant features of the Sutterby fluid. The obtained non-uniqueness solutions were classified based on how early the solution converged asymptotically. For example, the numerical solution that converged earlier asymptotically in the velocity/temperature profiles was labelled the first solution. The other solution that converged later, asymptotically, was labelled as the second solution. Before presenting the numerical results, we provide validation of our numerical method by solving the model presented in [41] and compare the numerical results with the results reported by [41]. Table 2 shows the comparison results, and there is a good agreement. Bhattacharyya et al. [41] employed the shooting method to solve the model, and Table 2 confirms that the bvp4c function is capable of precisely solving the boundary value problem.

Table 2. Numerical validation of $f''(0)$ when $S = 0$ in [41].

*c/a.	Present Result		Bhattacharyya et al. [41]	
	First Solution	Second Solution	First Solution	Second Solution
−0.250	1.40224078	–	1.40224051	–
−0.500	1.49566974	–	1.49566972	–
−0.625	1.50715589	–	1.50715673	–
−0.750	1.48929822	–	1.48929811	–
−1.000	1.32881685	0	1.32881689	0
−1.150	1.08223113	0.11670214	1.08223164	0.11670230
−1.200	0.93247330	0.23364972	0.93247277	0.23364910
−1.2465	0.58429940	0.55429554	0.58429146	0.55428565

*c/a is the stretching/shrinking parameter in [41].

Figure 2 shows the influence of the suction parameter (s) on the reduced skin friction coefficient $\left(C_{fx}\text{Re}_{\overline{x}}^{1/2}\right)$ and velocity profiles ($f'(\eta)$). Based on the first solution in Figure 2a, an increment in s increases the values of $C_{fx}\text{Re}_{\overline{x}}^{1/2}$ past a shrinking sheet. Primarily, an increment in s from 6 to 9 strengthens the impact of suction at the surface of the shrinking sheet. The act of suction encourages the laminar flow by trapping the low speed fluid molecules in the boundary layer region. This then leads to increasing of the fluid velocity past the shrinking sheet and is illuminated in Figure 2b. The increment of the fluid velocity reduces the momentum boundary layer thickness and increases the wall shear stress over the shrinking sheet. The high wall shear stress eventually increases the values of $C_{fx}\text{Re}_{\overline{x}}^{1/2}$ as s increases. Interestingly, the second solution in Figure 2a shows the opposite trend to the first solution, where an increment in s decreases the values of $C_{fx}\text{Re}_{\overline{x}}^{1/2}$. The state of suction, which was interpreted as enhancing the fluid velocity, is now seen to decrease the fluid velocity and increase the momentum boundary layer thickness (see Figure 2). The saturated state of the shrinking sheet may be the cause of these consequences. Later, the reducing fluid velocity lowers the wall shear stress and then decreases the values of $C_{fx}\text{Re}_{\overline{x}}^{1/2}$ as s increases.

Figure 3 demonstrates the impact of the Deborah number (De) on $C_{fx}\text{Re}_{\overline{x}}^{1/2}$ and velocity profiles. Both solutions in Figure 3a convey that an increment in De augments the values of $C_{fx}\text{Re}_{\overline{x}}^{1/2}$ as the sheet is shrinking. The Deborah number is used to enlighten the viscoelastic feature of a material. Here, an increment in De results in the increment of the shear thickening Sutterby fluid velocity. The valuable work of Azhar et al. [34] also reported a similar trend. The increment of the fluid velocity then enhances the wall shear stress past the shrinking sheet and increases the values of $C_{fx}\text{Re}_{\overline{x}}^{1/2}$. The increment in s and De assist in delaying the flow separation significantly.

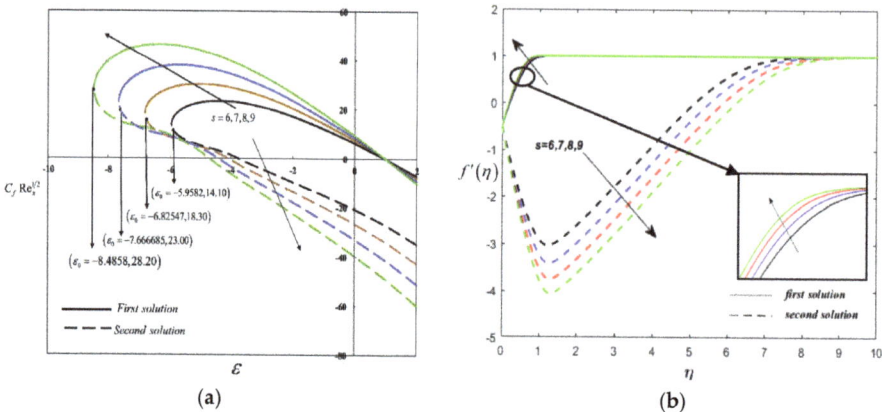

Figure 2. Impact of the suction parameter (s) on: (**a**) the reduced skin friction coefficient; (**b**) velocity profiles as s varies when $Rd = 1.2, m_0 = 1.5, \Pr = 4.36, De = 1.5, M = 0.5$, and $\phi = 0.02$.

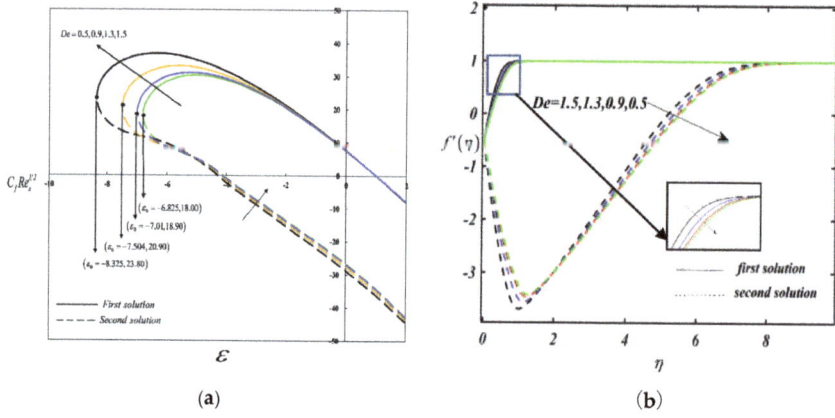

Figure 3. Impact of the Deborah number (De) on: (**a**) the reduced skin friction coefficient; (**b**) velocity profiles as De varies when $Rd = 1.2, m_0 = 1.5, \Pr = 4.36, s = 7, M = 0.5$, and $\phi = 0.02$.

The first and second solutions in Figure 4a lead to a decrement in $C_{fx}\text{Re}_x^{1/2}$ when M increases from 0.5 to 1.0. The magnetic field presents in an electrically conducting fluid as an electromagnetic force in the fluid flow region, which slows the fluid moving past the shrinking sheet. This is reflected by the velocity profiles in Figure 4b, where the fluid velocity decreases when M increases. The decrement in the fluid velocity then leads to an increase of the momentum boundary layer thickness and decreases the wall shear stress past the permeable shrinking sheet. Thus, the values of $C_{fx}\text{Re}_x^{1/2}$ decrease with the rising value of M. Unlike the shrinking case, different fluid flow behavior is perceived in the first solution when the Sutterby nanofluid flows towards a permeable stretching sheet. When the magnetic effect increases past a stretching sheet, the fluid velocity increases, although the increment is not significant. This is agreeable because the fluid flow is in the same direction as the stretching sheet and the action of the stretching sheet speeds up the fluid flow. The increment in fluid velocity reduces the momentum boundary layer thickness, increases the wall shear stress, and enhances the value of $C_{fx}\text{Re}_x^{1/2}$. The negative values of $C_{fx}\text{Re}_x^{1/2}$ indicate that the stretching sheet imposes a drag force on the fluid. Moreover, the reverse flow is observed through the second solution's presence when the permeable sheet is stretching in Figure 4a. The velocity profiles in Figure 4c support this by displaying the velocity overshoot (see the second solution profiles) when M varies. Thus, it is clear that reverse flow does exist in the stretching sheet case, and this may be due to the state of the sheet where the suction intensity is weak when the effect of M increases.

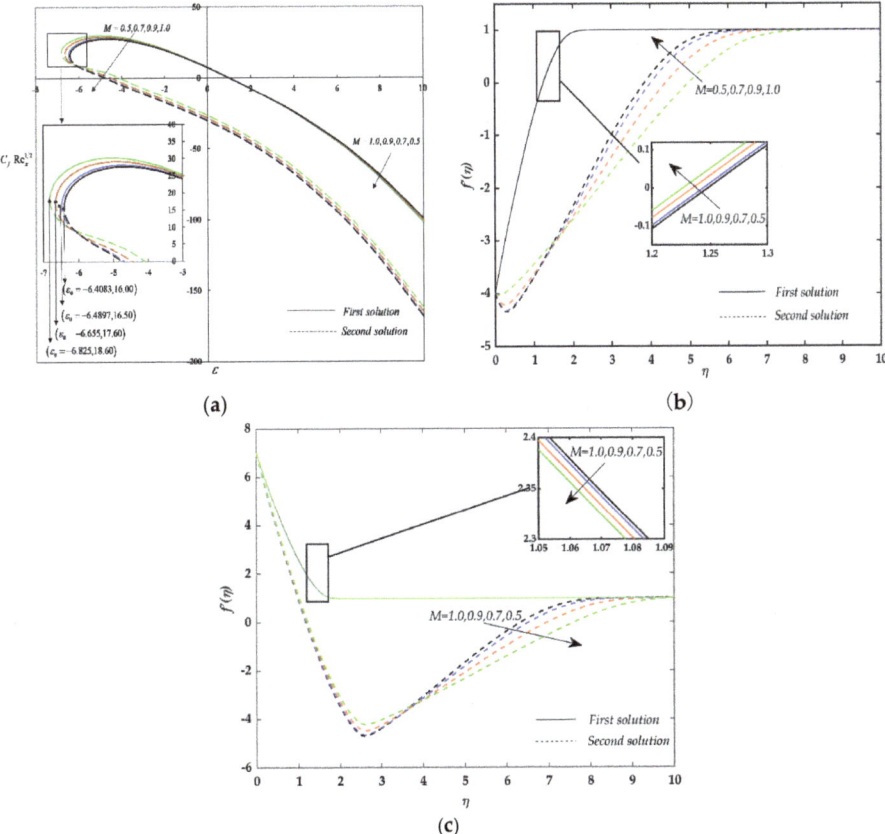

Figure 4. Impact of the magnetic parameter (M) on: (**a**) the reduced skin friction coefficient; (**b**) velocity profiles as M varies past a shrinking sheet ($\varepsilon = -4$); (**c**) velocity profiles as M varies past a stretching sheet ($\varepsilon = 7$) when $Rd = 1.2, m_0 = 1.5, \Pr = 4.36, De = 1.5, s = 7,$ and $\phi = 0.02$.

Figure 5 portrays the effect of the nanoparticle volume fraction or ϕ on $C_{fx}\text{Re}_{\overline{x}}^{1/2}$ and velocity profiles. The increment in ϕ increases the values of $C_{fx}\text{Re}_{\overline{x}}^{1/2}$ over a permeable shrinking sheet. An increased ratio of ϕ in the base fluid increases fluid viscosity, which then enhances the fluid velocity past the permeable shrinking sheet (see Figure 5b). These then affect the wall shear stress to increase and, consequently, raise the values of $C_{fx}\text{Re}_{\overline{x}}^{1/2}$. Velocity overshoots in the boundary layer are apparent in Figures 2b, 3b, 4b and 5b. These velocity overshoots near the permeable shrinking sheet indicate that the fluid velocity is higher than the shrinking sheet's velocity [42].

Table 3 exhibits the effect of the radiation parameter (Rd) on the reduced local Nusselt number $\left(\text{Re}_{\overline{x}}^{-1/2} Nu_{\overline{x}} x^{-2/5}\right)$ over the permeable shrinking surface. Both solutions allude to the enhancement of $\text{Re}_{\overline{x}}^{-1/2} Nu_{\overline{x}} x^{-2/5}$ when the impact of radiation grows in the fluid flow region. An increment in Rd hints at the release of energy in the form of heat from the fluid flow and decreases the fluid temperature profile. Thus, the thermal boundary layer becomes thinner and the wall heat flux increases. The depreciation in the thermal conductivity induces an increase in the rate of heat transfer or $\left(\text{Re}_{\overline{x}}^{-1/2} Nu_{\overline{x}} x^{-2/5}\right)$.

Furthermore, the magnetic parameter and the nanoparticle volume fraction have minimal effect in delaying flow separation. This is evident by the critical values (ε_0), as shown in Figures 4a and 5a.

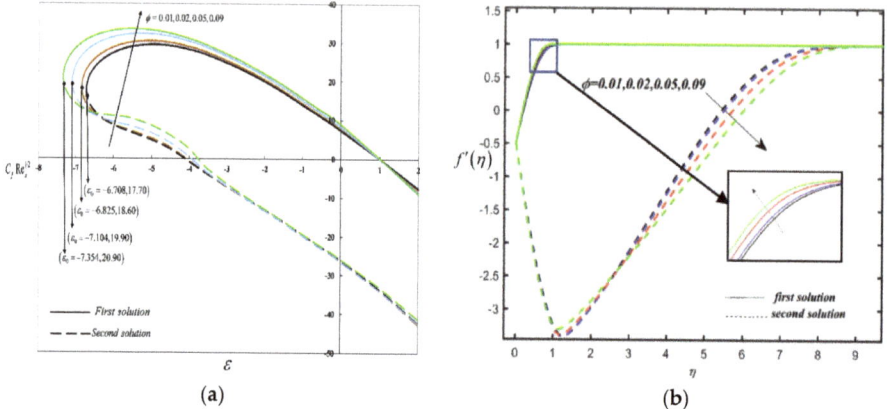

(a)   (b)

Figure 5. Impact of the nanoparticle volume fraction (ϕ) on: (a) the reduced skin friction coefficient; (b) velocity profiles as ϕ varies when $Rd = 1.2, m_0 = 1.5, \Pr = 4.36, De = 1.5, s = 7, M = 0.5$, and $\phi = 0.02$.

Table 3. The effect of the radiation parameter (Rd) on the reduced local Nusselt number when $M = 0.5, m_0 = 1.5, \Pr = 4.36, \phi = 0.02, s = 7$, and $De = 1.5$ as ε varies.

ε	Radiation Parameter (Rd)	$\operatorname{Re}_x^{-1/2} Nu_{\bar{x}} x^{-1/5}$	
		First Solution	Second Solution
−1.5	0.5	−1846.311663	−1846.31001
−3.5		−1845.804534	−1845.803095
−5.5		−1845.296838	−1845.296378
−6.5		−1845.042742	−1845.042486
−1.5	1.2	−1846.051576	−1846.046864
−3.5		−1845.195172	−1845.191067
−5.5		−1844.337149	−1844.335834
−6.5		−1843.907426	−1843.906694
−1.5	3.5	−1845.128211	−1845.099751
−3.5		−1844.064298	−1844.051593
−5.5		−1842.553219	−1842.549137
−6.5		−1841.795465	−1841.793188

The results of the stability analysis are presented in Table 4. The first solution achieves the positive eigenvalues while the second solution attains the negative eigenvalues. Based on the signs of eigenvalues, one can say that the positive eigenvalues specify the first solution as a stable solution; the stable solution can be understood as feasible and able to overcome the growth of an initially given disturbance. Furthermore, the negative eigenvalues reveal the second solution as an unstable solution associated with flow separation. The second solution promotes the growth of an initially given disturbance and hence achieves the negative eigenvalue. However, it is vital to identify and verify the stability of non-unique solutions so that the variety of possibilities of fluid flow behavior can be predicted.

Figure 6 depicts the streamlines of the Sutterby fluid under a number of settings. In particular, Figure 6a shows the streamlines when the sheet is impermeable and stretching at the rate of 1.4, while Figure 6b illustrates the streamlines when the sheet is impermeable and shrinking. The reverse flow in Figure 6b is noticeable and proves that the shrinking sheet's state instigates the reverse flow. Next, the streamlines for the fluid flow under the suction influence can be examined (see Figure 6c,d).

The reverse flow is now absent past the permeable shrinking sheet (Figure 6d). Thus, it is proved that mass suction succeeds in sustaining the laminar boundary layer flow over a shrinking surface. Figure 6e,f shows the behavior of fluid flow when the rate of stretching or shrinking increases; the fluid pattern being pulled at the surface of the sheet is clear, and again the reverse flow is absent.

Table 4. Lowest eigenvalues (γ_1) when $Rd = 1.2, m_0 = 1.5, \Pr = 4.36, De = 1.5, s = 7, M = 0.5$, and $\phi = 0.02$ as ε varies.

ε	γ_1	
	First Solution	Second Solution
−6.8	0.5844	−0.4163
−6.82	0.2842	−0.1794
−6.822	0.2339	−0.1354
−6.8250	0.1128	−0.0237
−6.82520	0.0961	−0.0077
−6.825250	0.0911	−0.0028

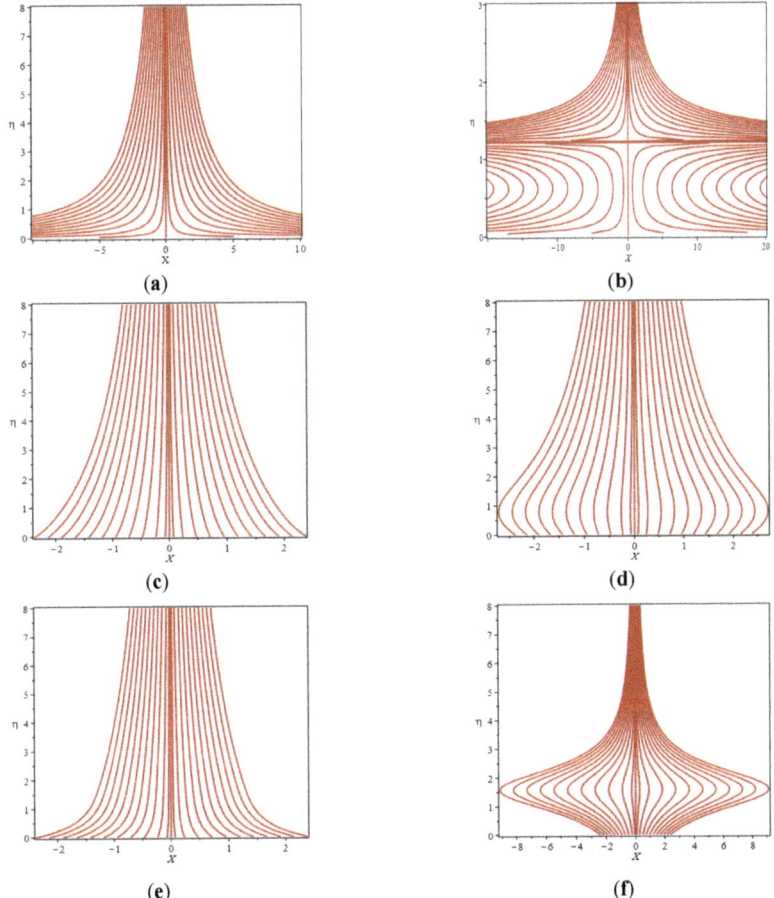

Figure 6. Streamlines when $Rd = 1.2, m_0 = 1.5, \Pr = 4.36, De = 1.5, \phi = 0.02, M = 0.5$; (**a**) $s = 0, \varepsilon = 1.4$; (**b**) $s = 0, \varepsilon = -1.4$; (**c**) $s = 7, \varepsilon = 1.4$; (**d**) $s = 7, \varepsilon = -1.4$; (**e**) $s = 7, \varepsilon = 4$; (**f**) $s = 7, \varepsilon = -4$.

7. Conclusions

The present numerical investigation aimed to reveal the Sutterby nanofluid fluid flow and heat transfer over a permeable stretching/shrinking surface together with the effects of thermal radiation and magnetohydrodynamics (MHD). The appropriate form of the similarity transformations for the present flow problem was derived using scaling group analysis. The newly derived similarity variables then transformed the mathematical model into a more straightforward form to solve the boundary value problem utilizing the solver function bvp4c in the MATLAB software. The significant results are summarized as follows:

- An increment in the suction parameter, the Deborah number, and the nanoparticle volume fraction delay flow separation.
- The dominance of the magnetic parameter in the fluid flow regime accelerates flow separation.
- Non-unique solutions are observed when governing parameters, such as the suction parameter, the Deborah number, the magnetic number, the radiation parameter, and the nanoparticle volume fraction, vary.
- The increment in the radiation parameter slightly enhances the convective heat transfer rate past a permeable shrinking sheet.
- Stability analysis elucidates the first solution as a stable solution, and the second solution as an unstable solution.

Author Contributions: Conceptualization, M.M.F., K.N., M.F.M.B., I.H. and R.R.; methodology, M.M.F., K.N. and M.F.M.B.; software, M.M.F.; validation, M.M.F., K.N., M.F.M.B., I.H. and R.R.; formal analysis, M.M.F., K.N. and M.F.M.B.; investigation, M.M.F., K.N. and M.F.M.B.; writing—original draft preparation, M.M.F., K.N., M.F.M.B., I.H. and R.R.; writing—review and editing, M.M.F., K.N., M.F.M.B., I.H. and R.R.; funding acquisition, I.H. All authors have read and agreed to the published version of the manuscript.

Funding: This research was funded by the Ministry of Education Malaysia (Project Code: FRGS/1/2019/STG06/UKM/01/2) and the APC was funded by the Ministry of Education Malaysia (Project Code: FRGS/1/2019/STG06/UKM/01/2). The author from Universiti Teknologi Malaysia (UTM) would like to acknowledge the Ministry of Education (MOE) and Research Management Centre-UTM for supported this work under the research grant (vote number 17J52).

Conflicts of Interest: The authors declare no conflict of interest.

Nomenclature

Roman letters

a	dimensional positive constant
A_1	first order Rivlin–Erickson tensor
B_0	magnetic field strength
C_p	specific heat capacity
$CH_3(CH_2)_4CH_3$	n-Hexane
De	Deborah number
E	material time constant
Fe_3O_4	magnetite
f_w	constant mass transfer parameter
I	identity tensor
k	thermal conductivity
k_1	Rosseland mean absorption coefficient
M	magnetic parameter
m	power-law index
p	pressure
Rd	radiation parameter
S	extra stress tensor
Pr	Prandtl number

Roman letters

T	Cauchy stress tensor
$\bar{t}$	dimensional time variable
T_w	wall temperature
T_∞	free stream temperature
T_0	reference temperature
u	dimensionless velocity
$\bar{u}$	dimensional velocity
u_1	reference velocity
V	velocity field
v_1	normal reference velocity
v_w	dimensionless surface mass flux velocity
$\bar{v}_w$	dimensional surface mass flux velocity
x, y	dimensionless Cartesian coordinates
$\bar{x}, \bar{y}$	dimensional Cartesian coordinates

Greek letters

α	thermal diffusivity
$\dot{\gamma}$	second invariant strain tensor
γ_1	smallest eigenvalue
ε	stretching/shrinking parameter
ε_0	critical value
η	similarity variable
θ	dimensionless temperature
μ_0	viscosity at low shear rates
μ	dynamic viscosity
ν	kinematic viscosity
ρ	density
σ	electrical conductivity
σ_1	Stefan Boltzmann constant
τ	dimensionless time variable
ϕ	nanoparticle volume fraction
ψ	stream function

Subscripts

bf	base fluid
e	condition at the free stream
nf	nanofluid
s	solid fractions
w	condition at the wall

Superscript

$'$	differentiation with respect to η

References

1. Sutterby, J.L. Laminar converging flow of dilute polymer solutions in conical sections: Part I. Viscosity data, new viscosity model, tube flow solution. *AIChE J.* **1966**, *12*, 63–68. [CrossRef]
2. Sutterby, J.L. Laminar converging flow of dilute polymer solutions in conical sections. II. *Trans. Soc. Rheol.* **1965**, *9*, 227–241. [CrossRef]
3. Batra, R.L.; Eissa, M. Helical flow of a Sutterby model fluid. *Polym. Plast. Technol. Eng.* **1994**, *33*, 489–501. [CrossRef]
4. Bergeron, V. Designing intelligent fluids for controlling spray applications. *C. R. Phys.* **2003**, *4*, 211–219. [CrossRef]
5. Giudice, D.; Haward, S.J.; Shen, A.Q. Relaxation time of dilute polymer solutions: A microfluidic approach. *J. Rheol.* **2017**, *61*, 327–337. [CrossRef]
6. Rehage, H.; Hoffmann, H. Viscoelastic surfactant solutions: Model systems for rheological research. *Mol. Phys.* **1991**, *74*, 933–973. [CrossRef]

7. Fujii, T.; Miyatake, O.; Fujii, M.; Tanaka, H.; Murakami, K. Natural convective heat transfer from a vertical isothermal surface to a non-Newtonian Sutterby fluid. *Int. J. Heat Mass Tranf.* **1973**, *16*, 2177–2187.
8. Fujii, T.; Miyatake, O.; Fujii, M.; Tanaka, H.; Murakami, K. Natural convective heat-transfer from a vertical surface of uniform heat flux to a non-Newtonian Sutterby fluid. *Int. J. Heat Mass Tranf.* **1974**, *17*, 149–154.
9. Choi, S.U.S.; Eastman, J.A. Enhancing thermal conductivity of fluids with nanoparticles. *ASME Publ. Fed.* **1995**, *231*, 99–103.
10. Mahian, O.; Kolsi, L.; Amani, M.; Estellé, P.; Ahmadi, G.; Kleinstreuer, C.; Marshall, J.S.; Taylor, R.A.; Abu-Nada, E.; Rashidi, S.; et al. Recent advances in modeling and simulation of nanofluid flows-Part II: Applications. *Phys. Rep.* **2019**, *791*, 1–59. [CrossRef]
11. Naganthran, K.; Basir, M.F.M.; Alharbi, S.O.; Nazar, R.; Alwatban, A.M.; Tlili, I. Stagnation point flow with time-dependent bionanofluid past a sheet: Richardson extrapolation technique. *Processes* **2019**, *7*, 722. [CrossRef]
12. Pop, I.; Naganthran, K.; Nazar, R.; Ishak, A. The effect of vertical throughflow on the boundary layer flow of a nanofluid past a stretching/shrinking sheet. *Int. J. Numer. Method Heat* **2017**, *27*, 1910–1927. [CrossRef]
13. Zainal, N.A.; Nazar, R.; Naganthran, K.; Pop, I. Unsteady three-dimensional MHD non-axisymmetric Homann stagnation point flow of a hybrid nanofluid with stability analysis. *Mathematics* **2020**, *8*, 784. [CrossRef]
14. Khan, M.I.; Qayyum, S.; Hayat, T.; Alsaedi, A. Stratified flow of Sutterby fluid with homogeneous-heterogeneous reactions and Catteneo-Christov heat flux. *Int. J. Numer. Method Heat* **2019**, *29*, 2977–2992. [CrossRef]
15. Mir, N.A.; Alqarni, M.S.; Farooq, M.; Malik, M.Y. Analysis of heat generation/absorption in thermally stratified Sutterby fluid flow with Cattaneo-Christov theory. *Microsyst. Technol.* **2019**, *25*, 3365–3373.
16. Mouli, G.B.C.; Gangadhar, K.; Raju, B.H.S. On spectral relaxation approach for Soret and Dufour effects on Sutterby fluid past a stretching sheet. *Int. J. Ambient Energy* **2019**. [CrossRef]
17. Nadeem, S.; Maraj, E.N. Peristaltic flow of Sutterby nanofluid in a curved channel with compliant walls. *J. Comput. Theor. Nanos.* **2015**, *12*, 226–233. [CrossRef]
18. Akbar, N.S.; Nadeem, S. Nano Sutterby fluid model for the peristaltic flow in small intestines. *J. Comput. Theor. Nanos.* **2013**, *10*, 2491–2499. [CrossRef]
19. Hayat, T.; Ayub, S.; Alsaedi, A.; Tanveer, A.; Ahmad, B. Numerical simulation for peristaltic activity of Sutterby fluid with modified Darcy's law. *Results Phys.* **2017**, *7*, 762–768. [CrossRef]
20. Ahmad, S.; Farooq, M.; Javed, M.; Anjum, A. Double stratification effects on chemically reactive squeezed Sutterby fluid flow with thermal radiation and mixed convection. *Results Phys.* **2018**, *8*, 1250–1259. [CrossRef]
21. Hayat, T.; Afzal, S.; Khan, M.I.; Alsaedi, A. Irreversibility aspects to flow of Sutterby fluid subject to nonlinear heat flux and Joule heating. *Appl. Nanosci.* **2019**, *9*, 1215–1226. [CrossRef]
22. Hayat, T.; Masood, F.; Qayyum, S.; Alsaedi, A. Sutterby fluid flow subject to homogeneous-heterogeneous reactions and nonlinear radiation. *Physica A* **2020**, *544*, 123439. [CrossRef]
23. Nawaz, M. Role of hybrid nanoparticles in thermal performance of Sutterby fluid, the ethylene glycol. *Physica A* **2020**, *537*, 122447. [CrossRef]
24. Wang, F.; Wang, N.; Yu, F.; Wang, X.; Cui, J. Study on micro-structure, solid solubility and tensile properties of 5A90 Al-Li alloy cast by low-frequency electromagnetic casting processing. *J. Alloys Compd.* **2020**, *820*, 153318. [CrossRef]
25. Beg, O.A.; Ferdows, M.; Karim, M.E.; Hasan, M.M.; Bég, T.A.; Shamshuddin, M.D.; Kadir, A. Computation of non-isothermal thermo-convective micropolar fluid dynamics in a wall MHD generator system with non-linear distending wall. *Int. J. Appl. Comput. Math.* **2020**, *6*, 42.
26. Raptis, A.; Perdikis, C.; Takhar, H.S. Effect of thermal radiation on MHD flow. *Appl. Math. Comput.* **2004**, *153*, 645–649. [CrossRef]
27. Sabir, Z.; Imran, A.; Umar, M.; Zeb, M.; Shoaib, M.; Raja, M.A.Z. A numerical approach for two-dimensional Sutterby fluid flow bounded at a stagnation point with an inclined magnetic field and thermal radiation impacts. *Therm. Sci.* **2020**, *186*. [CrossRef]
28. Bilal, S.; Sohail, M.; Naz, R.; Malik, M.Y.; Alghamdi, M. Upshot of ohmically dissipated Darcy-Forchheimer slip flow of magnetohydrodynamic Sutterby fluid over radiating linearly stretched surface in view of Cash and Crap method. *Appl. Math. Mech. Engl. Ed.* **2019**, *40*, 861–876. [CrossRef]

29. Prandtl, L. Über Flussigkeitsbewegungen bei sehr kleiner Reibung. *Verhandl. III Intern. Math.* **1904**, 484–491. Available online: https://ci.nii.ac.jp/naid/20000989592/ (accessed on 20 July 2020).
30. Schwarz, F. Symmetries of differential equations: From Sophus Lie to computer algebra. *SIAM Rev.* **1988**, *30*, 450–481. [CrossRef]
31. Pakdemirli, M.; Yurusoy, M. Similarity transformations for partial differential equations. *SIAM Rev.* **1998**, *40*, 96–101. [CrossRef]
32. Naganthran, K.; Basir, M.F.M.; Thumma, T.; Ige, E.O.; Nazar, R.; Tlili, I. Scaling group analysis of bioconvective micropolar fluid flow and heat transfer in a porous medium. *J. Therm. Anal. Calorim.* **2020**. [CrossRef]
33. Hamad, M.A.A.; Pop, I. Scaling transformations for boundary layer flow near the stagnation-point on a heated permeable stretching surface in a porous medium saturated with a nanofluid and heat generation/absorption effects. *Transp. Porous Media* **2011**, *87*, 25–39. [CrossRef]
34. Azhar, E.; Iqbal, Z.; Maraj, E.N. Impact of entropy generation on stagnation-point flow of Sutterby nanofluid: A numerical analysis. *Z. Naturforsch.* **2016**, *71*, 837–848. [CrossRef]
35. Azhar, E.; Iqbal, Z.; Ijaz, S.; Maraj, E.N. Numerical approach for stagnation point flow of Sutterby fluid impinging to Cattaneo-Christov heat flux model. *Pramana J. Phys.* **2018**, *91*, 61. [CrossRef]
36. Naganthran, K.; Basir, M.F.M.; Kasihmuddin, M.S.M.; Ahmed, S.E.; Olumide, F.B.; Nazar, R. Exploration of dilatant nanofluid effects conveying microorganism utilizing scaling group analysis: FDM Blottner. *Physica A* **2020**, *549*, 124040. [CrossRef]
37. Brinkman, H. The viscosity of concentrated suspensions and solutions. *J. Chem. Phys.* **1952**, *20*, 571. [CrossRef]
38. Hewitt, G.F.; Shires, G.L.; Bott, T.R. *Process Heat Transfer*; CRC Press: Boca Raton, FL, USA, 1994; p. 987.
39. Merkin, J.H. Mixed convection boundary layer flow on a vertical surface in a saturated porous medium. *J. Eng. Math.* **1980**, *14*, 301–313. [CrossRef]
40. Merkin, J.H. On dual solutions occurring in mixed convection in a porous medium. *J. Eng. Math.* **1985**, *20*, 171–179. [CrossRef]
41. Bhattacharyya, K.; Layek, G.C. Effects of suction/blowing on steady boundary layer stagnation-point flow and heat transfer towards a shrinking sheet with thermal radiation. *Int. J. Heat Mass Tranf.* **2011**, *54*, 302–307. [CrossRef]
42. Tie-Gang, F.; Ji, Z.; Shan-Shan, Y. Viscous flow over an unsteady shrinking sheet with mass transfer. *Chin. Phys. Lett.* **2009**, *26*, 014703. [CrossRef]

© 2020 by the authors. Licensee MDPI, Basel, Switzerland. This article is an open access article distributed under the terms and conditions of the Creative Commons Attribution (CC BY) license (http://creativecommons.org/licenses/by/4.0/).

Article

A Numerical Approach for the Heat Transfer Flow of Carboxymethyl Cellulose-Water Based Casson Nanofluid from a Solid Sphere Generated by Mixed Convection under the Influence of Lorentz Force

Firas A. Alwawi [1,2,*], Hamzeh T. Alkasasbeh [3], Ahmed M. Rashad [4] and Ruwaidiah Idris [1]

1. Faculty of Ocean Engineering Technology and Informatics, University Malaysia Terengganu, Kual Nerus 21030, Terengganu, Malaysia; ruwaidiah@umt.edu.my
2. Department of Mathematics, College of Sciences and Humanities in Al-Kharj, Prince Sattam bin Abdulaziz University, Al-Kharj 11942, Saudi Arabia
3. Department of Mathematics, Faculty of Science, Ajloun National University, P.O. Box 43, Ajloun 26810, Jordan; alkasasbehh@gmail.com
4. Department of Mathematics, Aswan University, Faculty of Science, Aswan 81528, Egypt; am_rashad@yahoo.com
* Correspondence: f.alwawi@psau.edu.sa

Received: 9 June 2020; Accepted: 2 July 2020; Published: 4 July 2020

Abstract: The heat transfer of a carboxymethyl cellulose aqueous solution (CMC-water) based Casson nanofluid, flowing under the impact of a variable-strength magnetic field in mixed convection around a solid sphere, has been examined in this work. Aluminum (Al), copper (Cu), and silver (Ag) nanoparticles were employed to support the heat transfer characteristics of the host fluid. A numerical approach called the Keller-box method (KBM) was used to solve the governing system for the present problem, and also to examine and analyze the numerical and graphic results obtained by the MATLAB program, verifying their accuracy through comparing them with the prior literature. The results demonstrate that a Al–CMC-water nanoliquid is superior in terms of heat transfer rate and skin friction. The velocity of CMC-water is higher with Ag compared to Al–CMC-water, and Ag–CMC-water possesses the lowest temperature. Growing mixed parameter values result in a rising skin friction, velocity and Nusselt number or decline in temperature.

Keywords: MHD; CMC-water; Casson fluid; mixed convection; solid sphere

1. Introduction

Carboxymethyl cellulose (CMC), also known as cellulose gum [1], has many features: a high solubility, clarity of its solutions, the ability to hold water, controlled crystal growth, and it can modify viscosity, in addition to its capacity to fit the required smooth texture or body. These multifunctional aspects of a non-toxic cellulose derivative are why it is utilized in many industries and technical applications. It is employed to enhance moisturizing impact due to its polymeric structure that works as a film-forming factor [2,3]. CMC is utilized in paper industries and pharmaceuticals and is also used to stabilize clay particles [2,4] and others [5–10]. In view of the massive uses of CMC, many researchers have devoted their time to studying it. Saqib et al. [11,12] employed a Caputo–Fabrizio fractional derivative (CFFD) approach and an Atangana–Baleanu fractional derivative (ABFD) approach alongside the Laplace technique to investigate the convection flow of CMC-water nanofluid. They confirmed that multiple wall carbon nanotubes are more effective in terms of improved heat transfer, and that the velocity of CMC-water is higher with multiple wall carbon nanotubes. Rahmati et al. [13] examined the laminar flow of a CMC-aqueous solution in a horizontal 2D microtube. Their findings revealed that the slip velocity coefficient contributed notably to the growth of the heat transfer rate, and significantly reduced the friction factor of the horizontal microtube wall.

The real reason for using nanotechnology is its capacity to work at the molecular level, atom-by-atom, to make large structures via essentially novel molecular organization. The actual birth of nanotechnology was at the end of 1959 when it was introduced by physicist Richard P Feynman [14]. He concluded that the physical properties of materials change depending on the scale of its molecules, and also posed two challenges: writing "Encyclopedia Britannica" on the head of a pin and making the nanometer. Two decades later, IBM Zurich scientists were able to invent the scanning tunneling microscope, which enabled scientists for the first time to observe materials at the atomic scale, a paradigm shift that had significantly contributed to the spread of nanotechnology in all industrialized countries by the 1990s. In the heat transfer field, Choi and Eastman [15] incorporated nanotechnology unprecedentedly through immersed metallic nanoparticles in a base fluid. These ultrafine particles possessed extraordinary properties that made them notably improve the thermal conductivity of the ordinary fluid. Buongiorno [16] developed a mathematical model that shows that the heat transfer rate is affected by several factors other than the thermal conductivity impact. Tiwari and Das [17] also developed a mathematical model to consider the solid volume fraction. Recently, many researchers have used the Tiwari and Das model to examine the nanofluid flow behavior of nanoparticles. Swalmeh et al. [18] used the Tiwari and Das model to investigate the behavior of micropolar nanofluid from a sphere. Selimefendigil et al. [19] analyzed the magnetohydrodynamic (MHD) combined convection flow of a nanofluid in a lid-driven triangular cavity by the use of the Tiwari and Das model. Alwawi et al. [20] employed the Tiwari and Das model to simulate the flow behavior of a sodium alginate based Casson nanofluid from a sphere. Metal nanoparticles are distinguished by excellent electrical and thermal conductivity, chemical stability, optical and magnetic distinct properties and also, they have a high surface-to-volume ratio. However, in this study aluminum (Al), copper (Cu), and silver (Ag) metal nanoparticles were used because of their similar thermo-physical properties and their common uses and many applications in polymers and pharmaceuticals [21–23], which may be due to their presence accompanied with the presence of CMC-water in these applications.

In real life, mixed convection plays a pivotal role in many engineering and industrial applications. It appears clearly in the cooling of electronic devices and nuclear reactors, food processing, and solar collectors. In addition, Lorentz forces, generated by the passage of a magnetic field via a flowing conducting fluid, has occupied a prominent place in several modern processes of metallurgy and metalworking. Makinde and Aziz [24] analyzed mixed convection on a vertical plate in a porous medium considering the MHD impact and convective boundary condition. Tham et al. [25] studied the boundary layer flow of nanofluid with the MHD effect. Chamkha et al. [26] investigated the magneto-mixed convection flow of ferrofluids in the presence of a partial slip. Here are some of the most important recently conducted studies related to MHD mixed convection [27–32].

Casson's model [33] was developed in 1959 to be able to predict the behavior of non-Newtonian fluids efficiently, and since then it has demonstrated its competence by foretelling the behavior of shear-thinning fluids, such as human blood, honey, concentrated fruit juice, ketchup, and others. Later a considerable number of articles employed this model. Malik et al. [34] employed the Runge–Kutta–Fehlberg technique to examine the flow of a Casson nanoliquid about a vertical cylinder. Mukhopadhyay et al. [35] emphasized that the flow separation could be curbed by raising the Casson parameter. Mustafa et al. [36] investigated the convection of Casson fluid from a stretching sheet taking into account viscous dissipation. See also these recent and efficient studies [37–41].

To the best of our knowledge, and judging by the prior literature, no study has been conducted on the heat transfer of a CMC-based Casson nanoliquid induced by combined convection past a solid sphere with a MHD influence via the KBM that has been investigated in this work. It is also an extension and development of these studies [20,25,42–44] which may be useful in academic studies, polymer processes, pharmaceutical and food industries, and others.

2. Basic Governing Equations

A MHD mixed convection flow of three types of metals (Al, Ag, Cu) in a host Casson fluid over an isothermal sphere of radius a with a prescribed wall l temperature T_w and ambient l temperature T_∞ were taken into account. Additionally, a heated and cooled sphere ($T_w > T_\infty$ & $T_w < T_\infty$, respectively) was considered.

Figure 1 depicts the schematic configuration and geometrical coordinates, where U_∞, and g are the free stream velocity, and the gravity vector, respectively. The ($\widetilde{\xi}, \widetilde{\eta}$) coordinates were measured along the circumference of the sphere at the stagnation point ($\widetilde{\xi} \approx 0$), and the distance normal to the surface of the sphere, respectively.

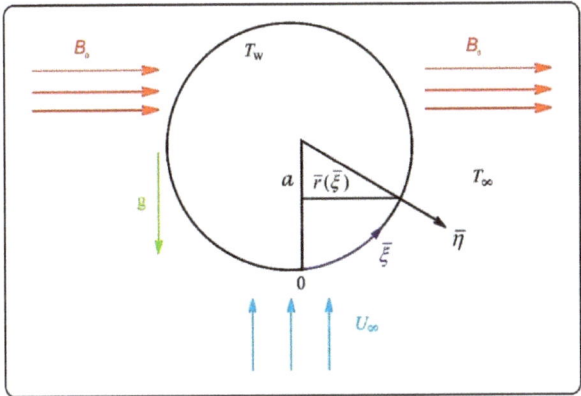

Figure 1. Schematic configuration of the problem.

Based on the previous assumption, the governing PDEs. for the Casson nanofluid are:

$$\frac{\partial}{\partial \widetilde{\xi}}(\widetilde{ru}) + \frac{\partial}{\partial \widetilde{\eta}}(\widetilde{rv}) = 0, \tag{1}$$

$$\widetilde{u}\frac{\partial \widetilde{u}}{\partial \widetilde{\xi}} + \widetilde{v}\frac{\partial \widetilde{u}}{\partial \widetilde{\eta}} = \widetilde{u}_e\frac{d\widetilde{u}_e}{d\widetilde{\xi}} + v_{nf}\left(1 + \frac{1}{\beta}\right)\frac{\partial^2 \widetilde{u}}{\partial \widetilde{\eta}^2} + \left(\frac{\chi \rho_s \beta_s + (1-\chi)\rho_f \beta_f}{\rho_{nf}}\right)g(T - T_\infty)\sin\left(\frac{\widetilde{\xi}}{a}\right) - \frac{\sigma_{nf} B_0^2}{\rho_{nf}}\widetilde{u}, \tag{2}$$

$$\widetilde{u}\frac{\partial T}{\partial \widetilde{\xi}} + \widetilde{v}\frac{\partial T}{\partial \widetilde{\eta}} = \alpha_{nf}\frac{\partial^2 T}{\partial \widetilde{\eta}^2}, \tag{3}$$

When they are associated with the boundary conditions:

$$\widetilde{u} = \widetilde{v} = 0, \ T = T_w, \ \text{as } \widetilde{\eta} = 0, \widetilde{u} \to \widetilde{u}_e(\widetilde{\xi}), \ T \to T_\infty, \ \text{as } \widetilde{\eta} \to \infty. \tag{4}$$

where $\widetilde{r}(\widetilde{\xi})$ and $\widetilde{u}_e(\widetilde{\xi})$ are given by:

$$\widetilde{r}(\widetilde{\xi}) = a\sin(\widetilde{\xi}/a), \text{ and } \widetilde{u}_e(\widetilde{\xi}) = \frac{3}{2}U_\infty \sin(\widetilde{\xi}/a), \tag{5}$$

The properties of the nanofluid (defined by [45]) are:

$$\frac{\sigma_{nf}}{\sigma_f} = 1 + \frac{3(\sigma-1)\chi}{(\sigma+2)-(\sigma-1)\chi}, \ \sigma = \frac{\sigma_s}{\sigma_f}, \ \frac{k_{nf}}{k_f} = \frac{(k_s+2k_f)-2\chi(k_f-k_s)}{(k_s+2k_f)+\chi(k_f-k_s)}, \ \mu_{nf} = \frac{\mu_f}{(1-\chi)^{2.5}},$$
$$(\rho c_p)_{nf} = (1-\chi)(\rho c_p)_f + \chi(\rho c_p)_s, \ \rho_{nf} = (1-\chi)\rho_f + \chi \rho_s, \ \alpha_{nf} = \frac{k_{nf}}{(\rho c_p)_{nf}}, \tag{6}$$

The following non-dimensional variables that are expressed by Rashad et al. [46] were used:

$$x = \frac{\tilde{\xi}}{a},\ y = \text{Re}^{1/2}\left(\frac{\tilde{\eta}}{a}\right),\ r(\tilde{\xi}) = \frac{\bar{r}(\tilde{\xi})}{a},\ u = \frac{\bar{u}}{U_\infty},$$

$$v = \text{Re}^{1/2}\left(\frac{\bar{v}}{U_\infty}\right),\ u_e(\xi) = \frac{\bar{u}_e(\tilde{\xi})}{U_\infty},\ \theta = \frac{T - T_\infty}{T_w - T_\infty}, \tag{7}$$

where $\text{Re} = U_\infty \frac{a}{v_f}$ is the Reynolds number.

By substituting Equation (7) into Equations (1)–(4) we get the following non-dimensional equations:

$$\frac{\partial}{\partial \xi}(ru) + \frac{\partial}{\partial \eta}(rv) = 0, \tag{8}$$

$$u\frac{\partial u}{\partial \xi} + v\frac{\partial u}{\partial \eta} = u_e(\xi)\frac{du_e}{d\xi} + \frac{\rho_f}{\rho_{nf}}\frac{1}{(1-\chi)^{2.5}}\left(1 + \frac{1}{\beta}\right)\frac{\partial^2 u}{\partial \eta^2}$$
$$+ \left(\frac{\chi \rho_s \beta_s + (1-\chi)\rho_f \beta_f}{\rho_{nf}}\right)\lambda\theta\sin\xi - \frac{\rho_f \sigma_{nf}}{\rho_{nf}\sigma_f}Mu, \tag{9}$$

$$u\frac{\partial \theta}{\partial \xi} + v\frac{\partial \theta}{\partial \eta} = \frac{1}{\text{Pr}}\left(\frac{k_{nf}/k_f}{(1-\chi) + \chi(\rho c_p)_s/(\rho c_p)_f}\right)\frac{\partial^2 \theta}{\partial \eta^2}, \tag{10}$$

here $M = \left(\frac{\sigma_f \beta_0^2 a}{\rho_f v_f}\right)$, $\text{Pr} = \frac{v_f}{\alpha_f}$, $\lambda = Gr/\text{Re}^2$, and $Gr = g\beta_f(T_w - T_\infty)\frac{a^3}{v_f^2}$ and the dimensionless boundary conditions are:

$$u = v = 0,\ \theta = 1,\ \text{at } \eta = 0,$$

$$u \to \frac{3}{2}\sin\tilde{\xi},\ \theta \to 0,\ \text{as } \eta \to \infty. \tag{11}$$

To solve the non-dimensional Equations (8)–(10), associated with the boundary conditions in Equation (11), defined the non-dimensional stream function ψ is defined as the following (defined by Nazar et al. [43]):

$$\psi = \tilde{\xi}r(\tilde{\xi})F(\tilde{\xi},\tilde{\eta}),\ \theta = \theta(\tilde{\xi},\tilde{\eta}),$$

$$u = \frac{1}{r}\frac{\partial \psi}{\partial \tilde{\eta}} \text{ and } v = -\frac{1}{r}\frac{\partial \psi}{\partial \tilde{\xi}} \tag{12}$$

By using Equation (12), the non-dimensional Equations (8)–(10) are reduced to:

$$\frac{\rho_f}{\rho_{nf}}\frac{1}{(1-\chi)^{2.5}}\left(1+\frac{1}{\beta}\right)\frac{\partial^3 F}{\partial \eta^3} + (1 + \xi\cot\xi)F\frac{\partial^2 F}{\partial \eta^2} - \left(\frac{\partial F}{\partial \eta}\right)^2 - \frac{\rho_f \sigma_{nf}}{\rho_{nf}\sigma_f}M\frac{\partial F}{\partial \eta}$$
$$+ \left(\frac{\chi \rho_s \beta_s/\beta_f + (1-\chi)\rho_f}{\rho_{nf}}\right)\lambda\theta\frac{\sin\xi}{\xi} + \frac{9}{4}\frac{\sin\xi\cos\xi}{\xi} = \xi\left(\frac{\partial F}{\partial \eta}\frac{\partial^2 F}{\partial \xi \partial \eta} - \frac{\partial F}{\partial \xi}\frac{\partial^2 F}{\partial \eta^2}\right), \tag{13}$$

$$\frac{1}{\text{Pr}}\left(\frac{k_{nf}/k_f}{(1-\chi) + \chi(\rho c_p)_s/(\rho c_p)_f}\right)\frac{\partial^2 \theta}{\partial \eta^2} + (1 + \xi\cot\xi)F\frac{\partial \theta}{\partial \eta} = \xi\left(\frac{\partial F}{\partial \eta}\frac{\partial \theta}{\partial \xi} - \frac{\partial F}{\partial \xi}\frac{\partial \theta}{\partial \eta}\right), \tag{14}$$

and the boundary conditions become:

$$\frac{\partial F}{\partial \eta} = F = 0,\ \theta = 1 \text{ at } \eta = 0,$$

$$\frac{\partial F}{\partial \eta} \to \frac{3}{2}\frac{\sin\xi}{\xi},\ \theta \to 0,\ \text{as } \eta \to \infty. \tag{15}$$

At the stagnation point of the sphere when ($\bar{\xi} \approx 0$), Equations (13)–(15) reduce to:

$$\frac{\rho_f}{\rho_{nf}}\frac{1}{(1-\chi)^{2.5}}\left(1+\frac{1}{\beta}\right)F''' + 2FF'' - (F')^2 - \frac{\rho_f \sigma_{nf}}{\rho_{nf}\sigma_f}MF' \\ + \left(\frac{\chi\rho_s\beta_s/\beta_f + (1-\chi)\rho_f}{\rho_{nf}}\right)\lambda\theta + \frac{9}{4} = 0, \quad (16)$$

$$\frac{1}{\Pr}\left(\frac{k_{nf}/k_f}{(1-\chi)+\chi(\rho c_p)_s/(\rho c_p)_f}\right)\theta'' + 2F\theta' = 0, \quad (17)$$

The subject to

$$F' = F = 0, \; \theta = 1 \text{ at } \eta = 0,$$

$$F' \to \frac{3}{2}, \; \theta \to 0, \text{ as } \eta \to \infty. \quad (18)$$

In this work two physical quantities were taken into consideration, specifically the local skin friction coefficient C_f and the local Nusselt number Nu, which are given by Molla et al. [47]:

$$C_f = \left(\frac{\tau_w}{\rho U_\infty^2}\right), \; Nu = \left(\frac{aq_w}{k_f(T_w - T_\infty)}\right), \quad (19)$$

where

$$\tau_w = \mu_{nf}\left(\frac{\partial \tilde{u}}{\partial \tilde{\eta}}\right)_{\tilde{\eta}=0}, \; q_w = -k_{nf}\left(\frac{\partial T}{\partial \tilde{\eta}}\right)_{\tilde{\eta}=0}. \quad (20)$$

Using Equations (7) and (11), C_f and Nu are turned into:

$$\text{Re}^{1/2}C_f = \frac{1}{(1-\chi)^{2.5}}\left(1+\frac{1}{\beta}\right)\xi\frac{\partial^2 F}{\partial \eta^2}(\xi,0), \; \text{Re}^{-1/2}Nu = \frac{-k_{nf}}{k_f}\left(\frac{\partial \theta}{\partial \eta}\right)_{\eta=0}. \quad (21)$$

3. Numerical Approach

In 1970 Keller [48] was first proposed the Keller-box method. About a decade later, this method became more popular when Jones [49] found a solution for boundary layer problems. Cebeci and Bradshaw [50] provided a detailed explanation of the Keller-box procedure, which we employed it in the current paper to construct the solution for the problem.

3.1. The Finite-Difference Method

In order to transform Equations (13) and (14) to first order equations, new independent unknowns will be defined as follows:

$w(\xi, \eta)$, $z(\xi, \eta)$, $p(\xi, \eta)$, and $g(\xi, \eta)$, where the temperature variable $\theta(\xi, \eta)$ is replaced by $g(\xi, \eta)$, and

$$F = w, \\ w' = z, \\ g' = p, \quad (22)$$

Thus, the Equations (13)–(15) are converted to:

$$\frac{\rho_f}{\rho_{nf}}\frac{1}{(1-\chi)^{2.5}}\left(1+\frac{1}{\beta}\right)z' + (1+\xi\cot\xi)Fz - w^2 - \frac{\rho_f \sigma_{nf}}{\rho_{nf}\sigma_f}Mw \\ + \left(\frac{\chi\rho_s\beta_s/\beta_f + (1-\chi)\rho_f}{\rho_{nf}}\right)\lambda g\frac{\sin\xi}{\xi} + \frac{9}{4}\frac{\sin\xi\cos\xi}{\xi} = \xi\left(w\frac{\partial w}{\partial \xi} - z\frac{\partial F}{\partial \xi}\right), \quad (23)$$

$$\frac{1}{\Pr}\left(\frac{k_{nf}/k_f}{(1-\chi)+\chi(\rho c_p)_s/(\rho c_p)_f}\right)p' + (1+\xi\cot\xi)Fp = \xi\left(w\frac{\partial g}{\partial \xi} - p\frac{\partial F}{\partial \xi}\right), \quad (24)$$

Subject to:
$$w(\xi,0) = F(\xi,0) = 0, \; g(\xi,0) = 1,$$
$$w(\xi,\infty) = \frac{3}{2}\frac{\sin\xi}{\xi}, \; g(\xi,\infty) = 0, \tag{25}$$

where the prime notation denotes the 1st derivative with respect to η,

Next the finite-difference form of Equation (22) for the midpoint $(\xi^n, \eta_{j-1/2})$ of the segment, and find the finite difference form of Equations (23) and (24) about the midpoint $(\xi^{n-1/2}, \eta_{j-1/2})$ of the rectangle have been obtained as:

$$F_j^n - F_{j-1}^n - \frac{h_j}{2}(w_j^n + w_{j-1}^n) = 0. \tag{26}$$

$$w_j^n - w_{j-1}^n - \frac{h_j}{2}(z_j^n + z_{j-1}^n) = 0. \tag{27}$$

$$g_j^n - g_{j-1}^n - \frac{h_j}{2}(p_j^n + p_{j-1}^n) = 0. \tag{28}$$

$$\frac{\rho_f}{\rho_{nf}}\frac{1}{(1-\chi)^{2.5}}\left(1+\frac{1}{\beta}\right)(z_j^n - z_{j-1}^n) + \left(\frac{A+\alpha}{4}\right)h_j(F_j^n + F_{j-1}^n)(z_j^n + z_{j-1}^n) - \left(\frac{1+\alpha}{4}\right)h_j(w_j^n + w_{j-1}^n)^2$$
$$+\left(\frac{\alpha}{2}\right)h_j z_{j-1/2}^{n-1}(F_j^n + F_{j-1}^n) + \frac{1}{2}\left(\frac{\chi\rho_s(\beta_s/\beta_f)+(1-\chi)\rho_f}{\rho_{nf}}\right)\frac{\sin x^{n-1/2}}{x^{n-1/2}}\lambda h_j(g_j^n + g_{j-1}^n) \tag{29}$$
$$-\frac{1}{2}\frac{\rho_f \sigma_{nf}}{\rho_{nf}\sigma_f}Mh_j(w_j^n + w_{j-1}^n) - \left(\frac{\alpha}{2}\right)h_j F_{j-1/2}^{n-1}(z_j^n + z_{j-1}^n) + \frac{9}{4}\frac{\sin x^{n-1/2}\cos x^{n-1/2}}{x^{n-1/2}}h_j = (R_1)_{j-1/2}^{n-1}$$

$$\frac{1}{\Pr}\frac{k_{nf}/k_f}{\left((1-\chi)(\rho C_p)_f + \chi(\rho c_p)_s/(\rho c_p)_f\right)}(p_j^n - p_{j-1}^n) - \frac{\alpha}{4}h_j(w_j^n + w_{j-1}^n)(g_j^n + g_{j-1}^n)$$
$$+\frac{A+\alpha}{4}h_j(F_j^n + F_{j-1}^n)(p_j^n + p_{j-1}^n) + \frac{\alpha}{2}h_j(w_j^n + w_{j-1}^n)g_{j-1/2}^{n-1} - \frac{\alpha}{2}h_j w_{j-1/2}^{n-1}(g_j^n + g_{j-1}^n) \tag{30}$$
$$-\frac{\alpha}{2}h_j(p_j^n - p_{j-1}^n)F_{j-1/2}^{n-1} + \frac{\alpha}{2}h_j p_{j-1/2}^{n-1}(F_j^n + F_{j-1}^n) = (R_2)_{j-1/2}^{n-1}$$

where

$$\alpha = \frac{x^{n-1/2}}{k_n}, \; A = \left(1 + x^{n-1/2}\cot x^{n-1/2}\right), \; k_n \text{ is } \Delta\xi, \text{ and } h_j \text{ is } \Delta\eta$$

$$(R_1)_{j-1/2}^{n-1} = -h_j \left\{ \begin{array}{l} \frac{\rho_f}{\rho_{nf}}\frac{1}{(1-\chi)^{2.5}}\left(1+\frac{1}{\beta}\right)\frac{(z_j^n - z_{j-1}^n)}{h_j} + (A-\alpha)F_{j-1/2}^n z_{j-1/2}^n \\ +(\alpha-1)\left(w_{j-1/2}^n\right)^2 - \frac{\rho_f \sigma_{nf}}{\rho_{nf}\sigma_f}Mw_{j-1/2}^n + \frac{9}{4}\frac{\sin x^{n-1/2}\cos x^{n-1/2}}{x^{n-1/2}} \\ +\left(\frac{\chi\rho_s(\beta_s/\beta_f)+(1-\chi)\rho_f}{\rho_{nf}}\right)\frac{\sin x^{n-1/2}}{x^{n-1/2}}\lambda g_{j-1/2}^n \end{array} \right\}^{n-1}$$

$$(R_2)_{j-1/2}^{n-1} = -h_j \left\{ \begin{array}{l} \frac{1}{\Pr}\frac{k_{nf}/k_f}{\left((1-\chi)(\rho C_p)_f + \chi(\rho c_p)_s/(\rho c_p)_f\right)}\frac{(p_j^n - p_{j-1}^n)}{h_j} \\ +(A-\alpha)F_{j-1/2}^n p_{j-1/2}^n + \alpha w_{j-1/2}^n g_{j-1/2}^n \end{array} \right\}^{n-1} \tag{31}$$

when $\xi = \xi^n$ the boundary conditions become:

$$F_0^n = w_0^n = 0, \; g_0^n = 1,$$
$$w_J^n = \frac{3}{2}\frac{\sin\xi}{\xi}, \; g_J^n = 0, \tag{32}$$

3.2. Newton's Method

Applying Newton's method on the system shown in Equations (26)–(30) to obtains:

$$\delta F_j - \delta F_{j-1} - \frac{1}{2}h_j(\delta w_j + \delta w_{j-1}) = (r_1)_{j-1/2} \tag{33}$$

$$\delta w_j - \delta w_{j-1} - \frac{1}{2}h_j(\delta z_j + \delta z_{j-1}) = (r_2)_{j-1/2} \tag{34}$$

$$\delta g_j - \delta g_{j-1} - \frac{1}{2}h_j(\delta p_j + \delta p_{j-1}) = (r_3)_{j-1/2} \tag{35}$$

$$(a_1)_j \delta z_j + (a_2)_j \delta z_{j-1} + (a_3)_j \delta F_j + (a_4)_j \delta F_{j-1} + (a_5)_j \delta w_j \\ +(a_6)_j \delta w_{j-1} + (a_7)_j \delta g_j + (a_8)_j \delta g_{j-1} = (r_4)_{j-1/2} \tag{36}$$

$$(b_1)_j \delta p_j + (b_2)_j \delta p_{j-1} + (b_3)_j \delta F_j + (b_4)_j \delta F_{j-1} + (b_5)_j \delta w_j \\ +(b_6)_j \delta w_{j-1} + (b_7)_j \delta g_j + (b_8)_j \delta g_{j-1} = (r_5)_{j-1/2} \tag{37}$$

where

$$(a_1)_j = \left[\frac{\rho_f}{\rho_{nf}}\frac{1}{(1-\chi)^{2.5}}\left(1+\frac{1}{\beta}\right) + h_j\left(\frac{(A+\alpha)}{2}F_{j-1/2} - \frac{\alpha}{2}F_{j-1/2}^{n-1}\right)\right]$$

$$(a_2)_j = \left[(a_1)_j - 2\frac{\rho_f}{\rho_{nf}}\frac{1}{(1-\chi)^{2.5}}\left(1+\frac{1}{\beta}\right)\right]$$

$$(a_3)_j = h_j\left[\frac{(A+\alpha)}{2}z_{j-1/2} + \frac{\alpha}{2}z_{j-1/2}^{n-1}\right]$$

$$(a_4)_j = (a_3)_j$$

$$(a_5)_j = h_j\left[-(1+\alpha)w_{j-1/2} - \frac{1}{2}\frac{\rho_f \sigma_{nf}}{\rho_{nf} \sigma_f}M\right]$$

$$(a_6)_j = (a_5)_j$$

$$(a_7)_j = h_j\left[\frac{\lambda}{2}\left(\frac{\chi\rho_s(\beta_s/\beta_f) + (1-\chi)\rho_f}{(1-\chi)\rho_f + \chi\rho_s}\right)\frac{\sin x^{n-1/2}}{x^{n-1/2}}\right]$$

$$(a_8)_j = (a_7)_j \tag{38}$$

$$(b_1)_j = \left[\frac{1}{\Pr}\frac{k_{nf}/k_f}{(1-\chi)(\rho C_p)_f + \chi(\rho c_p)_s/(\rho c_p)_f} + h_j\left(\frac{(A+\alpha)}{2}F_{j-1/2} - \frac{\alpha}{2}F_{j-1/2}^{n-1}\right)\right]$$

$$(b_2)_j = \left[\frac{2}{\Pr} - (b_1)_j\right]$$

$$(b_3)_j = h_j\left[\frac{(A+\alpha)}{2}p_{j-1/2} + \frac{\alpha}{2}p_{j-1/2}^{n-1}\right]$$

$$(b_4)_j = (b_3)_j$$

$$(b_5)_j = h_j\left[-\frac{\alpha}{2}g_{j-1/2} + \frac{\alpha}{2}g_{j-1/2}^{n-1}\right]h_j$$

$$(b_6)_j = (b_5)_j$$

$$(b_7)_j = h_j\left[-\frac{\alpha}{2}w_{j-1/2} - \frac{\alpha}{2}h_j w_{j-1/2}^{n-1}\right]$$

$$(b_8)_j = (b_7)_j \tag{39}$$

$$(r_1)_{j-1/2} = F_{j-1} - F_j + h_j w_{j-1/2}$$

$$(r_2)_{j-1/2} = w_{j-1} - w_j + h_j z_{j-1/2}$$

$$(r_3)_{j-1/2} = g_{j-1} - g_j + h_j p_{j-1/2}$$

$$(r_4)_{j-1/2} = \frac{\rho_f}{\rho_{nf}} \frac{1}{(1-\chi)^{2.5}} \left(1 + \frac{1}{\beta}\right)(z_{j-1} - z_j) - (A+\alpha)h_j F_{j-1/2} z_{j-1/2}$$
$$+ h_j \left(\alpha z_{j-1/2} F^{n-1}_{j-1/2} - \alpha z^{n-1}_{j-1/2} F_{j-1/2} - \frac{9}{4} \frac{\sin x^{n-1/2} \cos x^{n-1/2}}{x^{n-1/2}} \right)$$
$$- h_j \left(\frac{\chi \rho_s (\beta_s/\beta_f) + (1-\chi)\rho_f}{(1-\chi)\rho_f + \chi \rho_s} \right) \frac{A}{2} \frac{\sin x^{n-1/2}}{x^{n-1/2}} g_{j-1/2}$$
$$+ h_j \left((1+\alpha) w^2_{j-1/2} + \frac{\rho_f \sigma_{nf}}{\rho_{nf} \sigma_f} M w_{j-1/2} \right) + (R_1)^{n-1}_{j-1/2}$$

$$(r_5)_{j-1/2} = \frac{1}{\Pr} \frac{k_{nf}/k_f}{\left((1-\chi)(\rho C_p)_f + \chi(\rho c_p)_s/(\rho c_p)_f\right)} (p_{j-1} - p_j)$$
$$- (A+\alpha) h_j F_{j-1/2} p_{j-1/2} - \alpha h_j p^{n-1}_{j-1/2} F_{j-1/2} + \alpha h_j p_{j-1/2} F^{n-1}_{j-1/2} \tag{40}$$
$$+ \alpha h_j w_{j-1/2} g_{j-1/2} - \alpha h_j w_{j-1/2} g^{n-1}_{j-1/2} + \alpha h_j w^{n-1}_{j-1/2} g_{j-1/2} + (R_2)^{n-1}_{j-1/2}$$

3.3. The Block Tridiagonal Matrix

The matrix form of a linearized tridiagonal system is:

$$\Lambda \delta = r, \tag{41}$$

where

$$S = \begin{bmatrix} [A_1] & [C_1] & & & \\ [B_2] & [A_2] & [C_2] & & \\ & \ddots & \ddots & \ddots & \\ & & & & \\ & & [B_{J-1}] & [A_{J-1}] & [C_{J-1}] \\ & & & [B_J] & [A_J] \end{bmatrix}, \delta = \begin{bmatrix} [\delta_1] \\ [\delta_2] \\ \vdots \\ [\delta_{J-1}] \\ [\delta_J] \end{bmatrix}, r = \begin{bmatrix} [r_1] \\ [r_2] \\ \vdots \\ [r_{J-1}] \\ [r_J] \end{bmatrix}.$$

The boundary conditions in Equation (32) are satisfied precisely with no iteration. Due to these suitable values being maintained in every iterate, we assume $\delta F_0 = 0$, $\delta w_0 = 0$, $\delta p_0 = 0$, $\delta w_J = 0$, $\delta g_J = 0$, and let $d_J = -\frac{1}{2}h_J$.

The entries of the matrices are

$$[A_1] = \begin{bmatrix} 0 & 0 & 1 & 0 & 0 \\ d_1 & 0 & 0 & d_1 & 0 \\ 0 & -1 & 0 & 0 & d_1 \\ (a_2)_1 & (a_8)_1 & (a_3)_1 & (a_1)_1 & 0 \\ 0 & (b_8)_1 & (b_3)_1 & 0 & (b_1)_1 \end{bmatrix} \tag{42}$$

$$[A_j] = \begin{bmatrix} d_j & 0 & 0 & 0 & 0 \\ -1 & 0 & 0 & 0 & 0 \\ 0 & -1 & 0 & 0 & 0 \\ (a_6)_j & (a_8)_j & (a_3)_j & (a_1)_j & 0 \\ (b_6)_j & (b_8)_j & (b_3)_j & 0 & (b_1)_j \end{bmatrix}, 2 \leq j \leq J, \tag{43}$$

$$[B_j] = \begin{bmatrix} 0 & 0 & -1 & 0 & 0 \\ 0 & 0 & 0 & d_j & 0 \\ 0 & 0 & 0 & 0 & d_j \\ 0 & 0 & (a_4)_j & (a_2)_j & 0 \\ 0 & 0 & (b_4)_j & 0 & (b_2)_j \end{bmatrix}, 2 \leq j \leq J, \tag{44}$$

$$[C_j] = \begin{bmatrix} d_j & 0 & 0 & 0 & 0 \\ 1 & 0 & 0 & 0 & 0 \\ 0 & 1 & 0 & 0 & 0 \\ (a_5)_j & (a_7)_j & 0 & 0 & 0 \\ (b_5)_j & (b_7)_j & 0 & 0 & 0 \end{bmatrix}, 1 \leq j \leq J-1, \tag{45}$$

$$[\delta_1] = \begin{bmatrix} \delta z_0 \\ \delta g_0 \\ \delta F_1 \\ \delta z_1 \\ \delta p_1 \end{bmatrix}, [\delta_j] = \begin{bmatrix} \delta w_{j-1} \\ \delta g_{j-1} \\ \delta F_{j-1} \\ \delta z_{j-1} \\ \delta p_{j-1} \end{bmatrix}, 2 \leq j \leq J, [r_j] = \begin{bmatrix} (r_1)_{j-(1/2)} \\ (r_2)_{j-(1/2)} \\ (r_3)_{j-(1/2)} \\ (r_4)_{j-(1/2)} \\ (r_5)_{j-(1/2)} \end{bmatrix}, 1 \leq j \leq J \tag{46}$$

The final step is to solve the system in Equation (41) by the LU (lower–upper) factorization method, then implement numerical operations using MATLAB software (version 7, MathWorks, Natick, MA, USA). In this work the wall shear stress parameter $z(x,0)$ is considered as the convergence criterion (as it is usually considered, see Cebeci and Bradshaw [50]), so the calculations were repeated until the convergence criterion was satisfied, and stopped when $\left|\delta z_0^{(i)}\right| < \varepsilon_1$, where ε_1 is chosen to be 10^{-5} which give precise values up to four decimal places.

4. Results and Discussion

This section aims to predict and analyze graphically the behavior of a CMC-based Casson nanofluid under the impact of meaningfully related parameters with regard to the velocity, temperature, skin friction coefficient, and local Nusselt number. The ranges of parameters that are taken into consideration are the mixed parameter ($\lambda > 0$ & $\lambda < 0$), Casson parameter ($\beta > 0$), magnetic parameter ($M > 0$) and nanoparticles volume fraction ($0.1 \leq \chi \leq 0.2$).

Table 1 shows the thermo-physical properties of CMC-water and the nanoparticles. The numerical results obtained were in a close agreement with the literature and can be seen in comparative Tables 2 and 3.

Table 1. Thermo-physical properties of CMC-water (0.0–0.4%) and metals nanoparticles [51].

Thermo-Physical Property	CMC-Water	Al	Ag	Cu
ρ (kg/m^3)	997.1	2701	10,500	8933
C_p (J/kgk)	4179	902	235	385
K (w/mK)	0.613	237	429	401
$\beta \times 10^{-5}$ (K^{-1})	21	2.31	1.89	1.67
σ (s/m)	5.5×10^{-6}	35×10^6	63×10^6	95.6×10^6
Pr	6.2	-	-	-

Table 2. Comparison of $Re^{1/2}C_f$ with published findings by Nazar et al. [43] for several values of λ ($\beta \to \infty$, $M = 0$, $\chi = 0$, $Pr = 0.7$).

λ	−4		−1		0		0.74		1	
x	[43]	Present	[43]	Present	[43]	Present	[43]	Present	[43]	Present
0°	0.0000	0.0000	0.0000	0.0000	0.0000	0.0000	0.0000	0.0000	0.0000	0.0000
10°	0.0801	0.0780	0.3438	0.3443	0.4160	0.4167	0.4669	0.4545	0.4843	0.4851
20°	0.1149	0.1153	0.6564	0.6500	0.8014	0.8035	0.9031	0.8935	0.9380	0.9279
30°			0.9098	0.9076	1.1284	1.1244	1.2813	1.2759	1.3335	1.3277
40°			1.0790	1.0824	1.3733	1.3748	1.5775	1.5778	1.6471	1.6470
50°			1.1434	1.1537	1.5172	1.5253	1.7737	1.7806	1.8607	1.8672
60°			1.0866	1.1047	1.5477	1.5630	1.8580	1.8720	1.9627	1.9762
70°			0.8929	0.9202	1.4583	1.4811	1.8260	1.8470	1.9486	1.9691
80°			0.5280	0.5680	1.2480	1.2780	1.6800	1.7079	1.8216	1.8489
90°					0.9154	0.9530	1.4289	1.4656	1.5915	1.6284
100°					0.4308	0.4812	1.0847	1.1351	1.2732	1.3160
110°							0.6543	0.7241	0.8831	0.9559
120°									0.4220	0.5094

Table 3. Heat transfer coefficient $Q_w(\xi) = -(\partial\theta/\partial\eta)_{\eta=0}$ with published findings by Nazar et al. [43] for several values of λ ($\beta \to \infty$, $M = 0$, $\chi = 0$, $Pr = 0.7$).

λ	−4		−1		0		0.74		1	
x	[43]	Present	[43]	Present	[43]	Present	[43]	Present	[43]	Present
0°	0.6534	0.6519	0.7870	0.7858	0.8162	0.8150	0.8354	0.8342	0.8463	0.8406
10°	0.6440	0.6435	0.7818	0.7812	0.8112	0.8104	0.8307	0.8301	0.8371	0.8362
20°	0.6150	0.6158	0.7669	0.7670	0.7974	0.7974	0.8173	0.8174	0.8239	0.8239
30°			0.7422	0.7433	0.7746	0.7747	0.7955	0.7963	0.8024	0.8031
40°			0.7076	0.7097	0.7429	0.7447	0.7652	0.7669	0.7725	0.7741
50°			0.6624	0.6658	0.7022	0.7039	0.7267	0.7293	0.7345	0.7371
60°			0.6055	0.6103	0.6525	0.6565	0.6800	0.6837	0.6887	0.6922
70°			0.5224	0.5403	0.5934	0.5986	0.6253	0.6300	0.6352	0.6397
80°			0.4342	0.4432	0.5236	0.5287	0.5672	0.5671	0.5742	0.5784
90°					0.4398	0.4382	0.4920	0.4887	0.5060	0.5025
100°					0.3263	0.3197	0.4120	0.3978	0.4304	0.4152
110°							0.3179	0.3004	0.3458	0.3246
120°									0.2442	0.2314

Figures 2 and 3 display the influence of the mixed parameter in opposing and assisting flow cases ($\lambda > 0$ & $\lambda < 0$) on the skin friction coefficient and Nusselt number, respectively. From these figures, we found that the Al–CMC-water has the highest skin friction coefficient values in the case of assisting flow and the lowest in the case of the opposing flow. For the Nusselt number, Al–CMC-water has the highest value in both cases ($\lambda > 0$ & $\lambda < 0$) and this is due to the thermo-physical properties that the aluminum possesses. It can also be observed that, in both the cases of opposing and assisting flow, when λ increases, $Re^{1/2}C_f$ and $Re^{-1/2}Nu$ increase due to increase in the buoyancy force.

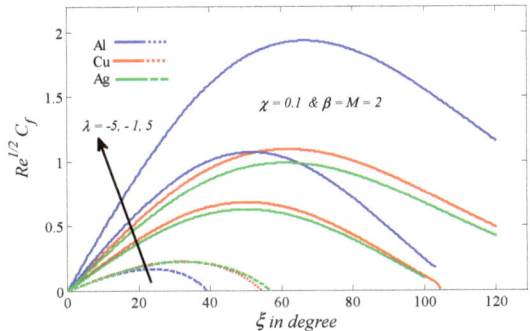

Figure 2. Mixed parameter versus the local skin friction.

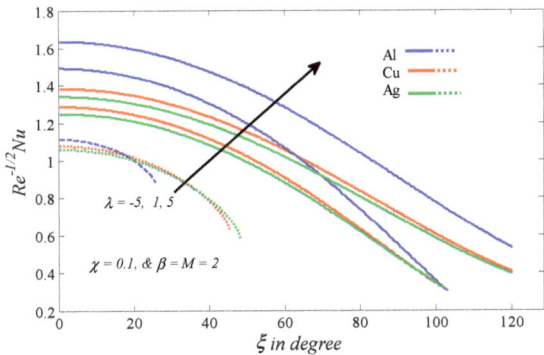

Figure 3. Mixed parameter versus the local Nusselt number.

In Figures 4 and 5 it can be seen that the increment in the value of nanoparticles volume fraction χ resulted in a noteworthy improvement in both the skin friction coefficient and Nusselt number. The improvement in the Nusselt number is caused by the enhancement of the density and thermal conductivity of CMC-water.

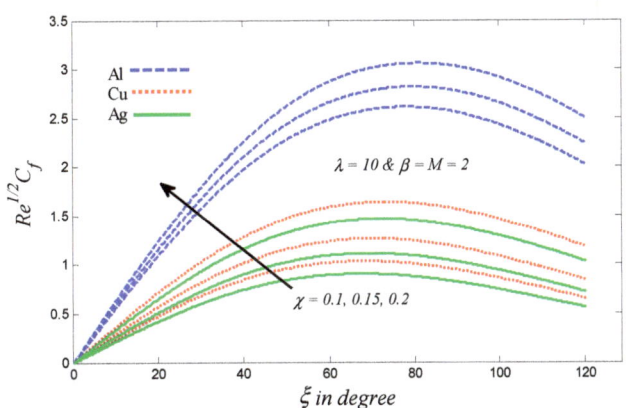

Figure 4. Nanoparticles volume fraction versus the local skin friction coefficient.

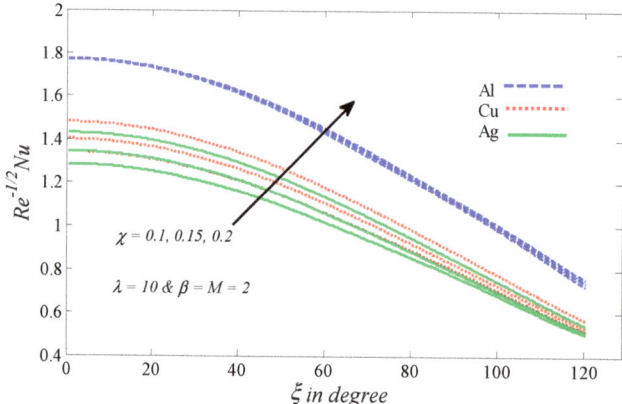

Figure 5. Nanoparticles volume fraction versus the local Nusselt number.

Figures 6 and 7 show the relationship between β and both the skin friction coefficient, and Nusselt number respectively. It's noticed that the Casson parameter β is inversely proportional to the skin friction coefficient, but it is directly proportional to the Nusselt number. Physically, when the values of β rise, the yield stress decreases and therefore the skin friction coefficient decreases.

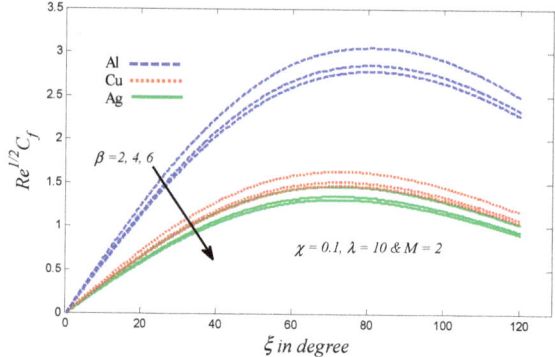

Figure 6. Casson parameter versus the local skin friction coefficient.

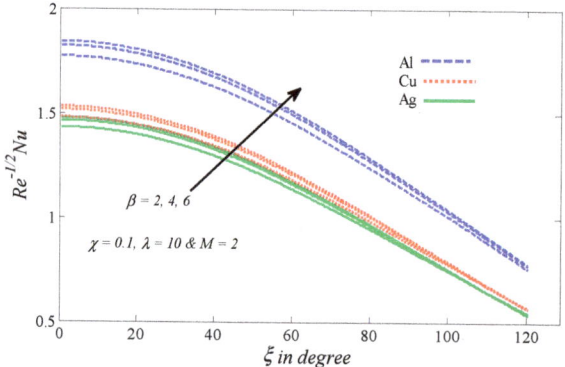

Figure 7. Casson parameter versus the local Nusselt number.

Figures 8 and 9 illustrate the graphical findings of $Re^{1/2}C_f$ and $Re^{-1/2}Nu$ respectively, with various values of the magnetic parameter (M). It is clear that as the values of M grow, both the skin friction coefficient and Nusselt number decline. In fact, this decline is a result of the restraining that occurred in the fluid flow, caused by the increase in intensity of the magnetic current which curbs convection and thereby reduces the skin fraction coefficient and Nusselt number. Furthermore, these figures demonstrate that, whatever the values of parameters λ, χ, β or M, Al–CMC-water has the highest $Re^{1/2}C_f$ and $Re^{-1/2}Nu$.

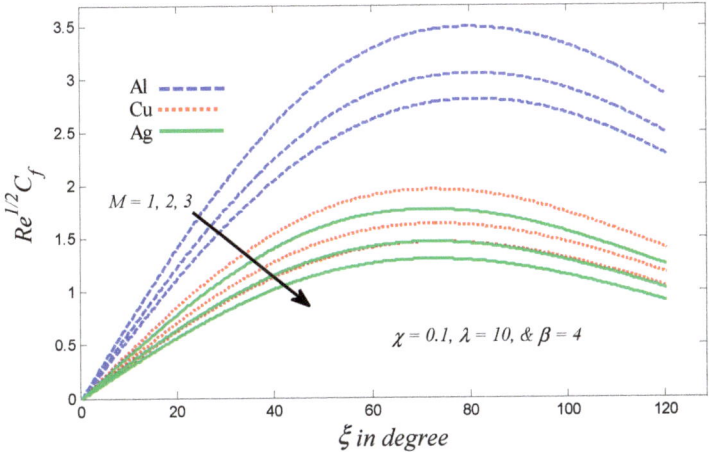

Figure 8. Magnetic parameter versus the local skin friction coefficient.

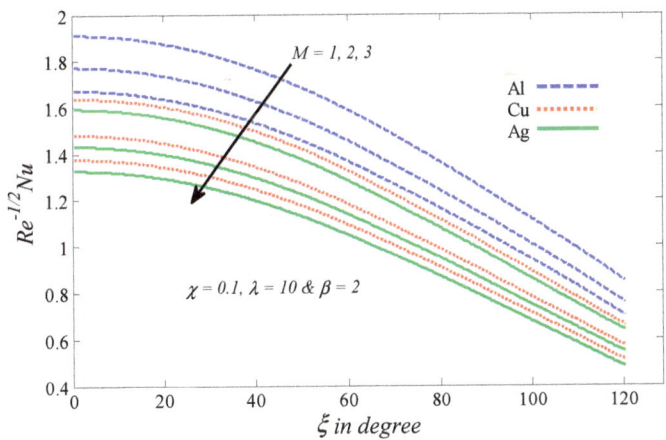

Figure 9. Magnetic parameter versus the local Nusselt number.

Figures 10 and 11 demonstrate the impact of the mixed parameter λ on the velocity and temperature in both cases opposing and assisting flow ($\lambda > 0$ & $\lambda < 0$). Both the cases of flow indicate that an increment in λ is accompanied by an improvement in the velocity or a decay in the temperature profiles. In fact, the growth in the mixed parameter enhances the thermal buoyancy force—and, hence the velocity increases.

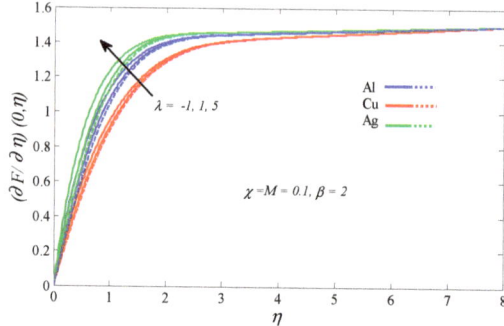

Figure 10. Mixed parameter versus velocity.

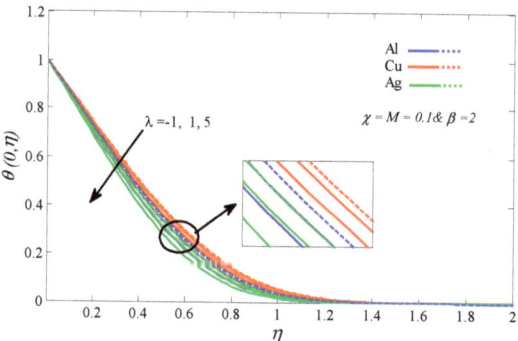

Figure 11. Mixed parameter versus temperature.

Figures 12 and 13 confirmed that the effect of the nanoparticles volume fraction (χ), on both velocity and temperature, is a positive effect. A rise in χ leads to a quicker transfer of heat from the outside of the sphere to the fluid and thus aids in the augmentation of the thickness of the thermal layer due to the increase in the temperature of the fluid. In addition, the increase in χ enhances energy transmission, which increases the fluid velocity. According to Figures 14 and 15, higher values of the Casson parameter (β) cause a curb in the velocity and temperature, which is verifiable because the augmentation in β creates a resistance force that restricts the flow of the fluid, which restrains the nanofluid velocity.

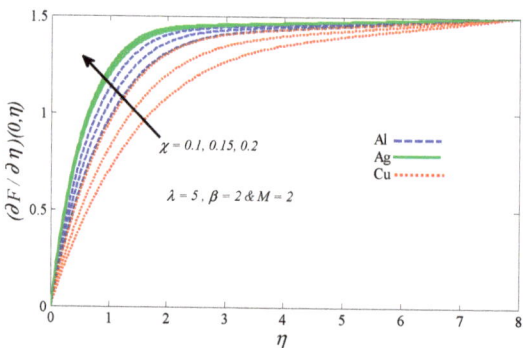

Figure 12. Nanoparticles volume fraction versus velocity.

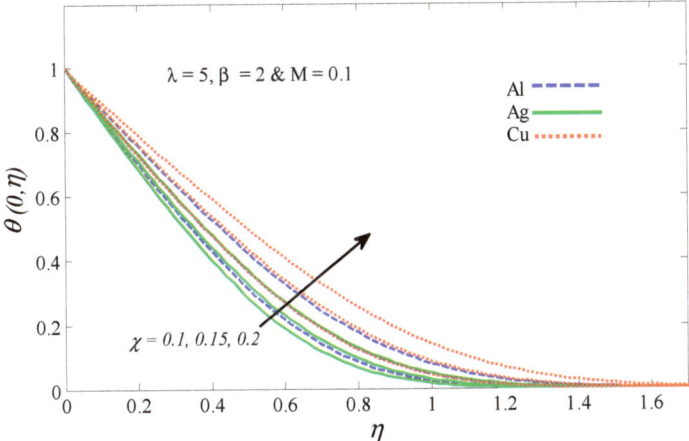

Figure 13. Nanoparticles volume fraction versus temperature.

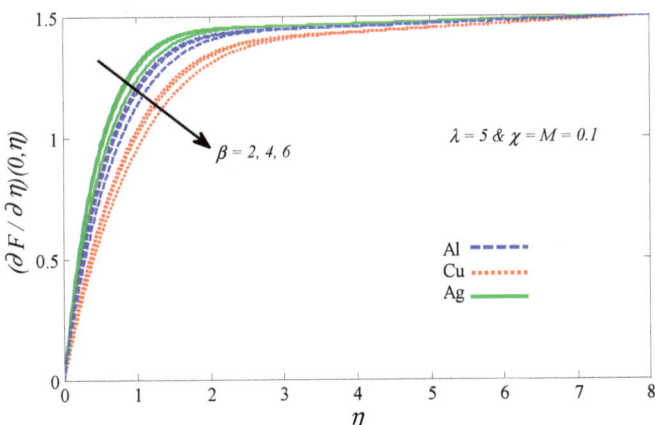

Figure 14. Casson parameter versus velocity.

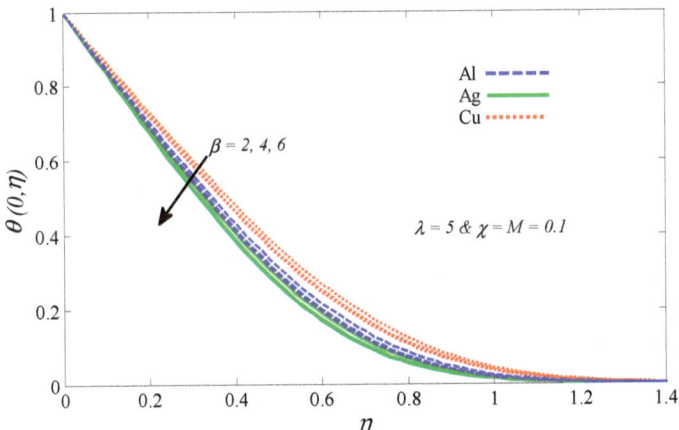

Figure 15. Casson parameter versus temperature.

Figures 16 and 17 depict the graphical findings of temperature and velocity versus the magnetic parameter (M), respectively. It is evident in these figures that as the value of M grows, the temperature increases but the velocity decreases. This phenomenon occurs when a magnetic current passes through a flowing nanofluid, which produces a kind of force known as the Lorentz force and, consequently, resists the nanofluid movement. It is worth noting that, whatever the values of parameters λ, χ, β or M, Silver–CMC-water is superior in terms of velocity, and we found that the Copper–CMC-water temperature was the highest.

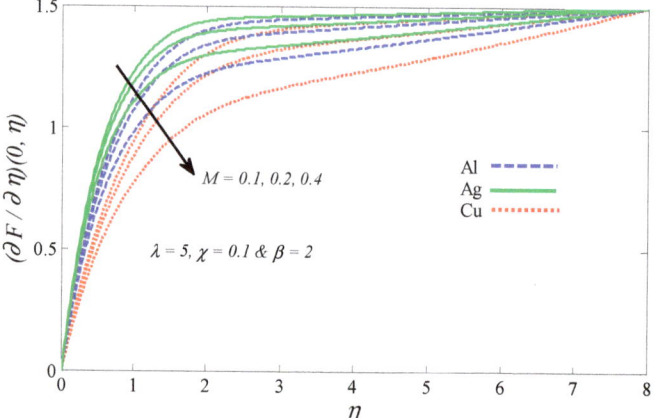

Figure 16. Magnetic parameter versus velocity.

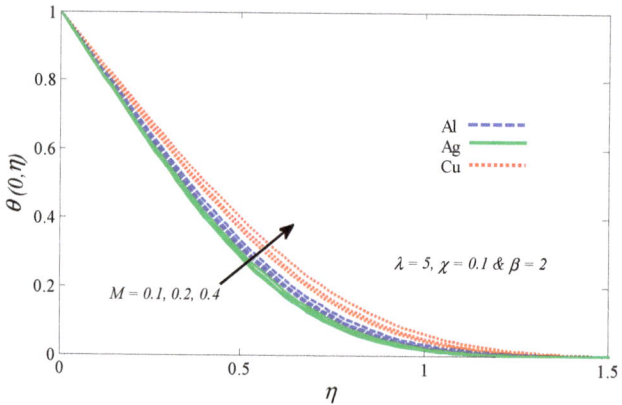

Figure 17. Magnetic parameter versus temperature.

5. Conclusions

In this research, we have explored the behavior of a CMC-water based Casson nanofluid from a solid sphere produced by mixed convection under a MHD influence. The following meaningful observations are worth mentioning:

1. The temperature profile increases when the values of each of χ or M parameters grow, and decreases as the values of β or λ increase.
2. The nanoparticles volume fraction has a positive relationship with all the physical quantities examined in this research.

3. The skin friction, velocity, and Nusselt number are decreasing functions of the magnetic field intensity, whereas temperature is an increasing function of it.

Regardless of the values of examined parameters, the values of temperature for Cu–CMC-water were the highest and had the lowest velocity.

Author Contributions: F.A.A.: Formal analysis, Investigation, and Methodology; F.A.A. and H.T.A.: Software; R.I., H.T.A. and A.M.R.: Supervision; H.T.A., A.M.R. and R.I.: Validation; H.T.A., A.M.R. and R.I.: Writing—review & editing. All authors have read and agreed to the published version of the manuscript.

Funding: This research was funded by University Malaysia Terengganu (UMT) through grant vote number 55193/4.

Conflicts of Interest: The authors declare no conflict of interest.

Nomenclature

a	Radius of Cylinder	α	Thermal diffusivity
B_0	Magnetic field strength	β	Casson parameter
C_f	Skin friction coefficient	β_f	Thermal expansion of base fluid
$r(\xi)$	Radial Distance	β_s	Thermal expansion of nanoparticles
Gr	Grashof number	θ	Temperature of nanofluid
g	Gravity vector	μ_β	Plastic Dynamic viscosity of base fluid
k	Thermal conductivity	μ_f	Dynamic viscosity of base fluid
M	Magnetic parameter	ρ	Density
Nu	Nusselt Number	(ρc_p)	Heat capacity
Pr	Prandtl number	τ_w	Wall shear stress
p_y	Yield stress	χ	Nanoparticle volume fraction
T	Temperature of the fluid	ψ	Stream function
T_w	Wall temperature	σ	Electrical conductivity
T_∞	Ambient temperature	λ	Mixed parameter
u	ξ- component of velocity	Subscript	
v	η- component of velocity	s	nanoparticles
v_f	Kinematic viscosity	nf	Nanofluid
u_e	Free stream velocity	f	Base fluid

References

1. Mondal, I.H. *Carboxymethyl Cellulose: Synthesis and Characterization*; Nova Science Publishers: Hauppauge, NY, USA, 2019.
2. Benchabane, A.; Bekkour, K. Rheological properties of carboxymethyl cellulose (CMC) solutions. *Colloid Polym. Sci.* **2008**, *286*, 1173. [CrossRef]
3. Bekkour, K.; Sun-Waterhouse, D.; Wadhwa, S.S. Rheological properties and cloud point of aqueous carboxymethyl cellulose dispersions as modified by high or low methoxyl pectin. *Food Res. Int.* **2014**, *66*, 247–256. [CrossRef]
4. Babazadeh, A.; Tabibiazar, M.; Hamishehkar, H.; Shi, B. Zein-CMC-PEG multiple nanocolloidal systems as a novel approach for nutra-pharmaceutical applications. *Adv. Pharm. Bull.* **2019**, *9*, 262. [CrossRef] [PubMed]
5. Karabinos, J.; Hindert, M. Carboxymethylcellulose. In *Advances in Carbohydrate Chemistry*; Elsevier: Amsterdam, The Netherlands, 1954; pp. 285–302.
6. Hollabaugh, C.; Burt, L.H.; Walsh, A.P. Carboxymethylcellulose. Uses and applications. *Ind. Eng. Chem.* **1945**, *37*, 943–947. [CrossRef]
7. Mondal, I.H. *Carboxymethyl Cellulose: Pharmaceutical and Industrial Applications*; Nova Science Publishers: Hauppauge, NY, USA, 2019.
8. Grządka, E.; Matusiak, J.; Bastrzyk, A.; Polowczyk, I. CMC as a stabiliser of metal oxide suspensions. *Cellulose* **2020**, *27*, 2225–2236. [CrossRef]
9. Chen, J.; Li, H.; Fang, C.; Cheng, Y.; Tan, T.; Han, H. Synthesis and structure of carboxymethylcellulose with a high degree of substitution derived from waste disposable paper cups. *Carbohydr. Polym.* **2020**, *2020*, 116040. [CrossRef]

10. Novak, U.; Bajić, M.; Kõrge, K.; Oberlintner, A.; Murn, J.; Lokar, K.; Triler, K.V.; Likozar, B. From waste/residual marine biomass to active biopolymer-based packaging film materials for food industry applications—A review. *Phys. Sci. Rev.* **2019**, *5*, e0099. [CrossRef]
11. Saqib, M.; Khan, I.; Shafie, S. Natural convection channel flow of CMC-based CNTs nanofluid. *Eur. Phys. J. Plus* **2018**, *133*, 549. [CrossRef]
12. Saqib, M.; Khan, I.; Shafie, S. Application of Atangana–Baleanu fractional derivative to MHD channel flow of CMC-based-CNT's nanofluid through a porous medium. *Chaos Solitons Fract.* **2018**, *116*, 79–85. [CrossRef]
13. Rahmati, A.R.; Akbari, O.A.; Marzban, A.; Toghraie, D.; Karimi, R.; Pourfattah, F. Simultaneous investigations the effects of non-Newtonian nanofluid flow in different volume fractions of solid nanoparticles with slip and no-slip boundary conditions. *Sci. Eng. Por.* **2018**, *5*, 263–277. [CrossRef]
14. Feynman, R.P. There's plenty of room at the bottom. California Institute of Technology. In *Engineering and Science Magazine*; California Institute of Technology: Pasadena, CA, USA, 1960.
15. Choi, S.U.; Eastman, J.A. *Enhancing Thermal Conductivity of Fluids with Nanoparticles*; Argonne National Lab.: Lemont, IL, USA, 1995.
16. Buongiorno, J. Convective transport in nanofluids. *J. Heat Transf.* **2006**, *128*, 240–250. [CrossRef]
17. Tiwari, R.K.; Das, M.K. Heat transfer augmentation in a two-sided lid-driven differentially heated square cavity utilizing nanofluids. *Int. J. Heat Mass Trans.* **2007**, *50*, 2002–2018. [CrossRef]
18. Swalmeh, M.Z.; Alkasasbeh, H.T.; Hussanan, A.; Mamat, M. Heat transfer flow of Cu-water and Al2O3-water micropolar nanofluids about a solid sphere in the presence of natural convection using Keller-box method. *Res. Phys.* **2018**, *9*, 717–724. [CrossRef]
19. Selimefendigil, F.; Chamkha, A.J. Magnetohydrodynamics mixed convection in a power law nanofluid-filled triangular cavity with an opening using Tiwari and Das' nanofluid model. *J. Anal. Calorim.* **2019**, *135*, 419–436. [CrossRef]
20. Alwawi, F.A.; Alkasasbeh, H.T.; Rashad, A.; Idris, R. MHD natural convection of Sodium Alginate Casson nanofluid over a solid sphere. *Res. Phys.* **2020**, *16*, 102818. [CrossRef]
21. Mathur, P.; Jha, S.; Ramteke, S.; Jain, N. Pharmaceutical aspects of silver nanoparticles. *Artif. Cells Nanomed. Biotechnol.* **2018**, *46*, 115–126. [CrossRef]
22. Bonde, H.C.; Fojan, P.; Popok, V.N. Controllable embedding of size-selected copper nanoparticles into polymer films. *Plasma Process Polym.* **2020**, *17*, 1900237. [CrossRef]
23. dos Santos, C.A.; Ingle, A.P.; Rai, M. The emerging role of metallic nanoparticles in food. *Appl. Microbiol. Biotechnol.* **2020**, *104*, 2373–2383. [CrossRef]
24. Makinde, O.; Aziz, A. MHD mixed convection from a vertical plate embedded in a porous medium with a convective boundary condition. *Int. J. Sci.* **2010**, *49*, 1813–1820. [CrossRef]
25. Tham, L.; Nazar, R.; Pop, I. Mixed convection boundary-layer flow about an isothermal solid sphere in a nanofluid. *Phys. Scr.* **2011**, *84*, 025403. [CrossRef]
26. Chamkha, A.J.; Rashad, A.; Alsabery, A.; Abdelrahman, Z.; Nabwey, H.A. Impact of Partial Slip on Magneto-Ferrofluids Mixed Convection Flow in Enclosure. *J. Sci. Eng. Appl.* **2020**, 1–25. [CrossRef]
27. Waqas, M.; Farooq, M.; Khan, M.I.; Alsaedi, A.; Hayat, T.; Yasmeen, T. Magnetohydrodynamic (MHD) mixed convection flow of micropolar liquid due to nonlinear stretched sheet with convective condition. *Int. J. Heat Mass Trans.* **2016**, *102*, 766–772. [CrossRef]
28. Makinde, O.; Mabood, F.; Khan, W.; Tshehla, M. MHD flow of a variable viscosity nanofluid over a radially stretching convective surface with radiative heat. *J. Mol. Liq.* **2016**, *219*, 624–630. [CrossRef]
29. Chamkha, A.; Rashad, A.; Mansour, M.; Armaghani, T.; Ghalambaz, M. Effects of heat sink and source and entropy generation on MHD mixed convection of a Cu-water nanofluid in a lid-driven square porous enclosure with partial slip. *Phys. Fluids* **2017**, *29*, 052001. [CrossRef]
30. Rashad, A.; Mansour, M.; Armaghani, T.; Chamkha, A. MHD mixed convection and entropy generation of nanofluid in a lid-driven U-shaped cavity with internal heat and partial slip. *Phys. Fluids* **2019**, *31*, 042006. [CrossRef]
31. Alkasasbeh, H.T.; Swalmeh, M.Z.; Hussanan, A.; Mamat, M. Effects of mixed convection on methanol and kerosene oil based micropolar nanofluid containing oxide nanoparticles. *CFD Lett.* **2019**, *11*, 55–68.
32. Swalmeh, M.Z.; Alkasasbeh, H.T.; Hussanan, A.; Mamat, M. Numerical investigation of heat transfer enhancement with Ag-GO water and kerosene oil based micropolar nanofluid over a solid sphere. *Adva. Res. Fluid. Mech. Sci.* **2019**, *59*, 269–282.

33. Casson, N. A flow equation for pigment-oil suspensions of the printing ink type. *Rheol. Disperse Syst.* **1959**, *2*, 84–102.
34. Malik, M.; Naseer, M.; Nadeem, S.; Rehman, A. The boundary layer flow of Casson nanofluid over a vertical exponentially stretching cylinder. *Appl. Nanosci.* **2014**, *4*, 869–873. [CrossRef]
35. Mukhopadhyay, S.; Mondal, I.C.; Chamkha, A.J. Casson fluid flow and heat transfer past a symmetric wedge. *Heat Transf. Asian Res.* **2013**, *42*, 665–675. [CrossRef]
36. Mustafa, M.; Hayat, T.; Ioan, P.; Hendi, A. Stagnation-point flow and heat transfer of a Casson fluid towards a stretching sheet. *Z. Nat. A* **2012**, *67*, 70–76. [CrossRef]
37. Ghadikolaei, S.; Hosseinzadeh, K.; Ganji, D.; Jafari, B. Nonlinear thermal radiation effect on magneto Casson nanofluid flow with Joule heating effect over an inclined porous stretching sheet. *Case Stud. Eng.* **2018**, *12*, 176–187. [CrossRef]
38. Rafique, K.; Anwar, M.I.; Misiran, M.; Khan, I.; Alharbi, S.; Thounthong, P.; Nisar, K. Numerical Solution of Casson Nanofluid Flow Over a Non-linear Inclined Surface With Soret and Dufour Effects by Keller-Box Method. *Front. Phys.* **2019**, *7*, 00139. [CrossRef]
39. Alwawi, F.A.; Alkasasbeh, H.T.; Rashad, A.M.; Idris, R. Natural convection flow of Sodium Alginate based Casson nanofluid about a solid sphere in the presence of a magnetic field with constant surface heat flux. *J. Phys. Conf. Ser.* **2019**, *1366*, 012005. [CrossRef]
40. Alwawi, F.A.; Alkasasbeh, H.T.; Rashad, A.; Idris, R. Heat transfer analysis of ethylene glycol-based Casson nanofluid around a horizontal circular cylinder with MHD effect. *Proc. Inst. Mech. Eng. Part C J. Mech. Eng. Sci.* **2020**. [CrossRef]
41. Alkasasbeh, H.; Swalmeh, M.; Bani Saeed, H.; Al Faqih, F.; Talafha, A. Investigation on CNTs-Water and Human Blood based Casson Nanofluid Flow over a Stretching Sheet under Impact of Magnetic Field. *Front. Heat Mass Transf.* **2020**, *14*, 15. [CrossRef]
42. Huang, M.; Chen, G. Laminar free convection from a sphere with blowing and suction. *J. Heat Transf.* **1987**, *109*, 6068496. [CrossRef]
43. Nazar, R.; Amin, N.; Pop, I. Mixed convection boundary layer flow about an isothermal sphere in a micropolar fluid. *Int. J. Sci.* **2003**, *42*, 283–293. [CrossRef]
44. Molla, M.M.; Rahman, A.; Rahman, L.T. Natural convection flow from an isothermal sphere with temperature dependent thermal conductivity. *J. Nav. Archit. Mar. Eng.* **2005**, *2*, 53–64. [CrossRef]
45. Ahmed, S.E.; Mansour, M.; Hussein, A.K.; Mallikarjuna, B.; Almeshaal, M.A.; Kolsi, L. MHD mixed convection in an inclined cavity containing adiabatic obstacle and filled with Cu–water nanofluid in the presence of the heat generation and partial slip. *J. Anal. Calorim.* **2019**, *138*, 1443–1460. [CrossRef]
46. Rashad, A.; Chamkha, A.J.; El-Kabeir, S. Effect of chemical reaction on heat and mass transfer by mixed convection flow about a sphere in a saturated porous media. *Int. J. Numer. Method H* **2011**, *21*, 418–433. [CrossRef]
47. Molla, M.M.; Hossain, M.; Taher, M. Magnetohydrodynamic natural convection flow on a sphere with uniform heat flux in presence of heat generation. *Acta Mech.* **2006**, *186*, 75. [CrossRef]
48. Keller, H.; Bramble, J. Numerical solutions of partial differential equations. *Mathematics* **1970**, *1*, 81–94.
49. Jones, E. An asymptotic outer solution applied to the Keller box method. *J. Comput. Phys.* **1981**, *40*, 411–429. [CrossRef]
50. Cebeci, T.; Bradshaw, P. *Physical and Computational Aspects of Convective Heat Transfer*; Springer Science & Business Media: New York, NY, USA, 2012; p. 487.
51. Das, S.; Banu, A.; Jana, R.; Makinde, O. Entropy analysis on MHD pseudo-plastic nanofluid flow through a vertical porous channel with convective heating. *AEJ* **2015**, *54*, 325–337. [CrossRef]

© 2020 by the authors. Licensee MDPI, Basel, Switzerland. This article is an open access article distributed under the terms and conditions of the Creative Commons Attribution (CC BY) license (http://creativecommons.org/licenses/by/4.0/).

Article

Lie Group Analysis of Unsteady Flow of Kerosene/Cobalt Ferrofluid Past A Radiated Stretching Surface with Navier Slip and Convective Heating

Hossam A. Nabwey [1,2,*], Waqar A. Khan [3] and Ahmed M. Rashad [4]

1. Department of Mathematics, College of Science and Humanities in Al-Kharj, Prince Sattam bin Abdulaziz University, Al-Kharj 11942, Saudi Arabia
2. Department of Basic Engineering Science, Faculty of Engineering, Menoufia University, Shebin El-Kom 32511, Egypt
3. Department of Mechanical Engineering, College of Engineering, Prince Mohammad Bin Fahd University, Al Khobar 31952, Saudi Arabia; wkhan@pmu.edu.sa
4. Department of Mathematics, Aswan University, Faculty of Science, Aswan 81528, Egypt; am_rashad@aswu.edu.eg
* Correspondence: h.mohamed@psau.edu.sa or eng_hossam21@yahoo.com

Received: 29 April 2020; Accepted: 17 May 2020; Published: 19 May 2020

Abstract: In this work, we identified the characteristics of unsteady magnetohydrodynamic (MHD) flow of ferrofluid past a radiated stretching surface. Cobalt–kerosene ferrofluid is considered and the impacts of Navier slip and convective heating are additionally considered. The mathematical model which describes the problem was built from some partial differential equations and then converted to self-similar equations with the assistance of the Lie group method; after that, the mathematical model was solved numerically with the aid of Runge–Kutta–Fehlberg method. Graphical representations were used to exemplify the impact of influential parameters on dimensionless velocity and temperature profiles; the obtained results for the skin friction coefficient and Nusselt number were also examined graphically. It was demonstrated that the magnetic field, Navier slip, and solid volume fraction of ferroparticles tended to reduce the dimensionless velocity, while the radiation parameter and Biot number had no effects on the dimensionless velocity. Moreover, the magnetic field and solid volume fraction increase skin friction whereas Navier slip reduces the skin friction. Furthermore, the Navier slip and magnetic field reduce the Nusselt number, whereas solid volume fraction of ferroparticles, convective heating, and radiation parameters help in increasing the Nusselt number.

Keywords: MHD; ferrofluid; Lie group framework; unsteady slip flow; stretching surface; thermal radiation

1. Introduction

Flow and convective heat transfer through a stretching surface play an essential role in research due to their presence in many engineering and industrial applications. Many authors have emphasized this and the details are found in [1–3]. To overcome the poor thermal conductivity and increase the other thermophysical properties of the conventional fluids, nanoparticles were suspended in a base fluid. These nanoparticles are called nanofluids and can be generated from diverse operations or chemical deposition mechanisms. Enchantment in the surface area and the rate of heat transfer occurred and many improvements have recently been performed for this issue [4]. This scheme of nanofluid is processed by integrating the pure fluid and classical equations of mass. Many investigations of nanofluid flow can be found in [5,6].

Heat transfer has been improved by adding nano-sized particles to a base fluid, as has been extensively enacted in heating and cooling methods in engineering and industries. The nano-scaled particles and the host fluid molecules are almost the same size and are identified as stable suspensions for an extended period. Convective thermal transport characteristics of nanofluids depend on the flow model, the volume fraction of nanofluid and shape of the particles [7–14]. Electronic gadgets, design of turbomachines, biomedicine, transportation, lubrication, enhanced oil recovery, lasers, petroleum drilling operations, and manufacturing process are some of the applications. The study of the combination of the fluid flow dynamic traits and the trait of electromagnetism is called magneto-hydrodynamics (MHD). It is a technique where the activities can be arrested electrically, associated with fluid flow in the presence of a magnetic flux field. The particles suspended in the fluid are controlled by the applied magnetic field and restructure their concentration; thus, the irregular heat transfer of the flow will be changed. A few situations with MHD issues are like the prediction of room climate, magneto-optical wavelength filters, estimations of stream rates of refreshments in the nourishment industry, optical switches and optical modulators. Magneto-nanoparticles are highly used in cancer therapy, MRI, magnetic drug targeting, hyperthermia, magnetic cell separation and drug delivery. They likewise have uses in geophysics; this is connected to thinking about stellar and solar structures, design of MHD pumps, etc. Several other significant investigations in this concern are due to [15–21].

Finally, Lie-group methods and their invariants offer a powerful, sophisticated, and methodical technique to obtain group-invariant solutions which are called self-similarity transformations. Self-similarity transformations achieved reduction of the independent variable numbers of a set of PDEs, leading to conversion of the non-linear governing PDEs into ODEs. Analysis using Lie groups has been executed by many scientists and applied mathematicians in many investigations [22–28].

In the current work, we analyze the unsteady MHD flow of ferrofluid and convective heat transfer confined by a radiate stretched sheet with the influence of Navier slip and convective heating. The mathematical model was solved numerically with the aid of Runge–Kutta–Fehlberg method. The aspects of various parameters such as velocity, temperature, shear stress fields and skin friction coefficient parameters associated with the current analysis are graphically examined. The recent advancements in modern technology have stimulated research interest in the analysis of boundary layer ferrofluid flow overstretching surfaces for its use in various engineering and industrial applications, such as paper production, fiberglass production, several engineering processes like solar power technology, etc.

2. Problem Formulation

In the current research, it is assumed that a 2D unsteady magneto-forced convective flow of ferrofluid past a radiate stretchable surface with impacts on Navier slip and convective heating are additionally considered. In this work, Cobalt is considered and is treated as a base nanoparticle, with kerosene as a base ferrofluid. The stretchable surface switches on from a fine slot, which is positioned at the starting point of a 2D coordinate system (x, y). At this point, the x-axis is considered all along the stretching direction of the sheet, having stretched velocity $U_w = ax$, which is applied vertically to the sheet externally. A constant magnetic strength B_0 is applied normal to the sheet. The mathematical model describing the system is (see Chamkha [29])

$$\frac{\partial u}{\partial x} + \frac{\partial v}{\partial y} = 0 \qquad (1)$$

$$\frac{\partial u}{\partial t} + u\frac{\partial u}{\partial x} + v\frac{\partial u}{\partial y} = \frac{\mu_{ff}}{\rho_{ff}}\frac{\partial^2 u}{\partial y^2} - \frac{\sigma_{ff} B_0^2}{\rho_{ff}} u \qquad (2)$$

$$\frac{\partial T}{\partial t} + u\frac{\partial T}{\partial x} + v\frac{\partial T}{\partial y} = \alpha_{ff}\left(\frac{\partial^2 T}{\partial y^2} - \frac{1}{k_{ff}}\frac{\partial q^r}{\partial y}\right) \qquad (3)$$

Subjected to the corresponding boundary conditions (see [30–34]):

$$u(t,x,0) = U_w + L\mu_{ff}\frac{\partial u}{\partial y}, v(t,x,0) = 0, -k_{ff}\frac{\partial T}{\partial y}(t,x,0) = h_f(T_f - T)$$
$$u(t,x,\infty) = 0, T(t,x,\infty) = T_\infty. \tag{4}$$

where t, u and v are the time and velocity components along the x and y axes and T is the temperature in the fluid phase. ρ_{ff} stands for the density. μ_{ff} stands for viscosity. β_{ff} Stands for the ferrofluid volumetric thermal expansion coefficient. σ_{ff} stands for electrical conductivity. $\alpha_{ff} = k_{ff}/(\rho C_p)_{ff}$ stands for the thermal diffusivity of the ferrofluid. L stands for the slip coefficient, which represents Navier slip, and h_f stands for the heat transfer coefficient. T_f stands for the uniform temperature of the stretchable surface. k_{ff} stands for the thermal conductivity of ferrofluid. $(\rho C_p)_{ff}$ stands for the specific heat of the ferrofluid at a constant pressure. The radiative heat flux qr is approached according to the Rosseland approximation (see [35,36]):

$$\frac{\partial q^r}{\partial y} = -\frac{4\sigma_1}{3\beta_R}\frac{\partial T^4}{\partial y} \tag{5}$$

where β_R and σ_1 stand for the mean absorption coefficient and the Stefan–Boltzmann constant. As carried out by Raptis [35], the fluid-phase temperature variations within the flow are approached to be adequately tiny so that T^4 may be obvious as a linear function of temperature. This is created by extending T^4 in a Taylor series on the free-stream temperature T_∞ and removing higher-order terms to yield

$$T^4 = 4T_\infty^3 T - 3T_\infty^4 \tag{6}$$

By applying Equations (5) and (6) in the last term of Equation (3), we obtain

$$\frac{\partial q^r}{\partial y} = -\frac{16\sigma_1 T_\infty^3}{3\beta_R}\frac{\partial^2 T}{\partial y^2} \tag{7}$$

In the current investigation, the following thermophysical relations are applied [37];

$$\rho_{ff} = (1-\chi)\rho_f + \chi\rho_s, \mu_{ff} = \frac{\mu_f}{(1-\chi)^{2.5}}, \alpha_{ff} = \frac{k_{ff}}{(\rho C_p)_{ff}},$$
$$(\rho C_p)_{ff} = (1-\chi)(\rho C_p)_f + \chi(\rho C_p)_s, (\rho\beta)_{ff} = (1-\chi)(\rho\beta)_f + \chi(\rho\beta)_s, \tag{8}$$
$$\frac{k_{ff}}{k_f} = \frac{(k_s+2k_f)-2\chi(k_f-k_s)}{(k_s+2k_f)+\chi(k_f-k_s)}, \frac{\sigma_{ff}}{\sigma_f} = 1 + \frac{3(\gamma-1)\chi}{(\gamma+2)-(\gamma-1)\chi} \text{ where } \gamma = \frac{\sigma_p}{\sigma_f}$$

Here, χ is nanoparticle volume fraction. Table 1 represents the thermophysical properties of ferrofluid.

Table 1. Thermophysical properties of kerosene, water and cobalt [37].

Property	Kerosene	Water	Cobalt
ρ (kg m^{-3})	780	997.1	8900
C_p (Jkg^{-1} K^{-1})	2090	4179	420
k (W m^{-1} K^{-1})	0.149	0.613	100
β (K^{-1})	9.9×10^{-4}	21×10^{-5}	1.3×10^{-5}
σ (Simens/m)	6×10^{-10}	0.05	1.602×10^7
μ (kg^{-1} m^{-1} s^{-1})	164×10^{-5}	625×10^{-6}	-

In this stage, the expressions for u, v, and θ will be defined as:

$$u = \frac{\partial \Psi}{\partial y}, v = -\frac{\partial \Psi}{\partial x}, \theta = \frac{(T - T_\infty)}{(T_f - T_\infty)}, \tag{9}$$

Substituting Equations (7)–(9) into Equations (1)–(4), we obtain

$$\frac{\partial^2 \Psi}{\partial t \partial y} + \frac{\partial \Psi}{\partial y}\frac{\partial^2 \Psi}{\partial x \partial y} - \frac{\partial \Psi}{\partial x}\frac{\partial^2 \Psi}{\partial y^2} = v_f \Xi_1 \frac{\partial^3 \Psi}{\partial y^3} - \frac{\sigma_{ff}}{\sigma_f}\frac{B_0^2 \sigma_f}{\rho_f}\frac{1}{1-\phi+\phi(\rho_s/\rho_f)}\frac{\partial \Psi}{\partial y} \qquad (10)$$

$$\frac{\partial \theta}{\partial t} + \frac{\partial \Psi}{\partial y}\frac{\partial \theta}{\partial x} - \frac{\partial \Psi}{\partial x}\frac{\partial \theta}{\partial y} = \frac{v_f}{\Pr}\Xi_2\left(\frac{k_{ff}}{k_f}+\frac{4}{3}Rd\right)\frac{\partial^2 \theta}{\partial y^2} \qquad (11)$$

$$\frac{\partial \Psi}{\partial y} = ax + L\mu_{ff}\frac{\partial^2 \Psi}{\partial y^2}, \frac{\partial \Psi}{\partial x} = 0, \frac{k_{ff}}{k_f}\frac{\partial \theta}{\partial y} = -\frac{h_f}{k_f}(1-\theta) \text{ at } y = 0,$$

$$\frac{\partial \Psi}{\partial y} = 0, \theta = 0, \text{ at } y \to \infty.$$

where $\Xi_1 = \dfrac{1}{(1-\chi)^{2.5}[1-\chi+\chi(\rho_s/\rho_f)]}$, $\Xi_2 = \dfrac{1}{\left[1-\chi+\chi(\rho C_p)_s/(\rho C_p)_f\right]}$

3. Lie Group Framework

Obtaining the solutions of the PDEs (partial differential equations) (10)–(12) governing the investigation understudy is equivalent to satisfying the constant solutions of these equations under a special continuous one-parameter group. The proposed technique is to search for a transformation group from the primary collection of one parameter scaling transformation. The facilitated form of Lie group framework, namely, the scaling group of transformations Δ (see [38–46]), will be presented here:

$$\Delta: \widehat{x} = x\ell^{\varepsilon\kappa_1}, \widehat{y} = y\ell^{\varepsilon\kappa_2}, \widehat{t} = t\ell^{\varepsilon\kappa_3}, \widehat{\Psi} = \Psi\ell^{\varepsilon\kappa_4}, \widehat{\theta} = \theta\ell^{\varepsilon\kappa_5} \qquad (13)$$

where κ_1, κ_2, κ_3, κ_4, and κ_5 are transformation parameters and ε is a small parameter whose interrelationship will be determined by our investigation. Equation (13) may be scrutinized as a point transformation, which transfers the coordinates (x, y, t, Ψ, θ) to $(\widehat{x}, \widehat{y}, \widehat{t}, \widehat{\Psi}, \widehat{\theta})$. Substituting transformations Equation (13) in Equations (10)–(12), we obtain;

$$\ell^{\varepsilon(\kappa_2+\kappa_3-\kappa_4)}\frac{\partial^2 \widehat{\Psi}}{\partial \widehat{t}\partial \widehat{y}} + \ell^{\varepsilon(\kappa_1+2\kappa_2-2\kappa_4)}\left(\frac{\partial \widehat{\Psi}}{\partial \widehat{y}}\frac{\partial^2 \widehat{\Psi}}{\partial \widehat{x}\partial \widehat{y}} - \frac{\partial \widehat{\Psi}}{\partial \widehat{x}}\frac{\partial^2 \widehat{\Psi}}{\partial \widehat{y}^2}\right) = v_f \Xi_1 \ell^{\varepsilon(3\kappa_2-\kappa_4)}\frac{\partial^3 \widehat{\Psi}}{\partial \widehat{y}^3}$$
$$-\frac{\sigma_{ff}}{\sigma_f}\frac{B_0^2 \sigma_f}{\rho_f}\frac{1}{1-\phi+\phi(\rho_s/\rho_f)}\ell^{\varepsilon(\kappa_2-\kappa_4)}\frac{\partial \widehat{\Psi}}{\partial \widehat{y}} \qquad (14)$$

$$\ell^{\varepsilon(\kappa_3-\kappa_5)}\frac{\partial \widehat{\theta}}{\partial \widehat{t}} + \ell^{\varepsilon(\kappa_1+\kappa_2-\kappa_4-\kappa_5)}\left(\frac{\partial \widehat{\Psi}}{\partial \widehat{y}}\frac{\partial \widehat{\theta}}{\partial \widehat{x}} - \frac{\partial \widehat{\Psi}}{\partial \widehat{x}}\frac{\partial \widehat{\theta}}{\partial \widehat{y}}\right) = \frac{v_f}{\Pr}\Xi_2\left(\frac{k_{ff}}{k_f}+\frac{4}{3}Rd\right)\ell^{\varepsilon(2\kappa_2-\kappa_5)}\frac{\partial^2 \widehat{\theta}}{\partial \widehat{y}^2} \qquad (15)$$

The following relations should be determined to reserve the system to be constant:

$$\kappa_2 + \kappa_3 - \kappa_4 = \kappa_1 + 2\kappa_3 - 2\kappa_4 = 3\kappa_2 - \kappa_4 = \kappa_2 - \kappa_4$$
$$\kappa_3 + \kappa_5 = \kappa_1 + \kappa_2 - \kappa_4 - \kappa_5 = 2\kappa_2 - \kappa_5 \qquad (16)$$

These relations give

$$\kappa_4 = \kappa_1, \kappa_2 = \kappa_3 = \kappa_5 = 0 \qquad (17)$$

and the one-parameter group of transformations can be obtained as

$$\widehat{x} = x\ell^{\varepsilon\kappa_1}, \widehat{y} = y, \widehat{t} = t, \widehat{\Psi} = \Psi\ell^{\varepsilon\kappa_1}, \widehat{\theta} = \theta \qquad (18)$$

Developing by Taylor's technique in powers of ε, we obtain:

$$\widehat{x} - x = x\varepsilon\kappa_1, \, \widehat{y} - y = 0, \, \widehat{t} - t = 0, \, \widehat{\Psi} - \Psi = \Psi\varepsilon\kappa_1, \, \widehat{\theta} - \theta = 0$$

which yields

$$\frac{dx}{x\kappa_1} = \frac{dy}{0} = \frac{dt}{0} = \frac{d\Psi}{\Psi\kappa_1} = \frac{d\theta}{0} \qquad (19)$$

$$\eta = \Gamma_1(x,t)y, \tau = \Gamma_2(x,y)t, \Psi = \Gamma_3(y,t)x, \theta = \theta(\tau,\eta), \qquad (20)$$

where $\Gamma_1, \Gamma_2,$ and Γ_3 are arbitrary functions which should be determined by its equations. η and τ are the similarity variable and dimensionless time.

To avert the fluid properties manifesting explicitly in the coefficients of the above equations, determining mass balance in Equation (1), with keeping generality, we have dropped three different convenient arbitrary constants based on the transformations performed previously by Nabwey [25] and Chamkha [29] as follows:

$$\Gamma_1(x,t) = \left(1/2a\sqrt{v_f t}\right), \Gamma_2(y,t) = a, \Gamma_3(y,t) = 2a\sqrt{v_f t} \qquad (21)$$

As a consequence, we find

$$t = \frac{\tau}{a}, y = 2\sqrt{v_f t}\eta, \Psi = 2ax\sqrt{v_f t}f(\tau,\eta) \qquad (22)$$

with the assistance of these formulations in Equation (22). Equations (10)–(12) are characterized as

$$\Xi_1 f''' + 2\eta f'' - 4\tau\left(f'^2 - ff''\right) + \frac{\sigma_{ff}}{\sigma_f}\frac{Ha^2}{1 - \phi + \phi(\rho_s/\rho_f)}f' - 4\tau\frac{\partial f'}{\partial \tau} = 0 \qquad (23)$$

$$\frac{\Xi_2}{Pr}\left(\frac{k_{ff}}{k_f} + \frac{4}{3}R\right)\theta'' + 2\eta\theta' + 4\tau(f\theta' - f'\theta) - 4\tau\frac{\partial\theta}{\partial\tau} = 0 \qquad (24)$$

subject to the following boundary conditions:

$$f(\tau,0) = 0, f'(\tau,0) = 1 + \frac{\delta/\sqrt{\tau}}{(1-\chi)^{2.5}}f''(\tau,0), \frac{k_{ff}}{k_f}\theta'(\tau,0) = -Bi\sqrt{\tau}(1-\theta(\tau,0))$$
$$f'(\tau,\infty) = 0, \theta(\tau,\infty) = 0 \qquad (25)$$

where $Ha = B_0\sqrt{\frac{\sigma_f}{a\rho_f}}$ stands for the Hartmann number. $Bi = \frac{2h_f}{k_f}\sqrt{\frac{v_f}{a}}$ stands for Biot number. $Rd = 4\sigma_1 T_\infty^3/k_f\beta_R$ stands for the radiation parameter. $\delta = L\mu_f/2\sqrt{v_f/a}$ stands for the velocity slip parameter.

The local skin-drag coefficient and local Nusselt number can be written respectively, as

$$C_f = -\mu_{ff}\left.\frac{\partial u}{\partial y}\right|_{y=0}/\left(\mu_f ax/2\sqrt{v_f t}\right) = -\frac{1}{(1-\chi)^{2.5}}f''(\tau,0) \qquad (26)$$

$$Nu = -\left[\left(k_{ff} + \frac{16\sigma_1 T_\infty^3}{3\beta_R}\right)\frac{\partial T}{\partial y}\right]_{y=0}/\left(k_f(T_f - T_\infty)/2\sqrt{v_f t}\right)$$
$$= -\left(\frac{k_{ff}}{k_f} + \frac{4Rd}{3}\right)\theta'(\tau,0) \qquad (27)$$

4. Numerical Method

Following [47,48], Equations (23) and (24), subject to (25), will be solved using the local similarity method, where the first derivatives with respect to τ are neglected and the Equations (23) and (24) with boundary conditions (25) can be re-written as

$$\Xi_1 f''' + 2\eta f'' - 4\tau \left(f'^2 - ff'' + \frac{\sigma_{ff}}{\sigma_f} \frac{Ha^2}{1 - \phi + \phi(\rho_s/\rho_f)} f' \right) = 0 \tag{28}$$

$$\frac{\Xi_2}{\Pr} \left(\frac{k_{ff}}{k_f} + \frac{4}{3} Rd \right) \theta'' + 2\eta \theta' + 4\tau (f\theta' - f'\theta) = 0 \tag{29}$$

The boundary conditions (25) remain the same. These ordinary differential equations with the boundary conditions (25) can be solved numerically by applying the Runge–Kutta–Fehlberg method (RKF7 45). Following [47,48], for the local non-similarity solution, now we hold all the terms by assuming the new auxiliary functions $F(\tau, \eta)$, and $\Theta(\tau, \eta)$, which are defined by

$$F = \frac{\partial f}{\partial \tau}, \Theta = \frac{\partial \theta}{\partial \tau} \tag{30}$$

Thus, Equations (23) and (24) can be expressed as

$$\Xi_1 f''' + 2\eta f'' - 4\tau \left(f'^2 - ff'' + \frac{\sigma_{ff}}{\sigma_f} \frac{Ha^2}{1 - \phi + \phi(\rho_s/\rho_f)} f' \right) - 4\tau F' = 0 \tag{31}$$

$$\frac{\Xi_2}{\Pr} \left(\frac{k_{ff}}{k_f} + \frac{4}{3} Rd \right) \theta'' + 2\eta \theta' + 4\tau (f\theta' - f'\theta) - 4\tau \Theta = 0 \tag{32}$$

subject to the same condition in (25). The new ODEs (31)–(32), subject to (25) represent a local non-similarity model for the problem under consideration. Equations (31) and (32) and the boundary conditions (25) are now differentiated w.r.t. τ, simplified and the derivatives w.r.t. τ are neglected again. These equations represent a local similarity model and can be expressed as;

$$\left. \begin{array}{l} \Xi_1 F''' + 2\eta F'' + 4ff'' - f'^2 + 4\tau Ff'' - f'^2 + 4\tau fF'' - 2f'F' - \left(\frac{\sigma_{ff}}{\sigma_f} \frac{Ha^2}{1 - \phi + \phi(\rho_s/\rho_f)} F' \right) \\ -4F' = 0 \\ \frac{\Xi_2}{\Pr} \left(\frac{k_{ff}}{k_f} + \frac{4}{3} Rd \right) \Theta'' + 2\eta \Theta' + 4(f\theta' - f'\theta) + 4\tau F\theta' + f\Theta' - \Theta f' - \theta F' - 4\Theta = 0 \end{array} \right\} \tag{33}$$

$$\left. \begin{array}{l} F(\tau, 0) = 0, F'(\tau, 0) = -\frac{\delta}{2\tau^{3/2}(1-\chi)^{2.5}} F''(\tau, 0) + \frac{\delta/\sqrt{\tau}}{(1-\chi)^{2.5}} F''(\tau, 0), F'(\tau, \infty) = 0, \\ \frac{k_{ff}}{k_f} \Theta'(\tau, 0) = -Bi \left[\frac{[1-\theta(\tau, 0)]}{2\sqrt{\tau}} - \sqrt{\tau} \Theta(\tau, 0) \right], \Theta(\tau, \infty) = 0 \end{array} \right\} \tag{34}$$

The ODEs (31)–(33) subject to (25) and (34) were solved numerically by employing the Runge–Kutta–Fehlberg technique (RKF45) using MAPLE-19 software (MAPLE 2019.0, Maplesoft, Waterloo, ON, Canada). This method is generally known as one of the most excellent methods available for obtaining the solutions of nonlinear differential equations and provides more accurate results. The step size was selected. For the similarity variable η_{max}, Equations (25) and (34), were replaced as

$$f'(5) = 0, \theta(5) = 0, F'(5) = 0, \Theta(5) = 0 \tag{35}$$

The selection of $\eta_{max} = 5$ guarantees that all numerical solutions approached the asymptotic values properly.

5. Results and Discussions

In this study, we investigated the unsteady magento-flow and heat transfer of Cobalt–kerosene ferrofluid past a stretchable surface. The influences of several key parameters on the dimensionless velocity $f'(\tau, \eta)$, temperature $\theta(\tau, \eta)$, skin friction $C_f(\tau, 0)$, and Nusselt number $Nu(\tau, 0)$ are examined. The Lie group method is employed to reduce partial differential equations and local similar and non-similar models are solved employing the RK-45 technique.

The effects of the magnetic field Ha and dimensionless time τ on the velocity are symbolized in Figure 1a and on the dimensionless temperature in Figure 1b, respectively. As the time increases, the velocity at the surface rises. The magnetic field generates Lorentz strength on the fluid particles, which resist the fluid and reduce the fluid velocity, as shown in Figure 1a. Consequently, the velocity boundary layer thickness decreases. Due to the decline in velocity, the temperature increases. In the thermal boundary layer, the temperature declines to the ambient temperature. The thermal boundary layer thickness reduces with an enlargement of the dimensionless time, as exhibited in Figure 1b. The influence of the solid volume fraction of nanoparticles χ and Navierslip δ on the velocity and temperature is depicted in Figure 2a,b when $\tau = 0.5$. In the absence of slip, the velocity is found to be higher for the pure regular fluid. At the surface, the velocity decreases with the increase in the slip and solid volume fraction, as shown in Figure 2a. No appreciable impact of χ could be observed at the surface as well as within the velocity boundary layer. The velocity boundary layer thickness enlarges with δ, which enlarges the thermal resistance and reduces the heat transfer rate; see Figure 2b. The variation of the dimensionless temperature with the solid volume fraction χ is depicted in Figure 2b. In the absence of slip, the temperature is lower at the wall and intensifies with δ. As expected, the temperature at the wall is higher for the regular fluid and dwindles with an intensify in the solid volume fraction χ. This is due to the higher thermal conductivity of Cobalt nanoparticles. With the addition of nanoparticles, the thermal conductivity of the ferrofluid increases and the heat transfer rate is enhanced.

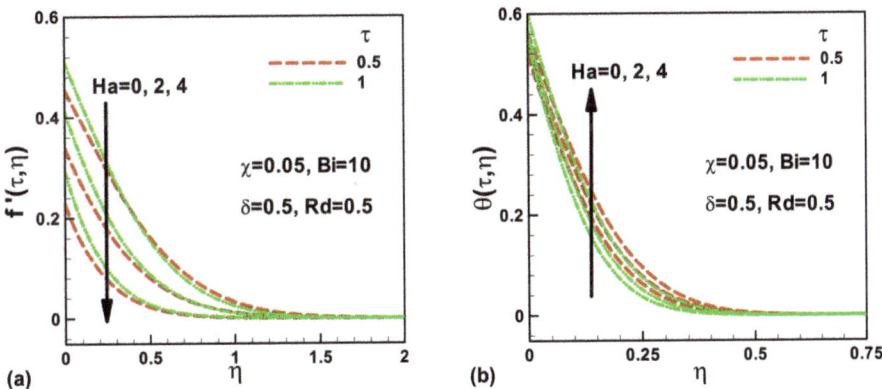

Figure 1. Effects of magnetic field Ha and dimensionless time τ on (**a**) dimensionless velocity, and (**b**) dimensionless temperature.

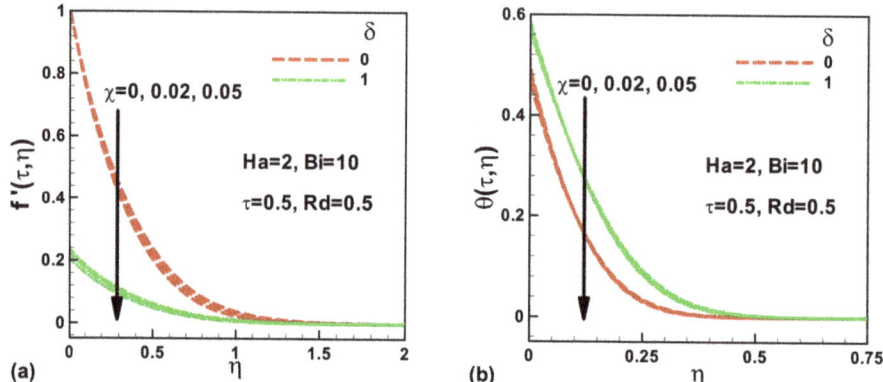

Figure 2. Effects of solid volume fraction of nanoparticles χ and Navier slip δ on (**a**) dimensionless velocity and (**b**) dimensionless temperature.

Figure 3a,b presents the effects of radiation parameter Rd and Biot number Bi on the velocity and temperature curves. It is important to note that equations of momentum and energy are independent of each other. The momentum equation and the velocity boundary conditions are independent of the radiation parameter Rd and convective heating parameter Bi. Therefore, there is no influence of these parameters on the velocity, which is obvious from Figure 3a. On the other side, the surface temperature increases significantly with a strengthen in both Rd and Bi. As a result, the thermal boundary layer thickness is boosted, with an increase in both parameters, as depicted in Figure 3b. The radiation parameter Rd reveals an enhancement in radiative heat, which improves the thermal state of fluid, causing its surface temperature to increase. Similarly, as the convective heating parameter increases and tends to infinity, the convective boundary condition changes to an isothermal boundary condition.

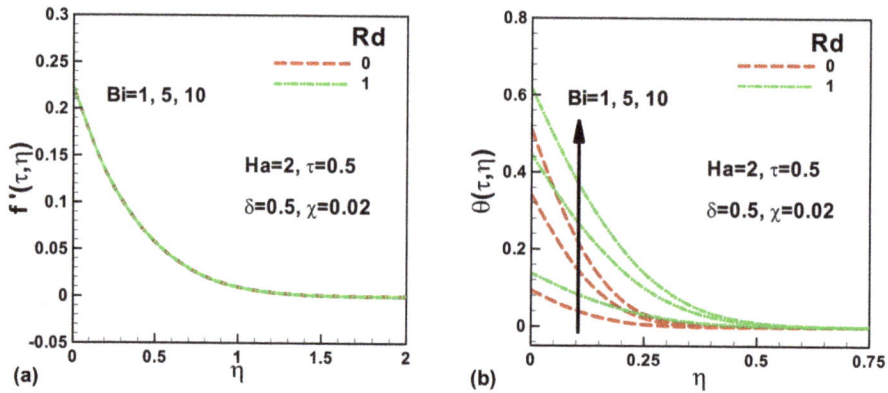

Figure 3. Effects of Biot number Bi and radiation parameter Rd on (**a**) dimensionless velocity and (**b**) dimensionless temperature.

The variations in skin friction and Nusselt number with the magnetic field Ha are depicted in Figures 4 and 5 for different values of the velocity slip δ and the solid volume fraction χ at $\tau = 1$ and $\tau = 2$, respectively. In the presence of magnetic strength, a Lorentz force is generated which resists the fluid and reduces the velocity curve. Therefore, the skin friction enhances with Ha, as shown in Figures 4 and 5a at different dimensionless times. As expected, the skin friction increases with dimensionless time τ. In the absence of velocity slip δ, the velocity curves are higher at the surface and decline with an increment in slip parameter δ. Consequently, the skin friction declines with the boosting of slip parameter δ. For the pure regular fluid, the skin friction is lower and increases with a rise in the solid volume fraction χ. This is due to an evolution in the ferrofluid density with the increased volume fraction of cobalt nanoparticles. Figures 4 and 5b illustrate the variation of Nusselt number with the magnetic field Ha and the volume fraction of ferroparticles χ at different dimensionless times. Like skin friction, Nusselt number also increases with dimensionless time. Due to Lorentz force, the dimensionless velocity decreases and, as a result, the Nusselt number is reduced with an increasing magnetic field. Similarly, the velocity decreases due to an intension in the slip and the Nusselt number reduces. The thermal conductivity of ferroparticles increases with an increase in the volume fraction of ferroparticles. Consequently, the Nusselt number increases with increasing χ.

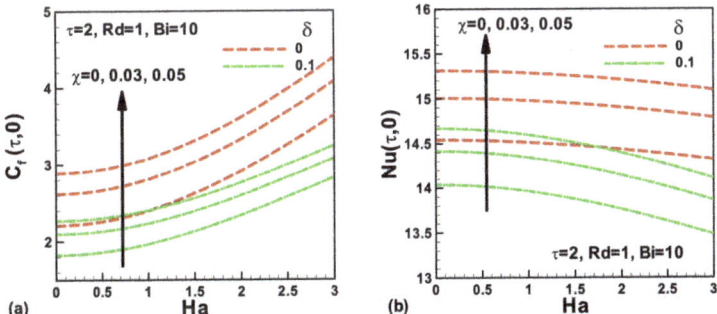

Figure 4. Effects of solid volume fraction of nanoparticles χ, magnetic Ha, and Navier slip δ parameters on (**a**) skin friction, and (**b**) Nusselt number when $\tau = 1$.

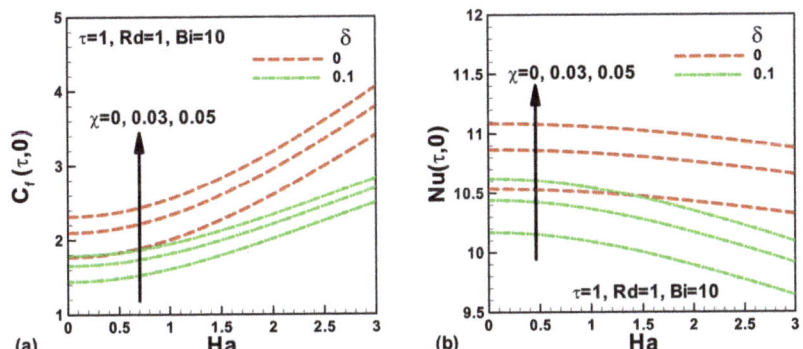

Figure 5. Effects of solid volume fractions of nanoparticles χ, magnetic Ha and Navierslip δ parameters on (**a**) skin friction, and (**b**) Nusselt number when $\tau = 2$.

Figure 6a, b presents the comparison of Nusselt numbers for kerosene oil and water for the same parameters. Due to the smaller Prandtl number Pr for water, the Nusselt numbers are found to be lower than kerosene. The Prandtl number Pr compares the rate of thermal diffusion in comparison to the rate of momentum diffusion. The higher the Prandtl number Pr, the higher the Nusselt number will be. It is also noticed that an increase in the Biot number Bi and radiation parameter Rd leads to an increase in the Nusselt number. These Nusselt numbers also become greater with increasing dimensionless time.

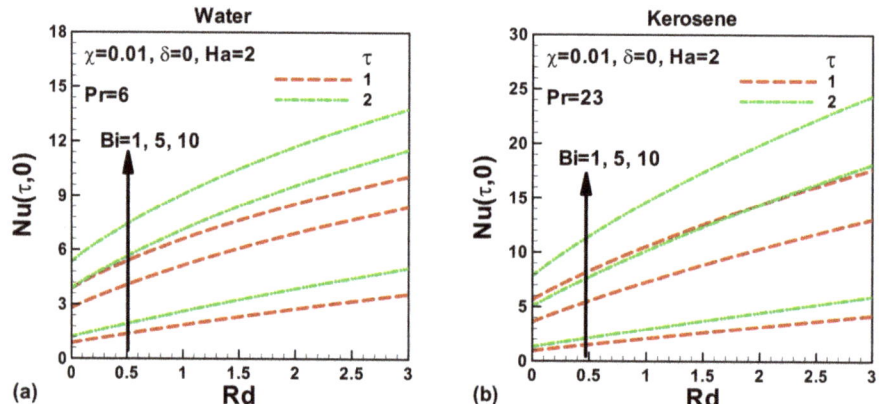

Figure 6. Effects of Biot number Bi, dimensionless time τ, and radiation parameter Rd on Nusselt number for (**a**) water and (**b**) kerosene oil as base fluid.

6. Conclusions

In this study, application of the scaling group of transformations to the unsteady magneto-flow of ferrofluid past a stretching surface was employed. The impacts of Navier slip, radiation and solid volume fraction of ferroparticles, as well as convective heating, were also investigated. From this study, it was concluded that:

- Employing the Lie group framework, the symmetries of the partial differential equations are presented exclusively in this investigation, these equations are reduced to self-similar equations utilizing translational and scaling symmetries. Numerical solutions for scaling symmetry are obtained applying the Runge–Kutta–Fehlberg method.
- The magnetic field, Navier slip, and solid volume fraction of ferroparticles tend to reduce the dimensionless velocity.
- The radiation parameter and Biot number have no effects on the dimensionless velocity.
- Magnetic field, radiation, Biot number, and Navier slip increase the surface temperature, whereas the solid volume fraction of ferroparticles reduces the surface temperature.
- The magnetic field, dimensionless time and solid volume fraction increase skin friction, whereas Navier slip reduces the skin friction.
- The magnetic field and Navier slip reduce the Nusselt number, whereas solid volume fraction of ferroparticles, convective heating, and radiation parameters help in increasing the Nusselt number.
- The Nusselt number for kerosene oil is higher than for water.

Author Contributions: Supervision, H.A.N.; Investigation, H.A.N., W.A.K. and A.M.R.; Methodology, H.A.N., W.A.K. and A.M.R.; Writing – original draft, H.A.N., W.A.K. and A.M.R.; Writing – review & editing, H.A.N., W.A.K. and A.M.R.; Funding acquisition, H.A.N. All authors have read and agreed to the published version of the manuscript.

Funding: This work was supported by the Deanship of Scientific Research at Prince Sattam bin Abdulaziz University under the research project No. 2020/01/14307.

Conflicts of Interest: The authors declare no conflict of interest.

References

1. Haq, R.; Nadeem, S.; Khan, Z.H.; Akbar, N.S. Thermal radiation and slip effects on MHD stagnation point over of nano fluid over a stretching sheet. *J. Phys.* **2015**, *6*, 7–13.
2. Kudenattia, R.B.; Kirsurb, S.R.; Achalab, L.N.; Bujurkec, N.M. MHD boundary layer flow over a non-linear stretching boundary with suction and injection. *Int. J. Non Linear Mech.* **2013**, *50*, 58–67. [CrossRef]

3. Yigra, Y.; Tesfay, D. Magneto-hydrodynamic flow of viscous fluid over a non-linearly stretching sheet. *J. Fluid Mech.* **2014**, *1*, 1–11.
4. Das, S.K.; Choi, S.U.S.; Yu, W.; Pradeep, T. *Nanofluids-Science and Technology*; John Wiley & Sons Publishers: Hoboken, NJ, USA, 2007.
5. Tlili, I.; Nabwey, H.A.; Ashwinkumar, G.P.; Sandeep, N. 3-D magnetohydrodynamic AA7072-AA7075/methanol hybrid nanofluid flow above an uneven thickness surface with slip effect. *Sci. Rep.* **2020**, *10*, 1–13. [CrossRef] [PubMed]
6. Abdelmalek, Z.; Khan, S.U.; Waqas, H.; Nabwey, H.A.; Tlili, I. Utilization of second order slip, activation energy and viscous dissipation consequences in thermally developed flow of third grade nanofluid with gyrotactic microorganisms. *Symmetry* **2020**, *12*, 309. [CrossRef]
7. Kumar, S.; Prasad, S.K.; Banerjee, J. Analysis of flow and thermal field in nanofluid using a single phase thermal dispersion model. *Appl. Math. Model.* **2010**, *34*, 573–592. [CrossRef]
8. Wang, F.C.; Wu, H.A. Enhanced oil droplet detachment from solid surfaces in charged nanoparticle suspensions. *Soft Matter.* **2013**, *9*, 7974–7980. [CrossRef]
9. Bég, O.A.; Espinoza, D.E.S.; Kadir, A.; Shamshuddin, M.; Sohail, A. Experimental study of improved rheology and lubricity of drilling fluids enhanced with nanoparticles. *Appl. Nanosci.* **2018**, *8*, 1069–1090. [CrossRef]
10. Zhu, J.; Cao, J. Effects of nanolayer and second order slip on unsteady nanofluid flow past a wedge. *Mathematics* **2019**, *7*, 1043. [CrossRef]
11. Avramenko, A.A.; Shevchuk, I.V.; Tyrinov, A.I.; Blinov, D.G. Heat transfer in stable film boiling of a nanofluid over a vertical surface. *Int. J. Therm. Sci.* **2015**, *92*, 106–118. [CrossRef]
12. Avramenko, A.A.; Shevchuk, I.V.; Tyrinov, A.I.; Blinov, D.G. Heat transfer at film condensation of stationary vapor with nanoparticles near a vertical plate. *Appl. Therm. Eng.* **2014**, *73*, 391–398. [CrossRef]
13. Avramenko, A.A.; Shevchuk, I.V.; Tyrinov, A.I.; Blinov, D.G. Heat transfer at film condensation of moving vapor with nanoparticles over a flat surface. *Int. J. Heat Mass Transf.* **2015**, *82*, 316–324. [CrossRef]
14. Avramenko, A.A.; Shevchuk, I.V. Lie group analysis and general forms of self-similar parabolic equations for fluid flow, heat and mass transfer of nanofluids. *J. Therm. Anal. Calorim.* **2019**, *135*, 223–235. [CrossRef]
15. Murthy, P.V.S.N.; RamReddy, C.; Chamkha, A.J.; Rashad, A.M. Magnetic effect on thermally stratified nanofluid saturated non-darcy porous medium under convective boundary condition. *Int. Commun. Heat Mass Transf.* **2013**, *47*, 41–48. [CrossRef]
16. Rashad, A.M. Impact of thermal radiation on MHD slip flow of a ferrofluid over a nonisothermal wedge. *J. Magn. Magn. Mater.* **2017**, *422*, 25–31. [CrossRef]
17. Chamkha, A.J.; Nabwey, H.A.; Abdelrahman, Z.M.A.; Rashad, A.M. Mixed bioconvective flow over a wedge in porous media drenched with a nanofluid. *J. Nanofluids* **2019**, *8*, 1692–1703. [CrossRef]
18. Nabwey, H.A.; Hashmi, M.S.; Khan, S.U.; Tlili, I. A theoretical analysis for mixed convection flow of Maxwell fluid between two infinite isothermal stretching disks with heat source/sink. *Symmetry* **2020**, *12*, 62. [CrossRef]
19. Kumar, K.G.; Ramesh, G.K.; Gireesha, B.J.; Rashad, A.M. On stretched magnetic flow of Carreau nanofluid with slip effects and nonlinear thermal radiation. *Nonlinear Eng.* **2019**, *8*, 340–349. [CrossRef]
20. Rashad, A.M.; Nabwey, H.A. Gyrotactic mixed bioconvection flow of a nanofluid past a circular cylinder with convective boundary condition. *J. Taiwan Inst. Chem. Eng.* **2019**, *99*, 9–17. [CrossRef]
21. Chamkha, A.J.; Rashad, A.M.; Alsabery, A.I.; Abdelrahman, Z.M.A.; Nabwey, H.A. Impact of partial slip on magneto-ferrofluids mixed convection flow in enclosure. *J. Therm. Sci. Eng. Appl.* **2020**, *12*. [CrossRef]
22. Ferdows, M.; Nabwey, H.A.; Rashad, A.M.; Uddin, M.J.; Alzahrani, F. Boundary layer flow of a nanofluid past a horizontal flat plate in a Darcy porous medium: A Lie group approach. *Proc. Inst. Mech. Eng. Part C J. Mech. Eng. Sci.* **2019**. [CrossRef]
23. Nabwey, H.A.; El-Mkyn, H.A. Lie group analysis of thermophoresis on a vertical surface in a porous medium. *J. King Saud Univ. Sci.* **2019**, *31*, 1048–1055. [CrossRef]
24. Nabwey, H.A.; Boumazgour, M.; Rashad, A.M. Group method analysis of mixed convection stagnation-point flow of non-Newtonian nanofluid over a vertical stretching surface. *Indian J. Phys.* **2017**, *91*, 731–742. [CrossRef]
25. Nabwey, H.A.; EL-Kabeir, S.M.M.; Rashad, A.M. Lie group analysis of effects of radiation and chemical reaction on heat and mass transfer by unsteady slip flow from a non-isothermal stretching sheet immersed in a porous medium. *J. Comput. Theor. Nanosci.* **2015**, *12*, 4056–4062. [CrossRef]

26. Bakier, A.Y.; Rashad, A.M.; Mansour, M.A. Group method analysis of melting effect on MHD mixed convection flow from radiate vertical plate embedded in a saturated porous media. *Commun. Nonlinear Sci. Numer. Simul.* **2009**, *14*, 2160–2170. [CrossRef]
27. EL-Kabeir, S.M.M.; EL-Hakiem, M.A.; Rashad, A.M. Group method analysis of combined heat and mass transfer by MHD non-Darcy non-Newtonian natural convection adjacent to horizontal cylinder in a saturated porous medium. *Appl. Math. Model.* **2008**, *32*, 2378–2395. [CrossRef]
28. EL-Kabeir, S.M.M.; EL-Hakiem, M.A.; Rashad, A.M. Lie group analysis of unsteady MHD three dimensional by natural convection from an inclined stretching surface saturated porous medium. *J. Comput. Appl. Math.* **2008**, *213*, 582–603. [CrossRef]
29. Chamkha, A.J. Unsteady hydromagnetic flow and heat transfer from a non-isothermal stretching sheet immersed in a porous medium. *Int. Comm. Heat Mass Transf.* **1998**, *25*, 899–906. [CrossRef]
30. Noghrehabadi, A.; Pourrajab, R.; Ghalambaz, M. Effect of partial slip boundary condition on the flow and heat transfer of nanofluids past stretching sheet prescribed constant wall temperature. *Int. J. Therm. Sci.* **2012**, *54*, 253–261. [CrossRef]
31. Rashad, A.M.; Ismael, M.A.; Chamkha, A.J.; Mansour, M.A. MHD mixed convection of localized heat source/sink in a nanofluid-filled lid-driven square cavity with partial slip. *J. Taiwan Inst. Chem. Eng.* **2016**, *68*, 173–186. [CrossRef]
32. Ismael, M.A.; Pop, I.; Chamkha, A.J. Mixed convection in a lid-driven square cavity with partial slip. *Int. J. Therm. Sci.* **2014**, *82*, 47–61. [CrossRef]
33. Hafidzuddin, E.H.; Nazar, R.; Arifin, N.M.; Pop, I. Effects of anisotropic slip on three-dimensional stagnation-point flow past a permeable moving surface. *Eur. J. Mech. B Fluids* **2017**, *65*, 515–521. [CrossRef]
34. Roşca, A.V.; Roşca, N.C.; Pop, I. Axisymmetric stagnation point flow and heat transfer towards a permeable moving flat plate with surface slip condition. *Appl. Math. Comput.* **2014**, *233*, 139–151. [CrossRef]
35. Raptis, A. Radiation and free convection flow through a porous medium. *Int. Comm. Heat Mass Transf.* **1998**, *25*, 289–295. [CrossRef]
36. EL-Hakiem, M.A.; Rashad, A.M. Effect of radiation on non-Darcy free convection from a vertical cylinder embedded in a fluid-saturated porous medium with a temperature-dependent viscosity. *J. Porous Media* **2007**, *10*, 209–218.
37. Sheikholeslami, M.; Gorji-Bandpy, M. Free convection of ferrofluid in a cavity heated from below in the presence of an external magnetic field. *Power Technol.* **2014**, *256*, 490–498. [CrossRef]
38. Mukhopadhyay, S.; Layek, G.C.; Samad, S.A. Study of MHD boundary layer flow over a heated stretching sheet with variable viscosity. *Int. J. Heat Mass Transf.* **2005**, *48*, 4460–4466. [CrossRef]
39. Reddy, G.V.R.; Chamkha, A.J. Lie group analysis of chemical reaction effects on MHD free convection dissipative fluid flow past an inclined porous surface. *Int. J. Numer. Methods Heat Fluid Flow* **2015**, *25*, 1557. [CrossRef]
40. Ibragimov, N.H. *CRC Handbook of Lie Group Analysis of Differential Equations*; CRC Press: Boca Raton, FL, USA, 1995; Volume 3.
41. Ovsiannikov, L.V. *Group Analysis of Differential Equations*; Academic Press: Cambridge, MA, USA, 2014.
42. Olver, P.J. *Applications of Lie Groups to Differential Equations*; Springer Science & Business Media: Berlin/Heidelberg, Germany, 2000; Volume 107.
43. Yürüsoy, M.; Pakdemirli, M.; Noyan, Ö.F. Lie group analysis of creeping flow of a second grade fluid. *Int. J. Non Linear Mech.* **2001**, *36*, 955–960. [CrossRef]
44. Yürüsoy, M.; Pakdemirli, M. Exact solutions of boundary layer equations of a special non-Newtonian fluid over a stretching sheet. *Mech. Res. Commun.* **1999**, *26*, 171–175. [CrossRef]
45. Sivasankaran, S.; Bhuvaneswari, M.; Kandaswamy, P.; Ramasami, E.K. Lie group analysis of natural convection heat and mass transfer in an inclined porous surface with heat generation. *Int. J. Appl. Math. Mech.* **2006**, *2*, 34–40.
46. Sivasankaran, S.; Bhuvaneswari, M.; Kandaswamy, P.; Ramasami, E.K. Lie group analysis of natural convection heat and mass transfer in an inclined surface. *Nonlinear Anal. Model. Control* **2006**, *11*, 201–212.

47. Sparrow, E.M.; Quack, H.; Boerner, C.J. Local non-similarity boundary-layer solutions. *AIAA J.* **1970**, *8*, 1936–1942. [CrossRef]
48. Minkowycz, W.J.; Sparrow, E.M. Numerical solution scheme for local non-similarity boundary-layer analysis. *Numer. Heat Transf. Part B Fundam. Int. J. Comput. Methodol.* **1978**, *1*, 69–85. [CrossRef]

© 2020 by the authors. Licensee MDPI, Basel, Switzerland. This article is an open access article distributed under the terms and conditions of the Creative Commons Attribution (CC BY) license (http://creativecommons.org/licenses/by/4.0/).

Article

Unsteady Three-Dimensional MHD Non-Axisymmetric Homann Stagnation Point Flow of a Hybrid Nanofluid with Stability Analysis

Nurul Amira Zainal [1,2], Roslinda Nazar [1], Kohilavani Naganthran [1,*] and Ioan Pop [3]

1. Department of Mathematical Sciences, Faculty of Science and Technology, Universiti Kebangsaan Malaysia, Bangi 43600, Selangor, Malaysia; nurulamira@utem.edu.my (N.A.Z.); rmn@ukm.edu.my (R.N.)
2. Fakulti Teknologi Kejuruteraan Mekanikal dan Pembuatan, Universiti Teknikal Malaysia Melaka, Hang Tuah Jaya, Durian Tunggal 76100, Melaka, Malaysia
3. Department of Mathematics, Babeş-Bolyai University, R-400084 Cluj-Napoca, Romania; popm.ioan@yahoo.co.uk
* Correspondence: kohi@ukm.edu.my

Received: 14 April 2020; Accepted: 3 May 2020; Published: 13 May 2020

Abstract: The hybrid nanofluid under the influence of magnetohydrodynamics (MHD) is a new interest in the industrial sector due to its applications, such as in solar water heating and scraped surface heat exchangers. Thus, the present study accentuates the analysis of an unsteady three-dimensional MHD non-axisymmetric Homann stagnation point flow of a hybrid Al_2O_3-Cu/H_2O nanofluid with stability analysis. By employing suitable similarity transformations, the governing mathematical model in the form of the partial differential equations are simplified into a system of ordinary differential equations. The simplified mathematical model is then solved numerically by the Matlab solver bvp4c function. This solving approach was proficient in generating more than one solution when good initial guesses were provided. The numerical results presented significant influences on the rate of heat transfer and fluid flow characteristics of a hybrid nanofluid. The rate of heat transfer and the trend of the skin friction coefficient improve with the increment of the nanoparticles' concentration and the magnetic parameter; however, they deteriorate when the unsteadiness parameter increases. In contrast, the ratio of the escalation of the ambient fluid strain rate to the plate was able to adjourn the boundary layer separation. The dual solutions (first and second solutions) are obtainable when the surface of the sheet shrunk. A stability analysis is carried out to justify the stability of the dual solutions, and hence the first solution is seen as physically reliable and stable, while the second solution is unstable.

Keywords: unsteady flow; non-axisymmetric flow; MHD; hybrid nanofluid; stagnation-point flow

1. Introduction

The stagnation point flow has attracted vast attention from many researchers because of its broad applications in both industrial and scientific applications. Some of the real-world applications of the stagnation point flow lie in the polymer industry, extrusion processes, plane counter-jets, and numerous forms of hydrodynamic modelling in engineering uses ([1–3]). An exact solution of the steady two-dimensional stagnation-point flow towards a solid surface in moving fluid was first discovered by Hiemenz [4] in 1911. In his particular study, the Navier–Stokes equations are reduced to non-linear ordinary differential equations by using a similarity transformation. The remarkable work done by Hiemenz [4] was extended by Homann [5], who started the classical work of three-dimensional stagnation point flow for the axisymmetric case. Meanwhile, the flow in the neighbourhood of a particular stagnation point on a surface was explored by Howarth [6], focusing on the non-axisymmetric three-dimensional flow near the stagnation region. Fast forward to 1961, the work of Howarth [6] was criticized by Davey [7], who indicated a mistake in Howarth's paper exposing that the results in the region $-1 \leq c \leq 0$ are unable to be achieved from those discovered for $0 \leq c \leq 1$ as reported in the study. In conjunction with these findings, Davey and Schofield [8] initiated the study of these saddle point solutions and justified the existence of the non-uniqueness solution. In another study, Weidman [9] modified the Homann's axisymmetric outer potential stagnation-point flow for non-axisymmetric stagnation flow of the strain rate. The study revealed a new clan of asymmetric viscous stagnation point flows liable on the shear rate ratio, $\gamma = b/a$ where $-\infty \leq \gamma \leq \infty$, a is the strain rate and b is the shear rate. An analysis of unsteady heat transmission in non-axisymmetric Homann stagnation-point flows of a viscous fluid over a rigid plate was investigated by Mahapatra and Sidui [10], and recently, an investigation on the non-axisymmetric Homann stagnation-point flows of a viscoelastic fluid towards a fixed plate was conducted by Mahapatra and Sidui [11].

Ever since the evolution study of the stagnation point flow with the presence of dual solutions by Davey [7], various works concerning the stagnation point flow towards a shrinking sheet were introduced. Wang [12] considered two-dimensional stagnation point flow on a two-dimensional shrinking sheet and axisymmetric stagnation point flow on an axisymmetric shrinking sheet, while Mahapatra and Sidui [13] assessed unsteady heat transfer in non-axisymmetric Homann stagnation-point flow towards a stretching/shrinking sheet with stability analysis. The continuous effort was carried out by Khashi'ie et al. [14] who examined the three-dimensional non-axisymmetric Homann stagnation point flow and heat transfer past a stretching/shrinking sheet using hybrid nanofluid. Meanwhile, Zaimi and Ishak [15] scrutinized the slip effects on the stagnation point flow towards a stretching vertical sheet. Nevertheless, explorations on the stagnation point flow keep evolving in various ways and have been working still because of its importance in massive engineering applications and also in the magnetohydrodynamics (MHD) flow field. A comprehensive study of the literature on the related works was reviewed by [16–19].

A fluid that is heated by electric energy in the occurrence of a vigorous magnetic field, such as crystal growth in melting, is essential in the industrial sector. The interaction of electrical currents and magnetic fields generates the divergence of Lorentz forces during the movement of fluid. In accordance with this phenomenon, MHD describes the hydrodynamics of a conducting fluid in the presence of a magnetic field. The examinations of MHD flow are very significant due to its massive number of uses implicating the magnetic effect in industrial and engineering areas, such as MHD electricity generators, sterilization tools, magnetic resonance graphs, MHD flow meters, and also in granular insulation (see [20,21]). The goods of the end product depend immensely on the rate of cooling involved in these processes, managed by the application of the magnetic field and electrically conducting fluids. The study of MHD flow in the Newtonian fluid was first carried out by Pavlov [22], who investigated the magnetohydrodynamic flow of an impressible viscous fluid caused by deformation of a surface. Chakrabarti and Gupta [23] broadened the study of hydromagnetic flow and heat transfer over a stretching sheet, followed by Vajravelu [24] who widened the hydromagnetic flow study over a continuous, moving, porous flat surface. Andersson, in 1995, introduced an exact solution of the Navier-Stokes equations for magnetohydrodynamic flow [25], and Lok et al. [26] analyzed the MHD stagnation-point flow towards a shrinking sheet using the Keller-box method and proved the existence of multiple (dual) solutions for small values of the magnetic field parameter for the shrinking case. Recently, Almutairi et al. [27] studied the influence of second-order velocity slip on the MHD flow of a nanofluid in a porous medium by considering the homogeneous-heterogeneous reactions. On the other hand, the impact of nonlinear and temperature jump on non-Newtonian MHD nanofluid flow and heat transfer past a stretched thin sheet was examined by Zhu et al. [28]

Over the last few decades, the researcher has experienced tremendous scholarly devotion to the study of heat transfer fluid. Recent demand for a high-efficiency refrigeration system and the ineffectiveness of traditional thermal conduction fluids encouraged analysts to discover another heat transfer fluid. Choi and Eastman [29] launched the exploration of nanofluids and illustrated the presence of suspended nanoparticles in a carrier fluid. This pioneering study led to the verdict of the colloidal suspension of intensely small-sized particles, for instance, carbon nanotubes, metals, oxides, and carbides, into the based fluid, which may ensure access to an advanced course of nanotechnology-based heat transfer media (see [30,31]). The eccentric features of nanofluids have gained great acknowledgement in various engineering, medical, and industrial applications like engine cooling, diesel generator efficiency, micro-manufacturing, solar water heating, cancer treatment, nuclear reactors, and diverse types of heat exchangers ([32–34]). Due to the massive potential for the applications of nanofluids, Choi and Eastmen [29] developed a mathematical model of nanofluids, which allowed Buongiorno [31] to contribute to heat transfer analysis in nanofluids by introducing the non-homogeneous model for transport and heat transfer phenomena in nanofluids with turbulence applications.

Recently, an expansion of new engineered nanofluids was achieved by dispersing composite nanopowder or dissimilar nanoparticles with sizes between 1 and 10nm in the base fluid [29]; it is known as a hybrid nanofluid. The hybrid nanofluid is a modern technology fluid that may offer better heat transfer performance and thermal physical properties. The progress related to the preparation methods of hybrid nanofluids, thermo-physical properties of hybrid nanofluids, and current applications of hybrid nanofluids was published by Sarkar et al. [35] and Sidik et al. [36]. In another study, Huminic and Huminic [37] highlighted the essential applications of hybrid nanofluids, such as in heat pipes, mini-channel heat sinks, plate heat exchangers, air conditioning systems, tubular heat exchangers, shell and tube heat exchangers, tube in a tube heat exchangers, and coiled heat exchangers. Turcu et al. [38], for the first time testified the hybrid nanocomposite particle synthesis, which consisted of two different hybrids, polypyrrole-carbon nanotube (PPY-CNT) nanocomposite and multi-walled carbon nanotube (MWCNT) on magnetic Fe_3O_4 nanoparticles. In the following year, Yen et al. [39] inspected the effect of hybrid nanofluids in channel flow numerically. Devi and Devi [40] analysed the problem of hydromagnetic hybrid nanofluid (Cu-Al_2O_3/water) flow on a permeable stretching sheet subject to Newtonian heating, and they continued the investigation to improve the heat transfer in hybrid nanofluid flow past a stretching sheet [41]. Subsequently, Yousefi et al. [42] reviewed on the stagnation point flow of an aqueous titania-copper hybrid nanofluid toward a wavy cylinder. At the same time, Khashi'ie et al. [43] performed a numerical study on the heat transfer and boundary layer flow of axisymmetric hybrid nanofluids driven by a stretching/shrinking disc. A detailed documentary on the numerical study of hybrid nanofluid flow and heat transfer is reviewed by [44–48].

Many practical situations, such as a sudden stretching of the plate or temperature change of the plate, involved unsteady conditions of the heat transfer flow. Cai et al. [49] explained that the flow in the viscous boundary layer near the plate would slowly be enlarged if the surface was extended unexpectedly, and hence converted into a steady flow after a certain interval. Technically, we believe that the consideration of physical quantities related to time is crucial in mathematical modeling and analysis, which is acknowledged in the formulation of this research problem.

Motivated by the work by Mahapatra and Sidui [13], this study aims to inspect the unsteady MHD non-axisymmetric Homann stagnation point of a hybrid nanofluid in three-dimensional flow. The proposed hybrid nanofluid model is adapted from Devi and Devi [40] and Hayat and Nadeem [50], recognized by suspending varied nanoparticles, namely alumina (Al_2O_3) and copper (Cu), in the base fluid (water). To the best of the authors' knowledge, no attempt has been made to examine the heat transfer and fluid flow of the hybrid nanofluid (Al_2O_3-Cu/H_2O) considering the unsteady parameter in non-axisymmetric Homann stagnation point flow. This possibly will benefit future works on choosing a significant parameter to enhance the heat transfer performance in the modern industry. The novelty of this study can also be seen in the discovery of dual solutions and the execution of stability analysis. Ultimately, this research is highly claimed to be authentic and original.

2. Mathematical Model

Consider the unsteady three-dimensional MHD non-axisymmetric Homann stagnation point flow of a hybrid Al_2O_3-Cu/water nanofluid with a stretching/shrinking sheet on the $x, y-$ plane where x, y and z are Cartesian coordinates with the $z-$ axis measured in the horizontal direction and the axes x and y are in the plane $z = 0$ as illustrated in Figure 1, respectively. We assume that the constant surface temperature T_w is stretched and shrunk in the x and y directions by the velocities $u_w = \frac{\varepsilon c x}{1+at}$ and $v_w = \frac{\varepsilon c y}{1+at}$. The uniform temperature is given by T_∞ and B_0 is introduced to the stretching/shrinking sheet in an orthogonal direction as a transverse uniform magnetic field. Meanwhile, the modified non-asymmetrically free streamflow along the x, y and z axes is described by ([9]):

$$u_e(x) = (a+b)x, \quad v_e(y) = (a-b)y, \quad w_e(z) = -2az. \tag{1}$$

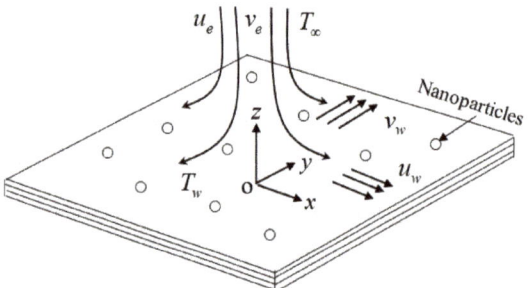

Figure 1. Flow model and coordinate systems of the physical model.

Here, a is the strain rate and b is the shear rate of stagnation point flow, correspondingly. By adapting the Tiwari and Das [30] nanofluid model, the continuity, momentum, and the energy equations of the hybrid nanofluid can be written as follows:

$$\frac{\partial u}{\partial x} + \frac{\partial v}{\partial y} + \frac{\partial w}{\partial z} = 0, \tag{2}$$

$$\frac{\partial u}{\partial t} + u\frac{\partial u}{\partial x} + v\frac{\partial u}{\partial y} + w\frac{\partial u}{\partial z} = \frac{\partial u_e}{\partial t} + u_e\frac{\partial u_e}{\partial x} + \frac{\mu_{hnf}}{\rho_{hnf}}\frac{\partial^2 u}{\partial z^2} - \frac{\sigma_{hnf}}{\rho_{hnf}}B^2(u - u_e), \tag{3}$$

$$\frac{\partial v}{\partial t} + u\frac{\partial v}{\partial x} + v\frac{\partial v}{\partial y} + w\frac{\partial v}{\partial z} = \frac{\partial v_e}{\partial t} + v_e\frac{\partial v_e}{\partial x} + \frac{\mu_{hnf}}{\rho_{hnf}}\frac{\partial^2 v}{\partial z^2} - \frac{\sigma_{hnf}}{\rho_{hnf}}B^2(v - v_e), \tag{4}$$

$$\frac{\partial T}{\partial t} + u\frac{\partial T}{\partial x} + v\frac{\partial T}{\partial y} + w\frac{\partial T}{\partial z} = \frac{k_{hnf}}{(\rho C_p)_{hnf}}\frac{\partial^2 T}{\partial z^2}. \tag{5}$$

The velocity component in $x-$ direction is given by u, while v is in $y-$ direction. Next, the boundary conditions are:

$$\begin{aligned}&t < 0: u = v = w = 0 \text{ for any } x, y, z,\\&t \geq 0: u = u_w, v = v_w, w = 0, T = T_w \text{ at } z = 0,\\&u \to u_e = \frac{(a+b)x}{1+at}, v \to v_e = \frac{(a-b)x}{1+at}, w_e \to \frac{-2az}{1+at}, T \to T_\infty \text{ as } z \to \infty.\end{aligned} \tag{6}$$

Note that T is the hybrid nanofluid temperature, μ_{hnf} is the dynamic viscosity of the hybrid nanofluid, ρ_{hnf} is the density of the hybrid nanofluid, k_{hnf} is the thermal conductivity of the hybrid nanofluid, $(\rho C_p)_{hnf}$ is the heat capacity of the hybrid nanofluid, σ_{hnf} is the electrical conductivity of the hybrid nanofluid, and the time-dependent of a transverse magnetic field is given by $B^2 = B_0^2/(1 + \alpha t)$ in detail.

The hybrid nanofluids thermophysical properties are specified in Table 1, as demonstrated by Devi and Devi [41,51]. At this point, ϕ is the volume fraction of nanoparticles, ρ_f and ρ_s are the base fluid density and hybrid nanoparticles, C_p is the heat capacity, $(\rho C_p)_f$ and $(\rho C_p)_s$ represent capacitance heating of the base fluid and hybrid nanoparticles, and finally k_f and k_s are the thermal conductivities of the base fluid and hybrid nanoparticles, respectively. Meanwhile, the thermophysical properties of the fluid and nanoparticles for aluminum oxide, copper, and the base fluid (water) are given in Table 2.

Table 1. Thermophysical properties of hybrid nanofluids (Devi and Devi [41,51]).

Properties	Hybrid Nanofluid
Density	$\rho_{hnf} = (1-\phi_2)[(1-\phi_1)\rho_f + \phi_1\rho_{s1}] + \phi_2\rho_{s2}$
Heat capacity	$(\rho C_p)_{hnf} =$ $(1-\phi_2)\left[(1-\phi_1)(\rho C_p)_f + \phi_1(\rho C_p)_{s1}\right] + \phi_2(\rho C_p)_{s2}$
Dynamic viscosity	$\mu_{hnf} = \dfrac{\mu_f}{(1-\phi_1)^{2.5}(1-\phi_2)^{2.5}}$
Thermal conductivity	$\dfrac{k_{hnf}}{k_{nf}} = \dfrac{k_{s2}+2k_{nf}-2\phi_2(k_{nf}-k_{s2})}{k_{s2}+2k_{nf}+\phi_2(k_{nf}-k_{s2})}$, where, $\dfrac{k_{nf}}{k_f} = \dfrac{k_{s1}+2k_f-2\phi_1(k_f-k_{s1})}{k_{s1}+2k_f+\phi_1(k_f-k_{s1})}$

Table 2. Thermophysical properties of nanoparticles and base fluid (Oztop and Abu-Nada [52]).

Properties	Al$_2$O$_3$	Cu	H$_2$O
ρ (kg/m^3)	3970	8933	997.1
C_p (J/kgK)	765	385	4179
k (W/mK)	40	400	0.613
$\beta \times 10^{-5}$ (mK)	0.85	1.67	21

Now, pursuing Mahapatra and Sidui [10], the resulting similarity transformation is proposed to achieve the similarity solutions:

$$u = \dfrac{cxf'(\eta)}{1+at}, v = \dfrac{cyg'(\eta)}{1+at}, w = -\sqrt{\dfrac{vc}{1+at}}(f+g), \theta(\eta) = \dfrac{(T-T_\infty)}{(T_w-T_\infty)}, \quad (7)$$
$$\eta = \sqrt{\dfrac{c}{v(1+at)}}z,$$

where the prime denotes differentiation with respect to η. By substituting (7) into the steady-state Equations (2)–(5), the following ordinary differential equations are obtained:

$$\dfrac{\mu_{hnf}/\mu_f}{\rho_{hnf}/\rho_f}f''' + A\left(\tfrac{1}{2}\eta + f + g\right)f'' + Af' - f'^2 - A(\lambda+\gamma) + (\lambda+\gamma)^2 \\ -M(f'-\lambda-\gamma) = 0, \quad (8)$$

$$\dfrac{\mu_{hnf}/\mu_f}{\rho_{hnf}/\rho_f}g''' + A\left(\tfrac{1}{2}\eta + f + g\right)g'' + Ag' - g'^2 - A(\lambda-\gamma) + (\lambda-\gamma)^2 \\ -M(g'-\lambda+\gamma) = 0, \quad (9)$$

$$\dfrac{1}{\Pr}\dfrac{k_{hnf}/k_f}{\rho C_{phnf}/\rho C_{pf}}\theta'' + \left(A\tfrac{1}{2}\eta + f + g\right)\theta' = 0, \quad (10)$$

with the boundary conditions (6) which are converted to:

$$f(0) = g(0) = 0, f'(0) = g'(0) = \varepsilon, \theta(0) = 1, \\ f'(\eta) \to \lambda + \gamma, g'(\eta) \to \lambda - \gamma, \theta(\eta) \to 0, \text{ as } \eta \to \infty. \quad (11)$$

In the equations mentioned above, $A = a/c$ represents the unsteadiness parameter, $\lambda = a/c$ is a ratio of the surrounding fluid strain rate to the surface strain rate, $\gamma = b/c$ is the surrounding fluid shear rate ratio to the strain rate of the sheet, $M = \frac{\sigma_{hnf}/\sigma_f}{\rho_{hnf}/\rho_f} \frac{B_0^2}{c}$ denotes the magnetic parameter and $\Pr = \nu_f/\alpha_f$ indicates the Prandtl number. The parameter of stretching/shrinking is meant by ε where $\varepsilon > 0$ determines the stretching sheet, while $\varepsilon < 0$ reflects the shrinking sheet. The related quantities of interest in this study are the skin friction coefficient, C_{fx} and C_{fy} along the $x-$ and $y-$ directions and the local Nusselt number Nu_x, which is specified as

$$C_{fx} = \frac{\tau_{wx}}{\rho_f u_e^2}, \quad C_{fy} = \frac{\tau_{wy}}{\rho_f v_e^2}, \quad Nu_x = \frac{xq_w}{k_f(T_w - T_\infty)}, \tag{12}$$

where τ_{wx}, τ_{wy} are the shear stresses along the $x-, y-$ axes and q_w represents the heat flux, correspondingly. Such terms can be defined by

$$\tau_{wx} = \mu_{hnf}\left(\frac{\partial u}{\partial z}\right)_{z=0}, \quad \tau_{wy} = \mu_{hnf}\left(\frac{\partial v}{\partial z}\right)_{z=0}, \quad q_w = -k_{hnf}\left(\frac{\partial T}{\partial z}\right)_{z=0}. \tag{13}$$

By prompting Equations (7), (12) and (13), we get:

$$(\lambda+\gamma)^{3/2}Re_x^{1/2}C_{fx} = \frac{\mu_{hnf}}{\mu_f}f''(0), (\lambda-\gamma)^{3/2}Re_y^{1/2}C_{fy} = \frac{\mu_{hnf}}{\mu_f}g''(0),$$
$$(\lambda+\gamma)^{1/2}Re_x^{-1/2}Nu_x = -\frac{k_{hnf}}{k_f}\theta'(0), \tag{14}$$

where $Re_x = \frac{(a+b)x^2}{(1+at)\nu_f}$ and $Re_y = \frac{(a+b)y^2}{(1+at)\nu_f}$ are the local Reynolds number along the $x-$ and $y-$ directions, respectively.

3. Stability Analysis

The system of Equations (8)–(10) along with the boundary conditions (11), is capable of generating more than one solution and ultimately permits the requirement analysis of the flow to identify the reliable and feasible solution. Going through the outstanding work done by Merkin [53] and Merrill et al. [54] in the stability analysis, the application of an unstable form of the boundary layer problem is analyzed by using the time variable and a dimensionless time variable denoted by τ [55]. Next, we consider the following new similarity variables:

$$u = \frac{cx}{1+at}\frac{\partial f}{\partial \eta}(\eta,\tau), \quad v = \frac{cy}{1+at}\frac{\partial g}{\partial \eta}(\eta,\tau), \quad w = -\sqrt{\frac{\nu c}{1+at}}[f(\eta,\tau) + g(\eta,\tau)],$$
$$\theta(\eta) = \frac{(T-T_\infty)}{(T_w-T_\infty)}, \quad \eta = \sqrt{\frac{c}{\nu(1+at)}}z, \quad \tau = \frac{ct}{1+at}. \tag{15}$$

Using the similarity variables of Equation (15) into Equations (8)–(10), the altered differential equations and the boundary conditions are as follows:

$$\frac{\mu_{hnf}/\mu_f}{\rho_{hnf}/\rho_f}\frac{\partial^3 f}{\partial \eta^3} + \left(A\frac{1}{2}\eta + f + g\right)\frac{\partial^2 f}{\partial \eta^2} + A\frac{\partial f}{\partial \eta} - \left(\frac{\partial f}{\partial \eta}\right)^2 - A(\lambda+\gamma) + (\lambda+\gamma)^2$$
$$-M\left(\frac{\partial f}{\partial \eta} - \lambda - \gamma\right) - (1-A\tau)\frac{\partial^2 f}{\partial \eta \partial \tau} = 0, \tag{16}$$

$$\frac{\mu_{hnf}/\mu_f}{\rho_{hnf}/\rho_f}\frac{\partial^3 g}{\partial \eta^3} + \left(A\frac{1}{2}\eta + f + g\right)\frac{\partial^2 g}{\partial \eta^2} + A\frac{\partial g}{\partial \eta} - \left(\frac{\partial g}{\partial \eta}\right)^2 - A(\lambda-\gamma) + (\lambda-\gamma)^2$$
$$-M\left(\frac{\partial g}{\partial \eta} - \lambda + \gamma\right) - (1-A\tau)\frac{\partial^2 g}{\partial \eta \partial \tau} = 0, \tag{17}$$

$$\frac{1}{\Pr}\frac{k_{hnf}/k_f}{\rho C_{phnf}/\rho C_{pf}}\frac{\partial^2 \theta}{\partial \eta^2} + \left(A\left(\frac{1}{2}\eta - \tau\right) + f + g + 1\right)\frac{\partial \theta}{\partial \eta} - (1-A\tau)\frac{\partial \theta}{\partial \tau} = 0, \tag{18}$$

262

$$f(0,\tau) = g(0,\tau) = 0, \frac{\partial f}{\partial \eta}(0,\tau) = \frac{\partial g}{\partial \eta}(0,\tau) = \varepsilon, \theta(0,\tau) = 1,$$
$$\frac{\partial f}{\partial \eta}(\eta,\tau) \to \lambda + \gamma, \frac{\partial g}{\partial \eta}(\eta,\tau) \to \lambda - \gamma, \theta(\eta,\tau) \to 0, \text{ as } \eta \to \infty. \quad (19)$$

Weidman et al. [55] highlighted that the stability of solutions is introduced by determining the system's decay or initial growth. This can be achieved via considering the following perturbation expressions of the primary flow $f = f_0(\eta), g = g_0(\eta)$ and $\theta = \theta_0(\eta)$ with the resulting equation:

$$f(\eta,\tau) = f_0(\eta) + e^{-\omega\tau}F(\eta), g(\eta,\tau) = g_0(\eta) + e^{-\omega\tau}G(\eta),$$
$$\theta(\eta,\tau) = \theta_0(\eta) + e^{-\omega\tau}H(\eta), \quad (20)$$

where ω is an unknown parameter of the eigenvalue, $F(\eta), G(\eta)$ and $H(\eta)$ are comparatively slight to $f_0(\eta), g_0(\eta)$ and $\theta_0(\eta)$. Substituting (20) into Equations (16)–(18) we attained the following system of equations:

$$\frac{\mu_{hnf}/\mu_f}{\rho_{hnf}/\rho_f}\frac{\partial^3 F}{\partial \eta^3} + \left(A\frac{1}{2}\eta + f_0 + g_0\right)\frac{\partial^2 F}{\partial \eta^2} + \left(A - M + (1-A\tau)\omega - 2\frac{\partial f_0}{\partial \eta}\right)\frac{\partial F}{\partial \eta}$$
$$+ (F+G)\frac{\partial^2 f_0}{\partial \eta^2} = 0, \quad (21)$$

$$\frac{\mu_{hnf}/\mu_f}{\rho_{hnf}/\rho_f}\frac{\partial^3 G}{\partial \eta^3} + \left(A\frac{1}{2}\eta + f_0 + g_0\right)\frac{\partial^2 G}{\partial \eta^2} + \left(A - M + (1-A\tau)\omega - 2\frac{\partial g_0}{\partial \eta}\right)\frac{\partial G}{\partial \eta}$$
$$+ (F+G)\frac{\partial^2 g_0}{\partial \eta^2} = 0, \quad (22)$$

$$\frac{1}{\Pr}\frac{k_{hnf}/k_f}{(\rho C_p)_{hnf}/(\rho C_p)_f}\frac{\partial^2 H}{\partial \eta^2} + (F+G)\frac{\partial \theta_0}{\partial \eta} + \left(A\frac{1}{2}\eta + f_0 + g_0\right)\frac{\partial H}{\partial \eta} + \omega H = 0, \quad (23)$$

subject to the boundary conditions:

$$F(0) = G(0) = 0, \frac{\partial F}{\partial \eta}(0) = \frac{\partial G}{\partial \eta}(0) = 0, H(0) = 0,$$
$$\frac{\partial F}{\partial \eta}(\infty) \to 0, \frac{\partial G}{\partial \eta}(\infty) \to 0, H(\infty) \to 0. \quad (24)$$

The stability of the steady-state flow and heat transfer solutions $f_0(\eta), g_0(\eta)$ and $\theta_0(\eta)$ was implemented by setting $\tau = 0$ with $F = f_0(\eta), G = g_0(\eta)$ and $H = \theta_0(\eta)$ in (21)–(24). Finally, the initial development or the solution decay of the solution (20) is identified. The value of ω is obtained by solving the following eigenvalue problem:

$$\frac{\mu_{hnf}/\mu_f}{\rho_{hnf}/\rho_f}F_0''' + \left(A\frac{1}{2}\eta + f_0 + g_0\right)F_0'' + \left(A - M + \omega - 2f_0'\right)F' + (F+G)f_0'' = 0, \quad (25)$$

$$\frac{\mu_{hnf}/\mu_f}{\rho_{hnf}/\rho_f}G_0''' + \left(A\frac{1}{2}\eta + f_0 + g_0\right)G_0'' + \left(A - M + \omega - 2g_0'\right)G' + (F+G)g_0'' = 0, \quad (26)$$

$$\frac{1}{\Pr}\frac{k_{hnf}/k_f}{(\rho C_p)_{hnf}/(\rho C_p)_f}H'' + (F+G)\theta_0' + \left(A\frac{1}{2}\eta + f_0 + g_0\right)H' + \omega H = 0, \quad (27)$$

along with the conditions:

$$F_0(0) = G_0(0) = 0, F_0'(0) = G_0'(0) = 0, H_0(0) = 0,$$
$$F_0'(\infty) \to 0, G_0'(\infty) \to 0, H_0(\infty) \to 0. \quad (28)$$

The range of possible eigenvalues can be determined by relaxing a boundary condition on $F'_0(\eta)$ according to the previous research by Harris et al. [56]. In this study, the condition $F'_0(\infty) \to 0$ is relaxed, and the linear eigenvalue problem (25)–(28) is solved together with the new boundary condition $F''_0(0) = 1$ for a fixed value of ω_1. The flow is considered unstable if the smallest eigenvalue ω_1 is negative, which indicates an initial growth of disturbances occurred, while a positive value of the smallest eigenvalue signifies that the flow is physically achievable and stable.

4. Results and Discussion

The bvp4c function in Matlab is adopted to produce the results of the nonlinear system of ordinary differential Equations (8)–(10) together with the boundary conditions (11). The relative error tolerance is set as 10^{-10} to gain the results of the numerical outcomes and stability analysis. Table 3 presents that the average central processing unit (CPU) time required for computing each result in Table 4 is approximately 1.4 s, and the dual solutions were obtained by indicating different initial guesses which must satisfy the far-field boundary conditions (11) asymptotically. The numerical results are validated with the numerical results produced by Mahapatra and Sidui [10] and Nawaz and Hayat [57] by collating the values of the shear stress $f''(0)$ of an axisymmetric ($\gamma = 0$) stagnation point flow of a viscous fluid with the exclusion of magnetic field and the unsteady parameter, which is depicted in Table 4. It demonstrates excellent agreement with the previous literature; hence, the practicality and effectiveness of the bvp4c method are verified. The estimated relative error, ε_r is also measured, and it shows that the calculated values of ε_r are relatively small between the present and previous results. Figure 2 exhibits the variation of the wall shear stress parameter $f''(0)$ and $g''(0)$ towards the difference value of γ when $\phi_1 = \phi_2 = M = A = 0, \lambda = 0.1, \varepsilon = 1.0$ and $\Pr = 1.0$ where the dotted lines correspond to the asymptotic behaviour of $f''(0)$ and $g''(0)$ which is in excellent agreement with Mahapatra and Sidui [10] who pursued a standard fourth-order Runge-Kutta integration technique in their study. This justifies the role of the bvp4c numerical technique as a dependable practice, and the present results are valid and correct.

Table 3. Computational time to generate the values of $f''(0)$ as λ varies when $\phi_1 = \phi_2 = 0$, $\gamma = 0, \varepsilon = 1, M = 0, A = 0$, and $\Pr = 6.2$.

λ	$f''(0)$	Time, t (s)
0.1	−1.12460540	0.851
0.2	−1.05562203	0.960
0.5	−0.75344581	0.877
1.0	0	1.474
2.0	2.20708771	1.419

The dimensionless velocity profiles $f'(\eta)$ and $g'(\eta)$ for different values of λ are illustrated in Figures 3 and 4. The figures prove that the profiles of the velocity comply with the far-field boundary conditions of Equation (11) asymptotically. The maximum value of the velocity gradient with the lowest thickness of the momentum boundary layer is preserved for the largest value of λ. Notably, the distance of two adjacent profiles increases remarkably with the increased amount of λ in both the first and second solutions. Figures 5 and 6 portray the variations of $f''(0)$ and $g''(0)$ towards ε for different values of λ in hybrid nanofluids with the existence of the magnetic field and the influences of the unsteadiness parameter. An enrichment in the amounts of λ embarking on the augmentation of both $\lambda + \gamma$ and $\lambda - \gamma$ provide a significant effect to the surface shear stresses. The variations of $f''(0)$ are expected to be higher by the increasing value of $\lambda + \gamma$ yet decrease when $\lambda + \gamma$ is decreased, and the same trend is expected in $g''(0)$ with $\lambda - \gamma$. On the other hand, the effect of the nanoparticles volume fraction is observed in Figures 7–9, respectively. Surprisingly, the critical values of the various usage of the nanoparticles' volume fraction give no significant effect to the trend of the nanofluid ($\phi_1 = 0.02, \phi_2 = 0.00$) and hybrid nanofluid flow ($\phi_1 = 0.02, \phi_2 = 0.002, 0.04$). It is observed that the reduced skin friction coefficient in x and y directions, as presented in Figures 7 and 8, and the reduced local Nusselt number in Figure 9 upsurges with the rising in the nanoparticles' volume fraction. This is due to the fact that more kinetic energy is produced with a higher concentration of nanoparticles and thus, enhances the heat transfer of the fluid particles. This finding also corresponds with several present particle-laden direct numerical studies (DNS), which reveal that the roughness components appear to redistribute the energy and, therefore, decrease the overall large-scale near-wall anisotropy of the flow pattern (see Yuan and Piomelli [58] and Ghodke and Apte [59]).

Table 4. Results of $f''(0)$ for specific values of λ when $\phi_1 = 0.0$, $\phi_2 = 0.0$, $\gamma = 0$, $\varepsilon = 1$, $M = 0$, $A = 0$, and $Pr = 6.2$.

λ	Present Result	Mahapatra and Sidui [10]	$\varepsilon_r = \lvert\frac{r-s}{s}\rvert \times 100\%$	Nawaz and Hayat [57]	$\varepsilon_r = \lvert\frac{r-s}{s}\rvert \times 100\%$
0.1	−1.12460540	−1.124000	0.053832%	−1.124600	0.000480%
0.2	−1.05562203	−1.054400	0.115764%	−1.055610	0.001139%
0.5	−0.75344581	−0.753400	0.006080%	−0.753100	0.045897%
1.0	0.00000000	0.000000	0.000000%	0.000000	0.000000%
2.0	2.20708771	2.190200	0.765158%	−	−

*ε_r is the estimate percentage relative error between the present result, r and previous result, s.

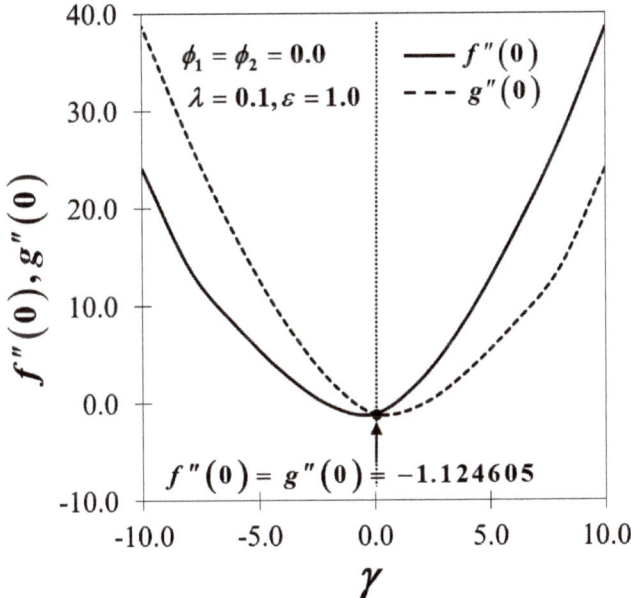

Figure 2. Variations of $f''(0)$ and $g''(0)$ for different values of γ.

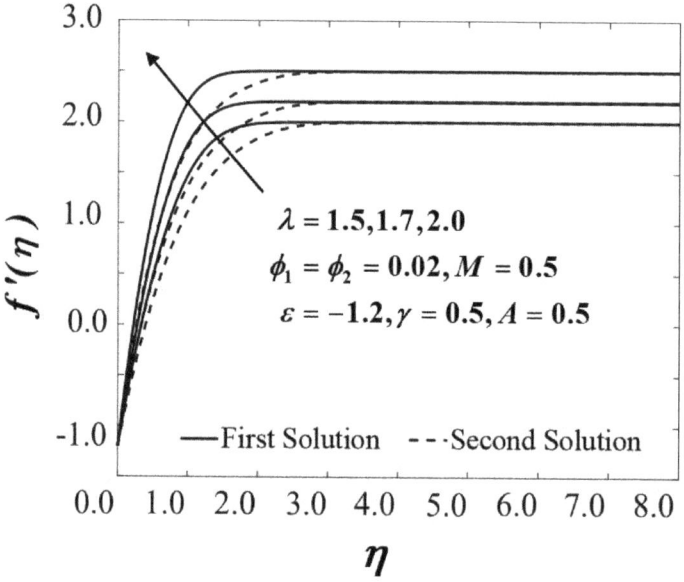

Figure 3. Velocity profiles of $f'(\eta)$ for different values of λ.

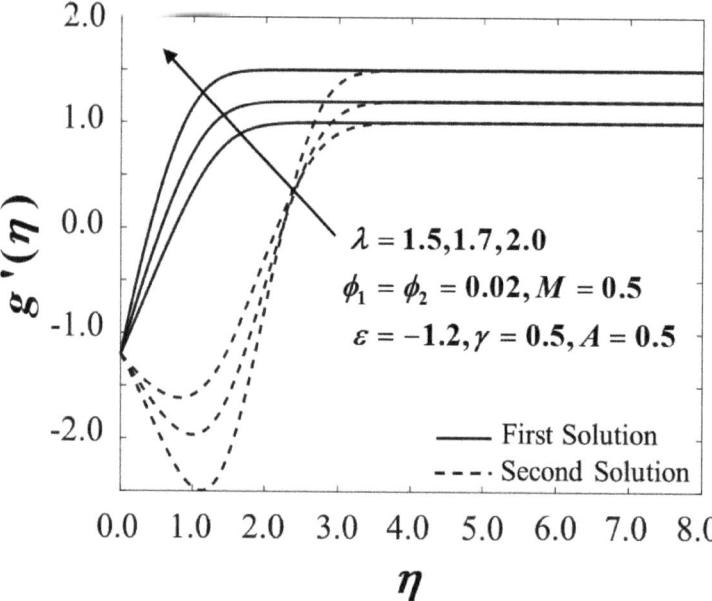

Figure 4. Velocity profiles of $g'(\eta)$ for different values of λ.

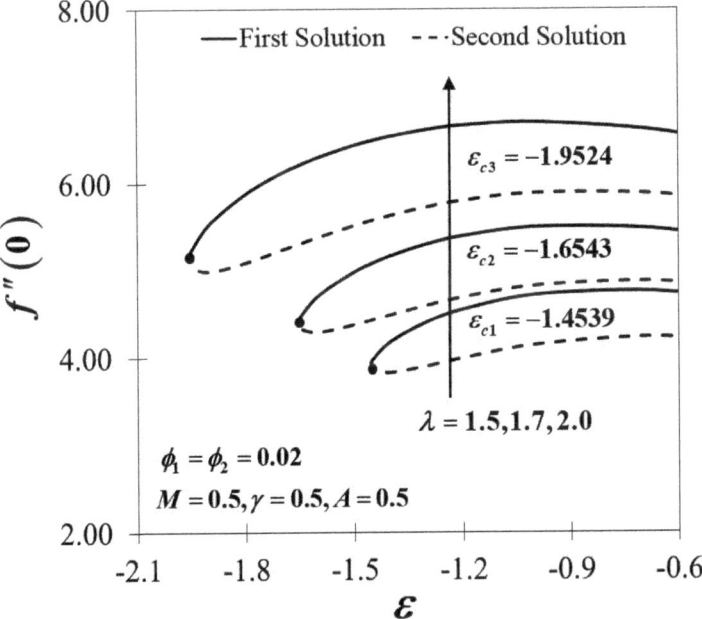

Figure 5. Variations of $f''(0)$ for different values of λ.

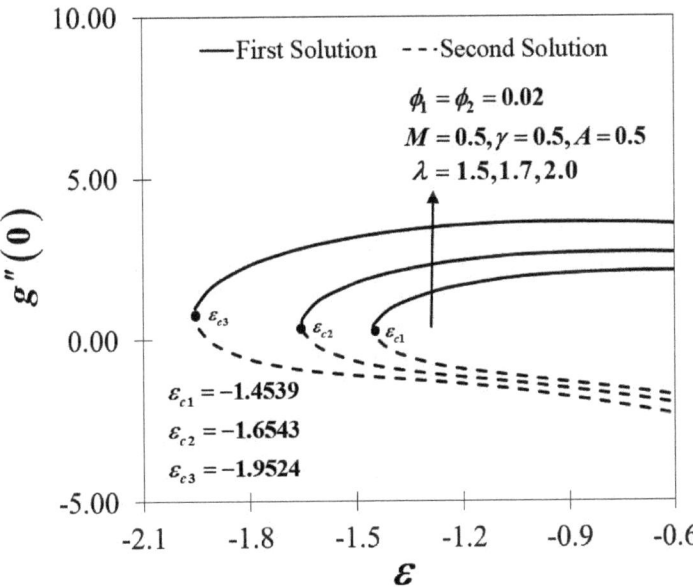

Figure 6. Variations of $g''(0)$ for different values of λ.

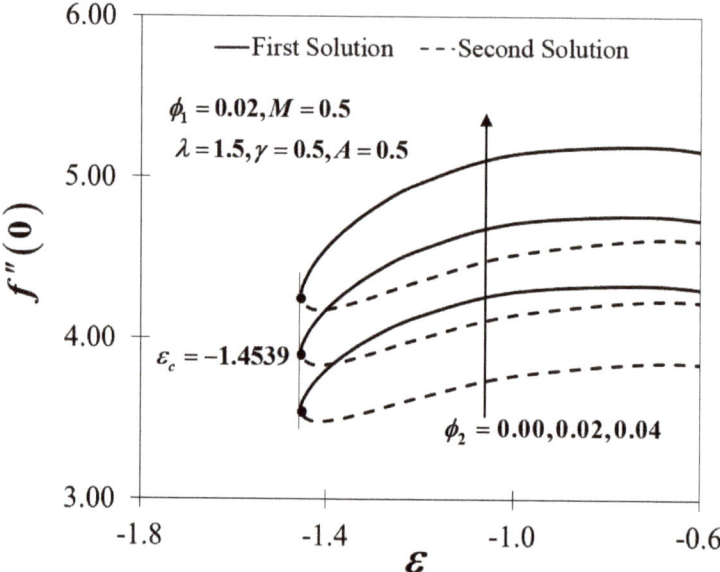

Figure 7. Variations of $f''(0)$ for different values of ϕ_2.

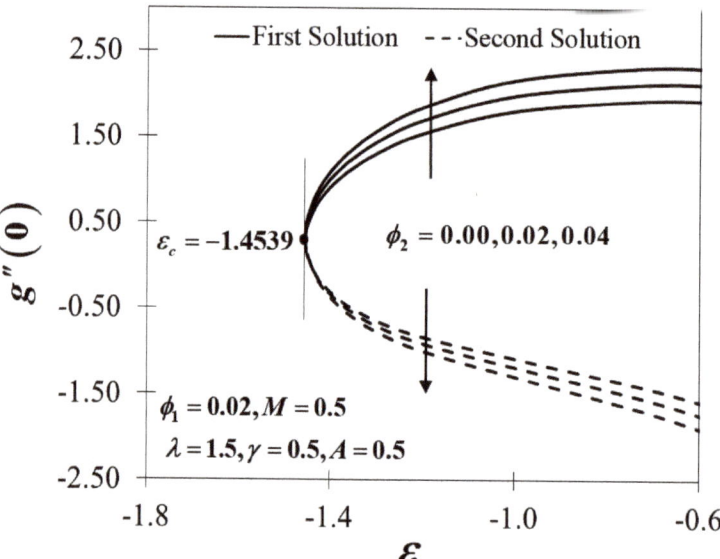

Figure 8. Variations of $g''(0)$ for different values of ϕ_2.

Figures 10–12 exhibit the influence of the magnetic parameter on $f''(0)$, $g''(0)$ and $-\theta'(0)$ which shows a prominent effect on the fluid flow of the shrinking sheet. The reduced skin friction coefficient in both the x- and y- directions rise and eventually increase the value of $-\theta'(0)$ with the escalation of M due to the occurrence of the Lorentz force, which acts to retard the fluid flow. The Lorentz force creates resistance to the motion of the fluid particles and then consequently reduces the fluid velocity. The synchronism of the magnetic and electric field that occurred from the formation of the Lorentz force tends to slow down the fluid movement. The boundary layer becomes thinner in both directions as proven in Figures 13 and 14 as M increases due to the delayed flow, hence contributing to the increment of $f''(0)$ and $g''(0)$. On the other hand, Figure 15 advertises the variations of the non-dimensional temperature profiles for different values of M which tend to decrease in the first solution but show a reverse trend in the second solution. The figures clearly reveal that the thicknesses of the thermal boundary layers decrease with the increase of M Practically, by restricting the magnetic field intensity, the progression of the thermal boundary layers' thicknesses can be managed and, thus, be able to reduce the temperature profile distributions.

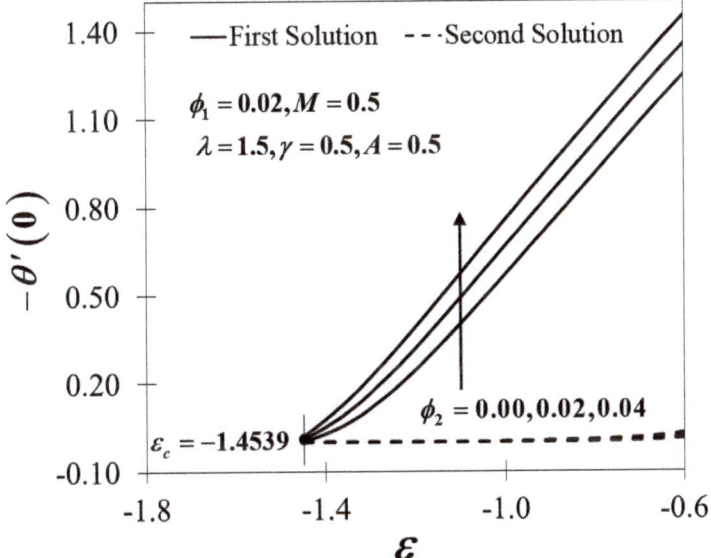

Figure 9. Variations of $-\theta'(0)$ for different values of ϕ_2.

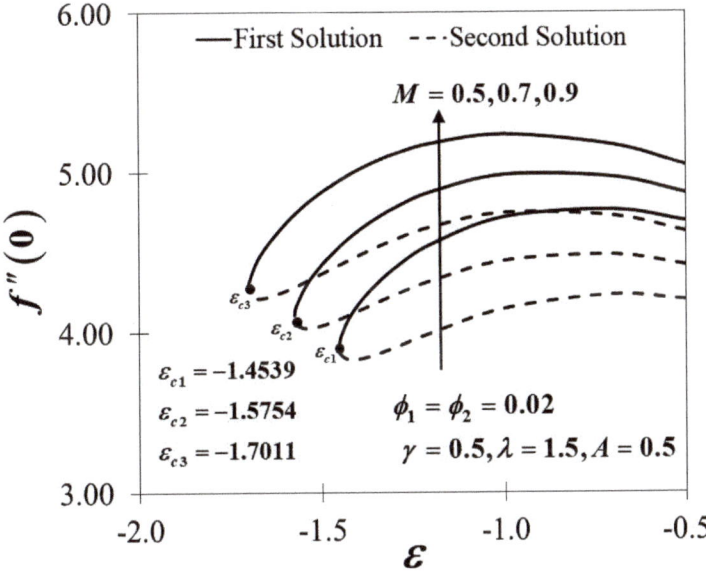

Figure 10. Variations of $f''(0)$ for different values of M.

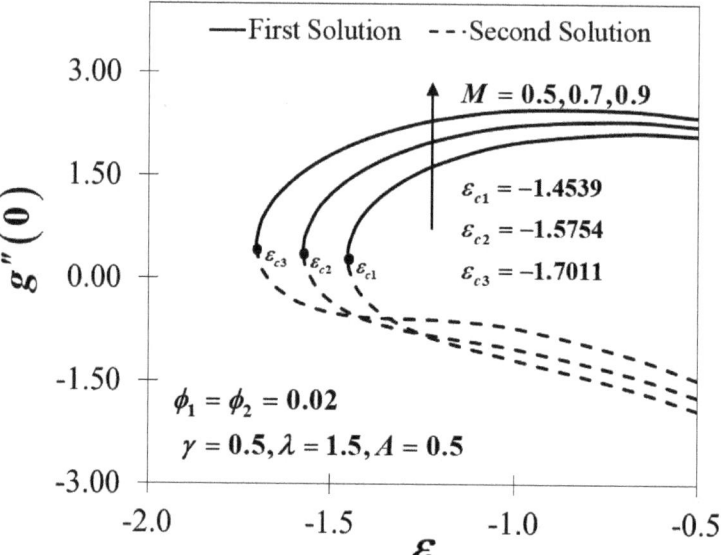

Figure 11. Variations of $g''(0)$ for different values of M.

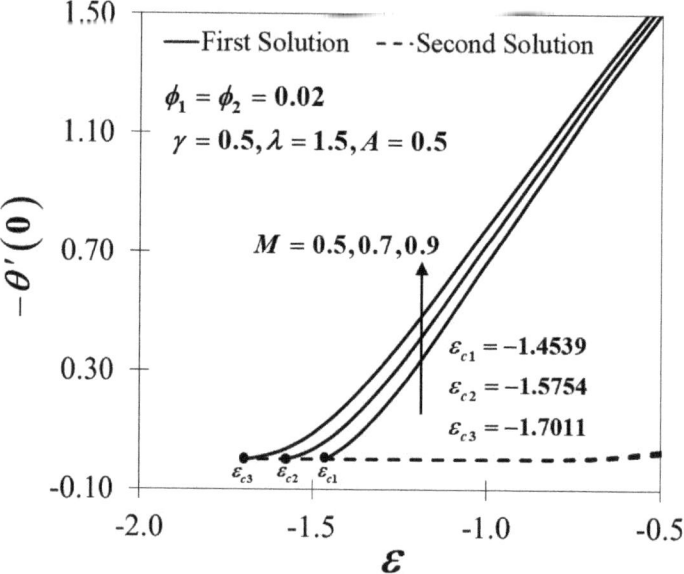

Figure 12. Variations of $-\theta'(0)$ for different values of M.

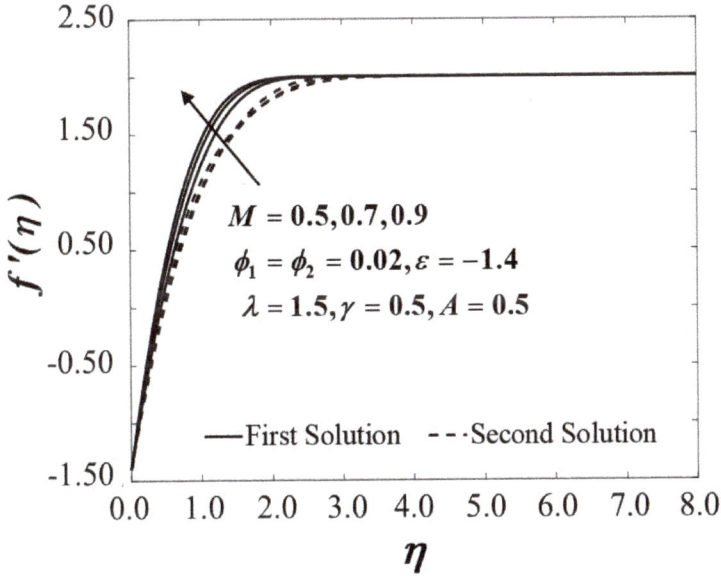

Figure 13. Velocity profiles of $f'(\eta)$ for different values of M.

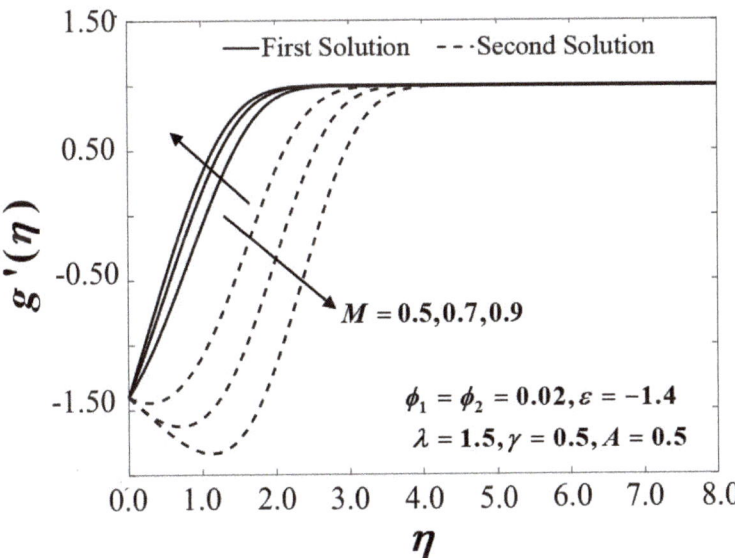

Figure 14. Velocity profiles of $g'(\eta)$ for different values of M.

The effect of the unsteadiness parameter A on $f''(0)$, $g''(0)$ and $-\theta'(0)$ is depicted in Figures 16–18, respectively. Increasing the values of A results in a reduction of the skin friction coefficients in both directions, as illustrated in Figures 16 and 17, which consequently decreases the reduced local Nusselt number. The fact that the unsteadiness parameter affects the velocity and temperature profile is proven in Figures 19–21. The velocity of the fluid is in decline along the surface of the sheet due to the increasing value of the shear stress and subsequently shrinks the thickness of the momentum boundary layer nearby the wall, as displayed in Figures 19 and 20, accordingly. The increasing value of A decreases in the temperature profile of the hybrid nanofluid, as shown in Figure 21, which is understandable because the spaces between the molecules become higher in unsteady flow and proportionately decrease the temperature profile and improve the cooling rates of the fluid. Therefore, the unsteadiness parameter should be highlighted and well considered for practical purposes.

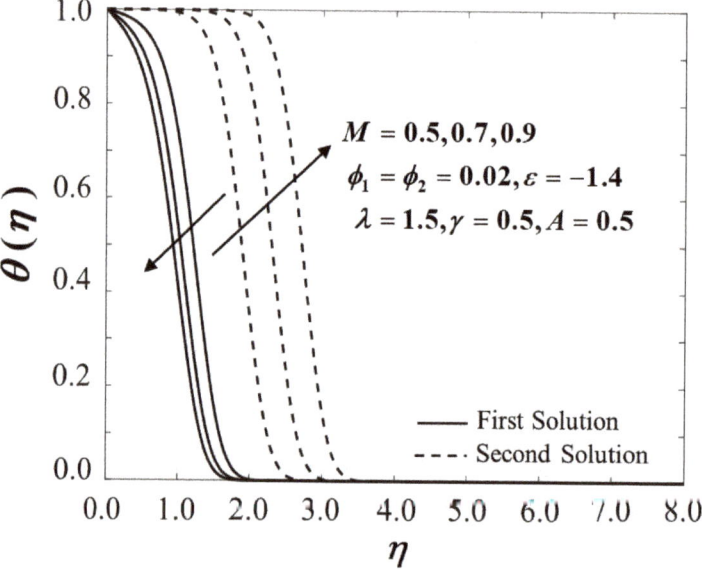

Figure 15. Temperature profiles of $\theta(\eta)$ for different values of M.

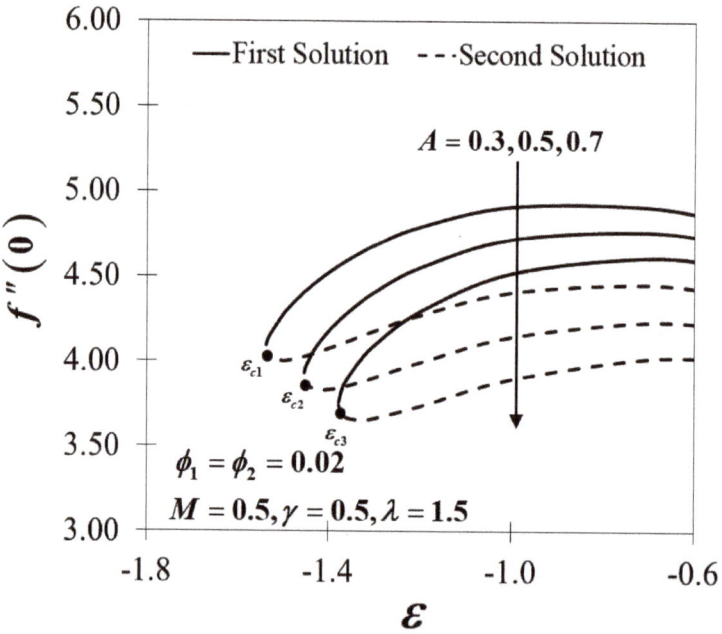

Figure 16. Variations of $f''(0)$ for different values of A.

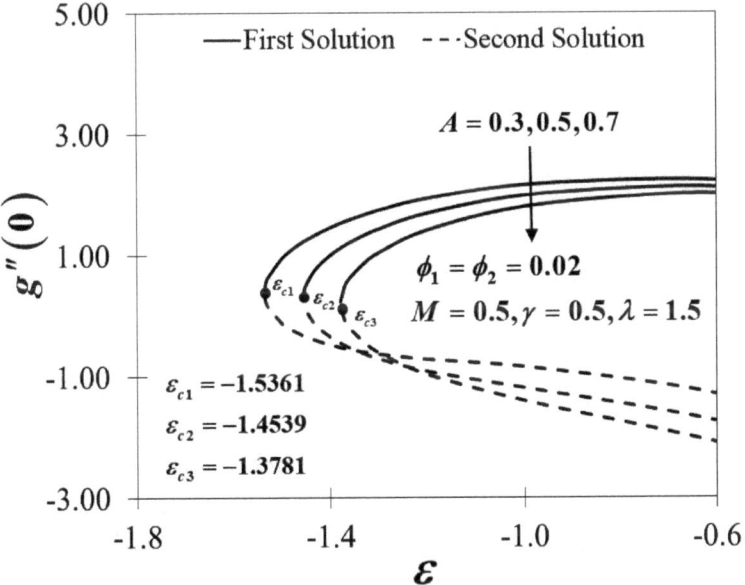

Figure 17. Variations of $g''(0)$ for different values of A.

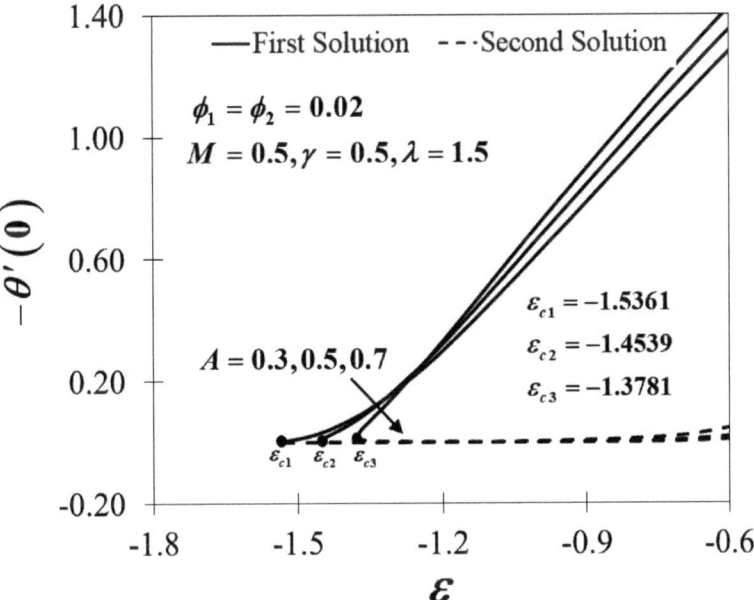

Figure 18. Variations of $-\theta'(0)$ for different values of A.

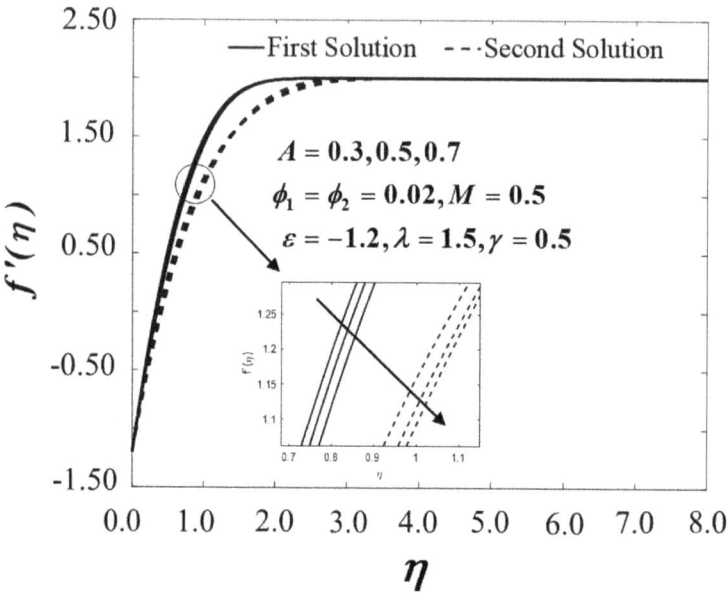

Figure 19. Velocity profiles of $f'(\eta)$ for different values of A.

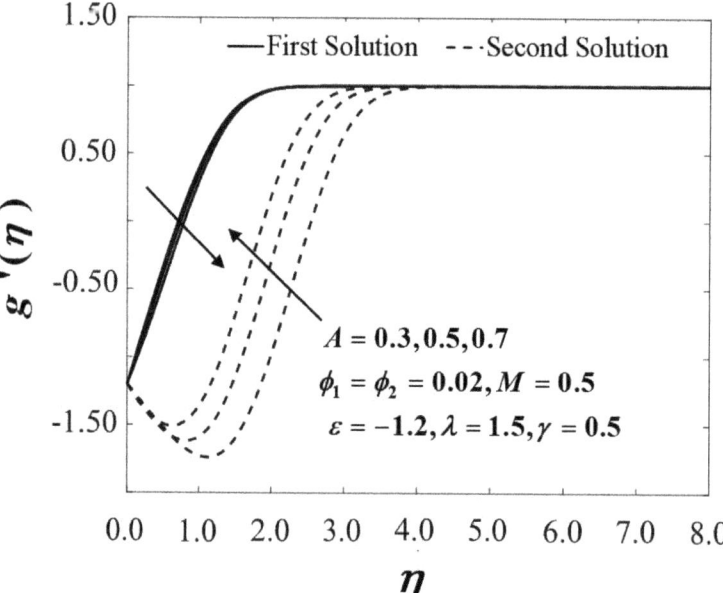

Figure 20. Velocity profiles of $g'(\eta)$ for different values of A.

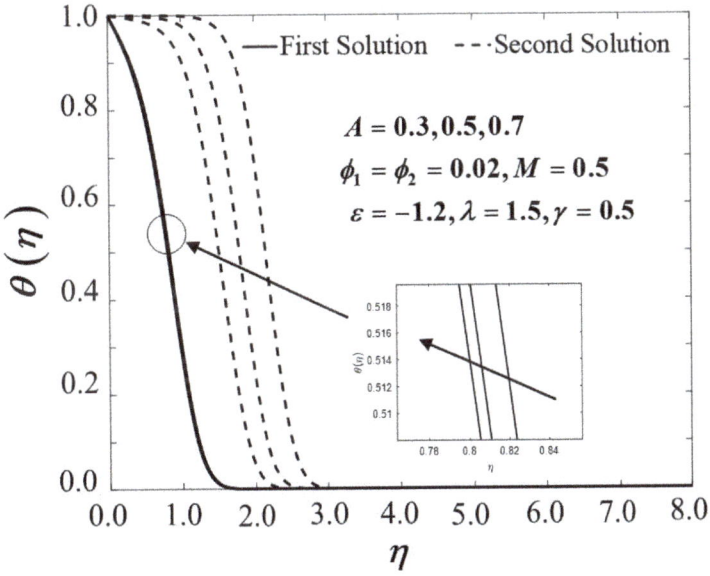

Figure 21. Temperature profiles of $\theta(\eta)$ for different values of A.

Since the dual solutions noticeably exist, as illustrated in Figures 3–21, a stability analysis is performed to discover a significant solution between the first and second solutions. This process is achievable by clarifying the eigenvalue problems in Equations (25)–(27) using bvp4c in Matlab, which produce an infinite set of $\omega_1 < \omega_2 < \omega_3$ .... The stability of the flow is reliant on the smallest positive eigenvalue ω_1 since there exists an initial decay of disturbances. In contrast, there is an initial growth of perturbations if the smallest eigenvalue ω_1 is negative, which signifies that the solution is unstable. As depicted in Table 5, the lowest eigenvalues for the first solution are positive, while the second solution is negative. In conclusion, the first solution is stable and physically reliable, while the second solution is unstable and unreal.

Table 5. Smallest eigenvalues ω_1 when $\phi_1 = \phi_2 = 0.02$, $\lambda = 1.5$, $\gamma = 0.5$, $M = 0.5$, $A = 0.5$ and $Pr = 6.2$.

ε	First Solution, ω_1	Second Solution, ω_1
−1.4	0.9181	−0.3952
−1.452	0.1712	−0.1469
−1.4532	0.1158	−0.0979
−1.4534	0.1040	−0.0869
−1.4536	0.0902	−0.0744
−1.4538	0.0741	−0.0593
−1.45388	0.0665	−0.0521

5. Conclusions

The current analysis is devoted to examining the unsteady three-dimensional non-axisymmetric Homann stagnation point flow of alumina (Al_2O_3) and copper (Cu) hybrid nanofluids in the presence of MHD. The governing partial differential equations are transformed into a system of ordinary differential equations by using a similarity transformation and appropriately solved by a bvp4c function in Matlab. An increment in λ may upsurge the velocity gradient and thus decline the momentum boundary layer thickness. A high concentration of the nanoparticle volume fraction speeds up the molecules' kinetic energy and then enhances the heat transfer process of the fluid particles. The increment in the intensity of the magnetic parameter M increases the local Nusselt number and the skin friction coefficient.

Further, the increasing value of A decreases the hybrid nanofluid temperature and eventually improves the cooling rates of the fluid. Dual solutions were disclosed in this study, and the analysis of solution stability confirmed that the first solutions are stable and physically reliable.

Author Contributions: Research design, N.A.Z., R.N., K.N., and I.P.; Formulation and methodology, N.A.Z.; Result analysis, N.A.Z.; Validation, R.N., and K.N.; Article preparation, N.A.Z.; Review and editing, N.A.Z., R.N., K.N., and I.P. All authors have read and agreed to the published version of the manuscript.

Funding: The author would like to express sincere appreciations to Universiti Kebangsaan Malaysia, Universiti Teknikal Malaysia Melaka, and the Ministry of Education Malaysia for the funding and continuous encouragement. This work is fully supported by the research university grant (GUP-2019-034) from the Universiti Kebangsaan Malaysia. The feedback and recommendations by the competent reviewers are very much appreciated.

Acknowledgments: The authors appreciate the valuable feedback and recommendations from the reviewers.

Conflicts of Interest: The authors declare no conflict of interest.

Nomenclature

Roman letters

a, b, c	positive constants $\left(s^{-1}\right)$
A	unsteadiness parameter $(-)$
B_0	transverse magnetic field $\left(kgA^{-1}s^{-2}\right)$
C_{fx}	skin friction coefficient along the $x-$ direction $(-)$
C_{fy}	skin friction coefficient along the $y-$ direction $(-)$
C_p	specific heat at constant pressure $\left(Jkg^{-1}K^{-1}\right)$
$\left(\rho C_p\right)$	heat capacitance of the fluid $\left(JK^{-1}m^{-3}\right)$
$f(\eta)$	dimensionless stream function in the $x-$ direction
$g(\eta)$	dimensionless stream function in the $y-$ direction
k	thermal conductivity of the fluid $\left(Wm^{-1}K^{-1}\right)$
M	magnetic parameter $(-)$
Nu_x	local Nusselt number $(-)$
Pr	Prandtl number $(-)$
Re_x, Re_y	local Reynolds number in the $x-$ and $y-$ directions, respectively $(-)$
t	time (s)
T	fluid temperature (K)
T_w	surface temperature (K)
T_∞	ambient temperature (K)
u, v, w	velocities component in the $x-, y-$ and $z-$ directions, respectively $\left(ms^{-1}\right)$
u_e, v_e	velocities of the free stream in the $y-$ and $y-$ directions $\left(ms^{-1}\right)$
u_w, v_w	velocities of the stretching/shrinking surface in the $y-$ and $y-$ directions $\left(ms^{-1}\right)$
x, y, w	Cartesian coordinates (m)

Greek symbols

α	positive constant $\left(s^{-1}\right)$
α_f	fluid thermal diffusivity $\left(m^2 s^{-1}\right)$
β	thermal expansion coefficient $\left(K^{-1}\right)$
γ	ratio of the ambient fluid shear rate to the plate strain rate $(-)$
ε	stretching/shrinking parameter $(-)$
ε_r	estimated relative error $(-)$
η	similarity variable $(-)$

θ	dimensionless temperature $(-)$
λ	ratio of the ambient fluid strain rate to the plate strain rate $(-)$
μ	dynamic viscosity $(N\,s\,m^{-2})$
ν	kinematic viscosity (m^2s^{-1})
ρ	density (kgm^{-3})
σ	electrical conductivity (Sm^{-1})
τ	dimensionless time variable $(-)$
τ_{wx}, τ_{wy}	wall shear stress along the $x-$ and $y-$ directions $(kgm^{-1}s^{-2})$
ϕ_1	nanoparticle volume fractions for Al_2O_3 (alumina) $(-)$
ϕ_2	nanoparticle volume fractions for Cu (copper) $(-)$
ω	eigenvalue $(-)$
ω_1	smallest eigenvalue $(-)$
Subscripts	
f	base fluid
nf	nanofluid
hnf	hybrid nanofluid
$s1$	solid component for Al_2O_3 (alumina)
$s2$	solid component for Cu (copper)
Superscript	
$\prime$	differentiation with respect to η

References

1. Fisher, E. *Extrusion of Plastics*; Wiley: New York, NY, USA, 1976.
2. Rauwendaal, C. *Polymer Extrusion*; Hanser Publication: Munich, Germany, 1985.
3. Ibrahim, W.; Shankar, B.; Nandeppanavar, M.M. MHD stagnation point flow and heat transfer due to nanofluid towards a stretching sheet. *Int. J. Heat Mass Transf.* **2013**, *56*, 1–9. [CrossRef]
4. Hiemenz, K. Die Grenzschicht an einem in den gleichförmingen Flüssigkeitsstrom eingetauchten geraden Kreiszylinder. *Dinglers Polytech J.* **1911**, *326*, 321–324.
5. Homann, F. Der Einfluss grosser Zähigkeit bei der Strömung um den Zylinder und um die Kugel. *Z. Angew. Math. Mech.* **1936**, *16*, 153–164. [CrossRef]
6. Howarth, L. CXLIV. The boundary layer in three-dimensional flow—Part II. The flow near a stagnation point. *Lond. Edinb. Dublin Philos. Mag. J. Sci.* **1951**, *42*, 1433–1440. [CrossRef]
7. Davey, A. Boundary-layer flow at a saddle point of attachment. *J. Fluid Mech.* **1961**, *10*, 593–610. [CrossRef]
8. Davey, A.; Schofield, D. Three-dimensional flow near a two-dimensional stagnation point. *J. Fluid Mech.* **1967**, *28*, 149–151. [CrossRef]
9. Weidman, P.D. Non-axisymmetric Homann stagnation-point flows. *J. Fluid Mech.* **2012**, *702*, 460–469. [CrossRef]
10. Mahapatra, T.R.; Sidui, S. Unsteady heat transfer in non-axisymmetric Homann stagnation-point flows. *Z. Angew. Math. Phys.* **2017**, *68*, 32. [CrossRef]
11. Mahapatra, T.R.; Sidui, S. Non-axisymmetric Homann stagnation-point flow of a viscoelastic fluid towards a fixed plate. *Eur. J. Mech. B/Fluids* **2020**, *79*, 38–43. [CrossRef]
12. Wang, C.Y. Stagnation flow towards a shrinking sheet. *Int. J. Non. Linear Mech.* **2008**, *43*, 377–382. [CrossRef]
13. Mahapatra, T.R.; Sidui, S. Unsteady heat transfer in non-axisymmetric Homann stagnation-point flows towards a stretching/shrinking sheet. *Eur. J. Mech. B/Fluids* **2019**, *75*, 199–208. [CrossRef]
14. Khashi'ie, N.S.; Arifin, N.M.; Pop, I.; Nazar, R.; Hafidzuddin, E.H.; Wahi, N. Non-axisymmetric Homann stagnation point flow and heat transfer past a stretching/shrinking sheet using hybrid nanofluid. *Int. J. Numer. Methods Heat Fluid Flow* **2020**. [CrossRef]
15. Zaimi, K.; Ishak, A. Stagnation-point flow towards a stretching vertical sheet with slip effects. *Mathematics* **2016**, *4*, 27. [CrossRef]
16. Arif Ullah Khan, S.; Saleem, S.; Nadeem, A.A.A. Analysis of unsteady non-axisymmetric Homann stagnation point flow of nanofluid and possible existence of multiple solutions. *Physica A* **2019**, 137567. [CrossRef]

17. Jusoh, R.; Nazar, R.; Pop, I. Impact of heat generation/absorption on the unsteady magnetohydrodynamic stagnation point flow and heat transfer of nanofluids. *Int. J. Numer. Methods Heat Fluid Flow* **2019**, *30*, 557–574. [CrossRef]
18. Khashi'ie, N.S.; Arifin, N.; Hafidzuddin, E.H.; Wahi, N. MHD mixed convective stagnation point flow with heat generation past a shrinking sheet. *ASM Sci. J.* **2019**, 71–81.
19. Ali, F.M.; Naganthran, K.; Nazar, R.; Pop, I. MHD mixed convection boundary layer stagnation-point flow on a vertical surface with induced magnetic field: A stability analysis. *Int. J. Numer. Methods Heat Fluid Flow* **2019**. [CrossRef]
20. Shehzad, S.A.; Abdullah, Z.; Alsaedi, A.; Abbasi, F.M.; Hayat, T. Thermally radiative three-dimensional flow of Jeffrey nanofluid with internal heat generation and magnetic field. *J. Magn. Magn. Mater.* **2016**, *397*, 108–114. [CrossRef]
21. Daniel, Y.S.; Aziz, Z.A.; Ismail, Z.; Salah, F. Double stratification effects on unsteady electrical MHD mixed convection flow of nanofluid with viscous dissipation and Joule heating. *J. Appl. Res. Technol.* **2017**, *15*, 464–476. [CrossRef]
22. Pavlov, K.B. Magnetohydrodynamic flow of an incompressible viscous fluid caused by deformation of a plane surface. *Magn. Gidrodin.* **1974**, *4*, 146–147.
23. Chakrabarti, A.; Gupta, A.S. Hydromagnetic flow and heat transfer over a stretching sheet. *Q. Appl. Math.* **1979**, *37*, 73–78. [CrossRef]
24. Vajravelu, K. Hydromagnetic flow and heat transfer over a continuous, moving, porous, flat surface. *Acta Mech.* **1986**, *185*, 179–185. [CrossRef]
25. Andersson, H.I. An exact solution of the Navier-Stokes equations for magnetohydrodynamic flow. *Acta Mech.* **1995**, *113*, 241–244. [CrossRef]
26. Lok, Y.Y.; Ishak, A.; Pop, I. MHD stagnation-point flow towards a shrinking sheet. *Int. J. Numer. Methods Heat Fluid Flow* **2011**, *21*, 61–72. [CrossRef]
27. Almutairi, F.; Khaled, S.M.; Ebaid, A. MHD flow of nanofluid with homogeneous-heterogeneous reactions in a porous medium under the influence of second-order velocity slip. *Mathematics* **2019**, *7*, 220. [CrossRef]
28. Zhu, J.; Xu, Y.; Han, X. A non-newtonian magnetohydrodynamics (MHD) nanofluid flow and heat transfer with nonlinear slip and temperature jump. *Mathematics* **2019**, *7*, 1199. [CrossRef]
29. Choi, S.U.S.; Eastman, J.A. Enhancing thermal conductivity of fluids with nanoparticles. *ASME Publ. Fed.* **1995**, *231*, 99–103.
30. Tiwari, R.K.; Das, M.K. Heat transfer augmentation in a two-sided lid-driven differentially heated square cavity utilizing nanofluids. *Int. J. Heat Mass Transf.* **2007**, *50*, 2002–2018. [CrossRef]
31. Buongiorno, J. Convective transport in nanofluids. *J. Heat Transf.* **2006**, *128*, 240–250. [CrossRef]
32. Mahian, O.; Kianifar, A.; Kalogirou, S.A.; Pop, I.; Wongwises, S. A review of the applications of nanofluids in solar energy. *Heat Mass Transf.* **2013**, *57*, 582–594. [CrossRef]
33. Saidur, R.; Leong, K.Y.; Mohammed, H.A. A review on applications and challenges of nanofluids. *Renew. Sustain. Energy Rev.* **2011**, *15*, 1646–1668. [CrossRef]
34. Thaker, R.; Patel, J.R. A review on application of nanofluids in solar energy. *Am. J. Nano Res. Appl.* **2015**, *2*, 53–61.
35. Sarkar, J.; Ghosh, P.; Adil, A. A review on hybrid nanofluids: Recent research, development and applications. *Renew. Sustain. Energy Rev.* **2015**, *43*, 164–177. [CrossRef]
36. Sidik, N.A.C.; Mahmud Jamil, M.; Aziz Japar, W.M.A.; Muhammad Adamu, I. A review on preparation methods, stability and applications of hybrid nanofluids. *Renew. Sustain. Energy Rev.* **2017**, *80*, 1112–1122. [CrossRef]
37. Huminic, G.; Huminic, A. Hybrid nanofluids for heat transfer applications—A state-of-the-art review. *Int. J. Heat Mass Transf.* **2018**, *125*, 82–103. [CrossRef]
38. Turcu, R.; Darabont, A.; Nan, A.; Aldea, N.; Macovei, D.; Bica, D.; Vekas, L.; Pana, O.; Soran, M.L.; Koos, A.A.; et al. New polypyrrole-multiwall carbon nanotubes hybrid materials. *J Optoelectron. Adv. Mater.* **2006**, *8*, 643–647.
39. Yen, T.H.; Soong, C.Y.; Tzeng, P.Y. Erratum: Hybrid molecular dynamics-continuum simulation for nano/mesoscale channel flows (Microfluid Nanofluid). *Microfluid. Nanofluidics* **2007**, *3*, 729. [CrossRef]

40. Devi, S.S.U.; Devi, S.P.A. Numerical investigation of three-dimensional hybrid Cu–Al$_2$O$_3$/water nanofluid flow over a stretching sheet with effecting Lorentz force subject to Newtonian heating. *Can. J. Phys.* **2016**, *94*, 490–496. [CrossRef]
41. Devi, S.U.; Devi, S.P.A. Heat transfer enhancement of Cu–Al$_2$O$_3$/water hybrid nanofluid flow over a stretching sheet. *J. Niger. Math. Soc.* **2017**, *36*, 419–433.
42. Yousefi, M.; Dinarvand, S.; Eftekhari Yazdi, M.; Pop, I. Stagnation-point flow of an aqueous titania-copper hybrid nanofluid toward a wavy cylinder. *Int. J. Numer. Methods Heat Fluid Flow* **2018**, *28*, 1716–1735. [CrossRef]
43. Khashi'ie, N.S.; Arifin, N.M.; Nazar, R.; Hafidzuddin, E.H.; Wahi, N.; Pop, I. Magnetohydrodynamics (MHD) axisymmetric flow and heat transfer of a hybrid nanofluid past a radially permeable stretching/shrinking sheet with Joule heating. *Chin. J. Phys.* **2020**, *64*, 251–263. [CrossRef]
44. Hayat, T.; Nadeem, S. Heat transfer enhancement with Ag–CuO/water hybrid nanofluid. *Results Phys.* **2017**, *7*, 2317–2324. [CrossRef]
45. Dinarvand, S.; Rostami, M.N.; Pop, I. A novel hybridity model for TiO$_2$-CuO/water hybrid nanofluid flow over a static/moving wedge or corner. *Sci. Rep.* **2019**, *9*, 16290. [CrossRef]
46. Waini, I.; Ishak, A.; Pop, I. Unsteady flow and heat transfer past a stretching/shrinking sheet in a hybrid nanofluid. *Int. J. Heat Mass Transf.* **2019**, *136*, 288–297. [CrossRef]
47. Waini, I.; Ishak, A.; Pop, I. Flow and heat transfer along a permeable stretching/shrinking curved surface in a hybrid nanofluid. *Phys. Scr.* **2019**, *94*, 105219. [CrossRef]
48. Waini, I.; Ishak, A.; Pop, I. Hybrid nanofluid flow and heat transfer over a nonlinear permeable stretching/shrinking surface. *Int. J. Numer. Methods Heat Fluid Flow* **2019**, *29*, 3110–3127. [CrossRef]
49. Cai, W.; Su, N.; Liu, X. Unsteady convection flow and heat transfer over a vertical stretching surface. *PLoS ONE* **2014**, *9*, e107229. [CrossRef]
50. Hayat, T.; Nadeem, S.; Khan, A.U. Rotating flow of Ag-CuO/H$_2$O hybrid nanofluid with radiation and partial slip boundary effects. *Eur. Phys. J. E* **2018**, *41*, 75. [CrossRef]
51. Devi, S.P.A.; Devi, S.S.U. Numerical investigation of hydromagnetic hybrid Cu—Al$_2$O$_3$/water nanofluid flow over a permeable stretching sheet with suction. *Int. J. Nonlinear Sci. Numer. Simul.* **2016**, *17*, 249–257. [CrossRef]
52. Oztop, H.F.; Abu-Nada, E. Numerical study of natural convection in partially heated rectangular enclosures filled with nanofluids. *Int. J. Heat Fluid Flow* **2008**, *29*, 1326–1336. [CrossRef]
53. Merkin, J.H. On dual solutions occurring in mixed convection in a porous medium. *J. Eng. Math.* **1986**, *20*, 171–179. [CrossRef]
54. Merrill, K.; Beauchesne, M.; Previte, J.; Paullet, J.; Weidman, P. Final steady flow near a stagnation point on a vertical surface in a porous medium. *Int. J. Heat Mass Transf.* **2006**, *49*, 4681–4686. [CrossRef]
55. Weidman, P.D.; Kubitschek, D.G.; Davis, A.M.J. The effect of transpiration on self-similar boundary layer flow over moving surfaces. *Int. J. Eng. Sci.* **2006**, *44*, 730–737. [CrossRef]
56. Harris, S.D.; Ingham, D.B.; Pop, I. Mixed convection boundary-layer flow near the stagnation point on a vertical surface in a porous medium: Brinkman model with slip. *Transp. Porous Media* **2009**, *77*, 267–285. [CrossRef]
57. Nawaz, M.; Hayat, T. Axisymmetric stagnation-point flow of nanofluid over a stretching surface. *Adv. Appl. Math. Mech.* **2014**, *6*, 220–232. [CrossRef]
58. Yuan, J.; Piomelli, U. Roughness effects on the Reynolds stress budgets in near-wall turbulence. *J. Fluid Mech.* **2014**, *760*, 1–12. [CrossRef]
59. Ghodke, C.D.; Apte, S.V. DNS study of particle-bed-turbulence interactions in an oscillatory wall-bounded flow. *J. Fluid Mech.* **2016**, *792*, 232–251. [CrossRef]

 © 2020 by the authors. Licensee MDPI, Basel, Switzerland. This article is an open access article distributed under the terms and conditions of the Creative Commons Attribution (CC BY) license (http://creativecommons.org/licenses/by/4.0/).

Article

Numerical Investigation on the Swimming of Gyrotactic Microorganisms in Nanofluids through Porous Medium over a Stretched Surface

Anwar Shahid [1], Hulin Huang [1], Muhammad Mubashir Bhatti [2], Lijun Zhang [2] and Rahmat Ellahi [3,4,5,*]

1. College of Astronautics, Nanjing University of Aeronautics & Astronautics, Nanjing 210016, China; anwar@nuaa.edu.cn (A.S.); hlhuang@nuaa.edu.cn (H.H.)
2. College of Mathematics and Systems Science, Shandong University of Science & Technology, Qingdao 266590, China; mmbhatti@sdust.edu.cn (M.M.B.); lijunzhang@zstu.edu.cn (L.Z.)
3. Department of Mathematics and Statistics, International Islamic University, Islamabad 44000, Pakistan
4. Fulbright Fellow Department of Mechanical Engineering University of California Riverside, Riverside, CA 92521, USA
5. Center for Modeling & Computer Simulation, Research Institute, King Fahd University of Petroleum and Minerals, Dhahran 31261, Saudi Arabia
* Correspondence: rellahi@alumni.ucr.edu

Received: 16 February 2020; Accepted: 5 March 2020; Published: 9 March 2020

Abstract: In this article, the effects of swimming gyrotactic microorganisms for magnetohydrodynamics nanofluid using Darcy law are investigated. The numerical results of nonlinear coupled mathematical model are obtained by means of Successive Local Linearization Method. This technique is based on a simple notion of the decoupling systems of equations utilizing the linearization of the unknown functions sequentially according to the order of classifying the system of governing equations. The linearized equations, that developed a sequence of linear differential equations along with variable coefficients, were solved by employing the Chebyshev spectral collocation method. The convergence speed of the SLLM technique can be willingly upgraded by successive applying over relaxation method. The comparison of current study with available published literature has been made for the validation of obtained results. It is found that the reported numerical method is in perfect accord with the said similar methods. The results are displayed through tables and graphs.

Keywords: successive local linearization method; swimming gyrotactic microorganisms; Darcy law; nanofluid

1. Introduction

The problems associated with the boundary layer mechanism and heat transfer flow through stretched subsurface have been eminently accepted through analysts as long as the presence in structures of enormous industrial and technologically significance. Few of the advanced spreading applications encompass the designing of plastic layers and copper cables, glass-fiber fabricating, food and polymer refining, geothermal power extraction, liquefying-spinning productions, polymer melting, hot roll glass blasting, in the formation of the final product, in the textile industry, and other abundant utilities. Sakiadis [1] performed the developing effort in the area of boundary layer flow on a continued stable subsurface flowing with steady velocities. Later, Crane [2] was the earliest who extended the conception for boundary layer flow through stretchable surfaces. He examined a closed mode result for the Newtonian fluid flow past a flat stretched subsurface. Banks [3] investigated the similarity solutions for the boundary layer flow through a stretched wall with non-Newtonian fluid. Gupta and Gupta [4] broadened the investigated idea by Crane with the heat and mass transfer past a stretchable surface, along with the influence of suction/blowing. Bujurke et al. [5] discussed the heat transfer phenomenon past a boundary layer, along with interval heat generation. Cortell [6] analyzed the viscous fluid flow numerically with heat transfer on a nonlinearly stretchable subsurface. Shahzad et al. [7] developed the exact solutions of the axisymmetric flow with heat transfer for MHD viscous fluid on a nonlinearly radial pervious stretched surface. Hayat et al. [8] explored the MHD axisymmetric flow for third-grade fluid with heat transfer over stretchable sheets. Shateyi and Makinde [9] recorded the heat transfer analysis for a viscous, electrically conducting fluid flow through a radial stretched and convectively heated disk. Khan et al. [10] discussed the mix convection heat transfer to Sisko fluid past radial stretchable surface together with the influence of convection boundary conditions, thermal radiation, and viscous dissipative terms. Since it is known that the standard of the final product relies on the rate of heat transfer as acknowledged, hence the nanofluids have a higher thermal conductivity of the nanoparticles utilized to enhance the rate for heat transfer [11,12]. For this purpose, distinct techniques are adopted to raise the thermal conductivity of the fluids by providing suspension of nano/micro or large-sized particles into liquids. An inventive approach to enhance the heat transfer rate is performed by utilizing nano-scale particles into the governing fluid by Choi et al. [13]. They recorded that by adding a tiny extent (less than 1% by volume) of the nanoparticles to regular heat transfer fluids enhanced the thermal conductivity for the fluids up to almost 4-times and higher. Kuznetsov and Nield [14] discussed the natural convection into a nanofluid through a vertical surface, along with the impact of thermophoresis and the Brownian-motion. Noghrehabadadi et al. [15] explored the heat transfer of nanofluids past a stretched subsurface with supposing of thermal convectively boundary conditions and partial slip. Zaraki et al. [16] analyzed the influence of the various shapes, sizes, and types of nanoparticles, and base-fluid flowing and heat transferring properties for a naturally convective boundary layer.

The investigations for magnetohydrodynamics have significant utilities, and also uses in cooling of nuclear reactors by the induction flow meter and liquid, depending on the capability variation into the fluid in order normal to the flow and the magnetic field. Ferdows et al. [17] explored the problem for magnetized nanofluid mixed convective flow past an exponential stretched plate. Bidin and Nazar [18] discussed the numerical investigation for boundary layer flow through an exponential stretchable surface, along with thermal emission. Khan et al. [19] studied the unsteady boundary layer flow for a nanofluid on a horizontal stretched plate together with the impact of MHD and thermal radiation. Mabood et al. [20] studied the MHD boundary layer nanofluid flowing with the influence of heat transfer and viscous dissipation through a nonlinear stretched surface. Freidoonimehr et al. [21] studied the magnetized stagnation point flow through a stretched/shrinkable surface alongside the impact of chemical reactions and heat absorption/generation. It is conclusive to mention here that Makinde and Animasaun [22] investigated an admirable work related to magnetized nanofluid flow alongside bioconvection with quartic autocatalysis chemical reaction. The results show that for a fixed numeric of a magnetic parameter, the local skin friction further develops at larger thickened parametric value, whereas the rate for local heat transfer turns lesser at a high-temperature parametric value past an uppermost subsurface of a paraboloid of an uprising. The possible developments and/or applications of the presented analysis to the same topic and to other related topics can be seen in [23–38].

The terminology bioconvection was first acknowledged in an article belonging to James Henry Platt to bring about other researchers to this consideration side towards the physics of streaming forms noticed in impenetrable fashions of free-floating microorganisms. In light of Platt [39], the movement of polygonal forms in impenetrable fashions of Tetrahymena (i.e., ciliate and flagellate), such as Benard cells, though not by thermo-convection. Since, it is well-known that the presence of microorganisms (bacteria) are everywhere, and it is illustrious evidence that a large number of bacteria may be accidentally suffered and sometimes can be shot down when periodically bared to a higher temperature, conflicting that thermophile is an organism that usually can be seen in different heated territories on the earth. The self-impelled motile microorganisms brought enhancement in the density of the base fluid in a peculiar way to produce a bioconvection kind of stream. Basing on the cause of propulsion, the motile microorganisms perhaps categorized into various types of microorganisms, counting oxytactic or chemotaxis, gyrotactic microorganisms, and negate gravitaxis. Contrasting the motile microorganisms, the nanoparticles are not self-propelled, and their migration is through the Brownian-motion and thermophoresis impact inward nanofluid. Ghorai and Hill [40] farther elucidated that bioconvection is a known terminology to indicate the phenomena for impromptu arrangement in the suspension of microorganisms, such as algae and bacteria. Bioconvection also can be explained as the macroscopic convective movement of fluid as a result of the density gradient, and is brought about by the jointly floating of motile microorganisms. Alike naturally convective process, bioconvection is induced by versatile stratification density. Kuznetsov and Avramenko [41] interpreted that when bioconvection takes place, it boosts mingling and diminishes the establishing of the particles that are decisive in medicine utilities. Khan and Makinde [42] examined nanofluid bioconvection caused by gyrotactic microorganisms and noticed that the self-propelled motile microorganisms enhance the density of the base-fluid as floating/swimming in a specific manner. Recently, Raees et al. [43] interpreted that bioconvection into nanofluids has enormous contributions in Colibri micro-volumes spectrometer and benefits the stability in nanofluids. Natural convection with double-diffusive effects over a boundary layer nanofluid flow has been examined by Kuznetsov and Nield [44]. Nonetheless, if the stimulators past the subsurface are more imperative and associate to the bulk-fluid, comprising 36 nm nanoparticles and gyrotactic microorganism, alike chemical backlash could be examined by applying the conception of homogeneous–heterogeneous quartic strategy. Sivaraj et al. [45] have discussed the gyrotactic microorganisms on the mechanism of 29 nm copper water nanofluids propagated through a horizontal surface of paraboloid. Amirsom et al. [46] have considered the movement of microorganisms on a magnetized nanofluid in the presence of second order slip conditions via bvp4c computational scheme. Waqas et al. [47,48] used a shooting method to discuss the propagation of nanoparticles and gyrotactic microorganisms through a stretching surface with magnetic and porous effects using non-Newtonian fluid models. A few other inquiries on gyrotactic microorganisms can be read here [49–51].

The impulsion of the current investigation is to explore the impact of a non-uniform magnetic field on the conduct of water suspension comprising nanoparticles and motile gyrotactic microorganisms flowing through a stretchable permeable sheet by employing Successive Local Linearization Method (SLLM) with the combination of Chebyshev spectral linearization method [52] not yet available in the existing literature. The governing flow equations and the boundary conditions were brought towards nonlinear ordinary differential equations by utilizing the similarity variable transformations, and are than solved numerically by spectral approach.

2. Mathematical Modeling

A two-dimensional, steady, incompressible viscous and electrically conducting nanofluid flow, comprising gyrotactic microorganisms through a stretched porous sheet by Darcy-Forchheimer relation is considered. It is also assumed that the flow field is under the effect of a varying magnetic field of strength $B(x) = B_0(\hat{x})$. The sheet is stretched vertically with velocity $\widetilde{U}_w = a\hat{x}$, with positive constant a. The induced magnetic field is ignored because it is minimal in comparison to the extraneous magnetic field, as can be seen in Figure 1. The concentration $\widetilde{C}_w$, temperature $\widetilde{T}_w$, and densities for motile microorganisms are $\widetilde{N}_w$ and $\widetilde{N}_\infty$ past the stretched subsurface are considered constant and bigger than the ambient concentration $\widetilde{C}_\infty$, temperature $\widetilde{T}_\infty$, respectively. It is further presumed that nanoparticles are not affecting the direction and velocity of microorganisms, and both the nanoparticles and the base fluid are in local thermal stability state; and the nanoparticles, motile microorganisms, and the base-fluid are having the equivalent velocities. Hence, for a suchlike problem, the governing equations for continuity, momentum, nanoparticle concentration, thermal energy, and microorganisms can be written as

$$\frac{\partial \widetilde{v}}{\partial \hat{y}} + \frac{\partial \widetilde{u}}{\partial \hat{x}} = 0, \tag{1}$$

$$\tilde{u}\frac{\partial \tilde{u}}{\partial \hat{x}} + \tilde{v}\frac{\partial \tilde{u}}{\partial \hat{y}} + \sigma B_0^2 \tilde{u} = -\frac{\partial \tilde{p}}{\partial \hat{x}} + v_f\left(\frac{\partial^2 \tilde{u}}{\partial \hat{x}^2} + \frac{\partial^2 \tilde{u}}{\partial \hat{y}^2}\right) + \overline{g}\beta\left(1 - \widetilde{C}_\infty\right)\left(\widetilde{T} - \widetilde{T}_\infty\right) - \overline{g}\left(\rho_p - \rho_f\right)\left(\widetilde{C} - \widetilde{C}_\infty\right) - \overline{g}\gamma\left(\rho_m - \rho_f\right)\left(\widetilde{N} - \widetilde{N}_\infty\right) - \frac{v_f}{k}\tilde{u}, \qquad (2)$$

$$\frac{\partial \tilde{p}}{\partial \hat{y}} = 0 \qquad (3)$$

$$\tilde{u}\frac{\partial \widetilde{T}}{\partial \hat{x}} + \tilde{v}\frac{\partial \widetilde{T}}{\partial \hat{y}} = \overline{\alpha}\left[\frac{\partial^2 \widetilde{T}}{\partial \hat{x}^2} + \frac{\partial^2 \widetilde{T}}{\partial \hat{y}^2}\right] + \overline{\tau}\left[D_B \frac{\partial \widetilde{C}}{\partial \hat{y}}\frac{\partial \widetilde{T}}{\partial \hat{y}} + \frac{D_T}{T_\infty}\left\{\left(\frac{\partial \widetilde{T}}{\partial \hat{y}}\right)^2 + \left(\frac{\partial \widetilde{T}}{\partial \hat{x}}\right)^2\right\}\right] + \frac{\mu_f \overline{\alpha}}{k_t}\left(\frac{\partial \tilde{u}}{\partial \hat{y}}\right)^2 + \frac{\sigma \overline{\alpha} B_0^2}{k_t}\tilde{u}^2 \qquad (4)$$

$$\tilde{u}\frac{\partial \widetilde{C}}{\partial \hat{x}} + \tilde{v}\frac{\partial \widetilde{C}}{\partial \hat{y}} = D_b\left[\frac{\partial^2 \widetilde{C}}{\partial \hat{x}^2} + \frac{\partial^2 \widetilde{C}}{\partial \hat{y}^2}\right] + \frac{D_T}{T_\infty}\left[\frac{\partial^2 \widetilde{T}}{\partial \hat{x}^2} + \frac{\partial^2 \widetilde{T}}{\partial \hat{y}^2}\right] \qquad (5)$$

$$\tilde{u}\frac{\partial \widetilde{N}}{\partial \hat{x}} - D_M\left(\frac{\partial^2 \widetilde{N}}{\partial \hat{x}^2} + \frac{\partial^2 \widetilde{N}}{\partial \hat{y}^2} + 2\frac{\partial^2 \widetilde{N}}{\partial \hat{x}\partial \hat{y}}\right) + \tilde{v}\frac{\partial \widetilde{N}}{\partial \hat{y}} + \frac{bW_C}{\left(\widetilde{C} - \widetilde{C}_\infty\right)}\left[\frac{\partial}{\partial \hat{y}}\left(N\frac{\partial \widetilde{C}}{\partial \hat{y}}\right) + \frac{\partial}{\partial \hat{x}}\left(\widetilde{N}\frac{\partial \widetilde{C}}{\partial \hat{x}}\right)\right] = 0 \qquad (6)$$

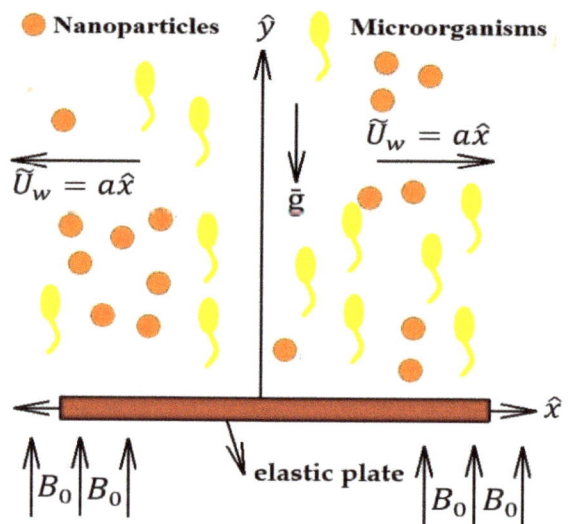

Figure 1. Flow structure through a stretch elastic plate.

Their respective boundary conditions can be read as

$$\tilde{u} = a\hat{x},\ \tilde{v} = 0, \widetilde{T} = \widetilde{T}_w, \widetilde{C} = \widetilde{C}_w, \widetilde{N} = \widetilde{N}_w \text{ at } \hat{y} = 0 \qquad (7)$$

$$\tilde{u} \to 0,\ \widetilde{C} \to \widetilde{C}_\infty, \tilde{v} \to 0,\ \widetilde{T} \to \widetilde{T}_\infty,\ \widetilde{N} \to \widetilde{N}_\infty \text{ as } \hat{y} \to \infty \qquad (8)$$

By cancelling Equation (3) from the momentum equations by cross-differentiation, only Equation (2) survives. In Equations (1)–(8), $\tilde{u}$ and $\tilde{v}$ are the velocity components for $\hat{x}$ and $\hat{y}$ directions correspondingly. Where $\widetilde{T}$ is the temperature, $\widetilde{C}$ is the concentration for nanoparticle, $\widetilde{N}$ is the density for motile microorganism, $\widetilde{p}$ is the pressure, ρ_f, ρ_m, ρ_p are the densities of nanofluid, microorganisms, and nanoparticles, D_b, D_m, D_T denote the Brownian-diffusion coefficient, diffusivity of microorganisms and thermophoresis-diffusion coefficient, k the porosity parameter, σ, k_t are the electrical and thermal conductivity for the fluid, γ indicates the average volume for a microorganism, respectively. $\overline{\alpha} = k_t/(\rho c_p)$ is the thermal diffusivity, bW_C are the constants, and the proportion of the effected heat capacitance of the nanoparticle to the base-fluid $\widetilde{\tau} = \frac{(\rho C)_p}{(\rho C)_f}$, respectively, are the other parametric quantities.

Invoking the following transformation

$$\left. \begin{array}{c} \widetilde{u} = a\hat{x}g'(\eta),\ \widetilde{v} = -\sqrt{av}\,g(\eta),\ \eta = \sqrt{\frac{a}{v}}\,\hat{y},\ \phi(\eta) = \frac{\widetilde{C}-\widetilde{C}_\infty}{\widetilde{C}_w - \widetilde{C}_\infty}, \\ \theta(\eta) = \frac{\widetilde{T}-\widetilde{T}_\infty}{\widetilde{T}_w - \widetilde{T}_\infty},\ \phi(\eta) = \frac{\widetilde{N}-\widetilde{N}_\infty}{\widetilde{N}_w - \widetilde{N}_\infty}. \end{array} \right\} \quad (9)$$

In Equations (1)–(8), the non-dimensional form of resulting equations, along with associated boundary conditions, can be written as

$$g''' + gg'' - g'^2 - Mg' - \beta_D g' + \frac{G_r}{R_e^2}(\theta - N_r \phi - R_b \phi) = 0 \quad (10)$$

$$\frac{1}{P_r}\theta'' + \theta'[g + N_b \phi'] + N_t \theta'^2 + E_c\{g''^2 + Mg'^2\} = 0 \quad (11)$$

$$\phi'' + L_e \phi' g + \frac{N_t}{N_b}\theta'' = 0 \quad (12)$$

$$\phi'' + L_b g \phi' - P_e([\phi + \Omega_d]\phi'' + \phi'\phi') = 0 \quad (13)$$

$$g(\eta) = 0, g'(\eta) = 1, \theta(\eta) = \phi(\eta) = \phi(\eta) = 1,\ \text{when}\ \eta = 0 \quad (14)$$

$$g'(\eta) = 0,\ \theta(\eta) = \phi(\eta) = \phi(\eta) = 0,\ \text{when}\ \eta \to \infty \quad (15)$$

In which

$$\beta_D = \frac{v}{a\rho_f k},\ M = \frac{\sigma B_0^2}{a\rho_f},\ \frac{G_r}{R_e^2} = \frac{\bar{g}\beta(1-\widetilde{C}_\infty)(\widetilde{T}-\widetilde{T}_\infty)}{a\widetilde{u}_w},\ N_r = \frac{(\rho_p - \rho_f)(\widetilde{C}_w - \widetilde{C}_\infty)}{\beta\rho_f(\widetilde{T}_w - \widetilde{T}_\infty)(1-\widetilde{C}_\infty)},\ P_r = \frac{v}{a},\ R_b =$$
$$\frac{\gamma(\rho_m - \rho_f)(\widetilde{N}_w - \widetilde{N}_\infty)}{\beta\rho_f(\widetilde{T}_w - \widetilde{T}_\infty)(1-\widetilde{C}_\infty)},\ N_T = \frac{\widetilde{\tau} D_T(\widetilde{T}_w - \widetilde{T}_\infty)}{v\widetilde{T}_\infty},\ N_b = \frac{\widetilde{\tau} D_B(\widetilde{C}_w - \widetilde{C}_\infty)}{v},\ E_c = \frac{\widetilde{u}_w^2}{c_p(\widetilde{T}_w - \widetilde{T}_\infty)},\ L_e = \frac{v}{D_B},\ L_b = \quad (16)$$
$$\frac{v}{D_M},\ \Omega_d = \frac{\widetilde{N}_\infty}{(\widetilde{N}_w - \widetilde{N}_\infty)},\ P_e = \frac{bW_C}{D_M},$$

These parametric quantities are permeability parameter β_D, Hartmann number M, the local Richardson number G_r/R_e^2, the buoyancy proportion parameter N_r, Prandtl number P_r, the bioconvection Rayleigh number R_b, the thermophoresis parameter N_t, the Brownian motion parameter N_b, Eckert number E_c, the conventional Lewis number and the bioconvection Lewis number L_e and L_b, the bioconvection Peclet number P_e, and Ω_d is the concentration of the microorganisms variance parametric quantity, respectively.

The motile density number, Sherwood, and Nusselt number for the present flow in dimensionless form are:

$$\frac{Nu_x}{Re_x^{1/2}} = -\theta'(0),\ \frac{Sh_x}{Re_x^{\frac{1}{2}}} = -\phi'(0),\ \frac{Nn_x}{Re_x^{\frac{1}{2}}} = -\phi'(0), \quad (17)$$

where $Re_x = \frac{U_0 \hat{x}}{v}$, the local Reynolds number.

3. Numerical Solutions

3.1. Spectral Local Linearization Scheme

Let us having a system of differential equations $G = [g_1(\xi), g_1(\xi), \ldots, g_p(\xi)]$ satisfying the system:

$$\mathcal{L}_j + \widetilde{N}_j = \mathcal{H}_j,\ j = 1, 2, \ldots, p, \quad (18)$$

where p describes the number of differential equations, each $\mathcal{H}_j$ is a function of $\xi \epsilon [A, B]$ and $\mathcal{L}_j, \widetilde{N}_j$ are the linear and nonlinear components in the system, respectively.

Usually, the SLLM is an iterative approach to solve the differential equations, starts from an initial approximation g_0, and then implements the SLLM successively, yielding the new approximations $g_1, g_2, \ldots$, where $G_t = [g_{1,t}, g_{2,t}, \ldots, g_{p,t}]$ for each $t = 0, 1, 2$. When once linearized, the nonlinear components are $\widetilde{N}_j$.

For this intention, the j-th differential Equation (18) after the first $t + 1$ iterations can be express as

$$\mathcal{L}_j\big|_{t+1} + \widetilde{N}_j\big|_{t+1} = \mathcal{H}_j, \tag{19}$$

The nonlinear components can be linearized by using Taylor series

$$\widetilde{N}_j\big|_{t+1} = \widetilde{N}_j\big|_t + \nabla \widetilde{N}_j\big|_t [V_{t+1} - V_t], \tag{20}$$

where V_t is an n-tuple of $G_{j,t}$ and its differentials. Now using Equations (19) and (20) in Equation (18), it becomes

$$\mathcal{L}_j\big|_{t+1} + \nabla \widetilde{N}_j\big|_t V_{t+1} = \mathcal{H}_j + \nabla \widetilde{N}_j\big|_t V_t - \widetilde{N}_j\big|_t. \tag{21}$$

3.2. Successive Local Linearization Method

For the implementation of Successive Local Linearization Method, first we have to reduce the order of Equation (24). To serve the purpose, a new transformation $g' = h$, leads Equation (10) to Equation (13) into the following form:

$$h'' + gh' - h^2 - Mh - \beta_D h + \frac{G_r}{R_e^2}(\theta - N_r \phi - R_b \phi) = 0, \tag{22}$$

$$\frac{1}{P_r}\theta'' + \theta'[g + N_b \phi'] + N_t \theta'^2 + E_c\{h'^2 + Mh^2\} = 0, \tag{23}$$

$$\phi'' + L_e \phi' g + \frac{N_t}{N_b}\theta'' = 0, \tag{24}$$

$$\phi'' + L_b g \phi' - P_e\{[\phi + \Omega_d]\phi'' + \phi'\phi'\} = 0. \tag{25}$$

Linearizing the non-linear term h^2 by applying Taylor series expansion can be written as

$$h^2{}_{t+1} = h^2{}_t + 2h_t[h_{t+1} - h_t] = 2h_t h_{t+1} - h^2{}_t \tag{26}$$

where the component having subscripts t and $t + 1$ stand for current previous and current approximated values. When Equation (26) is placed in Equation (22), then the non-linear system by means of Gauss-Seidel relaxation method can be decoupled as:

$$g'_{t+1} = h_t \tag{27}$$

$$h''_{t+1} + g_t h'_{t+1} - Mh_{t+1} - \beta_D h_{t+1} - 2h_t h_{t+1} = -h^2{}_t - \frac{G_r}{R_e^2}(\theta_t - N_r \phi_t - R_b \phi_t) \tag{28}$$

$$\frac{1}{P_r}\theta''_{t+1} + \theta'_{t+1}[g_t + N_b \phi'_t] + N_t \theta'^2{}_{t+1} = -E_c\{h'^2{}_{t+1} + Mh^2{}_{t+1}\} \tag{29}$$

$$\phi''_{t+1} + L_e g_t \phi'_{t+1} + \frac{N_t}{N_b}\theta''_{t+1} = 0 \tag{30}$$

$$\phi''_{t+1} + L_b g_t \phi'_{t+1} - P_e([\phi_{t+1} + \Omega_d]\phi''_{t+1} + \phi'_{t+1}\phi'_{t+1}) = 0 \tag{31}$$

The corresponding boundary conditions become

$$g_{t+1}(0) = 0, h_{t+1}(0) = 1 = \theta_{t+1}(0) = \phi_{t+1}(0) = \phi_{t+1}(0), \tag{32}$$

$$h_{t+1}(\infty) = 0 = \theta_{t+1}(\infty) = \phi_{t+1}(\infty) = \phi_{t+1}(\infty), \tag{33}$$

Writing a compact expression of Equations (27)–(31) as follows

$$g'_{t+1} = d_{00} \tag{34}$$

$$h''_{t+1} + d_{11}h'_{t+1} - d_{13}h_{t+1} - 2h_t h_{t+1} = d_{1,t} \tag{35}$$

$$\frac{1}{P_r}\theta''_{t+1} + d_{11}\theta'_{t+1} + N_b\,\phi'_t\theta'_{t+1} + N_t\theta'^2_{t+1} = d_{2,t} \tag{36}$$

$$\phi''_{t+1} + d_{32}\phi'_{t+1} + \frac{N_t}{N_b}\theta''_{t+1} = d_{3,t} \tag{37}$$

$$\phi''_{t+1} + d_{42}\phi'_{t+1} - P_e\{[\phi_{t+1} + \Omega_d]\phi''_{t+1} + \phi'_{t+1}\phi'_{t+1}\} = d_{4,t} \tag{38}$$

where

$$d_{00} = h_t, d_{11} = g_t, d_{12} = 2h_t, d_{13} = [M + \beta_D], d_{1,t} = -h^2_t - \frac{G_r}{R_e^2}(\theta_t - N_r\phi_t - R_b\phi_t) \tag{39}$$

$$d_{2,t} = -E_c\left(h'^2_{t+1} + Mh^2_{t+1}\right), d_{32} = L_e g_t, d_{42} = L_b g_t, d_{3,t} = d_{4,t} = 0 \tag{40}$$

Now, employing the Chebyshev spectral collocation method at the system of Equations (34)–(38), where the differentiation matrix $D = \frac{2}{l}D$ utilized to perform approximation for the derivatives of unknown variables in the above equations and our new system become

$$Dg_{t+1} = h_t \tag{41}$$

$$\{D^2 + diag[d_{11}]D - diag[d_{12}]I - d_{13}I\}H_{t+1} = d_{1,t} \tag{42}$$

$$\left\{\frac{1}{P_r}D^2 + diag[d_{11}]D + N_b diag[\phi'_t]D + N_t D^2\right\}\theta_{t+1} = d_{2,t} \tag{43}$$

$$\left\{D^2 + diag[d_{32}]D + \frac{N_t}{N_b}diag[\theta''_{t+1}]I\right\}\phi_{t+1} = d_{3,t} \tag{44}$$

$$\left\{\begin{array}{c} D^2 + diag[d_{42}]D - P_e\Omega_d diag[\phi''_{t+1}]I - P_e diag[\phi''_{t+1}]I \\ -diag[\phi'_{t+1}]D \end{array}\right\}\phi_{t+1} = d_{4,t} \tag{45}$$

With their respective boundary conditions

$$g_{t+1}(\eta_N) = 0, h_{t+1}(\eta_N) = 1 = \theta_{t+1}(\eta_N) = \phi_{t+1}(\eta_N) = \phi_{t+1}(\eta_N) \tag{46}$$

$$h_{t+1}(\eta_0) = 0 = \theta_{t+1}(\eta_0) = \phi_{t+1}(\eta_0) = \phi_{t+1}(\eta_0), \tag{47}$$

The system can be expressed in a more simplified way as

$$B_1 g_{t+1} = E_1 \tag{48}$$

$$B_2 h_{t+1} = E_2 \tag{49}$$

$$B_3 \theta_{t+1} = E_3 \tag{50}$$

$$B_4 \phi_{t+1} = E_4 \tag{51}$$

$$B_5 \phi_{t+1} = E_5 \tag{52}$$

where
$$B_1 = D, E_1 = h_t, \tag{53}$$

$$B_2 = D^2 + diag[d_{11}]D - diag[d_{12}]I - d_{13}I, E_2 = d_{1,t}, \tag{54}$$

$$B_3 = \frac{1}{P_r}D^2 + diag[d_{11}]D + N_b diag[\phi'_t]D + N_t D^2, E_3 = d_{2,t}, \tag{55}$$

$$B_4 = D^2 + diag[d_{32}]D + \frac{N_t}{N_b}diag[\theta''_{t+1}]I, E_4 = d_{3,t}, \tag{56}$$

$$B_5 = D^2 + diag[d_{42}]D - P_e\Omega_d diag[\phi''_{t+1}]I - P_e diag[\phi''_{t+1}]I \\ -diag[\phi'_{t+1}]D, E_5 = d_{4,t}, \tag{57}$$

$$diag[d_{11}] = \begin{bmatrix} d_{11}(\eta_0) & \cdots & \\ \vdots & \ddots & \vdots \\ & \cdots & d_{11}(\eta_N) \end{bmatrix}, diag[d_{12}] = \begin{bmatrix} d_{12}(\eta_0) & \cdots & \\ \vdots & \ddots & \vdots \\ & \cdots & d_{12}(\eta_N) \end{bmatrix}, \tag{58}$$

$$diag[d_{1,t}] = \begin{bmatrix} d_{1,t}(\eta_0) \\ \vdots \\ d_{1,t}(\eta_N) \end{bmatrix}, diag[d_{2,t}] = \begin{bmatrix} d_{2,t}(\eta_0) \\ \vdots \\ d_{2,t}(\eta_N) \end{bmatrix}, \tag{59}$$

$$diag[d_{32}] = \begin{bmatrix} d_{32}(\eta_0) & \cdots & \\ \vdots & \ddots & \vdots \\ & \cdots & d_{32}(\eta_N) \end{bmatrix}, diag[d_{42}] = \begin{bmatrix} d_{42}(\eta_0) & \cdots & \\ \vdots & \ddots & \vdots \\ & \cdots & d_{42}(\eta_N) \end{bmatrix}, \tag{60}$$

$$d_{3,t} = d_{4,t} = 0 = \begin{bmatrix} 0 \\ \vdots \\ 0 \end{bmatrix}, \tag{61}$$

$$g_{t+1} = [g(\eta_0), g(\eta_1), \ldots, g(\eta_N)]^T, h_{t+1} = [h(\eta_0), h(\eta_1), \ldots, h(\eta_N)]^T, \tag{62}$$

$$\theta_{t+1} = [\theta(\eta_0), \theta(\eta_1), \ldots, \theta(\eta_N)]^T, \phi_{t+1} = [\phi(\eta_0), \phi(\eta_1), \ldots, \phi(\eta_N)]^T \tag{63}$$

$\phi_{t+1} = [\phi(\eta_0), \phi(\eta_1), \ldots, \phi(\eta_N)]^T$ are vectors of sizes $(N+1) \times 1$ whereas 0 is a vector of order $(N+1) \times 1$ and I is an identity matrix of order $(N+1) \times (N+1)$.

In view of boundary conditions, the Equations (48)–(63) take the following form:

$$B_1 = \begin{bmatrix} B_1 & \\ \hline 0 & \cdots & 1 \end{bmatrix}, g_{t+1} = \begin{bmatrix} g_{t+1}(\eta_0) \\ g_{t+1}(\eta_1) \\ \vdots \\ g_{t+1}(\eta_N) \end{bmatrix}, E_1 = \begin{bmatrix} E_1 \\ \hline \overline{0} \end{bmatrix}, B_2 = \begin{bmatrix} 1 & \cdots & 0 \\ \hline & B_2 & \\ \hline 0 & \cdots & 1 \end{bmatrix}, h_{t+1} = \begin{bmatrix} h_{t+1}(\eta_0) \\ h_{t+1}(\eta_1) \\ \vdots \\ h_{t+1}(\eta_N) \end{bmatrix}, \tag{64}$$

$$E_2 = \begin{bmatrix} 0 \\ \hline E_2 \\ \hline \overline{1} \end{bmatrix}, B_3 = \begin{bmatrix} 1 & \cdots & 0 \\ \hline & B_3 & \\ \hline 0 & \cdots & 1 \end{bmatrix}, \theta_{t+1} = \begin{bmatrix} \theta_{t+1}(\eta_0) \\ \theta_{t+1}(\eta_1) \\ \vdots \\ \theta_{t+1}(\eta_N) \end{bmatrix}, E_3 = \begin{bmatrix} 0 \\ \hline E_3 \\ \hline \overline{1} \end{bmatrix} \tag{65}$$

$$\begin{bmatrix} 1 & \cdots & 0 \\ & B_4 & \\ 0 & \cdots & 1 \end{bmatrix} \phi_{t+1} = \begin{bmatrix} \phi_{t+1}(\eta_0) \\ \phi_{t+1}(\eta_1) \\ \vdots \\ \phi_{t+1}(\eta_N) \end{bmatrix}, \; E_4 = \begin{bmatrix} 0 \\ E_4 \\ \bar{1} \end{bmatrix}, \; \begin{bmatrix} 1 & \cdots & 0 \\ & B_5 & \\ 0 & \cdots & 1 \end{bmatrix} \phi_{t+1} = \begin{bmatrix} \phi_{t+1}(\eta_0) \\ \phi_{t+1}(\eta_1) \\ \vdots \\ \phi_{t+1}(\eta_N) \end{bmatrix}, \; E_5 = \begin{bmatrix} 0 \\ E_5 \\ \bar{1} \end{bmatrix} \quad (66)$$

The initial guesses are:

$$g_0(\eta) = (1 - e^{-\eta}), \; h_0(\eta) = e^{-\eta}, \; \theta_0(\eta) = \phi_0(\eta) = \phi_0(\eta) = e^{-\eta} \quad (67)$$

These initial assumptions approximation satisfying the boundary conditions (46)–(47) achieve subsequent approximations of $g_t, h_t, \theta_t, \phi_t, \phi_t$ for each $t = 1, 2, \ldots$ by employing the successive local linearization method.

4. Numerical Results and Discussion

4.1. Convergence Analysis

As the Gauss-Seidel method with the SOR parameter is utilized to enhance the convergence of the linear system of equations in the field of numerical linear algebra, therefore as matter of fact, an identical approach is applied to enhance the rate of convergence for successive local linearization method. If, for resolving function Z, the SLLM technique at the $(t + 1)$th iteration is

$$B_1 Z_{t+1} = E_1, \quad (68)$$

Then by revising, the new mode of the SLLM technique is indicated as

$$B_1 Z_{t+1} = (1 - \omega) B_1 Z_t + \omega E_1, \quad (69)$$

Here ω represents the convergence improving the parametric quantity, and B_1, E_1 are the matrices. This revised SLLM technique enlarges in improving the accuracy and efficiency of current results.

4.2. Graphical Illustrations

This section is dedicated to the numerical results, their validation, and the discussion. To examine the inclusion of all the leading parameters numerically, computational software MATLAB is used for the numerical simulations. Table 1 is drawn for the computed convergent outcomes of $Nu_x/Re_x^{1/2}$, $Sh_x/Re_x^{1/2}$, and $Nn_x/Re_x^{1/2}$ across the number of collocation points N, N_t, and N_b by fixing other parameters, whereas, Table 2 depicts the comparability of $-\theta'(0)$ with previously published data [53–55] across P_r with the preceding investigations by fixing other parameters of the governing equations. Table 3 is calculated to compare our computational results with the shooting method, and it can be observe that the results matched perfectly with the shooting method results. Figures 1–11 have been plotted against all the leading parameters for microorganism distribution, nanoparticle concentration, temperature, and velocity distribution, respectively.

Table 1. Numerical convergent values of Nusselt number, Sherwood Number, and the local density number of the motile microorganisms across N, N_t, and N_b by fixing $M = 1, \beta_D = E_c = 0, N_r = 0.5, R_b = 0.5, \frac{G_r}{R_e^2} = 0.5, P_r = 10, L_e = 10, L_b = 2, P_e = 0.5, \Omega_d = 1.0$.

N	N_t	N_b	$\frac{Nu_x}{Re_x^{1/2}}$	$\frac{Sh_x}{Re_x^{1/2}}$	$\frac{Nn_x}{Re_x^{1/2}}$
50	0.1	0.1	1.9320	7.2348	8.8760
60	0.1	0.1	1.9330	7.2357	8.8772
70	0.1	0.1	1.9334	7.2364	8.8778
80	0.1	0.1	1.9334	7.2364	8.8778
100	0.1	0.1	1.9334	7.2364	8.8778
50	0.5	0.5	1.1430	7.7175	9.3273
60	0.5	0.5	1.1446	7.7189	9.3290
70	0.5	0.5	1.1452	7.7198	9.3294
80	0.5	0.5	1.1452	7.7198	9.3294
100	0.5	0.5	1.1452	7.7198	9.3294

Table 2. Comparison of the current outcomes for Nusselt number with the previous investigations.

Current Results	Akbar and Khan [53]	Khan et al. [54]	Wang [55]
	$M = 1, \beta_D = 0, E_c = 0$	$M = \beta_D = E_c = 0$	$M = \beta_D, \frac{G_r}{R_e^2} = N_t = N_b = E_c = 0$
1.6045	1.6045		
0.3211		0.3211	
0.454072			0.4539

Table 3. Comparison of the present method with shooting technique.

N	$\frac{Nu_x}{Re_x^{1/2}}$	$\frac{Sh_x}{Re_x^{1/2}}$	$\frac{Nn_x}{Re_x^{1/2}}$
50	1.9320	7.2348	8.8760
60	1.9330	7.2357	8.8772
70	1.9334	7.2364	8.8778
80	1.9334	7.2364	8.8778
Shooting method	1.9334	7.2364	8.8778

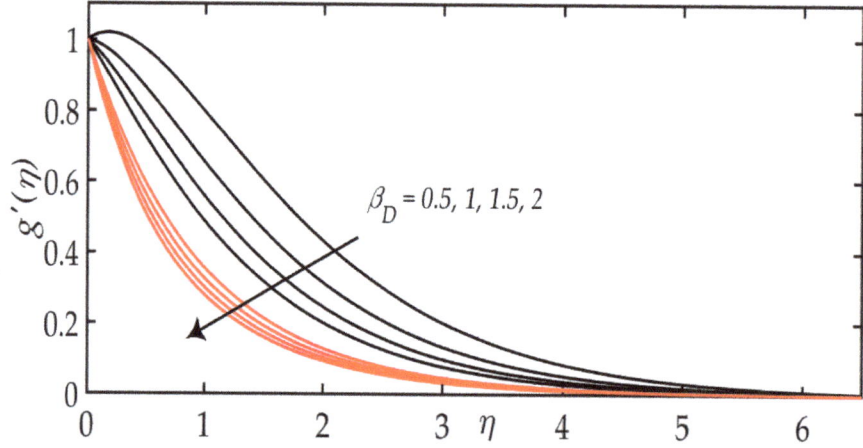

Figure 2. Variation of β_D and M on velocity distribution. Black line: $M = 0$, Red line: $M = 3$.

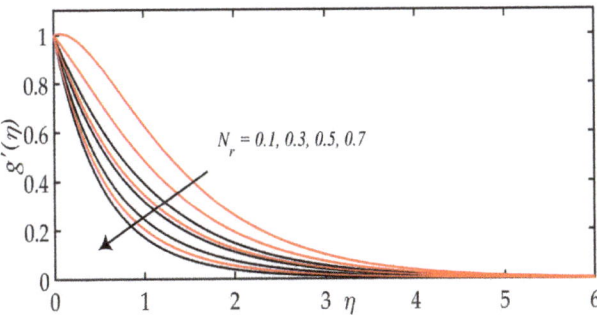

Figure 3. Variation of N_r and $\frac{G_r}{R_e^2}$ on velocity distribution. Black line: $\frac{G_r}{R_e^2} = 5.5$, Red line: $\frac{G_r}{R_e^2} = 10.5$.

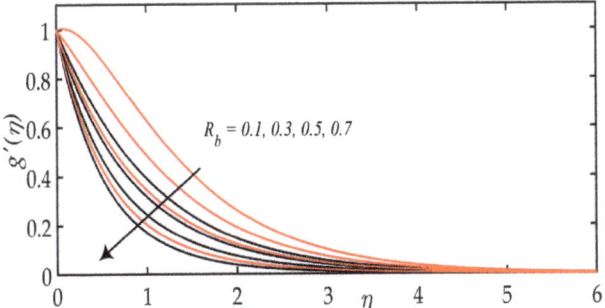

Figure 4. Variation of R_b and $\frac{G_r}{R_e^2}$ on velocity distribution. Black line: $\frac{G_r}{R_e^2} = 5.5$, Red line: $\frac{G_r}{R_e^2} = 10.5$.

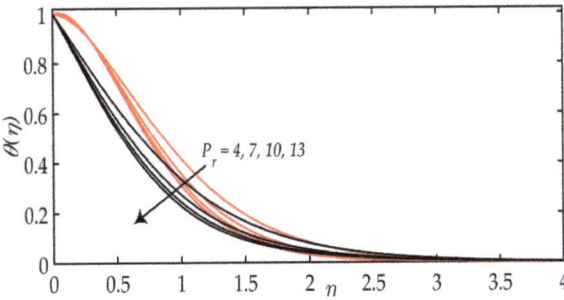

Figure 5. Variation of P_r and M on temperature profile. Black line: $M = 0$, Red line: $M = 3$.

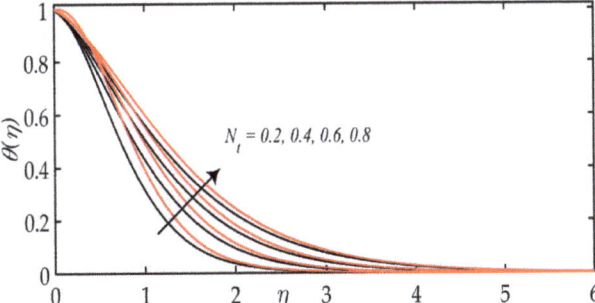

Figure 6. Variation of N_b and N_t on temperature profile. Black line: $N_b = 0.2$, Red line: $N_b = 0.5$.

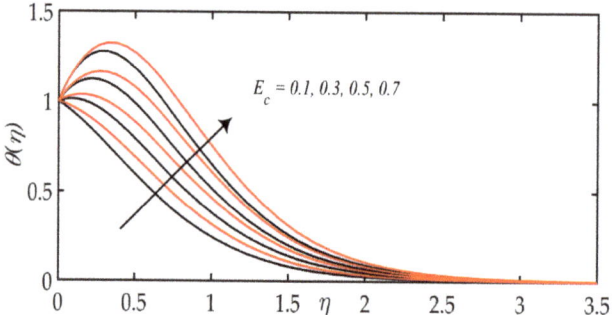

Figure 7. Variation of N_b and E_c on temperature profile. Black line: $N_b = 0.2$, Red line: $N_b = 0.5$.

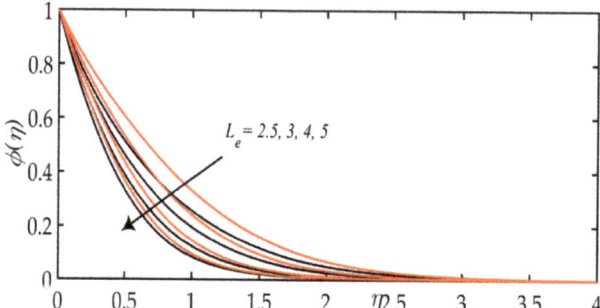

Figure 8. Variation of N_t and L_e on concentration profile. Solid line: $N_t = 5$, Dotted line: $N_t = 10$.

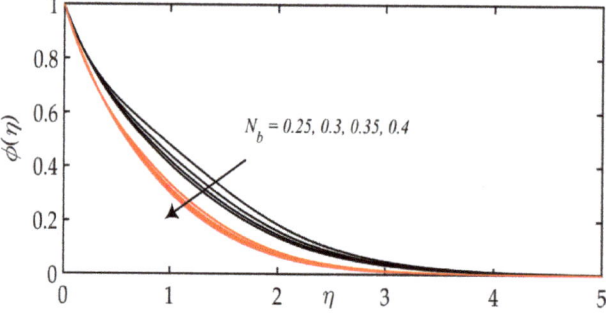

Figure 9. Variation of N_b and L_e on concentration profile. Solid line: $L_e = 1.5$, Dotted line: $L_e = 2.0$.

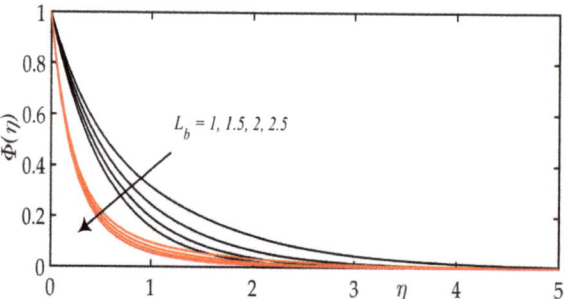

Figure 10. Variation of P_e and L_b on microorganism profile. Solid line: $P_e = 0.5$, Dotted line: $P_e = 2.0$.

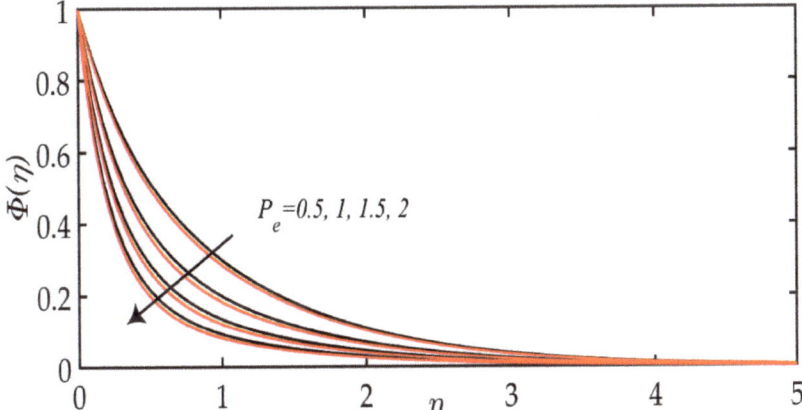

Figure 11. Variation of P_e and Ω_d on microorganism profile. Solid line: $\Omega_d = 0.1$, Dotted line: $\Omega_d = 0.3$.

Figure 2 shows that the velocity distribution decelerates by enhancing the permeability parameter β_D. It can be seen as a deceleration in momentum by taking increment in M, due to the existing body-force brought through the magnetic field. A well-known Lorentz force, causing a decrement for the velocity overshooting and momentum boundary-layer thickness. In Figure 3, it is recorded that by taking the increment in N_r, the velocity distribution decreases as a result of an increase in the negate buoyancy generated through the existence of nanoparticles, while for the Richardson number G_r/R_e^2, it is also found to be decreased by enhancing the values of the Richardson number. Figure 4 portrays that, by taking an increment in R_b, the velocity distribution falls because the power of convection due to bioconvection boosted against the convection of buoyancy force. In contrast, for the Richardson number G_r/R_e^2, it is found to be decreased by enlarging the values of the Richardson number.

The influence of Prandtl number P_r, Hartmann number M, the Brownian-motion parameter N_b, the thermophoresis parameter N_t, local Eckert number E_c, for various numeric values are drawn through Figures 5–7. From Figure 5, it is determined that by taking an increment in Prandtl number P_r, the temperature distribution slows down, although by enhancing the Hartmann number M, it accelerates the temperature distribution. Figure 6 is adorned for the effect of thermophoresis parameter N_t and the Brownian-motion parameter N_b of the temperature distribution, and also notice that the temperature distribution boosts for both parameters by enhancing the numeric value of these parameters. The influence of Eckert number E_c and the Brownian-motion parameter N_b of the temperature distribution is sketched in Figure 7, and it is noticed that the temperature distribution boosts for both parameters by enhancing the numeric value of these parameters. The further heating due to the interacting of the fluid to nanoparticles because of the Brownian-motion, thermophoresis impact, and viscous dissipation enhance the temperature. Therefore, the thickness of the thermal boundary layer turns into high-thicker across the larger numeric of N_t, N_b, and E_c, and temperature overshoots into the neighborhood of the stretched permeable sheet.

The impact of bioconvection Lewis number L_e, the Brownian-motion parameter N_b, the thermophoresis parameter N_t, the bioconvection L_b, Peclet number P_e, and the microorganisms concentration difference parameter Ω_d for concentration distribution and the density of motile microorganisms successively are shown through Figures 8–11. Figure 8 is adorned for the effect of bioconvection Lewis number L_e and thermophoresis parameter N_t of the concentration distribution, and also observed that the concentration distribution decelerates by enhancing the numeric value of Lewis number L_e because the convection of nanoparticles enhances by adding more immense value in Lewis number L_e, and also found decremented by taking increment in thermophoresis parameter N_t.

Therefore, the nanoparticles' boundary layer thickening has been developed to grow thicker with N_t. From Figure 9, it observed that by enlarging the Brownian-motion parameter N_b and the bioconvection Lewis number L_e, the concentration profile slows down for both the parameters. The graphical behavior of various values of the bioconvection L_b and Peclet number P_e in Figure 10 portrays that a decrement in the density for motile microorganisms quickly occurs by enhancing the bioconvection L_b and Peclet number P_e. That is, the density of motile microorganisms sharply slow downed, and indeed, by strengthening the bioconvection Lewis number L_b and Peclet number P_e, interprets the decrement of microorganisms diffusion, hence the density and boundary layer thickness together downturns for motile microorganisms by rising value in L_b and P_e. The influence of the Peclet number P_e and the concentration of the microorganisms varying parametric quantity Ω_d is sketched in Figure 11, and it is found that the density of motile microorganisms slowed down by enhancing both the parameters, i.e., the Peclet number P_e and the concentration of the microorganisms varying parametric quantity Ω_d.

5. Conclusions

The present analysis deals with the behavior of the swimming of the gyrotactic microorganisms in nanofluid propagating past a stretching permeable surface. The effects of porosity and magnetic field are also examined. The Successive Local Linearization Method is found very efficient in solving the nonlinear coupled equations. The SLLM is used across the shooting method, which utilizes the initial guesses for the missing slopes (a Newton-Raphson based iteration method for solving boundary value problem). To see the limitations and validations of this proposed computational methodology, the results are compared with previously published data and shooting method, and it is noticed that the obtained numerical results are in perfect accord with the other similar method. The significant findings of key parameters along with the performance of SLLM are.

1. It is observed that the permeability parameter and the magnetic field retard the velocity distribution while Richardson parameter boosts the velocity distribution.
2. Bioconvection Rayleigh number and Buoyancy proportion parametric quantity diminish the velocity distribution.
3. Prandtl number elevates the temperature distribution while it has been demoted by enlarging the values of the magnetic field.
4. The thermophoresis parameter and Eckert number significantly uplift the temperature distribution.
5. Brownian-motion parameter and Lewis number suppress the concentration distribution, whereas an enhancement in the thermophoresis parameter actively elevates the concentration profile.
6. Bioconvection Lewis number and Peclet number significantly demote the motile microorganism profile.
7. The SLLM algorithm is smooth to establish and employ because the scheme based on a simple univariate linearization of nonlinear functions.
8. The convergence speed of the SLLM technique can be willingly upgraded by applying successive over relaxation (SOR) method, the convergence was improved through relaxation parameter in the study.

Author Contributions: Conceptualization, H.H. Investigation, A.S. Methodology, M.M.B. Validation, L.Z. Writing-review & editing, R.E. All authors have read and agreed to the published version of the manuscript.

Funding: This research was funded by King Fahd University of Petroleum & Minerals, Dhahran, Saudi Arabia (ORCP2500).

Acknowledgments: R. Ellahi gratefully thanks to King Fahd University of Petroleum & Minerals, Dhahran, Saudi Arabia to honor him with the Chair Professor at KFUPM.

Conflicts of Interest: The authors declare no conflict of interest.

Nomenclature

a, b	Constants
B_0	Magnetic field
C_{gx}	Skin friction coefficient
$(c_p)_f$	Heat capacity of fluid (J/K)
$(c_p)_p$	Heat capacity of nanoparticles (J/K)
$(c_p)_s$	Heat capacity of solid fraction (J/K)
D_B	Brownian-diffusion coefficient (m²/s)
D_M	Diffusivity of microorganisms (m²/s)
D_T	Thermophoresis diffusion coefficient (m²/s)
E_c	Eckert number
g	Dimensionless stream function
k	Porosity parameter (H/m)
k_t	Thermal conductivity (W/m.K)
L_b	Bioconvection Lewis number
L_e	Lewis number
$\widetilde{N}$	Density for motile microorganism
N_b	Brownian motion parameter
N_r	Buoyancy proportion parameter
N_t	Thermophoresis parameter
Nu_x	Nusselt number
$\widetilde{p}$	Pressure (Pa)
P_e	Bioconvection Peclet number
P_r	Prandtl number (m²/s)
q_m	Local mass flux past the surface (kg/m²s)
q_w	Local heat flux past the surface (W/m²)
R_b	Bioconvection Rayleigh number
Re_x	Local Reynolds number
$Sh_{\widetilde{x}}$	Sherwood number
$\widetilde{T}_w$	Temperature of the wall (K)
$\widetilde{T}_\infty$	Ambient temperature (K)
$\widetilde{U}_w$	Stretching sheet velocity (m/s)
$\widetilde{u}, \widetilde{v}$	Components of velocity (m/s)
W_C	Heat capacitance of the nanoparticle (J/K)
$\hat{x}, \hat{y}$	Cartesian coordinates
Greek symbols	
$\overline{\alpha}$	Thermal diffusivity (m²/s)
β_D	Permeability parameter (m²)
γ	Average volume for a microorganism (m³)
θ	Temperature profile (K)
μ_{nf}	Dynamic viscosity (m²/s)
ν_{nf}	Kinematic viscosity of nanofluid (m²/s)
κ_{nf}	Thermal conductivity of nanofluid (W/mK)
ρ_f	Density of fluid (kg/m³)
ρ_p	Density of nanoparticles
ρ_m	Density of microorganisms
σ	Electrical conductivity (S/m)
$\overline{\sigma}$	Stefan-Boltzmann constant (J/K)
$\widetilde{\sigma}$	Dimensionless constant
τ_w	Shear stress (Pa)
Φ	Motile microorganism profile
ϕ	Nanoparticle volume fraction (m³/mol)
Ω_d	Microorganisms concentration variance parameter

References

1. Sakiadis, B.C. Boundary layer behavior on continuous solid surface, I: Boundary layer equations for two dimensional and axisymmetric flow. *AIChE J.* **1961**, *26–28*, 221–235. [CrossRef]
2. Crane, L.J. Flow past a stretching plane. *Z. Angew. Math. Phys.* **1970**, *21*, 645–647. [CrossRef]
3. Banks, W.H.H. Similarity solutions of the boundary layer equations for a stretching wall. *J. de Mécanique Théorique et Appliquée* **1983**, *2*, 92–375.
4. Gupta, P.S.; Gupta, A.S. Heat and mass transfer on a stretching sheet with suction or blowing. *Can. J. Chem. Eng.* **1977**, *55*, 744–746. [CrossRef]
5. Bujurke, N.M.; Biradar, S.N.; Hiremath, P.S. Second order fluid flow past a stretching sheet with heat transfer. *Z. Angew. Math. Phys.* **1987**, *38*, 890–892. [CrossRef]
6. Cortell, R. Viscous flow and heat transfer over a nonlinearly stretching sheet. *Appl. Math. Comput.* **2007**, *184*, 864–873. [CrossRef]
7. Shahzad, A.; Ali, R.; Khan, M. On the exact solution for axisymmetric flow and heat transfer over a nonlinear radially stretching sheet. *Chin. Phys. Lett.* **2012**, *29*, 084705. [CrossRef]
8. Hayat, T.; Shafiq, A.; Alsaedi, A.; Awais, M. MHD axisymmetric flow of third grade fluid between stretching sheets with heat transfer. *Comput. Fluids* **2013**, *86*, 103–108. [CrossRef]
9. Shateyi, S.; Makinde, O.D. Hydromagnetic stagnation-point flow towards a radially stretching convectively heated disk. *Math. Prob. Eng.* **2013**, *2013*, 616947. [CrossRef]
10. Khan, M.; Malik, R.; Munir, A. Mixed convective heat transfer to Sisko fluid over a radially stretching sheet in the presence of convective boundary conditions. *AIP Adv.* **2015**, *5*, 087178. [CrossRef]
11. Awais, M.; Hayat, T.; Irum, S.; Alsaedi, A. Heat Generation/Absorption Effects in a Boundary Layer Stretched Flow of Maxwell Nanofluid: Analytic and Numeric Solutions. *PLoS ONE* **2015**, *10*, e0129814. [CrossRef] [PubMed]
12. Ghalambaz, M.; Sheremet, M.A.; Pop, I. Free convection in a parallelogrammic porous cavity filled with a nanofluid using tiwari and das' nanofluid model. *PLoS ONE* **2015**, *10*, e0126486. [CrossRef] [PubMed]
13. Choi, S.; Zhang, Z.; Yu, W.; Lockwood, F.; Grulke, E. Anomalous thermal conductivity enhancement in nanotube suspensions. *Appl. Phys. Lett.* **2001**, *79*, 2252–2254. [CrossRef]
14. Kuznetsov, A.V.; Nield, D.A. Natural convective boundary-layer flow of a nanofluid past a vertical plate. *Int. J. Therm Sci.* **2010**, *49*, 243–247. [CrossRef]
15. Noghrehabadadi, A.; Pourrajab, R.; Ghalambaz, M. Flow and heat transfer of nanofluids over stretching sheet taking into account partial slip and thermal convective boundary conditions. *Heat Mass Transf.* **2013**, *49*, 1357–1366. [CrossRef]
16. Zaraki, A.; Ghalambaz, M.; Chamkha, A.J.; Ghalambaz, M.; De Rossi, D. Theoretical analysis of natural convection boundary layer heat and mass transfer of nanofluids: Effects of size, shape and type of nanoparticles, type of base fluid and working temperature. *Adv. Powder Technol.* **2015**, *26*, 935–946. [CrossRef]
17. Ferdows, M.; Khan, M.S.; Alam, M.M.; Sun, S. MHD mixed convective boundary layer flow of a nanofluid through a porous medium due to an exponentially stretching sheet. *Math. Probl. Eng.* **2012**, *2012*, 408528. [CrossRef]
18. Bidin, B.; Nazar, R. Numerical solution of the boundary layer flow over an exponentially stretching sheet with thermal radiation. *Eur. J. Sci. Res.* **2009**, *33*, 710–717.
19. Shakhaoath Khan, M.D.; Mahmud Alamb, M.D.; Ferdows, M. Effects of magnetic field on radiative flow of a nanofluid past a stretching sheet. *Procedia Eng.* **2013**, *56*, 316–322. [CrossRef]
20. Mabood, F.; Khan, W.A.; Ismail, A.I.M. MHD boundary layer flow and heat transfer of nanofluids over a nonlinear stretching sheet: A numerical study. *J. Magn. Magn. Mater.* **2015**, *374*, 569–576. [CrossRef]
21. Freidoonimehr, N.; Rashidi, M.M.; Jalilpour, B. MHD stagnation-point flow past a stretching/shrinking sheet in the presence of heat generation/absorption and chemical reaction effects. *J. Braz. Soc. Mech. Sci. Eng.* **2016**, *38*, 1999–2008. [CrossRef]
22. Makinde, O.; Animasaun, I. Bioconvection in MHD nanofluid flow with nonlinear thermal radiation and quartic autocatalysis chemical reaction past an upper surface of a paraboloid of revolution. *Int. J. Therm. Sci.* **2016**, *109*, 159–171. [CrossRef]

23. Pour, M.S.; Nassab, S.G. Numerical investigation of forced laminar convection flow of nanofluids over a backward facing step under bleeding condition. *J. Mech.* **2012**, *28*, N7–N12. [CrossRef]
24. Abu-Nada, E. Numerical prediction of entropy generation in separated flows. *Entropy* **2005**, *7*, 234–252. [CrossRef]
25. Marin, M.; Vlase, S.; Ellahi, R.; Bhatti, M.M. On the partition of energies for backward in time problem of the thermoelastic materials with a dipolar structure. *Symmetry* **2019**, *11*, 863. [CrossRef]
26. Almutairi, F.; Khaled, S.M.; Ebaid, A. MHD flow of nanofluid with homogeneous-heterogeneous reactions in a porous medium under the influence of second-order velocity slip. *Mathematics* **2019**, *7*, 220. [CrossRef]
27. Khan, U.; Zaib, A.; Kahn, I.; Nasar, K.S.; Baleanu, D. Insights into the stability of mixed convective Darcy–Forchheimer flows of cross liquids from a vertical plate with consideration of the significant impact of velocity and thermal slip conditions. *Mathematics* **2020**, *8*, 31. [CrossRef]
28. Wan, N.A.; Maleki, A.; Nazari, M.A.; Tlili, I.; Shadloo, M.S. Thermal conductivity modeling of nanofluids contain MgO particles by employing different approaches. *Symmetry* **2020**, *12*, 206.
29. Jamalabadi, M.Y.A.; Ghasemi, M.; Alamian, R.; Wongwises, S.; Afrand, M.; Shadloo, M.S. Modeling of subcooled flow boiling with nanoparticles under the influence of a magnetic field. *Symmetry* **2019**, *11*, 1275. [CrossRef]
30. Safaei, M.R.; Ahmadi, G.; Goodarzi, M.S.; Shadloo, M.S.; Goshayeshi, H.R.; Dahari, M. Heat transfer and pressure drop in fully developed turbulent flows of graphene nanoplatelets–silver/water nanofluids. *Fluids* **2016**, *1*, 20. [CrossRef]
31. Shahrestani, M.I.; Maleki, A.; Shadloo, M.S.; Tlili, I. Numerical investigation of forced convective heat transfer and performance evaluation criterion of Al_2O_3/water nanofluid flow inside an axisymmetric microchannel. *Symmetry* **2020**, *12*, 120. [CrossRef]
32. Ellahi, R.; Hussain, F.; Abbas, S.A.; Sarafraz, M.M.; Goodarzi, M.; Shadloo, M.S. Study of two-phase newtonian nanofluid flow hybrid with hafnium particles under the effects of slip. *Inventions* **2020**, *5*, 6. [CrossRef]
33. Alamri, S.Z.; Ellahi, R.; Shehzad, N.; Zeeshan, A. Convective radiative plane poiseuille flow of nanofluid through porous medium with slip: An application of Stefan blowing. *J. Mol. Liq.* **2019**, *273*, 292–304. [CrossRef]
34. Alamri, S.Z.; Khan, A.A.; Azeez, M.; Ellahi, R. Effects of mass transfer on MHD second grade fluid towards stretching cylinder: A novel perspective of cattaneo–christov heat flux model. *Phys. Lett. A* **2019**, *383*, 276–281. [CrossRef]
35. Riaz, R.; Ellahi, R.; Bhatti, M.M.; Marin, M. Study of heat and mass transfer in the eyring-powell model of fluid propagating peristaltically through a rectangular complaint channel. *Heat Transf. Res.* **2019**, *50*, 1539–1560. [CrossRef]
36. Bhatti, M.M.; Ellahi, R.; Zeeshan, A.; Marin, M.; Ijaz, N. Numerical study of heat transfer and hall current impact on peristaltic propulsion of particle-fluid suspension with complaint wall properties. *Mod. Phys. Lett. B* **2019**, *33*, 1950439. [CrossRef]
37. Ellahi, R.; Sait, S.M.; Shehzad, N.; Ayaz, Z. A hybrid investigation on numerical and analytical solutions of electro-magnetohydrodynamics flow of nanofluid through porous media with entropy generation. *Int. J. Numer. Methods Heat Fluid Flow* **2020**, *30*, 834–854. [CrossRef]
38. Sarafraz, M.M.; Pourmehran, O.; Yang, B.; Arjomandi, M.; Ellahi, R. Pool boiling heat transfer characteristics of iron oxide nano-suspension under constant magnetic field. *Int. J. Therm. Sci.* **2020**, *147*, 106131. [CrossRef]
39. Platt, J.R. "Bioconvection patterns" in cultures of free-swimming organisms. *Science* **1961**, *133*, 1766–1767. [CrossRef]
40. Ghorai, S.; Hill, N.A. Wavelengths of gyrotactic plumes in bioconvection. *Bull. Math. Biol.* **2000**, *62*, 429–450. [CrossRef]
41. Kuznetsov, A.V.; Avramenko, A.A. Effect of small particles on the stability of bioconvection in a suspension of gyrotactic microorganisms in a layer of finite depth. *Int. Commun. Heat Mass Transf.* **2004**, *31*, 1–10. [CrossRef]
42. Khan, W.A.; Makinde, O.D. MHD nanofluid bioconvection due to gyrotactic microorganisms over a convectively heat stretching sheet. *Int. J. Therm. Sci.* **2014**, *81*, 118–124. [CrossRef]
43. Raees, A.; Hang, X.U.; Qiang, S.U.N.; Pop, I. Mixed convection in gravity-driven nano-liquid film containing both nanoparticles and gyrotactic microorganisms. *Appl. Math. Mech.* **2015**, *36*, 163–178. [CrossRef]

44. Kuznetsov, A.V.; Nield, D.A. Double-diffusive natural convective boundary layer flow of a nanofluid past a vertical surface. *Int. J. Therm. Sci.* **2011**, *50*, 712–717. [CrossRef]
45. Sivaraj, R.; Animasaun, I.L.; Olabiyi, A.S.; Saleem, S.; Sandeep, N. Gyrotactic microorganisms and thermoelectric effects on the dynamics of 29 nm CuO-water nanofluid over an upper horizontal surface of paraboloid of revolution. *Multidiscip. Model. Mater. Struct.* **2018**, *14*, 695–721. [CrossRef]
46. Amirsom, N.A.; Uddin, M.; Basir, M.; Faisal, M.; Kadir, A.; Bég, O.A.; Ismail, M.; Izani, A. Computation of melting dissipative magnetohydrodynamic nanofluid bioconvection with second-order slip and variable thermophysical properties. *Appl. Sci.* **2019**, *9*, 2493. [CrossRef]
47. Waqas, H.; Khan, S.U.; Imran, M.; Bhatti, M.M. Thermally developed falkner–skan bioconvection flow of a magnetized nanofluid in the presence of a motile gyrotactic microorganism: Buongiorno's nanofluid model. *Phys. Scr.* **2019**, *94*, 115304. [CrossRef]
48. Waqas, H.; Khan, S.U.; Hassan, M.; Bhatti, M.M.; Imran, M. Analysis on the bioconvection flow of modified second-grade nanofluid containing gyrotactic microorganisms and nanoparticles. *J. Mol. Liq.* **2019**, *291*, 111231. [CrossRef]
49. Ferdows, M.; Reddy, M.G.; Alzahrani, F.; Sun, S. Heat and mass transfer in a viscous nanofluid containing a gyrotactic micro-organism over a stretching cylinder. *Symmetry* **2019**, *11*, 1131. [CrossRef]
50. Khan, W.A.; Rashad, A.M.; Abdou, M.M.M.; Tlili, I. Natural bioconvection flow of a nanofluid containing gyrotactic microorganisms about a truncated cone. *Eur. J. Mech. B Fluids* **2019**, *75*, 133–142. [CrossRef]
51. Ferdows, M.; Reddy, M.G.; Sun, S.; Alzahrani, F. Two-dimensional gyrotactic microorganisms flow of hydromagnetic power law nanofluid past an elongated sheet. *Adv. Mech. Eng.* **2019**, *11*. [CrossRef]
52. Trefethen, L.N. *Spectral Methods in MATLAB*; SIAM: Philadelphia, PA, USA, 2000; Volume 10.
53. Akbar, N.S.; Khan, Z.H. Magnetic field analysis in a suspension of gyrotactic microorganisms and nanoparticles over a stretching surface. *J. Magn. Magn. Mater.* **2016**, *410*, 72–80. [CrossRef]
54. Khan, W.A.; Pop, I. Boundary-layer flow of a nanofluid past a stretching sheet. *Int. J. Heat Mass Transf.* **2010**, *53*, 2477–2483. [CrossRef]
55. Wang, C.Y. Free convection on a vertical stretching surface. *J. Appl. Math. Mech.* **1989**, *69*, 418–420. [CrossRef]

© 2020 by the authors. Licensee MDPI, Basel, Switzerland. This article is an open access article distributed under the terms and conditions of the Creative Commons Attribution (CC BY) license (http://creativecommons.org/licenses/by/4.0/).

Article

Insights into the Stability of Mixed Convective Darcy–Forchheimer Flows of Cross Liquids from a Vertical Plate with Consideration of the Significant Impact of Velocity and Thermal Slip Conditions

Umair Khan [1], Aurang Zaib [2], Ilyas Khan [3,*], Kottakkaran Sooppy Nisar [4] and Dumitru Baleanu [5,6,7]

1. Department of Mathematics and Social Sciences, Sukkur IBA University, Sukkur 65200, Sindh, Pakistan; umairkhan@iba-suk.edu.pk
2. Department of Mathematical Sciences, Federal Urdu University of Arts, Science & Technology, Karachi 75300, Gulshan-e-Iqbal, Pakistan; aurangzaib@fuuast.edu.pk
3. Faculty of Mathematics and Statistics, Ton Duc Thang University, Ho Chi Minh City 72915, Vietnam
4. Department of Mathematics, College of Arts and Sciences, Prince Sattam bin Abdulaziz University, Wadi Aldawaser 11991, Saudi Arabia; n.sooppy@psau.edu.sa
5. Department of Mathematics, Cankaya University, Ankara 06790, Turkey; Baleanu@mail.cmuh.org.tw
6. Department of Medical Research, China Medical University Hospital, China Medical University, Taichung 40447, Taiwan
7. Institute of Space Sciences, Magurele-Bucharest 077125, Romania
* Correspondence: ilyaskhan@tdtu.edu.vn

Received: 30 November 2019; Accepted: 19 December 2019; Published: 24 December 2019

Abstract: This paper reflects the effects of velocity and thermal slip conditions on the stagnation-point mixed convective flow of Cross liquid moving over a vertical plate entrenched in a Darcy–Forchheimer porous medium. A Cross liquid is a type of non-Newtonian liquid whose viscosity depends on the shear rate. The leading partial differential equations (PDEs) are altered to nonlinear ordinary differential equations (ODEs) via feasible similarity transformations. These transmuted equations are computed numerically through the bvp4c solver. The authority of sundry parameters on the temperature and velocity distributions is examined graphically. In addition, the characteristics of heat transfer are analyzed in the presence of the impact of drag forces. The outcomes reveal that the permeability parameter decelerates the drag forces and declines the rate of heat transfer in both forms of solutions. Moreover, it is found that the drag forces decline with the growing value of the Weissenberg parameter in the upper branch solutions, while a reverse trend is revealed in the lower branch solutions. However, the rate of heat transfer shows a diminishing behavior with an increasing value of the Weissenberg parameter.

Keywords: slip effects; mixed convection flow; cross fluid; Darcy–Forchheimer model

1. Introduction

Many liquids such as detergents, printer ink, animal blood, foodstuff, paints, polymer fluids, etc., transform their properties of flow subjected to operating shear stress, and thus diverge from viscous fluids. These fluids are identified as non-Newtonian substances. Numerous researchers have reported different non-Newtonian fluid models and a few of them are micropolar, Casson, Burgers, Sisko, Maxwell, Oldroyd-B, generalized Burgers, and Cross models, etc. In this paper, we report the Cross liquid [1] model, which states features of stress. In addition, this model sufficiently distinguishes the flow in the region of the power law and high, as well as low, regions of shear rates. In this study, unlike the fluid of power law, first, we achieve a finite viscosity as the rate of shear disappears which also involves a time constant owing to the importance of this model in numerous industrial and engineering computations. Utilization of Cross fluid in industries comprises the polymer latex of the aqueous solution and blood, as well as solutions of synthesis polymeric. Khan et al. [2] inspected the flow of Cross liquid through heat transfer from a planar stretched sheet and found the numerical solution through the shooting technique. The impact of electric field with the characteristic of heat transfer involving Cross liquid from a stretched sheet was scrutinized by Hayat et al. [3] who found that the liquid velocity grew with a rising Weissenberg parameter while temperature distribution decayed due to the Pr. Khan et al. [4] scrutinized the axisymmetric flow and the characteristic of heat transfer containing Cross liquid using a radial stretched sheet and observed that the power-law index raised the structure of the velocity boundary layer. Ijaz Khan et al. [5] scanned the activation energy impact on the magnetic flow of Cross liquid from a stretched surface. Another study, by Ijaz Khan et al. [6], surveyed the magnetic influence on mixed convective flow involving Cross nanofluid with activation energy. Recently, Azam et al. [7] applied the concept of solar energy on time-dependent flow in the presence of Cross nanofluid from a stretched sheet with nonlinear radiation.

The impact of non-Newtonian liquids in the porous media is significant in the fields of engineering and industries due to its numerous applications such as mud injections, cement or slurry grouts to strengthen soils, blood circulation through the kidney, insulation of fibrous, electrochemistry, and drilling liquid injection in rocks for ornamental oil recovery, or for the fortification of the well, etc. Bejan et al. [8], Vafai [9], and Vadasz [10] discussed further applications in their books. Darcy's law has been utilized generally to inspect the behavior of flow in a porous medium. However, the connection between the velocity of flow and pressure gradient at rates of high flow cannot be modeled through Darcy's law (Spivey et al. [11]). There is further indication that at a high rate of flow, the non-Darcy involve several subsurface systems of biological porous and engineering porous flow [12–14]. Forchheimer [15] included a term of velocity squared in the Darcy to analyze the boundary and inertia aspects. This term is constantly applied to larger Reynolds numbers. Rashidi et al. [16] discovered the influence of electric field on fluid flow with the characteristic of heat transfer in a Darcy–Brinkman–Forchheimer medium. The impact of variable thermal conductivity of Darcy–Forchheimer flow in the presence of Cattaneo–Christov heat-flux was considered by Hayat et al. [17]. In another paper, Hayat et al. [18] examined the non-Newtonian viscoelastic fluid involving nanoliquid through nonlinear stretched surface engrossed in the Darcy–Forchheimer porous medium. Kang et al. [19] employed finite difference technique to discuss the Neumann condition for the general Darcy–Forchheimer problem. Hayat et al. [20] explored the homogenous-heterogeneous reaction of viscous liquid in a Darcy–Forchheimer porous medium through a curved stretched surface. They scrutinized that the porosity and inertia parameters produce larger temperature. Recently, Rasool [21] considered the Darcy–Forchheimer flow to investigate electric field containing nanoparticle through a nonlinear stretched surface. They observed that the mass and heat flux decline due to porosity while drag force is enhanced. A few other similar studies are given in [22–24].

As mentioned above, the present literature is packed with works comprising the heat transfer characteristics of boundary-layer flow involving Newtonian and non-Newtonian liquids. In addition, the research regarding the Darcy–Forchheimer flow through heat transport comprising Cross liquid has disclosed a vital pledge in industrial and environmental systems, such as the process of fermentation, petroleum resources, usage of geothermal energy, production of crude oil, grain storage, etc. However, the review of literature revealed that no one has considered the impact of slip effects on mixed convection flow of Cross liquid in the porous media. Therefore, in this research, we focus our attention to the Darcy–Forchheimer flows involving non-Newtonian Cross liquids from a vertical plate with mixed convection and slip effects. Similarity variables are employed to metamorphose the PDEs into nonlinear ODE's. The metamorphosed system is then exercised through bvp4c solver. The dual nature of solutions is acquired in opposing flow. The vital constraints in the flow field are discussed via graphical portraits.

2. Formulation of the Problem

Consider a steady incompressible flow of Cross liquid past a vertical plate in a porous medium with slip impacts. The x-axis is taking along the plate and the y-axis perpendicular to it, as illustrated in Figure 1.

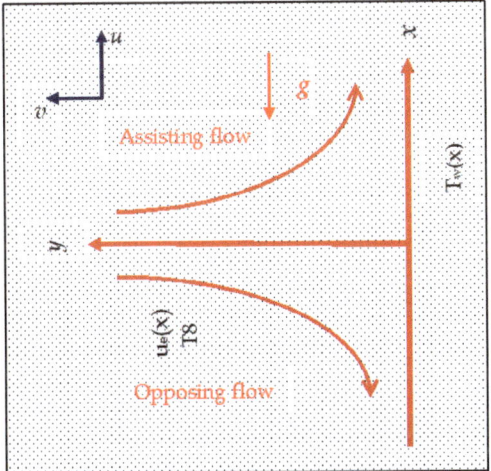

Figure 1. Physical diagram of the problem.

It is presumed that the free stream velocity $u_e(x) = bx$ and the wall temperature $T_w(x) = T_\infty + cx$ vary linearly, where b and c are two constants and T_∞ is the temperature away from the plate. We utilize the Darcy–Forchheimer model in which the square of the velocity factor is included. In addition, the rheology equations of Cross liquid in term of viscosity are

$$\begin{cases} \tau = -\breve{p}I + A_1 \breve{\mu}(\dot{\gamma}), \\ \breve{\mu} - \breve{\mu}_\infty = \dfrac{\breve{\mu}_0 - \breve{\mu}_\infty}{1 + \left(\breve{\Gamma}\dot{\gamma}\right)^n}, \end{cases} \quad (1)$$

Here, n the power-law index, $\breve{\Gamma}$ the time constant, A_1 the first tensor of Rivlin–Ericksen and defined as $A_1 = (\text{grad } V) + (\text{grad } V)^T$, $\breve{p}$ the pressure, I the identity vector, $\dot{\gamma}$ the rate of shear for the current model is taken as

$$\dot{\gamma} = \left[4\left(\frac{\partial u}{\partial x}\right)^2 + \left(\frac{\partial u}{\partial y} + \frac{\partial v}{\partial x}\right)\right]^{1/2}, \quad (2)$$

whereas $\breve{\mu}_0$ and $\breve{\mu}_\infty$ represent the zero and infinite shear rates, respectively. In the present study, $\breve{\mu}_\infty$ is considered to be zero. Therefore, Equation (1) can be written as

$$\breve{\mu} = \frac{\breve{\mu}_0}{1 + \left(\breve{\Gamma}\dot{\gamma}\right)^n}. \quad (3)$$

Keeping in mind that the temperature and velocity of the two-dimensional (2D) fluid flow are considered in the forms $T = T(x,t)$ and $V = [u(x,y), v(x,y), 0]$, then the governing equations become

$$\frac{\partial v}{\partial y} + \frac{\partial u}{\partial x} = 0 \quad (4)$$

$$\frac{1}{\varepsilon^2}\left(u\frac{\partial u}{\partial x} - u_e\frac{du_e}{dx} + v\frac{\partial u}{\partial y}\right) = v_{eff}\frac{\partial}{\partial y}\left(\frac{\frac{\partial u}{\partial y}}{1 + \left\{\Gamma\frac{\partial u}{\partial y}\right\}^{1-n}}\right) - \frac{v(u - u_e)}{K_1} - \frac{C_F(u^2 - u_e^2)}{K^{1/2}} + g\beta_T(T - T_\infty) \quad (5)$$

$$\alpha_m\frac{\partial^2 T}{\partial y^2} - u\frac{\partial T}{\partial x} - v\frac{\partial T}{\partial y} = 0 \quad (6)$$

The physical boundary conditions are

$$u = L_1\frac{\partial u}{\partial y}, \; v = 0, \; T = T_w(x) + S_1\frac{\partial T}{\partial y} \text{ at } y = 0,$$
$$u \to u_e(x), \; T \to T_\infty \text{ as } y \to \infty. \quad (7)$$

Here, (v, u) signify, respectively, the velocity components in x- and y-directions, μ_{eff} the effective (or "apparent") viscosity, $v_{eff} = \mu_{eff}/\rho$ the effective kinematic viscosity, ρ the density, ε the porosity parameter, K_1 the porous medium permeability, k the thermal conductivity of fluid, C_F drag coefficient, α_m the thermal diffusivity, T the temperature, L_1 length of slip, and S_1 proportionality constant.

Following Rosali et al. [22], we set up the similarity transformation

$$\eta = y\sqrt{\frac{b}{\alpha_m}}, \; \psi = \sqrt{b\alpha_m}xf(\eta), \; \theta(\eta) = \frac{T - T_\infty}{T_w - T_\infty}. \quad (8)$$

Using the similarity transformation in the above PDEs we obtain

$$\varepsilon_1 f'''\left(1 + n(We f'')^{1-n}\right) + \left(\begin{array}{c} 1 + ff'' - (f')^2 + K(1 - f') + \\ B(1 - (f')^2) + \lambda K\theta \end{array}\right)\left(1 + (We f'')^{1-n}\right)^2 = 0, \quad (9)$$

$$\theta'' + \theta' f - \theta f' = 0. \quad (10)$$

The physical boundary conditions are

$$f'(0) = \gamma_1 f''(0), \; f(0) = 0, \; \theta(0) = 1 + \gamma_2\theta'(0) \text{ at } \eta = 0,$$
$$f'(\eta) \to 1, \; \theta(\eta) \to 0 \text{ as } \eta \to \infty. \quad (11)$$

Here, the parameters are used in the above ODE's are modified porosity, dimensionless permeability, mixed convection, inertia coefficient, velocity slip, and thermal slip. These are defined as $\varepsilon_1 = \frac{\varepsilon^2 v_{eff}}{\alpha_m} = \varepsilon^2 \Pr$, $K = \frac{\varepsilon^2 v}{K_1 b}$, $\lambda = \frac{\varepsilon^2 g\beta_T c}{b^2} = \frac{Ra_x}{Pe_x^2}\Pr$, $B = \frac{\varepsilon^2 u_e C_F}{b\sqrt{K}}$, $\gamma_1 = L_1\sqrt{\frac{b}{\alpha_m}}$, $\gamma_2 = \sqrt{\frac{b}{\alpha_m}}S_1$. Here, $Ra_x = \frac{\varepsilon^2 g\beta_T(T_w - T_\infty)x^3}{v_{eff}\alpha_m}$ is the Rayleigh number and $Pe_x = \frac{xu_e}{\alpha_m}$ is the Peclet number.

3. Skin Friction and Nusselt Number

The coefficients of skin friction C_f and Nusselt number Nu_x are identified as

$$C_f = \frac{2\tau_w}{\rho u_e^2} \text{ and } Nu_x = \frac{xq_w}{k(T_w - T_\infty)}, \quad (12)$$

where q_w and τ_w are identified as the heat flux and the shear stress, respectively, which are specified as

$$q_w = -k\frac{\partial T}{\partial y}\bigg|_{y=0} \text{ and } \tau_w = \left[\frac{\mu\frac{\partial u}{\partial y}}{\left(1 + \left(\Gamma\frac{\partial u}{\partial y}\right)^{1-n}\right)}\right]_{y=0}. \quad (13)$$

Utilizing Equation (8), we have

$$\frac{1}{2}\sqrt{\frac{Re_x}{Pr}} C_f = \frac{f''(0)}{\left(1 + (We f''(0))^{1-n}\right)} \text{ and } -\theta'(0) = \frac{Nu_x}{\sqrt{Pe_x}}. \quad (14)$$

4. Numerical Procedure

The nonlinear coupled ODEs (9) and (10) with boundary constraint (11) through the bvp4c by converting the leading ODEs to an initial value problem (IVP). In this method, it is further helpful to provide a fixed value to $\eta \to \infty$, say η_∞. The above-mentioned higher order equations are converted into a first-order system as follows:

$$f' = p, \quad (15)$$

$$p' = q, \quad (16)$$

$$q' = \frac{\left(1 + (We f'')^{1-n}\right)^2}{\varepsilon_1\left(1 + n(We f'')^{1-n}\right)} \left[p^2 - fq - 1 + K(1-p) - B\left(1 - p^2\right) - \lambda \theta\right], \quad (17)$$

$$\theta' = z, \quad (18)$$

$$z' = p\theta - fz, \quad (19)$$

with

$$f(0) = 0, \; p(0) = \gamma_1 q(0), \; \theta(0) = 1 + \gamma_2 z(0). \quad (20)$$

Numerically grip the system of Equations (15)–(20) as an IVP, we require that the values for $q(0)$ and $\theta(0)$ are needed, however these values are not mentioned. The initial estimated values for $q(0)$ and $\theta(0)$ are conjectured and bvp4c is pertained on MATLAB software to achieve accurate results. It is also noted that the multiple solutions are attained by setting different guesses. After that, the considered values of $\theta(\eta)$ and $f'(\eta)$ at $(\eta_\infty = 8)$ are evaluated with the boundary conditions $\theta(\eta_\infty) = 0$ and $f'(\eta_\infty) = 1$, in which the predictable values of $q(0)$ and $\theta(0)$ are prescribed by the Secant method to achieve a better guess for the solutions. The step size is considered as $\Delta \eta = 0.01$. The procedure is iteratively repeated until required solutions with an acceptable level of accuracy (i.e., up to 10^{-5}) to fulfill the criterion of convergence.

5. Physical Explanation

In this study, the dimensionless parameters that were appearing in the momentum and the energy equations and the value of these parameters were taken to be fixed for the computational purpose are given as $\lambda = -3.5$, $n = 0.5$, $We = 0.5$, $\gamma_1 = \gamma_2 = 0.5$, $\varepsilon_1 = 0.5$, $B = 0.1$. The graphical features of the embedded flow of fluids were captured in Figures 2–21 on the velocity, temperature profiles, the skin friction, and the local Nusselt number against the enormous distinct parameters. The numerical results with accessible conclusions are referenced in Table 1, which shows the authenticity of our problem by comparing the results with the available results in the literature. Additionally, the green lines throughout the study demonstrate the first solution, which is also called the upper branch solution while the red lines exhibit the second solution called the lower branch in all the invoked figures.

Table 1. Comparison of the values of $f''(0)$ for distinct values of ε_1 and λ when $K = 1$ and rest of variables are absent.

ε_1	λ	Rosali et al. [22]	Present Results
	−0.5	4.1508	4.1389
0.1	1	6.4874	6.4864
	2	7.7611	7.7614
	−0.5	1.8821	1.8838
0.5	1	2.8597	2.8453
	2	3.3944	3.3944
	−0.5	1.5967	1.6008
0.7	1	2.4074	2.4124
	2	2.8514	2.8499
	−0.5	1.3418	1.3438
1	1	2.0050	2.0050
	2	2.3690	2.3620

6. Deviation of the Skin Coefficient and the Local Nusselt Number

The graphical behavior of our solutions for the skin friction coefficient $0.5(\mathrm{Re}_x)^{\frac{1}{2}}(\mathrm{Pr})^{\frac{-1}{2}} C_f$ and the local Nusselt number $Nu_x(Pe_x)^{\frac{-1}{2}}$ by exercising the different parameters against the mixed convection parameter λ are shown in all invoked Figures 2–7. The existing of dual solutions is marked in all the aforementioned figures in the case of mixed convection opposing flow ($\lambda < 0$) while the outcome is unique for the phenomenon of mixed convection assisting flow ($\lambda > 0$). The influence of the modified porosity parameter ε_1 on the skin friction and the local Nusselt number versus λ is depicted in Figures 2 and 3, respectively. Figure 2 shows that the values of the skin friction decelerate in the first solution with enhancing ε_1 in the range of $(-4 \leq \lambda)$, while the reverse trend is seen in the range of $(\lambda < -4)$. Figure 3 scrutinizes that the values of the Nusselt number accelerate due to ε_1. It is also observed from these sketches that the physical realizable solution is represented by the green solid lines and the decline of the unstable solution is displayed by the red dotted lines. The critical values $|\lambda|$ enhance as ε_1 augments, suggesting that the modified porosity parameter delays the boundary-layer separation. In addition, it can be clearly observed from these figures that the skin friction as well as the Nusselt number augments as λ increases in the assisting flow, while the contrary behavior is observed in the opposing flow. Physically, in the assisting flow case, the favorable pressure gradient produces which augments the motion of liquid, which consequently raises the shear stress and heat transfer rate. In contrast, opposing flow guides to an adverse pressure gradient that delays the motion of liquid. The impacts of the Weissenberg number We and the inertia parameter B against λ on $\left(0.5(\mathrm{Re}_x)^{\frac{1}{2}}(\mathrm{Pr})^{\frac{-1}{2}} C_f\right)$ and $\left(Nu_x(Pe_x)^{\frac{-1}{2}}\right)$ are depicted in Figures 4–7. For the upper branch solution, both the momentum boundary layer and the thermal boundary layer become lower by changing the value of We, while the opposite behavior is marked for the lower branch solution as shown in Figures 4 and 5. Figures 6 and 7 suggest that the fall trend with augmenting B in the lower branch solution, while the upper branch solution is enhanced for the similar choice of B.

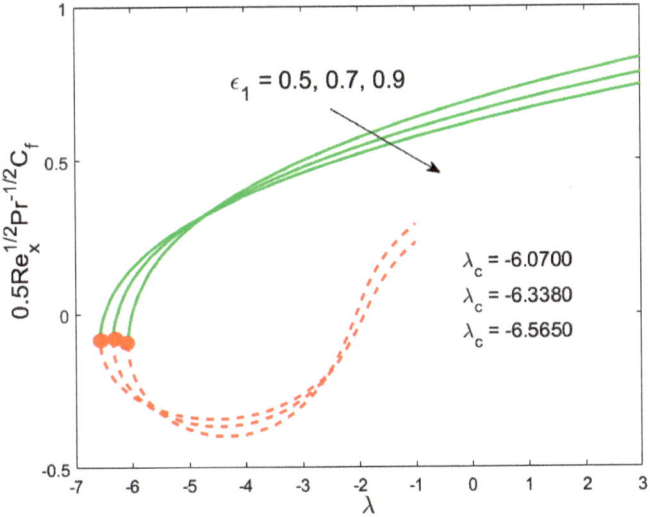

Figure 2. Influence of ε_1 on $0.5\mathrm{Re}_x^{1/2}\mathrm{Pr}^{-1/2}C_f$.

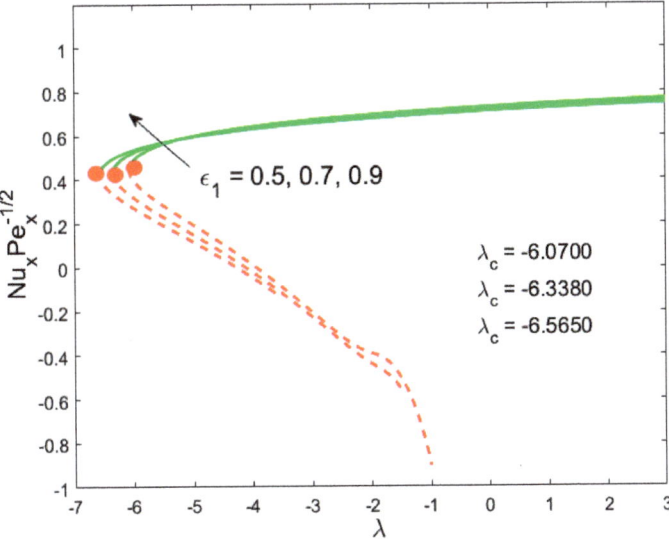

Figure 3. Influence of ε_1 on $Nu_x Pe_x^{-1/2}$.

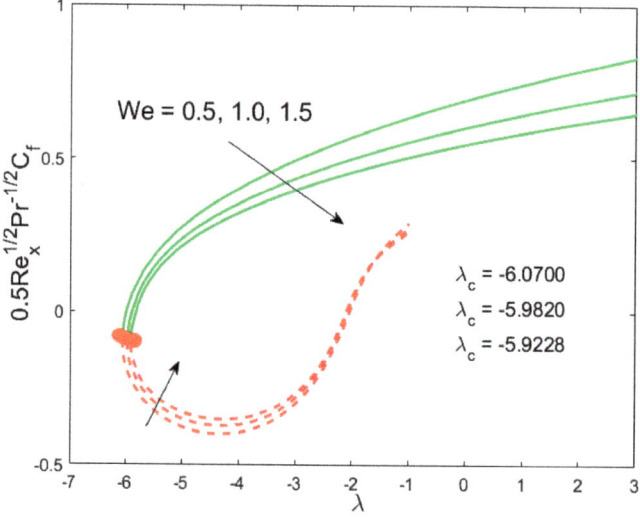

Figure 4. Influence of We on $0.5\text{Re}_x^{1/2}\text{Pr}^{-1/2}C_f$.

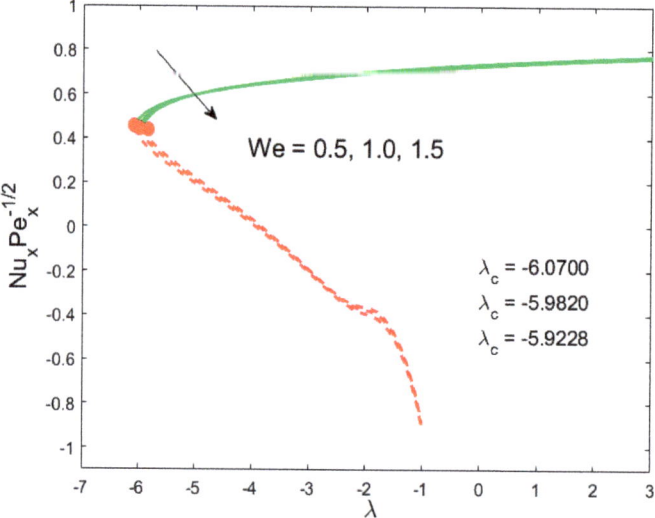

Figure 5. Influence of We on $Nu_x Pe_x^{-1/2}$.

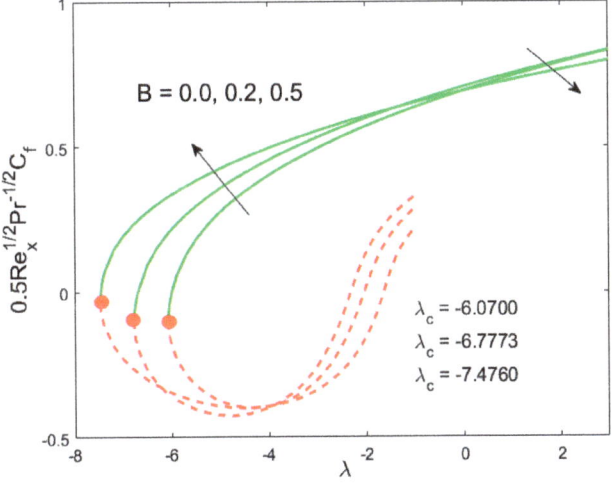

Figure 6. Influence of B on $0.5\text{Re}_x^{1/2}\text{Pr}^{-1/2}C_f$.

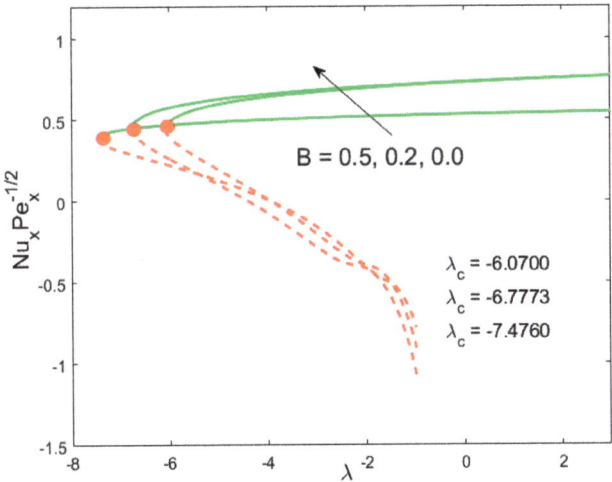

Figure 7. Influence of B on $Nu_x Pe_x^{-1/2}$.

7. Deviation of the Velocity and Temperature Fields

The analyses and the behavior are captured in Figures 8 and 9, respectively, showing the tasters of the $f'(\eta)$ and $\theta(\eta)$ profiles for the distinct values of the slip parameter γ_1 for both branches of the solutions, while the effects of the thermal slip parameter γ_2 on the velocity and temperature for various selected values are portrayed in Figures 10 and 11, respectively. Physically, when augmenting the values of γ_1, the wall shear stress insignificant and as a result, the momentum boundary layer (Figure 8) becomes larger and larger for both the upper branch and lower branch solutions, while the reverse trend is scrutinized for the temperature profile (Figure 9) due to escalating the γ_1. Figure 10 shows that the velocity of fluid rises with γ_2 in the first solution and declines in the second solution, while the opposite behavior is observed in the sketch of temperature, as shown in Figure 11. This is due to fact that the extra flow penetrates through the thermal boundary layer which consequently transmitted the additional heat and this guides in the decline of temperature distribution. Thus, for the authenticity of our solutions, it is clearly visible from behavior of momentum and temperature profiles in Figures 8–11 that these solutions satisfied the boundary conditions asymptotically. As shown in Figures 12–15, the behavior of the fluid flow is explored by exercising the dimensionless parameters We and n on $f'(\eta)$ and $\theta(\eta)$, respectively. The increment in the local Weissenberg number We and power-law index n, both the green solid lines, as well as the red dotted lines (first and second solution) are rising in Figures 12 and 13, while the contrary flow of fluid motion is noticeable corresponding to these parameters in Figures 13 and 15, respectively. From the physical view, the additional relaxation time is needed when the values of We increases and as a result, the velocity boundary layer and the fluid temperature was shrunk and declined in Figures 12 and 13, respectively. Figure 14 exhibits that the velocity profile increases due to the augmenting values of n in case of shear thinning and vice versa for the temperature profile which is invoked in Figure 15. Figure 16 shows the behavior of the permeability parameter K on $f'(\eta)$ as we enhance the parameter K, the upper solution is decelerated while the lower solution shows increasing behavior, whereas for the same parameter, the reverse behavior is noted in the temperature profile, as presented in Figure 17. Figure 18 illustrates that the liquid velocity enhances in both upper and lower solutions by changing the values of the modified porosity parameter ε_1, while the temperature profile behavior is shown in Figure 19, which decelerates in both branches of solutions as we boost up the value of ε_1. In Figure 20, we plotted the velocity profile for various values of inertia coefficient B, which shows that the first solution is enhanced and the second solution is declined. The temperature profile declines in the upper branch solution and rises in the second branch solution as the value of B augments and this behavior is captured in Figure 21. The cause for this trend is that the inertia of the porous medium offers an extra confrontation to the mechanism of the liquid flow, which grounds the liquid to progress at a retarded rate with reduced temperature.

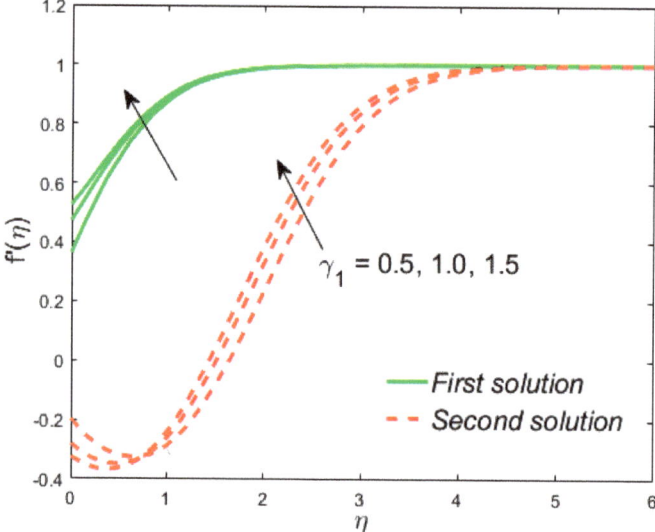

Figure 8. Influence of γ_1 on $f'(\eta)$.

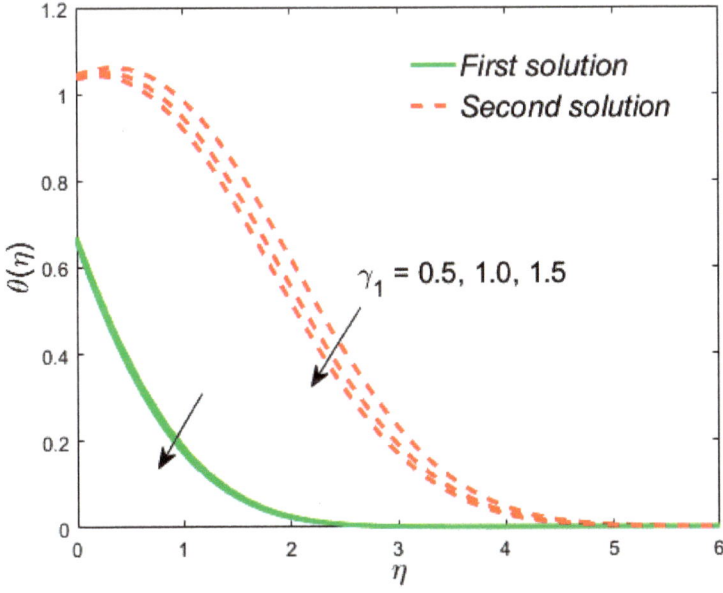

Figure 9. Influence of γ_1 on $\theta(\eta)$.

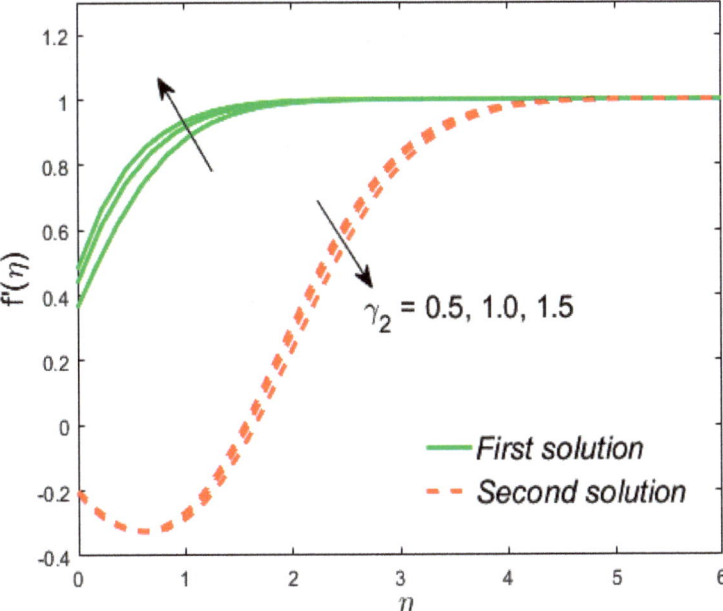

Figure 10. Influence of γ_2 on $f'(\eta)$.

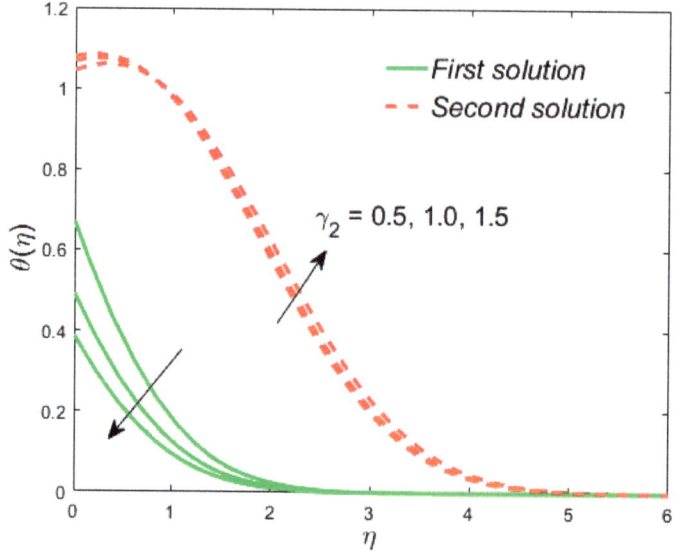

Figure 11. Influence of γ_2 on $\theta(\eta)$.

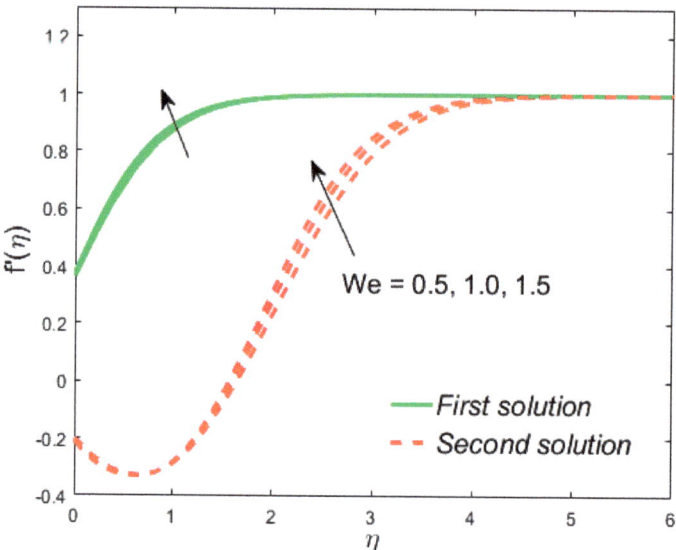

Figure 12. Influence of We on $f'(\eta)$.

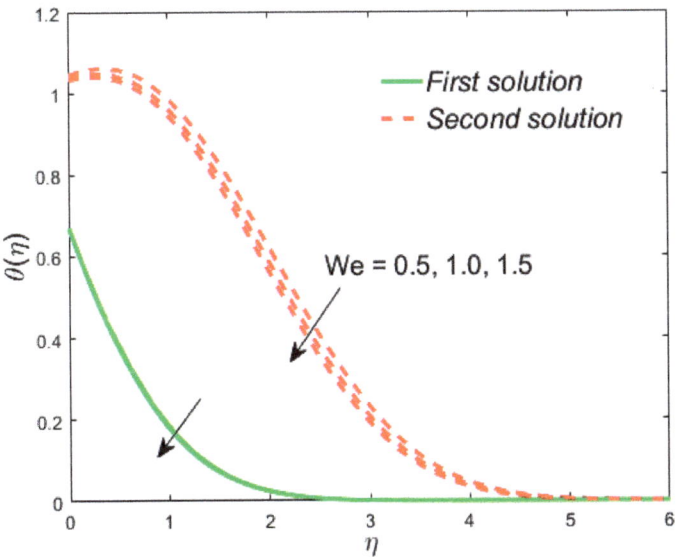

Figure 13. Influence of We on $\theta(\eta)$.

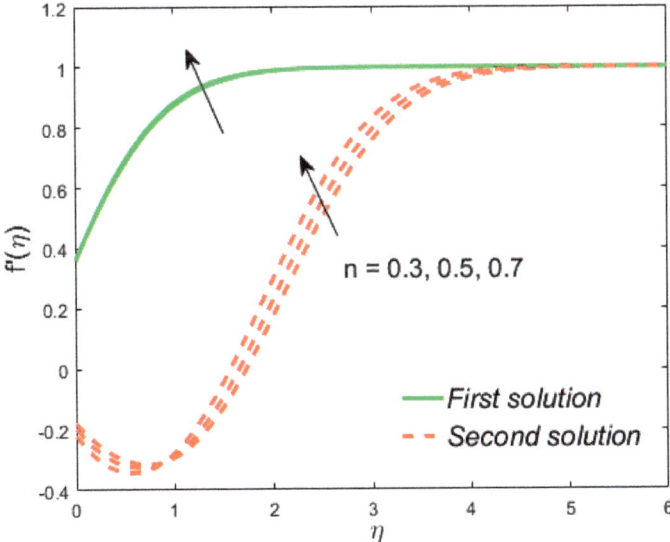

Figure 14. Influence of n on $f'(\eta)$.

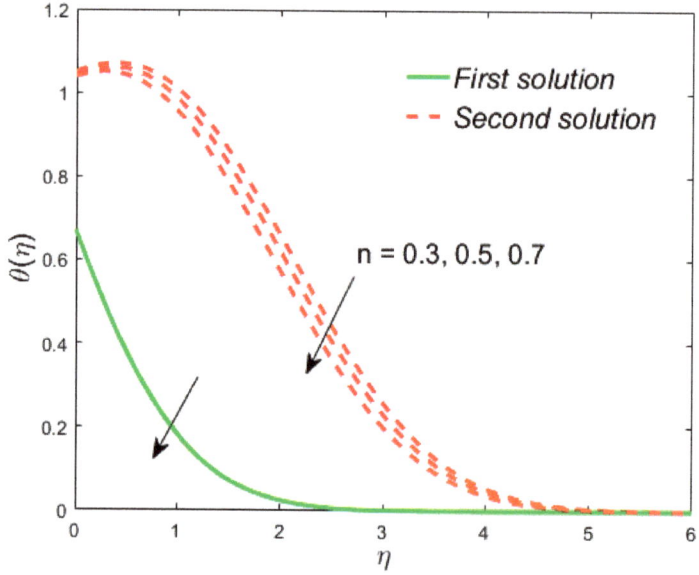

Figure 15. Influence of n on $\theta(\eta)$.

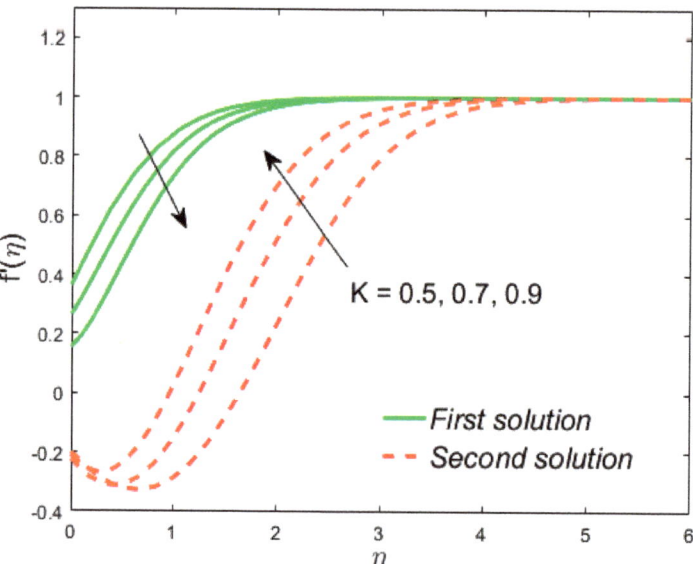

Figure 16. Influence of K on $f'(\eta)$.

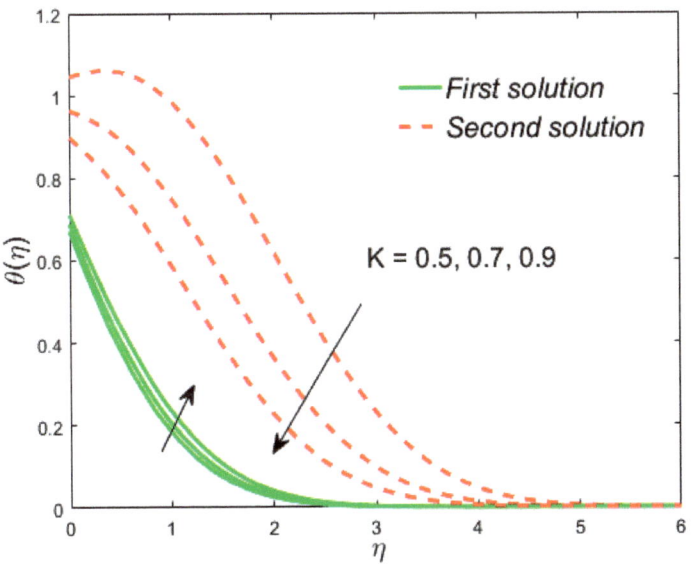

Figure 17. Influence of K on $\theta(\eta)$.

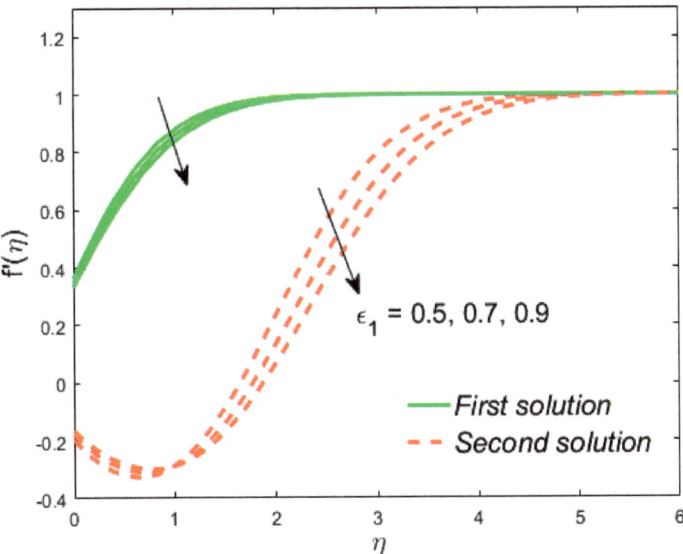

Figure 18. Influence of ε_1 on $f'(\eta)$.

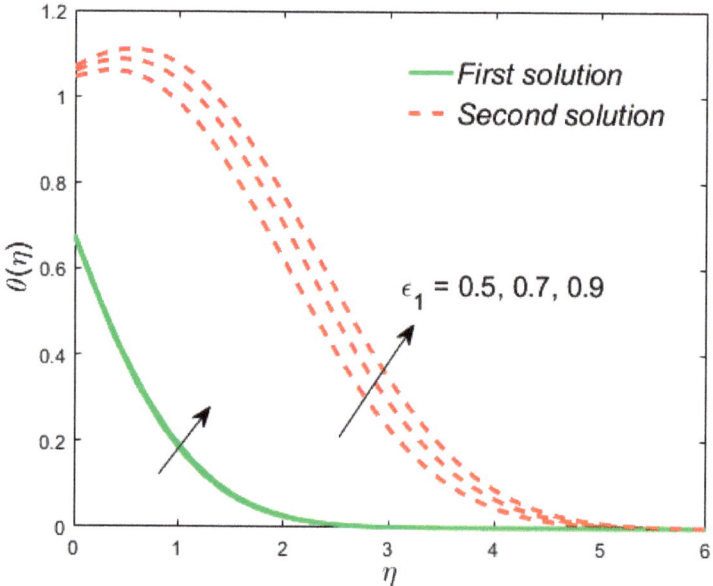

Figure 19. Influence of ε_1 on $\theta(\eta)$.

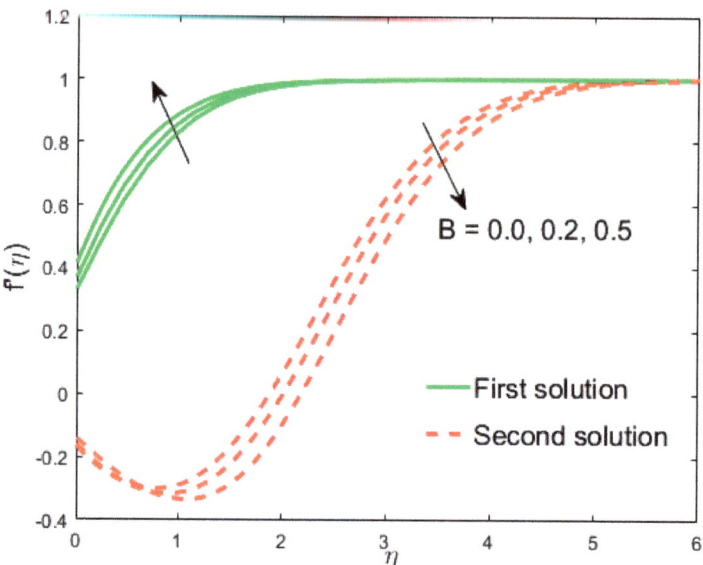

Figure 20. Influence of B on $f'(\eta)$.

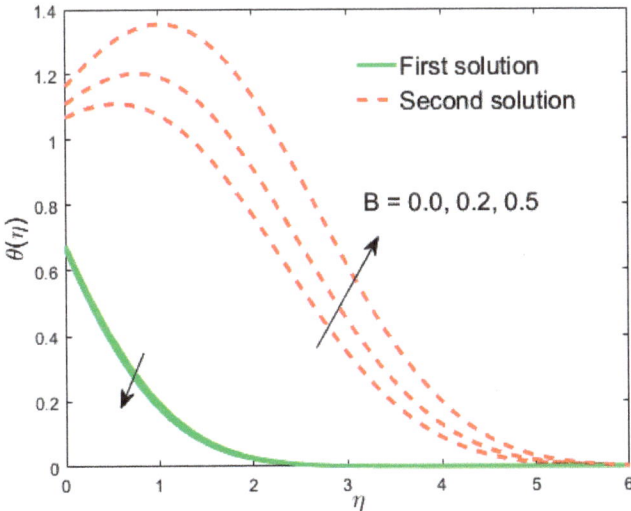

Figure 21. Influence of B on $\theta(\eta)$.

8. Closing Remarks

The impact of slip factors on the mixed convective flow of Cross liquid has been examined from a vertical plate immersed in Darcy–Forchheimer porous medium. The similarity variables are adopted to convert the PDEs to nonlinear ODEs. The transmuted system is numerically deciphered through the bvp4c solver. Core verdicts of the current research are stated as follows:

- Permeability parameter decelerates the drag force, as well as the rate of heat transfer in both forms of solutions;
- Due to the porosity parameter, the drag force slows down in upper and lower branch solutions, while the rate of heat transfer accelerates;
- The drag forces decline with the growing values of the Weissenberg parameter in the upper branch solutions, while a reverse trend is observed in the lower branch solutions. However, the rate of heat transfer is diminished with the Weissenberg parameter;
- The drag forces are declined initially and then enhance due to the inertia coefficient, while the rate of heat transfer increases in both solutions;
- Liquid velocity increases due to γ_1 in both solutions, while the temperature distribution behaves in a contrary direction;
- The temperature of the liquid is decreased due to γ_2 in the upper branch solutions and augmented in the lower branch solutions. The repeal tendency is scrutinized for the velocity;
- The velocity of the liquid has an enhancing behavior with the increasing values of We in both solutions, while the temperature is a declining function of We;
- The power-law index accelerates the velocity and reduces the temperature of the liquid in both solutions.

It is expected that the current numerical results provide significant knowledge for computer routines for further complex problems involving mixed convection of non-Newtonian fluids in porous media and stimulate curiosity for experimental work. In addition, the influence of slip effects in Darcy–Forchheimer flow with mixed convection has been of great interest especially in the utilization of geothermal energy and petroleum reservoir, etc.

Author Contributions: Conceptualization, A.Z. and K.S.N.; formal analysis, A.Z. and D.B.; funding acquisition, U.K. and A.Z., and D.B.; investigation, U.K.; methodology, U.K. and I.K.; software, I.K.; supervision, I.K. and K.S.N. All authors have read and agreed to the published version of the manuscript.

Funding: This research received no external funding.

Conflicts of Interest: The authors declare no conflict of interest.

Nomenclature

A_1	the first tensor of Rivlin–Ericksen
(b, c)	positive constants
B	inertia coefficient
C_F	drag coefficient
C_{fx}	skin friction coefficient
g	gravity acceleration
Gr_x	Grashof number
I	the identity vector
k	thermal conductivity of fluid
K_1	porous medium permeability
K	dimensionless permeability
L_1	length of slip
n	power-law index
Nu_x	Nusselt number
$\widetilde{p}$	the pressure
Pe_x	the Peclet number
q_w	the heat flux
Ra_x	the Rayleigh number
Re_x	local Reynolds number
S_1	proportionality constant
T	temperature (K)
T_∞	free-stream temperature (K)
T_w	wall temperature (K)
u_e	free-stream velocity (m s^{-1})
(u, v)	velocity components (m s^{-1})
We	Weissenberg number
(x, y)	Cartesian coordinates (m)

Greek Symbols

α_m	thermal diffusivity
β	thermal expansion
ε_1	modified porosity
$\dot{\gamma}$	the rate of shear
γ_1	velocity slip
γ_2	thermal slip
Γ	time constant
λ	mixed convective parameter
$\widetilde{\mu}_0$	zero shear rate
$\widetilde{\mu}_\infty$	infinite shear rate
μ_{eff}	the effective (or "apparent") viscosity
θ	dimensionless temperature
ν_{eff}	effective kinematic viscosity
ρ	density
ψ	stream function
τ_w	the shear stress
η	similarity variable

Subscripts

w	wall boundary condition
∞	free-stream condition

Superscripts

$'$	derivative w.r.t. η

References

1. Cross, M.M. Rheology of non-Newtonian fluids: A new flow equation for pseudoplastic systems. *J. Colloid Sci.* **1965**, *20*, 417–437. [CrossRef]
2. Khan, M.; Manzur, M.; Rahman, M.U. Boundary layer flow and heat transfer of Cross fluid over a stretching sheet. *arXiv* **2016**, arXiv:1609.01855. [CrossRef]
3. Hayat, T.; Khan, M.I.; Tamoor, M.; Waqas, M.; Alsaedi, A. Numerical simulation of heat transfer in MHD stagnation point flow of Cross fluid model towards a stretched surface. *Results Phys.* **2017**, *7*, 1824–1827. [CrossRef]
4. Khan, M.; Manzur, M.; Rahman, M.U. On axisymmetric flow and heat transfer of Cross fluid over a radially stretching sheet. *Results Phys.* **2017**, *7*, 3767–3772. [CrossRef]
5. Ijaz Khan, M.; Waqas, M.; Hayat, T.; Alsaedi, A. Magneto-hydrodynamical numerical simulation of heat transfer in MHD stagnation point flow of Cross fluid model towards a stretched surface. *Phys. Chem. Liq.* **2018**, *56*, 584–595. [CrossRef]
6. Ijaz Khan, M.; Hayat, T.; Khan, M.I.; Alsaedi, A. Activation energy impact in nonlinear radiative stagnation point flow of Cross nanofluid. *Int. Commun. Heat Mass Transf.* **2018**, *91*, 216–224. [CrossRef]
7. Azam, M.; Shakoor, A.; Rasool, H.F.; Khan, M. Numerical simulation for solar energy aspects on unsteady convective flow of MHD Cross nanofluid: A revised approach. *Int. J. Heat Mass Transf.* **2019**, *131*, 495–505. [CrossRef]
8. Bejan, A.; Dincer, I.; Lorente, S.; Miguel, A.F.; Reis, A.H. *Porous and Complex Flow Structures in Modern Technologies*; Springer: New York, NY, USA, 2004.
9. Vafai, K. *Handbook of Porous Media*, 2nd ed.; Taylor & Francis: New York, NY, USA, 2005.
10. Vadasz, P. *Emerging Topics in Heat and Mass Transfer in Porous Media*; Springer: New York, NY, USA, 2008.
11. Spivey, J.; Brown, K.; Sawyer, W.; Frantz, J.H. Estimating non-Darcy flow coefficient from buildup-test data with wellbore storage. *SPE Reserv. Eval. Eng.* **2004**, *7*, 256–269. [CrossRef]
12. Schafer, P.; Lohnert, G. Boiling experiments for the validation of dry out models used in reactor safety. *Nucl. Eng. Des.* **2006**, *236*, 1511–1519. [CrossRef]
13. Vafai, K. *Porous Media: Applications in Biological Systems and Biotechnology*; CRC Press: Boca Raton, FL, USA, 2010.
14. Wu, Y.S.; Lai, B.; Miskimins, J.L.; Fakcharoenphol, P.; Yuan, D. Analysis of multiphase non-Darcy flow in porous media. *Transp. Porous Med.* **2011**, *88*, 205–223. [CrossRef]
15. Forchheimer, P. Wasserbewegung durch boden. *Z. Ver. Dtsch. Ing.* **1901**, *45*, 1782–1788.
16. Rashidi, S.; Dehghan, M.; Ellahi, R.; Riaz, M.; Jamal-Abad, M.T. Study of stream wise transverse magnetic fluid flow with heat transfer around an obstacle embedded in a porous medium. *J. Magn. Magn. Mater.* **2015**, *378*, 128–137. [CrossRef]
17. Hayat, T.; Muhammad, T.; Al-Mezal, S.; Liao, S.J. Darcy-Forchheimer flow with variable thermal conductivity and Cattaneo-Christov heat flux. *Int. J. Numer. Methods Heat Fluid Flow* **2016**, *26*, 2355–2369. [CrossRef]
18. Hayat, T.; Haider, F.; Muhammad, T.; Alsaedi, A. On Darcy-Forchheimer flow of viscoelastic nanofluids: A comparative study. *J. Mol. Liq.* **2017**, *233*, 278–287. [CrossRef]
19. Kang, Z.; Zhao, D.; Rui, H. Block-centered finite difference methods for general Darcy–Forchheimer problems. *Appl. Math Comput.* **2017**, *307*, 124–140. [CrossRef]
20. Hayat, T.; Saif, R.S.; Ellahi, R.; Muhammad, T.; Ahmad, B. Numerical study for Darcy-Forchheimer flow due to a curved stretching surface with Cattaneo-Christov heat flux and homogeneous-heterogeneous reactions. *Results Phys.* **2017**, *7*, 2886–2892. [CrossRef]

21. Rasool, G.; Shafiq, A.; Khalique, C.M.; Zhang, T. Magneto-hydrodynamic Darcy-Forchheimer nanofluid flow over nonlinear stretching sheet. *Phys. Scr.* **2019**. [CrossRef]
22. Rosali, H.; Ishak, A.; Pop, I. Mixed convection stagnation-point flow over a vertical plate with prescribed heat flux embedded in a porous medium: Brinkman-extended Darcy formulation. *Trans. Porous Media* **2011**, *90*, 709–719. [CrossRef]
23. Faraz, N.; Khan, Y.; Anjum, A.; Kahshan, M. Three-Dimensional hydro-magnetic flow arising in a long porous slider and a circular porous slider with velocity slip. *Mathematics* **2019**, *7*, 748. [CrossRef]
24. Jamalabadi, M.Y.A. Optimal design of isothermal sloshing vessels by entropy generation minimization method. *Mathematics* **2019**, *7*, 380. [CrossRef]

© 2019 by the authors. Licensee MDPI, Basel, Switzerland. This article is an open access article distributed under the terms and conditions of the Creative Commons Attribution (CC BY) license (http://creativecommons.org/licenses/by/4.0/).

Article

Significance of Double Stratification in Stagnation Point Flow of Third-Grade Fluid towards a Radiative Stretching Cylinder

Anum Shafiq [1], Ilyas Khan [2,*], Ghulam Rasool [3], Asiful H. Seikh [4] and El-Sayed M. Sherif [4,5]

1. School of Mathematics and Statistics, Nanjing University of Information Science and Technology, Nanjing 210044, China; anumshafiq@ymail.com
2. Faculty of Mathematics and Statistics, Ton Duc Thang University, Ho Chi Minh City 72915, Vietnam
3. School of Mathematical Sciences, Zhejiang University, Hangzhou 310027, China; grasool@zju.edu.cn
4. Center of Excellence for Research in Engineering Materials (CEREM), King Saud University, P.O. Box 800, Al-Riyadh 11421, Saudi Arabia; aseikh@ksu.edu.sa (A.H.S.); esherif@ksu.edu.sa (E.-S.M.S)
5. Electrochemistry and Corrosion Laboratory, Department of Physical Chemistry, National Research Centre, El-Buhouth St., Dokki, Cairo 12622, Egypt
* Correspondence: ilyaskhan@tdtu.edu.vn

Received: 14 October 2019; Accepted: 7 November 2019; Published: 14 November 2019

Abstract: The present article is devoted to examine the significance of double stratification in third grade stagnation point flow towards a radiative stretching cylinder. The stagnation point is discussed categorically. Analysis is scrutinized in the presence of Thermophoresis, Brownian diffusion, double stratification and heat source/sink. Suitable typical transformations are used to drive the system of ordinary differential equation. The governing system is subjected to optimal homotopy analysis method (OHAM) for convergent series solutions. The impact of pertinent fluid parameters on the velocity field, temperature distribution and concentration of the nanoparticles is shown graphically. Numerical data is compiled in tabulare form for skin friction, Nusselt and Sherwood numbers to analyze the variation caused by the present model and to see the impact for industrial and engineering point of view.

Keywords: third-grade liquid; heat generation/absorption; stretched cylinder; series solution

1. Introduction

The study of stratification analyzes the variations and effects in thermal stratified object (medium) for the so-called common fluids. In industrial as well as natural processes, stratification plays an important role. Reason behind the existence of this phenomena is variation in temperature, variation of densities in different fluids and the concentration differences. Transfer of heat and mass simultaneously, doubles the stratification that belongs to the context of thermal stratification. Thermal stratification can be seen very often in the reservoirs and oceans. Another type of stratification is salinity stratification that is witnessed in rivers, estuaries, reservoirs storing the ground water, atmospheric heterogeneous mixtures, food industries and various manufacturing processes etc. A very few researchers in the past have made a significant contribution in investigating the effect of mass and thermal stratification over heat as well as mass transfer by a naturally convective flow. Keeping in view the above mentioned facts, double stratification gained a significant importance in the eyes of some researchers like Srinivasacharya and RamReddy [1,2] who investigated the double stratification's effect numerically. The medium was first considered non-porous and afterwards Darcian (porous) as well. Mixing process of oxygen with water in the bottom of reservoirs through biological processes can be controlled by using the tool of thermal stratification (see [3]). Stratification has also major contribution in environmental sciences. It can be very helpful in balancing the temperature differences and concentrations of oxygen and hydrogen to control the growth rates of various species in naturally unbalanced and less productive environments, Ibrahim and Makinde [4]. Various engineering processes occurring at a very high temperature direly depend upon a deep understanding and knowledge of thermal radiation. Combustion energy processes happening in fossil fuel, flows in astrophysics, harnessing the energy of sun in solar technology, turbines, devices for converting mechanical energy into propulsive force in aircraft, missiles and space-ships etc. are best examples of the importance and usage of the study of thermal radiations (see [5]). In some objects, fluid flow encounters a certain point where fluid motion becomes zero. In Geop physical setups, physical models and fluid mechanics, the point is called stagnation point. This stagnation point can be anywhere on the surface of object. However, the fluid continues flowing in neighborhood of this point, called stagnation point flow. Such an object is termed as impermeable object (see [6]). Stagnation point is sub-divided into two main categories (i) orthogonal and (ii) slanted stagnation point. In first case, the fluid particles act orthogonal to a rigid/solid surface and consequently, the resulting velocity is zero. The orthogonality of fluid particles at certain point makes it a perpendicular or orthogonal stagnation point. In second case, the fluid particles act on the rigid body through some random arbitrary angle of incidence. One can say that this point is a dual of orthogonal and shear stagnation point flow flowing parallel to the object. Numerous researches has been carried out on stagnation point flow. Describing the fluid motion near stagnation regions of a solid surface, the stagnation point flow was first studied by Hiemenz [7] using a similarity transformation for reducing the Navier-Stokes equations to Non-linear ODEs. Accordingly, stagnation flow can be categorized in various types depending upon the behavior of flow. Analyzing the density one can characterize it as inviscid or viscous flow, steady or unsteady flow, geometrically it can be two or three dimensional flow. The stagnation point flow can also be characterized according to the symmetry. Therefore, it can be symmetric or asymmetric, normal or slanted. Analyzing the flow behavior, it can be treated as homogeneous or immiscible fluid and forward or reverse fluid (see [8,9]). Importance of stagnation point flow can be witnessed in natural and industrial phenomena. Fluid striking the tips of submarines, oil-ships and air-crafts are best examples of stagnation point flow. The blood flowing through a junction in an artery is another biological example of stagnation point flow. Mabood et al. [10] investigated the radiation effects on stagnation point flow with melting heat transfer. Meanwhile, stagnation point flow of Tangent-hyperbolic liquid visualized by Shafiq et al. [11] witnesses its importance and significance in different aspects.

Process of natural convection can be witnessed in various physical phenomenon especially fire and heat engineering, nuclear science, reservoirs used for petroleum etc. The presence of heat (source/sink) and thermal radiation is a key factor in natural convection process. Such processes has been studied extensively because of naturally frequent existence. Ghoshdastidar [12] has explained various areas witnessing the applications of free convection. For example, the transfer of heat from heater to the neighborhood or heat dissipation through coil of refrigerator unit to the neighborhood etc. The encounter of such phenomena is common in wide range of thermal applications. Cheng [13,14] studied the boundary layer flow as natural convection. The medium was a vertical surface with Newtonian heating. The chemical reaction and thermal radiation are important aspects in engineering setups involving Riga patterns (see [15]). Boundary layer flow and the study of heat transfer in fluid mechanics and engineering is a contemporary research area (see [16]). Furthermore, Rasool et al. [17] reported MHD nanofluid flow over stretching surface with simple temperature attributes whereas, Rasool et al. [18] reported a study in the same representation using Cattaneo Christov heat and mass flux model over a stretching surface. Many researchers in the past have remained focused on this area and their work have been published. For example Kuznetsov and Nield [19] studied this phenomena of boundary layer flow analytically using the Brownian motion model. The effects of thermophoresis were taken into account. The results proved that Nusselt number is a decreasing function of the parameters of Brownian motion. Presence of gravity is a key element for density differences which plays a vital role in the mixing of heterogeneous fluids and their dynamics. A similar kind of boundary layer flow through a porous medium was investigated by Lesnic et al. [20]. Recently, Shafiq et al. [21]

investigated a boundary-Layer flow of Walters' B fluid in Newtonian heating depicted the heat transfer phenomena. The study highlights usefulness of boundary layer flow. The Newtonian heating, its effects and applications has been discussed in this research in detail. The study of two-dimensional boundary layer flow using an unsteady and permeable stretching surface is yet another recent improvement linking the effects of thermal radiations in boundary layer flows (see [22]). In this study Shafiq et al. investigated the effects of electric and magnetic fields. In present study the analysis is carried out by finding the optimal convergence. For details one can read the optimal control convergence procedure adopted in solving linearized Navier-Stokes equations in netlike domain [23] and pipeline flow [24].

In the literature mentioned above, the studies have been mainly reported on stretching surfaces with various assumptions including the porosity factor, Brownian diffusion and thermophoresis using HAM [25–31]. However, no research is found emphasizing the role of stagnation point in third grade fluid towards stretching surface (cylinder) which affirms the novelty of the present problem. Here the objective is to discuss the stagnation point and boundary layer flow, to analyze the corresponding results in the presence of sink/source and to graphically interpret various physical parameters involved in model using Optimal Homotopy approach.

2. Formulation

We consider a third grade stagnation point flow towards a radiative stretching cylinder in the context of double stratification. The stagnation point is discussed categorically. Analysis is scrutinized in the presence of Thermophoresis, Brownian diffusion, double stratification and heat source/sink. Suitable typical transformations are used to drive the system of ordinary differential equation. The governing system is subjected to optimal homotopy analysis method (OHAM) for convergent series solutions.. The effect of double stratification and thermal radiation is accounted. We assume that z-axis is directed along the given stretching cylinder whereas the radial r-axis goes perpendicular to it. Thus,

$$\frac{\partial u}{\partial r} + \frac{u}{r} + \frac{\partial w}{\partial z} = 0, \tag{1}$$

$$\begin{aligned}
\rho \left(u \frac{\partial w}{\partial r} + w \frac{\partial w}{\partial z} \right) &= W_e \frac{dw_e}{dz} + \mu \left(\frac{\partial^2 w}{\partial r^2} + \frac{1}{r} \frac{\partial w}{\partial r} \right) + \alpha_1 \left[\frac{w}{r} \frac{\partial^2 w}{\partial r \partial z} + \frac{u}{r} \frac{\partial^2 w}{\partial r^2} + \frac{3}{r} \frac{\partial w}{\partial r} \frac{\partial w}{\partial z} \right. \\
&\quad + \frac{1}{r} \frac{\partial u}{\partial r} \frac{\partial w}{\partial r} + 4 \frac{\partial w}{\partial r} \frac{\partial^2 w}{\partial r \partial z} + w \frac{\partial^3 w}{\partial r^2 \partial z} + 2 \frac{\partial u}{\partial r} \frac{\partial^2 w}{\partial r^2} \\
&\quad + u \frac{\partial^3 w}{\partial r^3} + 3 \frac{\partial^2 w}{\partial r^2} \frac{\partial w}{\partial z} + \frac{\partial^2 u}{\partial r^2} \frac{\partial w}{\partial r} \right] + \alpha_2 \left[\frac{2}{r} \frac{\partial u}{\partial r} \frac{\partial w}{\partial r} + \frac{2}{r} \frac{\partial w}{\partial r} \frac{\partial w}{\partial z} \right. \\
&\quad + 2 \frac{\partial^2 u}{\partial r^2} \frac{\partial w}{\partial r} + 2 \frac{\partial u}{\partial r} \frac{\partial^2 w}{\partial r^2} + 2 \frac{\partial^2 w}{\partial r^2} \frac{\partial w}{\partial z} \\
&\quad \left. + 4 \frac{\partial w}{\partial r} \frac{\partial^2 w}{\partial r \partial z} \right] + \beta_3 \left[\frac{2}{r} \left(\frac{\partial w}{\partial r} \right)^3 + 6 \left(\frac{\partial w}{\partial r} \right)^2 \frac{\partial^2 w}{\partial r^2} \right],
\end{aligned} \tag{2}$$

$$u \frac{\partial T}{\partial r} + w \frac{\partial T}{\partial z} = \frac{k}{\rho c_p} \left(\frac{\partial^2 T}{\partial r^2} + \frac{1}{r} \frac{\partial T}{\partial r} \right) + \frac{16\sigma T_\infty^3}{3k^* \rho c_p} \left(\frac{\partial^2 T}{\partial r^2} + \frac{1}{r} \frac{\partial T}{\partial r} \right) + \frac{Q_0}{\rho c_p} (T - T_\infty), \tag{3}$$

$$u \frac{\partial C}{\partial r} + w \frac{\partial C}{\partial z} = D \left(\frac{\partial^2 C}{\partial r^2} + \frac{1}{r} \frac{\partial C}{\partial r} \right), \tag{4}$$

with

$$\begin{aligned}
w(r,z) &= W_w(z) = \frac{W_0 z}{l}, \ u(r,z) = 0, \ T(r,z) = T_0 + b\left(\frac{z}{l}\right), \ C(r,z) = C_0 + d\left(\frac{z}{l}\right) \text{ at } r = R, \\
w(r,z) &\longrightarrow W_e(z) = \frac{W_\infty z}{l}, \ T(r,z) \to T_0 + c\left(\frac{z}{l}\right), \ C(r,z) \to C_0 + e\left(\frac{z}{l}\right) \text{ at } r \longrightarrow \infty.
\end{aligned} \tag{5}$$

where r represents radial distance while z is assigned to axial distance. u, v, correspond to r and z component of fluid velocity, T, T_∞ represent surface & ambient temperature while C and C_∞ represent surface & ambient concentration, respectively. Here ρ, ν correspond to fluid density and kinematic viscosity while c_p and k are the specific heat at constant pressure and thermal conductivity, respectively. The constants b, d, e and c are dimensionless. k^* is designated as coefficient of mean absorption, σ is Stephen Boltzman constant, reference length is represented by l, W_w is stretching velocity while W_e is the free stream velocity. Q_0 is used to represent the coefficient of heat generation as well as absorption. It is pertinent to mention that Q_0^+ (the positive values) behaves as source (heat generation) while Q_0^- (the negative values) behaves as sink (heat absorption). Using suitable transformations

$$\begin{aligned} w(r,z) &= \tfrac{W_0 z}{l} f'(\eta), \ u(r,z) = -\sqrt{\tfrac{\nu W_0}{l}} \tfrac{R}{r} f(\eta), \ \eta = \sqrt{\tfrac{W_0}{\nu l}} \left(\tfrac{r^2-R^2}{2R}\right), \\ \theta(\eta) &= \tfrac{T-T_\infty}{T_W-T_0}, \ \phi(\eta) = \tfrac{C-C_\infty}{C_w-C_0}. \end{aligned} \qquad (6)$$

It can easily be verified that the balance of mass given by Equation (1) is identically satisfied. On substituting Equation (6) into Equations (2)–(5) and then rearranging we have:

$$\begin{aligned} &(1+2\gamma\eta)\,f''' + A^2 + 2\gamma f'' - f'^2 + f f'' \\ &+\alpha_1 \left[(1+2\gamma\eta)\left\{2f'f''' - ff^{(iv)} + 3f''^2\right\} + \gamma\left(6f'f'' - 2ff'''\right)\right] \\ &+\alpha_2 \left[2(1+2\gamma\eta)\,f''^2 + \gamma\left(2f'f'' + 2ff'''\right)\right] \\ &+\beta Re \left[6(1+2\gamma\eta)^2 f''^2 f''' + 8\gamma(1+2\gamma\eta)\,f''^3\right], \end{aligned} \qquad (7)$$

$$f'(0) = 1, \ f(0) = 0, \ f'(\infty) = A, \qquad (8)$$

$$\left(1 + \tfrac{4}{3}R_d\right)(1+2\gamma\eta)\theta'' + 2\gamma\left(1 + \tfrac{4}{3}R_d\right)\theta' - \Pr(f'\theta + f'\theta) - S\Pr f' + Qg = 0, \qquad (9)$$

$$\theta(0) = 1 - S, \ \theta(\infty) \longrightarrow 0, \qquad (10)$$

$$(1+2\gamma\eta)\phi'' + 2\gamma\phi' + Sc(f\phi' - f'\phi) - ScSt(f') = 0, \qquad (11)$$

$$\phi(0) = 1 - St, \ \phi(\infty) \longrightarrow 0, \qquad (12)$$

where

$$\begin{aligned} \alpha_1 &= \tfrac{a_1^* W_0}{l\mu}, \ \alpha_2 = \tfrac{a_2^* W_0}{l\mu}, \ \beta = \tfrac{\beta_3 W_0^2}{l^2 \mu}, \ Re = \tfrac{Wz}{\nu}, \\ R_d &= \tfrac{4\sigma^* T_\infty^3}{k^* k}, \ \Pr = \tfrac{\mu c_p}{\rho}, \ \gamma = \left(\tfrac{\nu l}{W_0 R^2}\right)^{1/2}, \\ Q &= \tfrac{Q_0}{\rho c_p W_w}, \ S = \tfrac{c}{b}, \ St = \tfrac{e}{d}, \end{aligned} \qquad (13)$$

which respectively indicate the dimensionless third-grade parameters $(\alpha_1, \alpha_2, \beta)$, the Reynolds number (R_e), thermal radiation parameter (R_d), the Prandtl number $(\Pr)$, the curvature parameter (γ), heat generation/absorption parameter (Q), thermal stratification parameter (St) and solute stratification parameter (Sc). Expression of physical quantities are,

$$Nu_z = \tfrac{zq_w}{k(T_w - T_0)}, \ \text{where} \ q_w = -k\left(\tfrac{\partial T}{\partial r}\right)_{r=R}, \qquad (14)$$

$$C_f = \tfrac{T_{rz}}{\rho W_e^2}, \qquad (15)$$

$$Sh_z = \tfrac{zh_w}{D_w(C_w - C_0)}, \ \text{where} \ h_w = -D\left(\tfrac{\partial C}{\partial r}\right)_{r=R}, \qquad (16)$$

Such that

$$Nu_z Re_z^{-1/2} = -(1 + \tfrac{4}{3}Rd)(1 - S)\theta'(0), \qquad (17)$$

$$Re_z^{-1/2} C_f = [1 + 3\alpha_1 + 3\beta Re(f''(0))^2] f''(0), \qquad (18)$$

$$Sh_z Re_z^{-1/2} = -(1 - St)\phi'(0), \qquad (19)$$

where $Re_z = W_w l/\nu$ is the local Reynolds number.

3. Optimal Homotopic Solutions

The initial guesses and linear operators for the construction of series solutions are

$$f(\eta) = A\eta + (1-A)(1-Exp(-\eta)), \quad \theta_0(\eta) = (1-S)Exp(-\eta), \quad \phi_0(\eta) = (1-St)Exp(-\eta), \tag{20}$$

$$\mathcal{L}_f(f) = \frac{d^3 f}{d\eta^3} - \frac{df}{d\eta}, \quad \mathcal{L}_\theta(\theta) = \frac{d^2\theta}{d\eta^2} - \theta, \quad \mathcal{L}_\phi(\theta) = \frac{d^2\theta}{d\eta^2} - \theta, \tag{21}$$

with

$$\mathcal{L}_f [A_1 + A_2 \exp(\eta) + A_3 \exp(-\eta)] = 0, \tag{22}$$

$$\mathcal{L}_\theta [A_4 \exp(\eta) + A_5 \exp(-\eta)] = 0, \tag{23}$$

$$\mathcal{L}_\phi [A_6 \exp(\eta) + A_7 \exp(-\eta)] = 0, \tag{24}$$

where A_i ($i = 1, 2, \ldots, 7$) are the arbitrary constants.

4. Optimal Convergence Control Parameters

The parameters $\hbar_f$ and $\hbar_\theta$ are called convergence control parameters that are computed using the numerical BVPh2.0 package. Resulting optimal numerical values of these parameters are usually determined by the min of the average error. To significantly reduce the processing time of CPU, the tactic of average residual error is used at the mth-order of approximation, such that,

$$\varepsilon_m^f\left(\hbar_f\right) = \frac{1}{N+1} \sum_{j=0}^{N} \left[\sum_{i=0}^{k} (f_i)_{\eta=j\pi}\right]^2, \tag{25}$$

$$\varepsilon_m^\theta\left(\hbar_f, \hbar_\theta\right) = \frac{1}{N+1} \sum_{j=0}^{N} \left[\sum_{i=0}^{k} (f_i)_{\eta=j\pi}, \sum_{i=0}^{k} (\theta_i)_{\eta=j\pi}\right]^2, \tag{26}$$

and

$$\varepsilon_m^\theta\left(\hbar_f, \hbar_\phi\right) = \frac{1}{N+1} \sum_{j=0}^{N} \left[\sum_{i=0}^{k} (f_i)_{\eta=j\pi}, \sum_{i=0}^{k} (\theta_i)_{\eta=j\pi}\right]^2. \tag{27}$$

The optimal values of the convergence control parameters are $\hbar_f = -0.32677$, $\hbar_\theta = -0.56129$ and $\hbar_\phi = -0.46129$, when $\alpha_1 = \alpha_2 = \beta = 0.1$, $Re = \gamma = S = 0.2$, $A = 1.5$, $R_d = 0.4$, $Q = 0.2$, $Sc = 1.2$, $St = 0.3$, $Pr = 1$. The values of convergence control parameters are choosen very carefully. The admissible ranges of parameters are taken. The results are convergent within the ranges of these values. The values assigned to the fluid parameters are chosen carefully to satisfy the convergence criteria of OHAM. Beyond these values, the solution might not converge. One can see the total residual error in Figure 1.

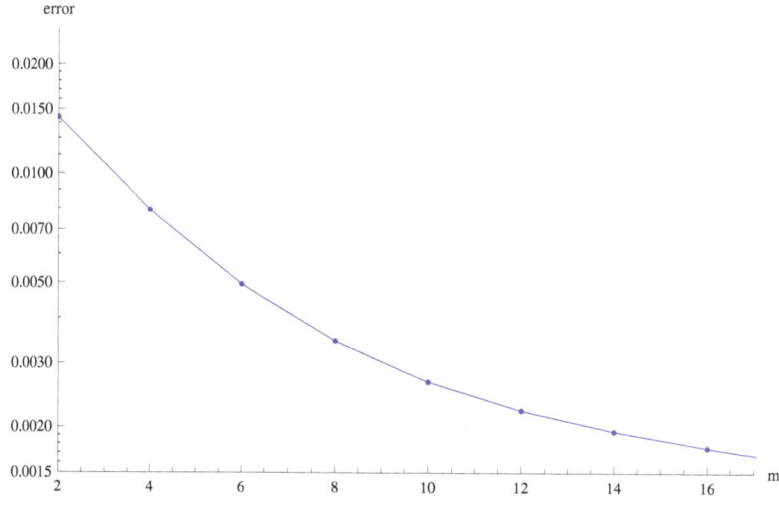

Figure 1. Total error vs order of approximations.

5. Discussions

The physical insights of parameters, used in this research, on velocity, concentration distribution and temperature are key aspects to be discussed in this section. Significance and impact of γ (curvature parameter) over the velocity field is shown in Figure 2. It is witnessed from Figure 2 that initially an inverse proportion between γ and velocity field as well as boundary layer thickness converts into a direct proportion at far away from the cylinder. The thickness of boundary layer and the velocity distribution with a certain decrease near the cylinder gradually starts increasing in fluid far away from cylinder. More the fluid is near to the cylinder, more is the affect of resistance. Figure 3 shows the behavior of ratio parameter on velocity distribution. Therefore, it can be noticed that velocity field goes higher and higher in both cases for $A > 1$ and $A < 1$ while the boundary layer shows a different behavior. Even At $A = 1$, there are no visuals of boundary layer. The significance of β, a third grade parameter, in the fluid velocity is depicted in Figure 4. The more the value of β, the low is the viscosity that causes enhancement in velocity distribution. Consequently, the velocity profile is enhanced. Figure 5, depicts the the impact of Reynolds Re on the velocity field. Certain decrease in the velocity field is noticed in moving from near the cylinder to away and finally, it vanishes at far away from surface. The reason behind this vanishing is the high value of Reynolds number that reduces the friction in between the surface and fluid. Figure 6 shows the variation of A, the ratio parameter, on temperature distribution. Higher is the value of A, the lesser is thickness of thermal and temperature boundary layer. Curvature parameter γ on $\theta(\eta)$ is analyzed in Figure 7. Both, the thermal field and connected/associated boundary layer are found as increasing functions of the γ. The Impact of well known Prandtl number on $\theta(\eta)$ is plotted in Figure 8. There is an inverse relation seen in thermal distributions for Prandtl. The smaller is the Prandtl factor, the higher is the temperature and thermal boundary layer thickness. The decremented thermal diffusion due to increment in Prandtl number forces the temperature distribution to decrease. One can conclude that fluids having low Prandtl numbers normally have high thermal diffusivity. The influence of heat generation and absorption on $\theta(\eta)$ is shown in Figure 9. An increase in heat generation parameter $Q > 0$ and decrease in heat absorption parameter $Q < 0$ ensures the increase in temperature field. Further, the increase in heat generation increases the thickness of thermal boundary layer because the heat generation produces more heat that certainly allows a temperature hike. R_d on $\theta(\eta)$ in Figure 10 shows the variation in temperature distribution due to thermal radiation. More is the thermal radiation R_d, lesser is the temperature distribution. The stratification parameter S over $\theta(\eta)$ is analyzed in

Figure 11. An increase in S certainly decreases the $\theta(\eta)$ significantly whereas the thickness of thermal boundary layer goes higher and higher over a decrease in S. This is justified with the reason that a hike in S reduces the difference between surface of cylinder and corresponding temperature. The effect of curvature parameter γ over $\phi(\eta)$ is analyzed in Figure 12. Nearby the cylinder, the concentration profile attains a decrement and goes on increasing away from the cylinder. Figure 13 shows the behavior and variation of Schmidt number Sc, ratio of momentum and mass diffusivity, over $\phi(\eta)$. The more is Sc, $\phi(\eta)$ goes on increasing while thickness of solute boundary layer decreases. Higher is the value of Sc, smaller is the mass diffusivity and therefore, $\phi(\eta)$ achieves an increment. Finally, the variation and behavior of solute stratified parameter St over the concentration profile is displayed in Figure 14. Decrements in concentration profile are noted for high values of St. Hence, an increase in St is responsible for decreasing concentration distribution existent between surface and ambient fluid. Consequently, the concentration field decreases. Optimal convergence control parameters are enlisted in Table 1. It shows the individually calculated average squared r-errors in momentum and energy equations at different order of approximation. A decrease in squared residual errors is noted as compared to the order of approximation. Behavior of the coefficient of skin-friction is enlisted in Table 2. It is evident that for large values of α_1, β, γ and Re, skin friction increases. The skin friction decreases for augmented values of α_2 and A. Table 3 enlists the variation in Nusselt number due to different parameters. Higher the values of α_1, α_2, β, A, γ, Q, St and R_d, higher is the Nusselt number. However, it decreases with Pr. Table 4 shows the influence of numerous parameters on Sherwood number. Higher the values of S, α_1, α_2, and β, higher is the Sherwood number. However, it experiences a decrease in values with γ, St and Sc.

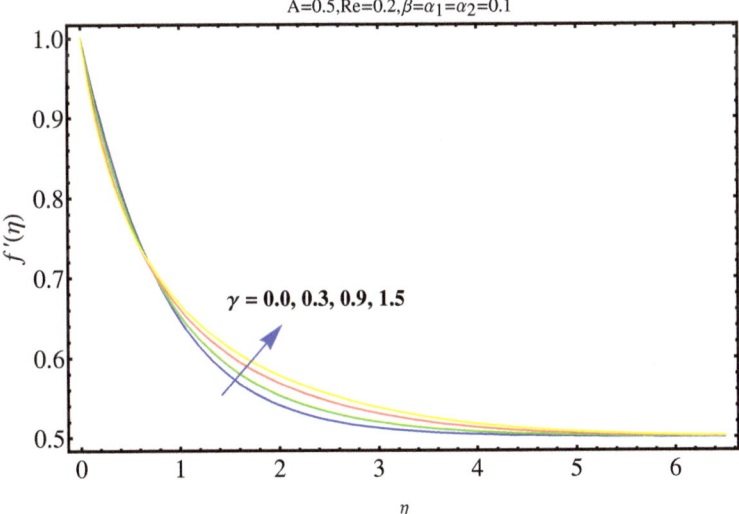

Figure 2. Impact of γ on $f'(\eta)$.

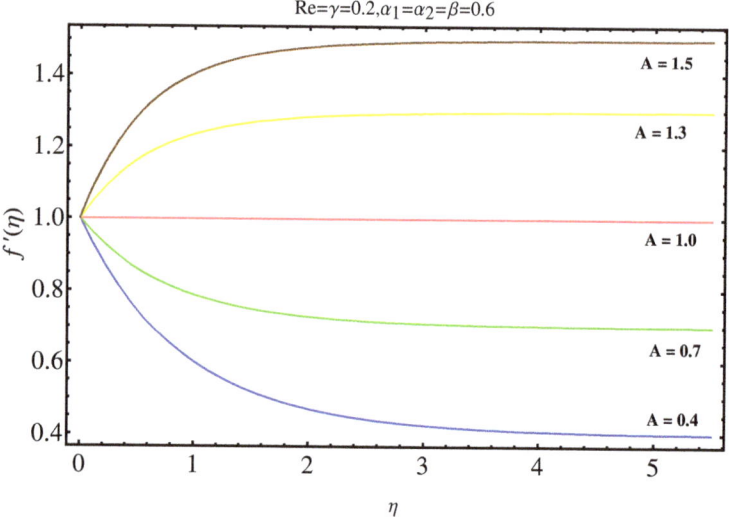

Figure 3. Impact of A on $f'(\eta)$.

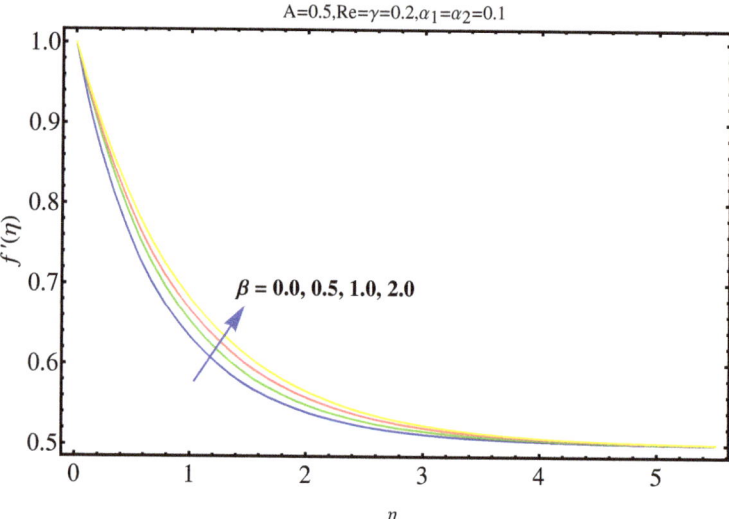

Figure 4. Impact of β on $f'(\eta)$.

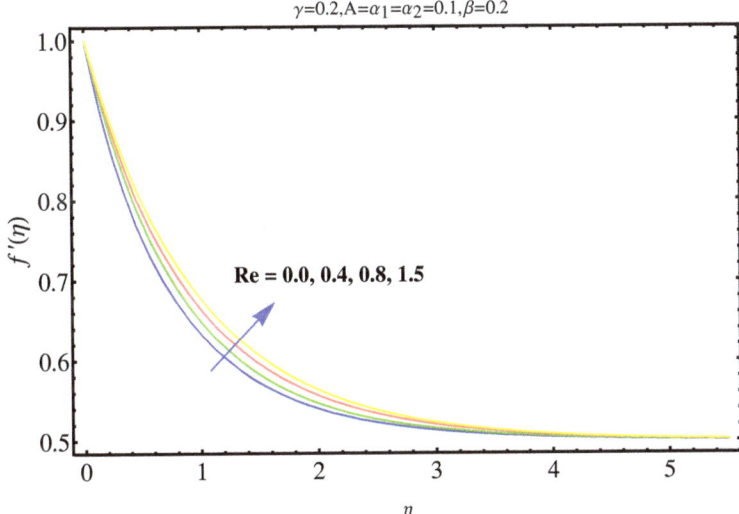

Figure 5. Impact of Re on $f'(\eta)$.

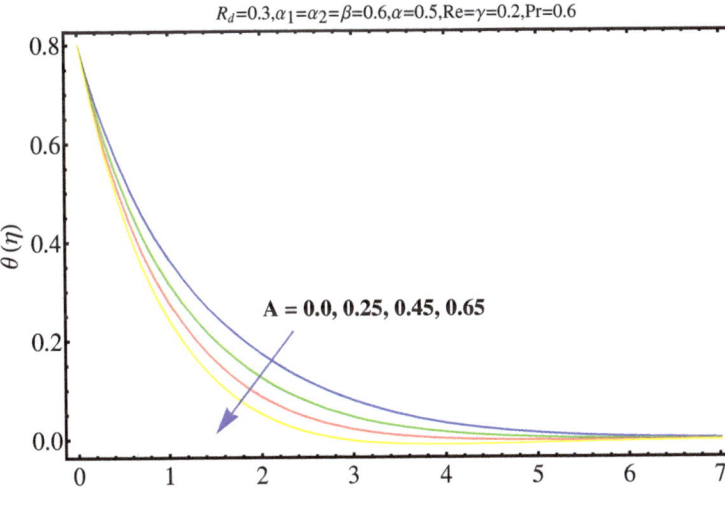

Figure 6. Impact of A on $\theta(\eta)$.

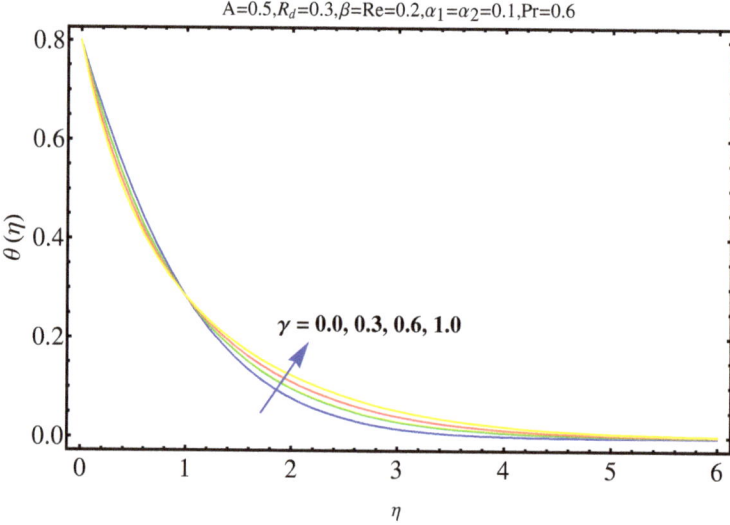

Figure 7. Impact of γ on $\theta(\eta)$.

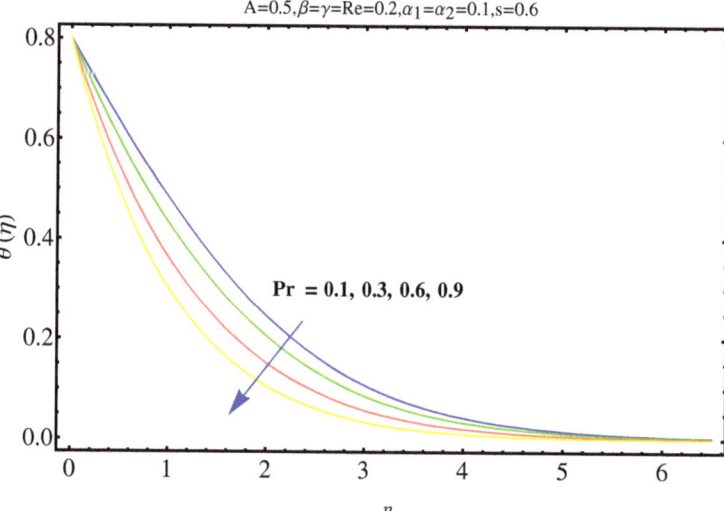

Figure 8. Impact of Pr on $\theta(\eta)$.

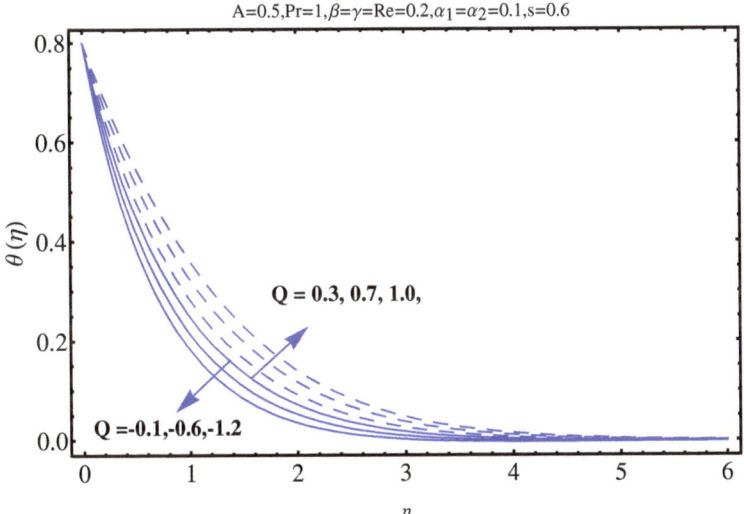

Figure 9. Impact of Q on $\theta(\eta)$.

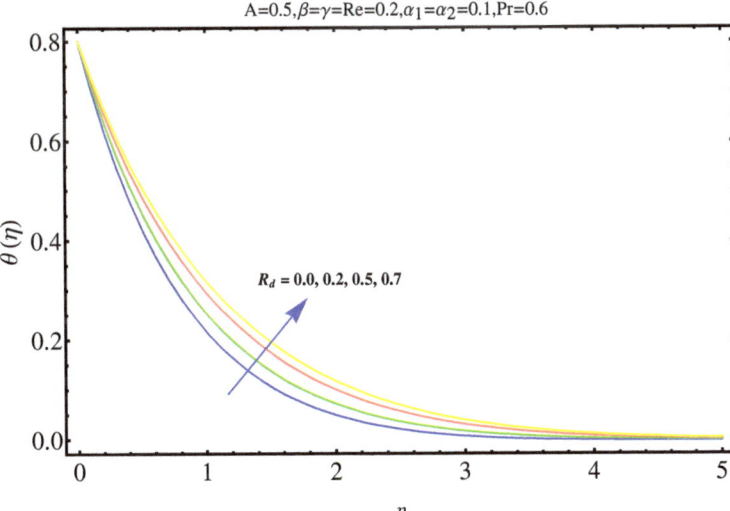

Figure 10. Impact of R_d on $\theta(\eta)$.

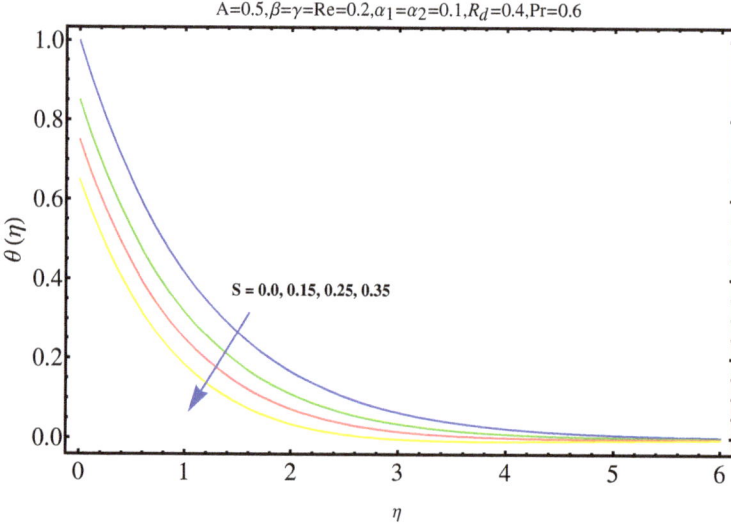

Figure 11. Impact of S on $\theta(\eta)$.

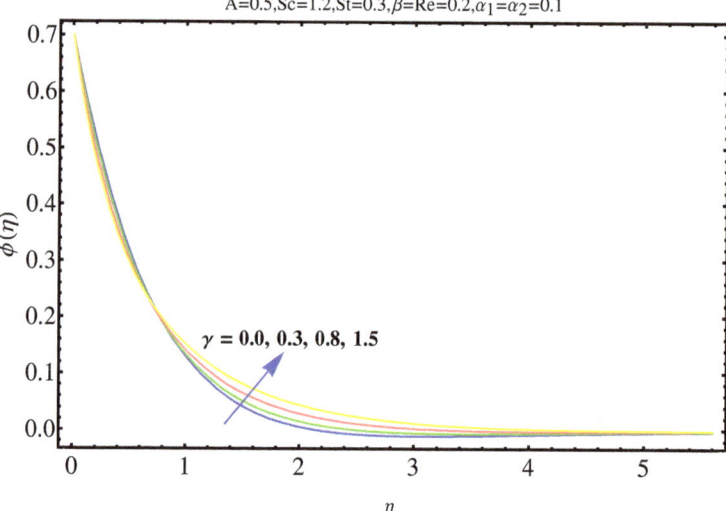

Figure 12. Impact of γ on $\phi(\eta)$.

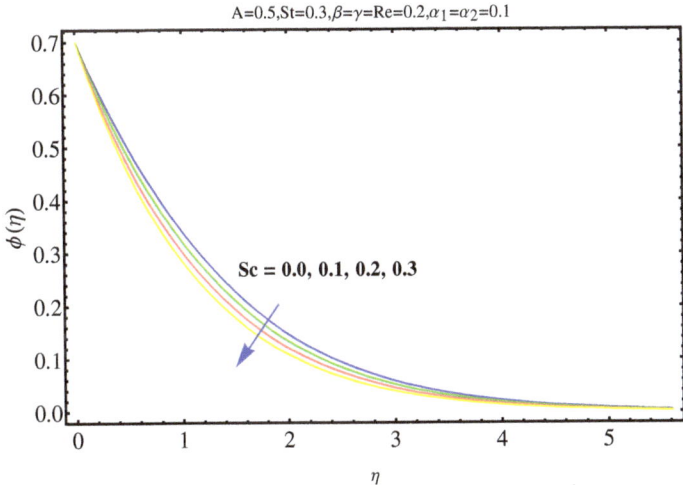

Figure 13. Impact of Sc on $\phi(\eta)$.

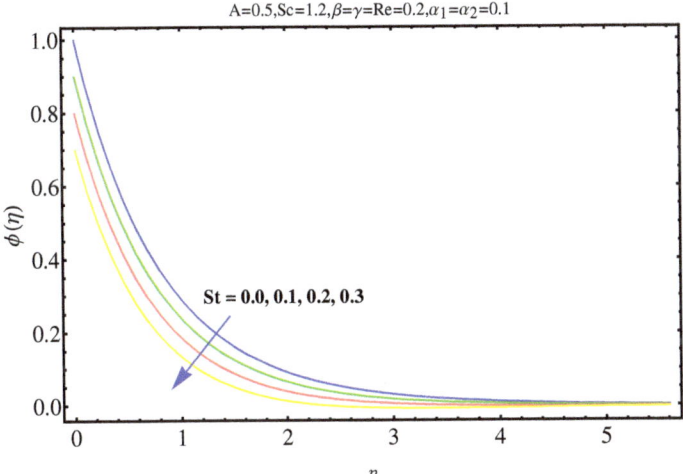

Figure 14. Impact of St on $\phi(\eta)$.

Table 1. Residual errors at $\alpha_1 = \alpha_2 = \beta = 0.1$, $Re = \gamma = S = 0.2$, $A = 1.5$, $R_d = 0.4$, $Q = 0.2$, $Sc = 1.2$, $St = 0.3$, $Pr = 1$ by means of optimal control parameters $\hbar_f = -0.32677$, $\hbar_\theta = -0.56129$ and $\hbar_\phi = -0.46129$.

m	ϵ_m^f	ϵ_m^θ	ϵ_m^ϕ	CPU Time [s]
2.0	0.00108904	0.0161511	0.136643	3.84396
4.0	7.87184×10^{-6}	0.0145533	0.117591	20.2511
6.0	1.15481×10^{-6}	0.0135712	0.105814	48.3618
8.0	2.90966×10^{-7}	0.012861	0.0975497	111.334
10	1.64734×10^{-7}	0.0123001	0.0915131	229.121
14	7.54581×10^{-8}	0.0114404	0.0838574	822.825
18	4.06748×10^{-8}	0.0107921	0.081305	2398.66
20	3.10657×10^{-8}	0.0105197	0.0803804	3958.69

Table 2. Numerical values of Skin friction for various physical parameters.

α_1	α_2	β	γ	A	Re	$-Re_r^{1/2}C_f$
0.0	0.1	0.1	0.2	0.6	0.4	1.04320
0.1						1.16269
0.2						1.19544
0.1	0.0	0.1	0.2	0.6	0.4	0.58384
	0.1					0.57578
	0.2					0.56811
0.1	0.1	0.0	0.2	0.6	0.4	0.57440
		0.1				0.57718
		0.2				0.57985
0.1	0.1	0.1	0.0	0.6	0.4	0.57310
			0.1			0.57440
			0.2			0.57578
0.1	0.1	0.1	0.2	0.0	0.4	1.21411
				0.1		1.16269
				0.2		1.09544
0.0	0.1	0.1	0.2	0.6	0.0	0.58381
					0.1	0.64578
					0.2	0.76811

Table 3. Numerical values of local Nusselt number for various physical parameters.

R_d	Pr	St	Q	γ	α_1	α_2	A	β	$Re_r^{-1/2}Nu_r$
0.4	1.0	0.3	0.3	0.2	0.1	0.1	0.6	0.1	1.01341
0.7									1.36421
1.0									1.65732
0.3	0.1	0.3	0.3	0.2	0.1	0.1	0.6	0.1	1.41321
	0.5								1.36969
	1.0								1.12294
0.3	1.0	0.0	0.3	0.2	0.1	0.1	0.6	0.1	1.11411
		0.5							1.26969
		1.0							1.39294
0.3	1.0	0.3	0.0	0.2	0.1	0.1	0.6	0.1	1.38941
			0.5						1.62315
			1.0						1.89561
0.3	1.0	0.3	0.3	0.0	0.1	0.1	0.6	0.1	1.12303
				0.1					1.35344
				0.5					1.67423
0.0	1.0	0.3	0.3	0.2	0.0	0.1	0.6	0.1	1.12141
					0.5				1.24922
					1.0				1.41034
0.3	1.0	0.3	0.3	0.2	0.1	0.0	0.6	0.1	1.30312
						0.5			1.32601
						1.0			1.38923
0.3	1.0	0.3	0.3	0.2	0.1	1.0	0.0	0.1	1.12423
							0.5		1.24921
							1.0		1.41235
0.3	1.0	0.3	0.3	0.2	0.1	1.0	0.6	0.0	1.25309
								0.5	1.31985
								1.0	1.51225

Table 4. Numerical values of local Sherwood number for various physical parameters when $A = 0.6$ and Re = 0.2.

γ	St	S	Sc	α_1	α_2	β	$Re_r^{-1/2} Sh_r$
0.0	0.3	0.2	1.2	0.1	0.1	0.1	0.98740
0.1							0.92745
0.2							0.90462
0.3	0.0	0.2	1.2	0.1	0.1	0.1	1.98741
	0.1						1.74501
	0.2						1.25019
0.3	0.3	0.0	1.2	0.1	0.1	0.1	1.11523
		0.1					1.32751
		0.2					1.89154
0.3	0.3	0.2	0.0	0.1	0.1	0.1	0.54315
			0.1				0.38612
			0.2				0.18612
0.3	0.3	0.2	1.2	0.0	0.1	0.1	0.81913
				0.1			0.83997
				0.2			0.91251
0.3	0.3	0.2	1.2	0.1	0.0	0.1	1.28712
					0.1		1.59874
					0.2		1.98717
0.3	0.3	0.2	1.2	0.1	0.1	0.0	1.71231
						0.1	1.82127
						0.2	1.92351

6. Concluding Remarks

Here we have considered an axisymmetric stagnation point third grade fluid flow over a radiative stretching surface/cylinder. The stagnation point is discussed in detail. Analysis is scrutinized in the presence of double stratification, heat generation/absorption and Brownian motion. Optimal homotopy method (OHAM) is used for final solutions. Salient features are listed below:

- Due to the effect of temperature, smaller values of stratified parameters results in higher values of velocity and temperature distributions.
- Velocity profile enhances with β and Re.
- Radiation parameter enhances while ratio parameter reduces the temperature distribution.
- The coefficient of skin friction is higher for higher values of α_1, β, γ and Re.
- An increase in Q and γ results in more convenient heat transfer.

Author Contributions: Data curation, A.H.S.; Formal analysis, E.-S.M.S.; Investigation, G.R.; Methodology, A.S.; Software, I.K. All authors contributed equally.

Funding: The authors would like to extend their sincere appreciation to the Deanship of Scientific Research at King Saud University for its funding of this research through the Research Group Project No. RGP-160.

Conflicts of Interest: The authors declare no conflict of interest.

References

1. Srinivasachariya, D.; RamReddy, C. Effect of double stratification on free convection in a micropolar fluid. *ASME J. Heat Transf.* **2011**, *133*, 1–7. [CrossRef]
2. Srinivasachariya, D.; RamReddy, C. Free convective heat and mass transfer in a doubly stratified non-Darcy micro polar fluid. *Korea J. Chem. Eng.* **2011**, *9*, 1924–1932.
3. Srinivasachariya, D.; RamReddy, C. Heat and mass transfer by natural convection in a doubly stratified non-Darcy micro polar fluid. *Int. Commun. Heat Mass Transf.* **2010**, *37*, 873–880. [CrossRef]
4. Ibrahim, W.; Makinde, O.D. The effect of double stratification on boundary-layer flow and heat transfer of nanofluid over a vertical plate. *Comput. Fluids* **2013**, *86*, 433–441. [CrossRef]

5. Shafiq, A.; Hammouch, Z.; Turab, A. Impact of Radiation in a Stagnation Point Flow of Walters' B Fluid Towards a Riga Plate. *Therm. Sci. Eng. Prog.* **2018**, *6*, 27–33. [CrossRef]
6. Tooke, R.M.; Blyth, M.G. A note on oblique stagnation-point flow. *Phys. Fluids* **2008**, *20*, 033101. [CrossRef]
7. Hiemenz, K. Die Grenzschicht an einem in dengleichfiirmingen Flussigkeits stromeing etauchten geraden Kreiszy linder. *Dinglers Polym. J.* **1911**, *326*, 321–410.
8. Khan, W.A.; Pop, I. Flow near the two-dimensional stagnation-point on an infinite permeable wall with a homogeneous–heterogeneous reaction. *Commun. Nonlinear Sci. Numer. Simul.* **2010**, *15*, 3435–3443. [CrossRef]
9. Ja'fari, M.; Rahimi, A.B. Axisymmetric stagnation-point flow and heat transfer of a viscous fluid on a moving plate with time-dependent axial velocity and uniform transpiration. *ScientiaIranica* **2013**, *20*, 152–161. [CrossRef]
10. Mabood, F.; Shafiq, A.; Hayat, T.; Abelman, S. Radiation effects on stagnation point flow with melting heat transfer and second order slip. *Results Phys.* **2017**, *7*, 31–42. [CrossRef]
11. Hayat, T.; Shafiq, A.; Alsaedi, A. Characteristics of magnetic field and melting heat transfer in stagnation point flow of Tangent-hyperbolic liquid. *J. Magn. Magn. Mater.* **2016**, *405*, 97–106. [CrossRef]
12. Ghoshdastidar, P.S. *Heat Transfer*; Oxford University Press: Oxford, UK, 2004.
13. Cheng, P. Heat transfer in geothermal systems. *Adv. Heat Transf.* **1978**, *14*, 1–105.
14. Cheng, P. Natural convection in a porous medium: External flow. In Proceedings of the NATO Advanced Study in Natural Convection, Izmir, Turkey, 1985.
15. Rasool, G.; Zhang, T. Characteristics of chemical reaction and convective boundary conditions in Powell-Eyring nanofluid flow along a radiative Riga plate. *Heliyon* **2019**, *5*, e01479. [CrossRef] [PubMed]
16. Vasu, B.; Reddy, R.; Murthy, P.V.S.N. Thermophoresis on boundary layer heat and mass transfer flow of Walters-B fluid past a radiate plate with heat sink/source. *Heat Mass Transf.* **2017**, *53*, 1553–1570. [CrossRef]
17. Rasool, G.; Shafiq, A.; Khalique, C.M.; Zhang, T. Magnetohydrodynamic Darcy Forchheimer nanofluid flow over nonlinear stretching sheet. *Phys. Scr.* **2019**, *94*, 105221. [CrossRef]
18. Rasool, G.; Zhang, T. Darcy-Forchheimer nanofluidic flow manifested with Cattaneo-Christov theory of heat and mass flux over non-linearly stretching surface. *PLoS ONE* **2019**, *14*, e0221302. [CrossRef]
19. Kuznetsov, A.V.; Nield, D.A. Natural convective boundary-layer flow of a nanofluid past a vertical plate. *Int. J. Therm. Sci.* **2010**, *49*, 243–307. [CrossRef]
20. Lesnic, D.; Ingham, D.B.; Pop, I. Free convection boundary layer flow along a vertical surface in a porous medium with Newtonian heating. *Int. J. Heat Mass Transf.* **1999**, *42*, 2621–2627. [CrossRef]
21. Hayat, T.; Shafiq, A.; Mustafa, M.; Alsaedi, A. Boundary-Layer Flow of Walters' B Fluid with Newtonian Heating. *Z. Naturforschung A* **2015**, *70*, 301–395. [CrossRef]
22. Hayat, T.; Shafiq, A.; Alsaedi, A. Hydromagnetic boundary layer flow of Williamson fluid in the presence of thermal radiation and Ohmic dissipation. *Alex. Eng. J.* **2016**, *55*, 2229–2240. [CrossRef]
23. Provotorov, V.V.; Provotorova, E.N. Optimal control of the linearized Navier-Stokes system in a netlike domain. *Vestnik Sankt-Peterburgskogo Universiteta Prikladnaya Matematika Informatika Protsessy Upravleniya* **2017**, *13*, 431–443.
24. Artemov, M.A.; Baranovskii, E.S.; Zhabko, A.P.; Provotorov, V.V. On a 3D model of non-isothermal flows in a pipeline network. *J. Phys. Conf. Ser.* **2019**, *1203*, 012094. [CrossRef]
25. Rasool, G.; Zhang, T.; Shafiq, A. Second grade nanofluidic flow past a convectively heated vertical Riga plate. *Phys. Scr.* **2019**, *94*, 125212. [CrossRef]
26. Rasool, G.; Zhang, T.; Shafiq, A.; Durur, H. Influence of chemical reaction on Marangoni convective flow of nanoliquid in the presence of Lorentz forces and thermal radiation: A numerical investigation. *J. Adv. Nanotech.* **2019**, *1*, 32–49. [CrossRef]
27. Rasool, G.; Zhang, T.; Shafiq, A. Marangoni effect in second grade forced convective flow of water based nanofluid. *J. Adv. Nanotech.* **2019**, *1*, 50–61. [CrossRef]
28. Shafiq, A.; Nawaz, M.; Hayat, T.; Alsaedi, A. Magnetohydrodynamic axisymmetric flow of a third-grade fluid between two porous disks. *Brazi. J. Chem. Eng. (USA)* **2013**, *3*, 599–609. [CrossRef]
29. Rasool, G.; Shafiq, A.; Tlili, I. Marangoni convective nano-fluid flow over an electromagnetic actuator in the presence of first order chemical reaction. *Heat Transf.-Asian Res.* **2019**. [CrossRef]

30. Rasool, G.; Shafiq, A.; Durur, H.Darcy-Forchheimer relation in Magnetohydrodynamic Jeffrey nanofluid flow over stretching surface, Discrete and Continuous Dynamical Systems - Series S. *Am. Inst. Math. Sci.* **2019**, accepted.
31. Rasool, G.; Shafiq, A.; Khalique, C.M. Marangoni forced convective Casson type nanofluid flow in the presence of Lorentz force generated by Riga plate, Discrete and Continuous Dynamical Systems - Series S. *Am. Inst. Math. Sci.* **2019**, accepted.

Sample Availability: Samples of the compounds ...... are available from the authors.

© 2019 by the authors. Licensee MDPI, Basel, Switzerland. This article is an open access article distributed under the terms and conditions of the Creative Commons Attribution (CC BY) license (http://creativecommons.org/licenses/by/4.0/).

Article

On the MHD Casson Axisymmetric Marangoni Forced Convective Flow of Nanofluids

Anum Shafiq [1], Islam Zari [2], Ghulam Rasool [3], Iskander Tlili [4,5,*] and Tahir Saeed Khan [2]

1. School of Mathematics and Statistics, Nanjing University of Information Science and Technology, Nanjing 210044, China; anumshafiq@ymail.com
2. Department of Mathematics, University of Peshawar, Khybar Pakhtunkhwa 25000, Pakistan; zari145@yahoo.com (I.Z.); tsk7@uop.edu.pk (T.S.K.)
3. School of Mathematical Sciences, Yuquan Campus, Zhejiang University, Hangzhou 310027, China; grasool@zju.edu.cn
4. Department for Management of Science and Technology Development, Ton Duc Thang University, Ho Chi Minh City 758307, Vietnam
5. Faculty of Applied Sciences, Ton Duc Thang University, Ho Chi Minh City 758307, Vietnam
* Correspondence: iskander.tlili@tdtu.edu.vn

Received: 26 September 2019; Accepted: 8 November 2019; Published: 11 November 2019

Abstract: The proposed investigation concerns the impact of inclined magnetohydrodynamics (MHD) in a Casson axisymmetric Marangoni forced convective flow of nanofluids. Axisymmetric Marangoni convective flow has been driven by concentration and temperature gradients due to an infinite disk. Brownian motion appears due to concentration of the nanosize metallic particles in a typical base fluid. Thermophoretic attribute and heat source are considered. The analysis of flow pattern is perceived in the presence of certain distinct fluid parameters. Using appropriate transformations, the system of Partial Differential Equations (PDEs) is reduced into non-linear Ordinary Differential Equations (ODEs). Numerical solution of this problem is achieved invoking Runge–Kutta fourth-order algorithm. To observe the effect of inclined MHD in axisymmetric Marangoni convective flow, some suitable boundary conditions are incorporated. To figure out the impact of heat/mass phenomena on flow behavior, different physical and flow parameters are addressed for velocity, concentration and temperature profiles with the aid of tables and graphs. The results indicate that Casson fluid parameter and angle of inclination of MHD are reducing factors for fluid movement; however, stronger Marangoni effect is sufficient to improve the velocity profile.

Keywords: Casson nanoliquid; Marangoni convection; inclined MHD; Joule heating; heat source

1. Introduction

The theory of magnetohydrodynamics (MHD) is highly appreciated for the industrial purposes. It is based on magnetic properties of electrically conducting liquids. The characteristic of MHD field is to generate currents in moving liquid and produce forces that act upon the liquid flow and reconstruct the magnetic field itself. To modify flow features of heat and mass analysis, the applied magnetic field impacts the deferred nanoparticles and reforms their absorption inside the liquid. This efficient phenomenon was first utilized for astrophysical and geophysical related problems. Recently, heat transportation and MHD flows have played significant roles in agricultural engineering, petroleum industries and medical treatment such as MHD strategy used for reduction of blood during surgeries, magnetic cell separation and treatment of certain arterial diseases. Basically, the MHD parameter is not only working as a significant parameter to control the cooling/heating rate but also to achieve desired quality of product for different flows. Further, MHD can be used in continuous casting of metal processing to suppress instabilities and control flow field. In this context, Hayat et al. [1,2] explored the MHD flow through moving surfaces and concluded that enhancement in magnetic parameter shows increase in nanoparticles concentration and temperature profiles. Hayat et al. [3,4] numerically studied heat transfer impact on MHD axisymmetric third grade liquid flow. Shafiq et al. [5] presented the study of bioconvective MHD tangent hyperbolic nanoliquid flow with Newtonian heating. Shateyi and Makinde [6] prepared MHD stagnant point flow through a radially stretching convectively heated disk. Hayat et al. [7] investigated the third grade axisymmetric MHD flow over a stretched cylinder and showed that momentum layer thickness and velocity profile are increasing when the curvature parameter increases. Moreover, Shafiq et al. [8] discussed magnetohydrodynamics axisymmetric third grade liquid flow between two porous disks.

The novelty of Marangoni convection is generally the edge dissipative layer between two phase fluid flows such as gas–liquid and liquid–liquid interfaces. It depends upon the variation of surface tension driven by temperature, chemical concentration and applied magnetic field. These gradients can occur only when fluid interfaces contain different fluid properties from each other. Due to the viscosity of interacting liquids, external forces such as gravitational and shear forces come into action. Most researchers have focused their interest on simulating these external forces by utilizing governing equations due to its widespread application in the fields of space processing, industrial manufacturing processes and microgravity science. The significance of Marangoni convective flows in the transportation process of heat and mass into different systems have been thoroughly scrutinized in [9–11]. Kumar et al. [12] discussed Marangoni convective Casson nanoliquid flow in the presence of chemical reaction and uniform heat source/sink and observed that Marangoni parameter showed dominant behavior in terms of velocity as well as temperature fields. Din et al. [13] examined the effect of Marangoni convection on based nanoliquid with thermal radiation and demonstrated that decreasing behavior of velocity profile depends on suction parameter, whereas the temperature distribution and boundary layer thickness increased with an increase in nanofluid volume fraction. Sheikholeslami and Ganji [14] studied the impact of magnetic field on nanoliquid flow by Marangoni convection by Runge–Kutta technique and observed that an increment in heat transfer depended on an increment in solid volume fraction of nanofluid. Hayat et al. [15] investigated the impact of radiation and Joule heating on Marangoni mixed convective flow.

For the last few decades, survey of non-Newtonian fluid flows has been the center of attraction for researchers, engineers and scientists. This is due to the application of non-Newtonian liquid flows in the real world, e.g., in bio-engineering, drilling operations, plastic polymers, paint, optical fibers, coated sheets, cosmetics, salt solutions, food item, etc. The existing problems in nature related with larger diameter and higher shear rates can be solved easily; however, when these flows are related to small diameter with low shear rates, the importance of non-Newtonian fluids (see [16,17]) are non-negligible. The deviation from classical Newton's law of viscosity and flow behavior under shear stress to the non-Newtonian fluids become complex. These flows are challenging task for researchers due to their non-linear rheological behavior. Casson liquid model is one of simplest models of non-Newtonian fluids. The idea of Casson fluid administrated by Casson (see [18]) is to build up the blood flow problems. Due to its rheological properties, Casson liquid behaves as a soft solid when yield stress is higher than shear stress, whereas, if shear stress approaches to infinity, then it starts to deform (see [19]). This structure is widely used for different materials, such as jelly, chocolate, honey, blood, tomato sauce and condensed fruit juices. Charm and Kurland [20] used Casson fluid model and investigated the viscosity of human blood. Bhattacharyya and Hayat [21] analyzed the Casson fluid on MHD boundary layer flow through shrinking sheet. Kumar et al. [22] investigated the viscous dissipation phenomenon in Casson nanoliquid over a moving radiative surface. Moreover, Casson fluid flow model [23–25] has been considered for different geometries and various effects in the literature.

The introduction of nanoparticles in different systems is most favorable to intensify thermal conductivity of classical liquid flows, convection heat transfer coefficient and to control loss in energy. The advantages of nanosize particles in fluid systems is to increase surface area, capacity of heat transfer, intensify the flow interface after collision and interact fluid particles with each other. Thus, this phenomenon is a backbone of the industrial processes and is also beneficial for solar energy resources and bio-medical treatment (see [26–34]). The proficiency of the solar systems [35] can be improved by incorporating the nanoparticles as working fluid into the systems. The iron based nanoparticles may be utilized as drug and radiation transportation for the treatment of cancer patient (see [36,37]). Using magnets the particles can be enter through blood stream to tumor. This type of cancer treatment permits high local doses of drugs into the body without any significant side effect. Further, micelles nanoparticles have been recently introduced to target the kidney cells diseases. These particles can pass into the kidney and remain there. Similarly, magnetic based nanoparticles are also used for cell separation, hyperthermia therapies and for the increment in Magnetic Resonance Imaging (MRI) with contrast behavior. Hayat et al. [38] judged that nanofluid enhanced the temperature and associated boundary layer width of Casson flow. Naseem et al. [39] numerically investigated third grade nanoliquid flow using the Cattaneo–Christov model over a Riga plate and observed that, with an increment in thermal and concentration relaxation parameters, a reduction occurred in concentration and temperature distribution, respectively. Rasool et al. [40] examined the MHD Darcy–Forchheimer nanoliquid flow under the nonlinear stretched surface. Rashid et al. [41] investigated the entropy generation in Darcy–Forchheimer flow of nanofluid with five nanomaterials due to stretching cylinder. Naseem et al. [42] considered the MHD biconvective flow of a Powell–Eyring nanoliquid over a stretching plate. Rasool et al. [43–48] reported some interesting results involving the role of nanoparticles in typical base fluids flowing over different surfaces.

In the studies mentioned above, one can see that an utmost attention is given to natural convection and heat and mass transfer analysis but less importance has been given to the convection through Marangoni phenomena especially in nanofluid flows. The thermo-capillary and soluto-capillary affects are the main factors in Marangoni convection of fluids and nanofluids. Furthermore, flat surfaces with linear stretching are assumed frequently but axisymmetric analyses are less reported. The main contribution of this research is to examine the process of heat and mass transportation for axisymmetric Marangoni convective flow with an inclined MHD by taking Casson nanofluid flowing towards an infinite disk. Brownian motion and thermophoresis are deliberated on account of nanoparticles structure. Finally, the problem is solved by an accurate numerical technique known as Runge–Kutta fourth-order algorithm, whereas previous studies are given mostly by HAM.

2. Problem Formulation and Coordinate System

The geometry of the problem (see Figure 1) is based on the MHD effect for axisymmetric Marangoni convective, incompressible, steady and laminar flow utilizing the electrically conducting Casson nanoliquid model. Marangoni convective flow is caused due to concentration and temperature gradients on surfaces generated by surface tension. A uniform magnetic field is applied in such a way that it makes an angle α_1 in non-vertical direction. The cylindrical coordinates system is considered along and normal to the interface of flow problem. Concentration and temperature interfaces of the flow structure are altered at the surface of the disk. The analysis of heat transfer is examined through Joule heating and viscous dissipation. The formulated governing equations for the MHD effect on Marangoni convective flow structure are given as (see, for example, [4–15]):

$$\frac{\partial \tilde{u}}{\partial \tilde{r}} + \frac{\partial \tilde{w}}{\partial \tilde{z}} + \frac{\tilde{u}}{\tilde{r}} = 0, \quad (1)$$

$$\tilde{u}\frac{\partial \tilde{u}}{\partial \tilde{r}} + \tilde{w}\frac{\partial \tilde{u}}{\partial \tilde{z}} = \frac{\mu}{\rho}\left(1 + \frac{1}{\beta_1}\right)\frac{\partial^2 \tilde{u}}{\partial \tilde{z}^2} - \frac{\sigma B_0^2}{\rho}\sin^2\alpha_1 \tilde{u}, \quad (2)$$

$$\tilde{u}\frac{\partial \tilde{T}}{\partial \tilde{r}} + \tilde{w}\frac{\partial \tilde{T}}{\partial \tilde{z}} = \frac{k}{\rho c_\omega}\frac{\partial^2 \tilde{T}}{\partial \tilde{z}^2} + \tau\frac{D_{\tilde{T}}}{\tilde{T}_\infty}\left(\frac{\partial \tilde{T}}{\partial \tilde{z}}\right)^2 + \tau D_B \frac{\partial \tilde{C}}{\partial \tilde{z}}\frac{\partial \tilde{T}}{\partial \tilde{z}} + \frac{1}{\rho c_\omega}Q_1(\tilde{T}-\tilde{T}_\infty)$$
$$+\frac{\mu}{\rho c_\omega}\left(1+\frac{1}{\beta_1}\right)\left(\frac{\partial \tilde{u}}{\partial \tilde{z}}\right)^2 + \frac{\sigma B_0^2}{\rho c_\omega}\sin^2\alpha_1 \tilde{u}^2, \tag{3}$$

$$\tilde{u}\frac{\partial \tilde{C}}{\partial \tilde{r}} + \tilde{w}\frac{\partial \tilde{C}}{\partial \tilde{z}} = D_B \frac{\partial^2 \tilde{C}}{\partial \tilde{z}^2} + \frac{D_{\tilde{T}}}{\tilde{T}_\infty}\frac{\partial^2 \tilde{T}}{\partial \tilde{z}^2}. \tag{4}$$

In Equations (1)–(4), ρ characterizes fluid density; μ signifies dynamic viscosity; β indicates parameter of Casson fluid; σ symbolizes surface tension; $\tilde{C}$ and $\tilde{T}$ represent fluid concentration and temperature, respectively; $\tilde{C}_\infty$ and $\tilde{T}_\infty$ characterize fluid ambient concentration and temperature far away from the surface, respectively; τ shows shear stress; D_B is the coefficient of Brownian diffusion; k indicates coefficient of absorption; c_ω denotes specific heat; $D_{\tilde{T}}$ characterizes coefficient of thermophoretic diffusion; Q_1 represents heat source sink coefficient; and α_1 signifies angle of inclination.

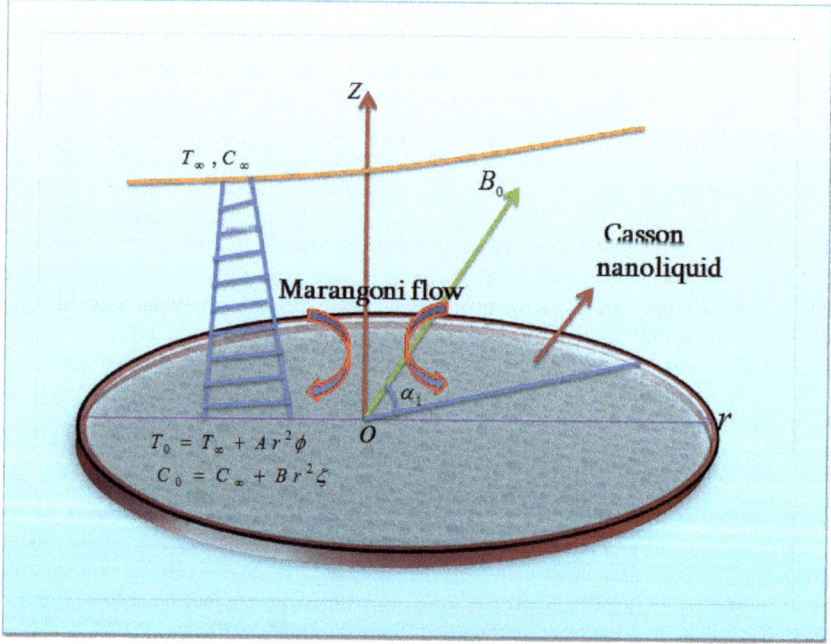

Figure 1. Physical diagram of the flow model.

The subjected boundary conditions are (see, for example, [10]):

$$\mu\left(1+\frac{1}{\beta_1}\right)\frac{\partial \tilde{u}}{\partial \tilde{z}}\bigg|_{\tilde{z}=0} = \frac{\partial \sigma}{\partial \tilde{T}}\frac{\partial \tilde{T}}{\partial \tilde{r}}\bigg|_{\tilde{z}=0} + \frac{\partial \sigma}{\partial \tilde{C}}\frac{\partial \tilde{C}}{\partial \tilde{r}}\bigg|_{\tilde{z}=0}, \quad \tilde{w}|_{\tilde{z}=0}=0,$$
$$\tilde{T}|_{\tilde{z}=0} = \tilde{T}_\infty + A\tilde{r}^2\phi, \quad \tilde{C}|_{\tilde{z}=0} = \tilde{C}_\infty + B\tilde{r}^2\zeta,$$
$$\tilde{u}|_{\tilde{z}\to\infty} \longrightarrow 0, \quad \tilde{T}|_{\tilde{z}\to\infty} \longrightarrow \tilde{T}_\infty, \quad \tilde{C}|_{\tilde{z}\to\infty} \longrightarrow \tilde{C}_\infty. \tag{5}$$

The suitable transformations incorporated in the proposed flow structure are (see, for example, [6]):

$$\eta = \sqrt{\frac{b}{\nu}}\tilde{z}, \ \tilde{u} = b\tilde{r}\, g'(\eta), \ \tilde{w} = -2\sqrt{b\nu}g(\eta), \ \zeta = \frac{\tilde{C} - \tilde{C}_\infty}{\tilde{C}_g - \tilde{C}_\infty}, \ \phi = \frac{\tilde{T} - \tilde{T}_\infty}{\tilde{T}_g - \tilde{T}_\infty}. \tag{6}$$

Moreover, assumptions indicate that surface tension is a linear function of concentration and temperature, which may be represented as (see, for example, [10]):

$$\sigma = \sigma_0 - \gamma_{\tilde{T}}(\tilde{T} - \tilde{T}_\infty) - \gamma_{\tilde{C}}(\tilde{C} - \tilde{C}_\infty), \tag{7}$$

where σ_0, $\gamma_{\tilde{T}}$ and $\gamma_{\tilde{C}}$ represent the positive constants. After incorporating the above-mentioned transformations into Equations (1)–(4), we obtain

$$\left(1 + \frac{1}{\beta_1}\right)g''' + 2gg'' - (g')^2 - M_1^2 \sin^2\alpha_1\, g' = 0, \tag{8}$$

$$g(0) = 0, \ (1 + \frac{1}{\beta_1})g''(0) = -2\, M_a(1 + R_a\, \zeta(0)), \ g'(\infty) = 0, \tag{9}$$

$$\phi'' + 2\Pr g\, \phi' + \Pr N_1 \phi' \zeta + \Pr N_2 (\phi')^2 + \Pr Ec \left(1 + \frac{1}{\beta_1}\right)(g'')^2 + \Pr Ec\, M_1^2 \sin^2\alpha_1 (g')^2 + \Pr B_1 \phi = 0, \tag{10}$$

$$\phi(0) = 1, \ \phi(\infty) = 0, \ N_1' \zeta(0) + N_2\, \phi'(0) = 0, \tag{11}$$

$$\zeta'' + 2Le\, g\zeta' + \frac{N_2}{N_1}\phi'' = 0, \tag{12}$$

$$\zeta(0) = 1, \ \zeta(\infty) \to 0. \tag{13}$$

In Equations (8)–(13), $N_1 = \frac{\tau D_B(\tilde{C}_g - \tilde{C}_\infty)}{\nu}$ indicates Brownian motion parameter, $N_2 = \frac{\tau D_T(\tilde{T}_g - \tilde{T}_\infty)}{\nu \tilde{T}_\infty}$ characterizes thermophoresis parameter, $M_1 = \frac{\sigma B_0^2}{8\rho b}$ shows magnetic number, $M_a = \frac{\gamma_{\tilde{T}}}{\mu \Omega}\sqrt{\frac{\Omega}{\gamma}}$ signifies Marangoni number, $R_a = \frac{\gamma_{\tilde{T}} B}{\gamma_{\tilde{T}} A}$ shows Marangoni ratio parameter, $B_1 = \frac{Q}{b\rho c_w}$ represents heat source sink, $\Pr = \frac{\rho c_w \mu}{k}$ signifies Prandtl number, $Ec = \frac{\tilde{u}_\infty^2}{c_w(\tilde{T}_w - \tilde{T}_\infty)}$ indicates Eckert number, and $Le = \frac{\nu}{D_B}$ denotes Lewis number. Additionally, Nu the local Nusselt number is given as

$$Nu = \frac{-\tilde{r}\left(\frac{\partial \tilde{T}}{\partial \tilde{z}}\right)\big|_{\tilde{z}=0}}{k(\tilde{T}_\infty - \tilde{T}_w)}, \tag{14}$$

and in dimensionless form becomes

$$R_d^{-1/2} Nu = \frac{Nu}{\sqrt{R_d}} = -\phi'(0), \tag{15}$$

where $R_d = \frac{\tilde{u}_w \tilde{r}}{\nu}$ is local Reynold's parameter.

3. Computational Scheme

We now solve the governing Equations (8)–(13), numerically by employing Runge–Kutta fourth-order technique. For different sundry parameters, we perform numerical computation.

4. Physical Interpretation and Analysis

The main objective of this segment is to communicate the physical importance of heat and mass transportation phenomenon in axisymmetric Marangoni convective flow with the impact of inclined MHD on Casson nanoliquid over an infinite disk. To clearly check the insight of proposed model, the impact of different parameters (Casson fluid parameter β_1, magnetic number M_1, angle of inclination α_1, Marangoni number M_a, Brownian motion parameter N_1, thermophoresis parameter N_2, Prandtl number Pr, heat source sink B_1 and Lewis number Le) are considered on velocity field $g(\eta)$, temperature profile $\phi(\eta)$, concentration distribution $\zeta(\eta)$ and local Nusselt number Nu.

4.1. Assessment of Velocity Distribution

The performance of Casson fluid parameter β_1 on velocity field $g'(\eta)$ is demonstrated in Figure 2. In this figure, one can see that, for enhancement in β_1, velocity profile increases near the wall but decreases when $\eta > 1.4$ and vanishes far away from the surface. This is because an increment in Casson fluid parameter produces a decrease in yield stress and the fluid adopts rheological behavior and associated boundary layer width reduces. In Figure 3, it is analyzed that a rise in magnetic parameter M_1 drops the fluid velocity. This logic is dependent on the fact that an increment in magnetic field M_1, which causes an increase in the resistive nature of Lorentz force, and consequently decreases the velocity field. Figure 4 demonstrates the influence of inclination angle α_1 on $g'(\eta)$. It is apparent from the sketch that velocity profile $g'(\eta)$ reduces when the angle of inclination α_1 increases. This is because, when angle of inclination increases, the impact of magnetic field rises on liquid and as a result Lorentz force increases, which in turn decreases the velocity profile. In addition, for $\alpha_1 = 0$, there is no effect of magnetic field on velocity profile, whereas, for $\alpha_1 = \pi/2$, maximum resistance is noted. In Figure 5, graphical representation signifies that velocity field is mounting function of Marangoni number M_a. This behavior is because of Marangoni number, as it is the ratio between tangential stress and viscosity. Therefore, the fluid with higher surface tension acts more strongly on the surrounding liquid and consequently it enhances velocity of the fluid.

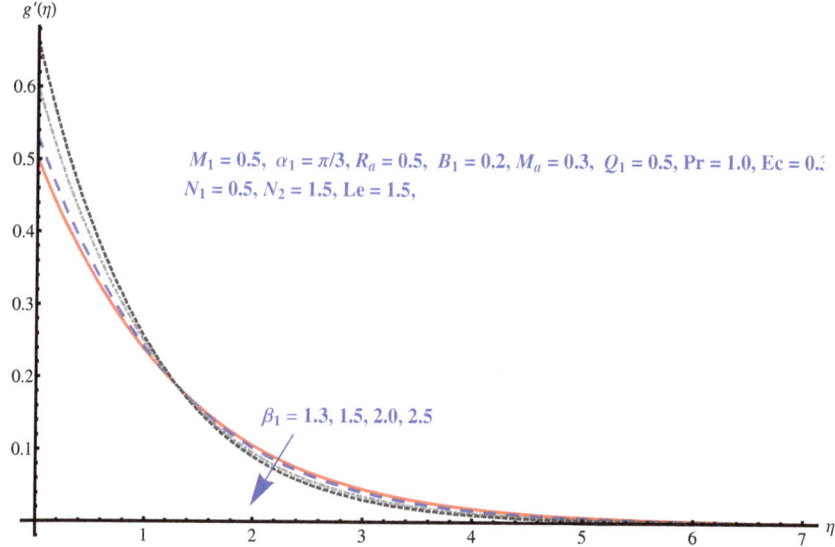

Figure 2. Influence of β_1 on velocity field.

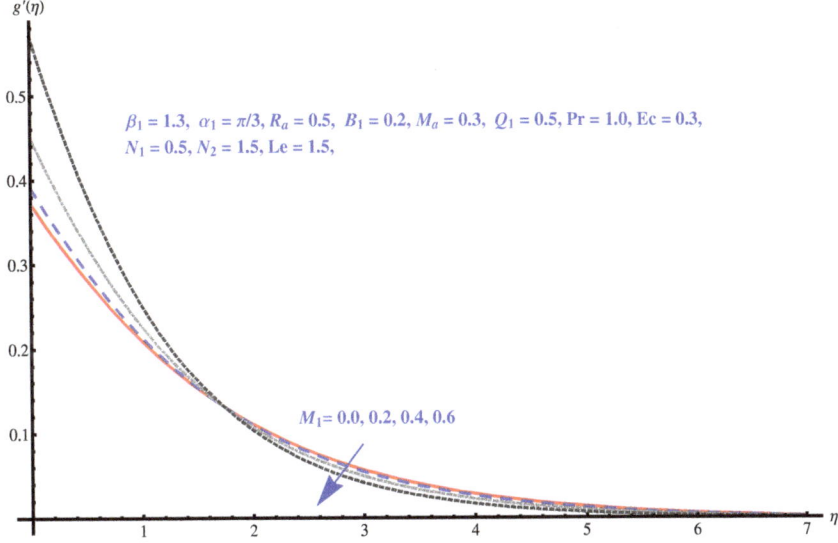

Figure 3. Influence of M_1 on velocity field.

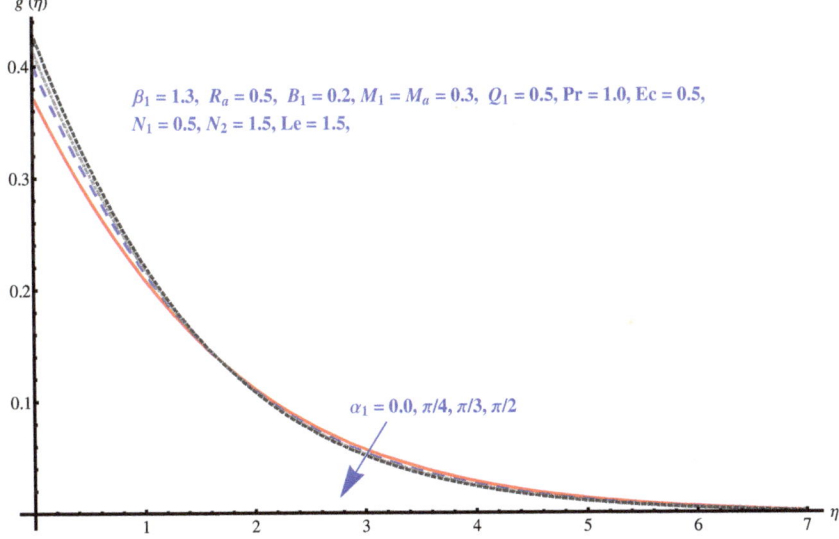

Figure 4. Influence of α_1 on velocity field.

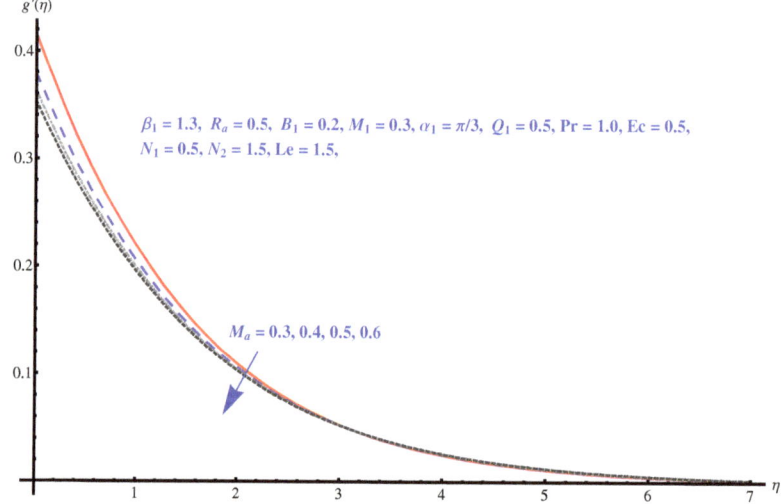

Figure 5. Influence of M_a on velocity field.

4.2. Assessment of Temperature Distribution

In this subsection, the temperature field $\phi(\eta)$ corresponds to various sundry parameter such as thermophoresis parameter N_2, Brownian motion parameter N_1 and heat source parameter B_1 are plotted in Figures 6–8. The behavior of N_1 on $\phi(\eta)$ is sketched in Figure 6. The temperature field $\phi(\eta)$ is increased with a rise in Brownian motion parameter N_1. With the Increment in Brownian motion parameter, the fluid molecules becomes more energetic. As a result, the temperature field is enhanced. Figure 7 shows the significance of Thermophoresis parameter N_2 on $\phi(\eta)$. It is noted that the temperature field is a mounting function of N_2. Figure 8 shows the variation of heat source B_1 via temperature field $\phi(\eta)$. It is examined that temperature as well as the associated boundary layer is increased by increment in heat source parameter B_1. Physically, the rise in rate of heat source parameter B_1 leads to the thermal boundary layer thickness becoming greater, as does the temperature field.

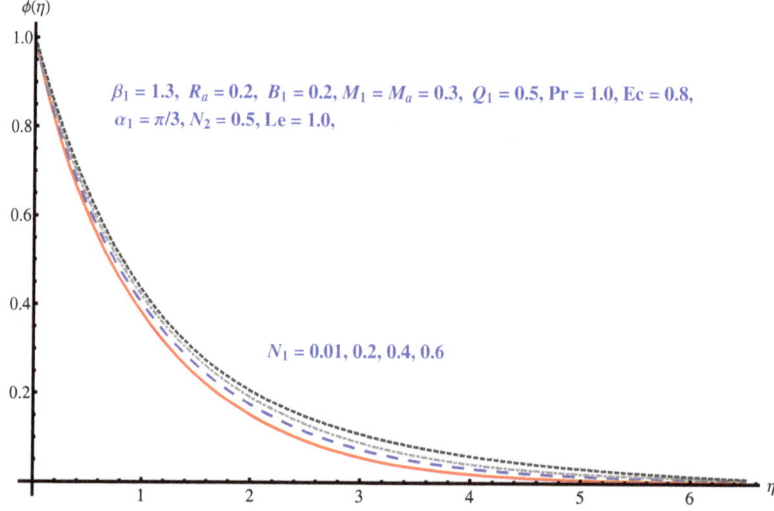

Figure 6. Influence of N_1 on temperature field.

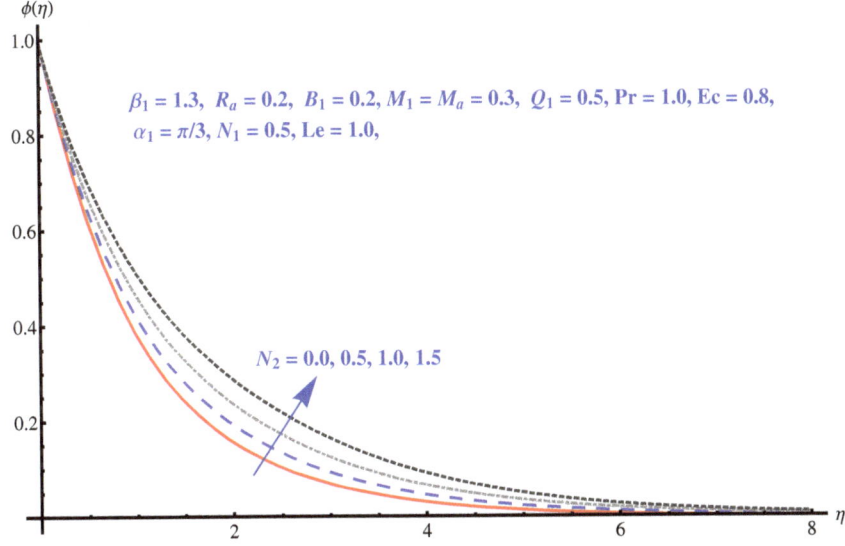

Figure 7. Influence of N_2 on temperature field.

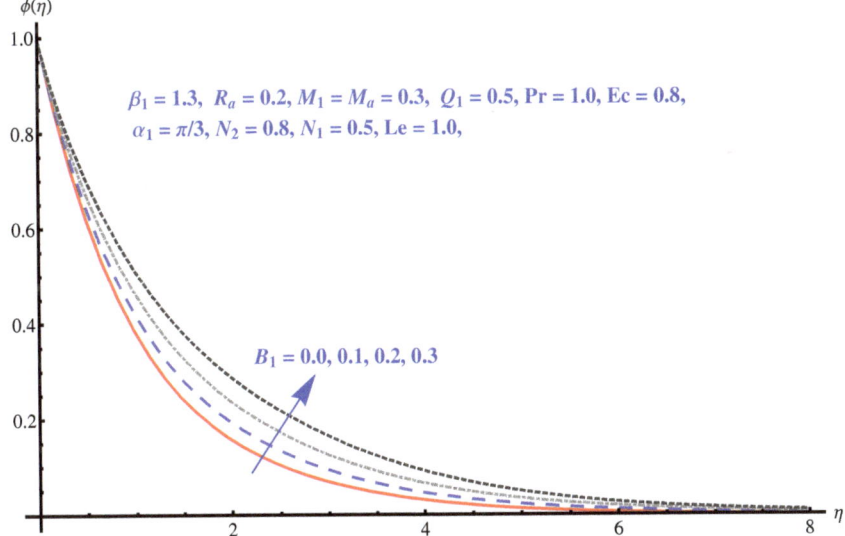

Figure 8. Influence of B_1 on temperature field.

4.3. Assessment of Concentration Distribution

The significance of Brownian motion N_1 on concentration profile $\zeta(\eta)$ is displayed in Figure 9. It is observed from the sketch that larger values of N_1 fluid concentration reduce far away from the surface and vanish after $\eta \geq 5$. On the other side, it increases near the surface. This is due to the existence of slip mechanisms of fluid particles, which influence the hydrodynamic and thermal bounce. Hence, the presence of this terminology does not have significant impact on flow concentration. Further, both thermophoresis parameter N_2 and Lewis number Le show increasing impact for concentration profiles (see Figures 10 and 11). The improvement in fluid concentration profile via Lewis number Le

is due to the fact that it is characterized by fluid flows where simultaneously mass and heat transport are involved. Therefore, an improvement is found in fluid concentration.

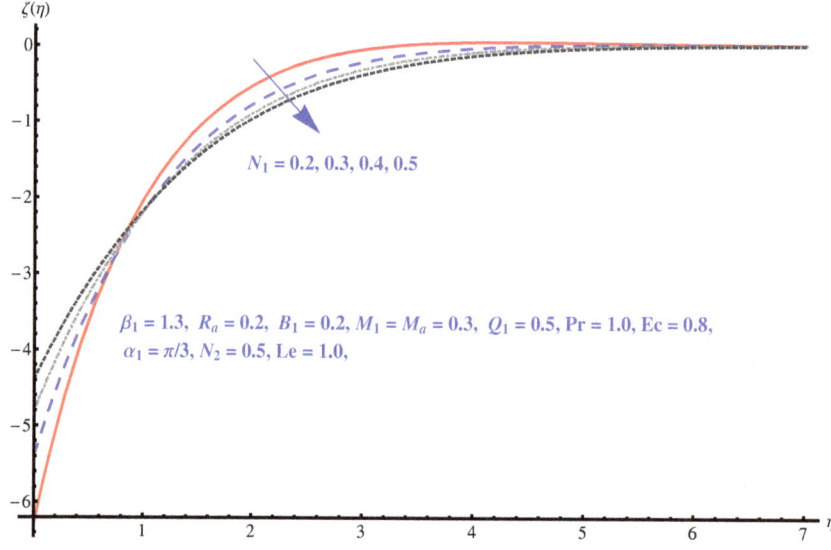

Figure 9. Influence of N_1 on concentration field.

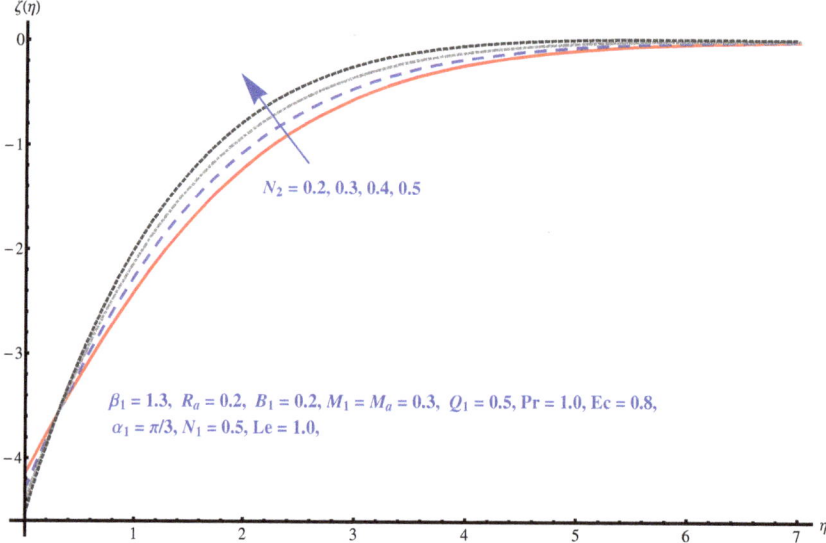

Figure 10. Influence of N_2 on concentration field.

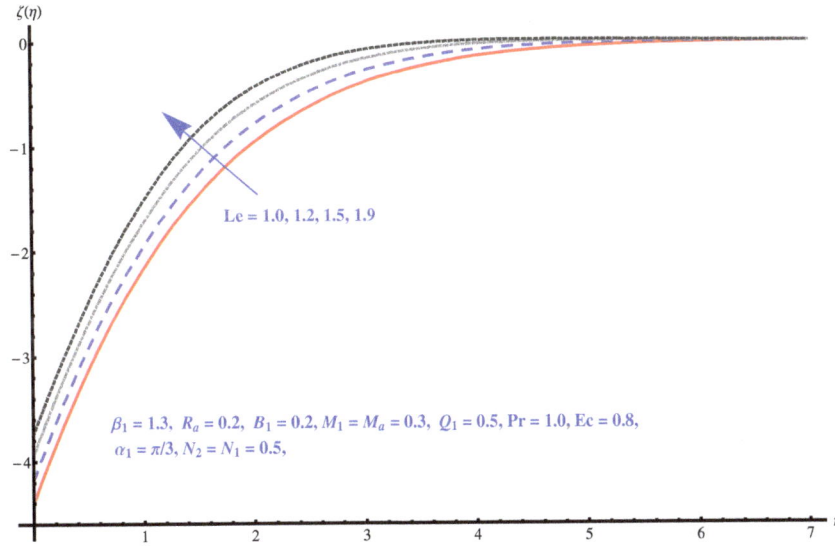

Figure 11. Influence of Le on concentration field.

4.4. Assessment of local Nusselt Number

In Table 1, one can see that a good agreement is found between the present results and previous literature. A very good agreement is found in RK-45 results; however, the results of HAM are a little different but the variation trend is similar in all the cases. Table 2 presents the significance of physical parameters through local Nusselt number Nu for axisymmetric Marangoni convective flow of Casson liquid over an infinite disk with the impact of an inclined MHD. Near the wall or boundary, Nusselt number has a dominant role, for the computation of thermal profile variations. The numerical quantities of Nusselt number is supportive to convey the cumulative tendency of temperature gradient in flow domain. It is observed in the table that rises in the Marangoni convective fluid parameter, M_a, Marangoni ratio, R_a, and N_2 monotonically decrease the Nusselt number Nu by keeping other fluid parameters fixed. On the other hand, the parameters N_1, Pr, B_1 and Le manifest rises in heat flux behavior Nu. The small increase on average Nusselt number indicates that greater heat exchange rate occurs near boundary of the disk due to these parameters.

Table 1. Comparison table of current results with previously published literature setting the additional parameters equals to zero.

r	$N_1 = N_b$	$N_2 = N_t$	Pr	Nu_x (Present)	Nu_x ([12])	Nu_x ([15])
0.0				1.488649	1.488646	–
0.1				1.551383	1.551382	–
0.2				1.609960	1.609962	–
	0.5			1.893601	1.893601	–
	1.0			1.609966	1.609962	–
	1.5			1.376585	1.376584	–
		0.5		1.893141	1.893141	–
		1.0		1.737723	1.737723	–
		1.5		1.609960	1.609962	–
			4.0	1.764112	1.764332	1.535191
			5.0	1.767000	1.767004	1.696162
			6.0	1.768490	1.768493	1.808222

Table 2. Numeric values of local Nusselt parameter Nu for distinct values of sundry parameter where $\beta_1 = 1.5$, $M_1 = 1.0$, $Ec = 0.8$ and $\alpha_1 = \pi/5$.

M_a	R_a	N_1	N_2	Pr	B_1	Le	$-R_d^{-1/2} Nu$
0.0	0.2	0.5	0.5	0.5	0.1	1.0	1.00000
0.3							0.99797
0.6							0.99772
0.3	0.1	0.5	0.5	0.5	0.1	1.0	0.99966
	0.3						0.99690
	0.5						0.99588
0.3	0.2	0.1	0.5	0.5	0.1	1.0	0.99657
		0.5					0.99797
		1.0					0.99966
0.0	0.2	0.5	0.1	0.5	0.1	1.0	0.99962
			0.5				0.99797
			1.0				0.99549
0.3	0.2	0.5	0.5	0.1	0.1	1.0	0.99755
				0.7			0.99797
				1.2			0.99857
0.3	0.2	0.5	0.5	0.5	0.1	1.0	0.99829
					0.4		0.99979
					0.8		1.00009
0.3	0.2	0.5	0.5	0.5	0.1	1.0	0.99829
						1.4	0.99883
						1.8	0.99923

5. Conclusions

In the present research work, we used RK45 scheme to simulate the two-dimensional Marangoni convective flow along with MHD effect and related heat and mass transfer problem over an infinite disk. The efficiency of proposed model was observed numerically and graphically, and found in good agreement for heat transportation process. The influence of distinct parameters on proposed flow problem are discussed in detail above. Further, the main findings of the present study are highlighted below:

- Increase in Brownian motion parameter enhances the flow temperature field, however the same goes for a declination of concentration field.
- Rise in thermophrases parameter improves the fluid temperature as well as concentration field.
- Larger values of Lewis number corresponds to the high concentration profile.
- Casson fluid parameter is found to be a reducing factor for fluid movement; therefore, admitting the higher quantity of Casson fluid parameter causes a reduction in fluid velocity.
- Increment in magnetic parameter and angle of inclination are reducing factors for the motion of fluid; however, the opposite performance in terms of heat transfer rate via Nusselt number is noted for the two parameters.
- The higher amount of Marangoni number condenses the active connectivity, which leads to improve the velocity profile.
- Temperature distribution rises up for the larger values of heat source sink.
- Increase in the Marangoni and Prandtl numbers show high increment on average Nusselt number, which leads to the conclusion that less heat exchange happens near the disk, while small values of fractional and physical parameters β_1, M_1, α_1, R_a, N_1, N_2, Ec, B_1, and Le manifest the high heat exchange rate near the boundary of the disk.

Author Contributions: Conceptualization, I.Z.; methodology, I.Z.; software, A.S.; validation, G.R.; formal analysis, T.S.K.; investigation, I.Z.; resources, I.T.; data curation, G.R.; writing—original draft preparation, I.Z.; writing—review and editing, A.S., G.R. and I.T.; visualization, A.S.; supervision, T.S.K.; project administration, A.S.; and funding acquisition, I.T.

Funding: This research received no external funding.

Conflicts of Interest: The authors declare no conflict of interest.

References

1. Hayat, T.; Aziz, A.; Muhammad, T.; Ahmad, B. On magnetohydrodynamic flow of second grade nanofluid over a nonlinear stretching sheet. *J. Mag. Mag. Mater.* **2016**, *408*, 99–106. [CrossRef]
2. Hayat, T.; Mumtaz, M.; Shafiq, A.; Alsaedi, A. Stratified magnetohydrodynamic flow of tangent hyperbolic nanofluid induced by inclined sheet. *Appl. Math. Mech.* **2017**, *38*, 271–288. [CrossRef]
3. Hayat, T.; Shafiq, A.; Alsaedi, A.; Awais, M. MHD axisymmetric flow of third grade fluid between stretching sheets with heat transfer. *Comput. Fluids* **2013**, *86*, 103–108. [CrossRef]
4. Hayat, T.; Shafiq, A.; Nawaz, M.; Alsaedi, A. MHD axisymmetric flow of third grade fluid between porous disks with heat transfer. *Appl. Math. Mech.* **2012**, *33*, 749–764. [CrossRef]
5. Shafiq, A.; Hammouch, Z.; Sindhu, T.N. Bioconvective MHD flow of tangent hyperbolic nanofluid with newtonian heating. *Int. J. Mech. Sci.* **2017**, *133*, 759–766. [CrossRef]
6. Shateyi, S.; Makinde, D. Numerical analysis of MHD stagnation point flow towards a radially stretching convectively heated disk. In Proceedings of the International Conference on Mechanics, Fluids, Heat, Elasticity and Electromagnetic Fields (MFHEEF 2013), Venice, Italy, 28–30 September 2013; pp. 1–8.
7. Hayat, T.; Shafiq, A.; Alsaedi, A. MHD axisymmetric flow of third grade fluid by a stretching cylinder. *Alex. Eng. J.* **2015**, *54*, 205–212. [CrossRef]
8. Shafiq, A.; Nawaz, M.; Hayat, T.; Alsaedi, A. Magnetohydrodynamic axisymmetric flow of a third-grade fluid between two porous disks. *Braz. J. Chem. Eng.* **2013**, *30*, 599–609. [CrossRef]
9. Rasool, G.; Zhang, T.; Shafiq, A.; Durur, H. Influence of Chemical Reaction on Marangoni Convective Flow of Nanoliquid in the Presence of Lorentz Forces and Thermal Radiation: A Numerical Investigation. *J. Adv. Nanotechnol.* **2019**, *1*, 32–49. [CrossRef]
10. Rasool, G.; Zhang, T.; Shafiq, A. Marangoni Effect in Second Grade Forced Convective Flow of Water Based Nanofluid. *J. Adv. Nanotechnol.* **2019**, *1*, 50–61. [CrossRef]
11. Mahanthesh, B.; Gireesha, B.J.; Shashikumar, N.S.; Shehzad, S.A. Marangoni convective MHD flow of SWCNT and MWCNT nanoliquids due to a disk with solar radiation and irregular heat source. *Phys. E Low-Dim. Syst. Nanostrut.* **2017**, *94*, 25–30. [CrossRef]
12. Kumar, G.; Gireesha, K.; Prasannakumara, B.J.; Makinde, O.D. Impact of Chemical Reaction on Marangoni Boundary Layer Flow of a Casson Nano Liquid in the Presence of Uniform Heat Source Sink. *Diffus. Found.* **2017**, *11*, 22–32. [CrossRef]
13. Din, M.; Usman, S.T.; Afaq, M.; Hamid, K.; Wang, W. Examination of carbon-water nanofluid flow with thermal radiation under the effect of Marangoni convection. *Eng. Comput.* **2017**, *34*, 2330–2343.
14. Sheikholeslami, M.; Ganji, D.D. Influence of magnetic field on CuO–H_2O nanofluid flow considering Marangoni boundary layer. *Int. J. Hydrogen Energy* **2017**, *42*, 2748–2755. [CrossRef]
15. Hayat, T.; Shaheen, U.; Shafiq, A.; Alsaedi, A.; Asghar, S. Marangoni mixed convection flow with Joule heating and nonlinear radiation. *AIP Adv.* **2015**, *5*, 1–15. [CrossRef]
16. Besthapu, P.P.; Haq, R.U.; Bandari, S.; Al-Mdallal, Q.M. Thermal radiation and slip effects on MHD stagnation point flow of non-Newtonian nanofluid over a convective stretching surface. *Neural Comput. Appl.* **2019**, *31*, 207–217. [CrossRef]
17. Lakshmi, K.B.; Kumar, K.A.; Reddy, J.V.; Sugunamma, V. Influence of nonlinear radiation and cross diffusion on MHD flow of Casson and Walters-B nanofluids past a variable thickness sheet. *J. Nanofluids* **2019**, *8*, 73–83. [CrossRef]
18. Casson, N. A flow equation for Pigment-Oil suspensions of the printing ink type. In *Rheology Of Disperse Systems*; Mill, C.C., Ed.; Pergamon Press: Oxford, London, UK, 1959.
19. Eldabe, N.T.M.; Saddeck, G.; El-Sayed, A.F. Heat transfer of MHD non-Newtonian Casson fluid flow between two rotating cylinders. *Mech. Mech. Eng.* **2001**, *5*, 237–251.

20. Charm, S.E.; Kurland, G. Viscometry of human blood for shear rates of 0–100,000 sec^{-1}. *Nature* **1965**, *206*, 617. [CrossRef] [PubMed]
21. Bhattacharyya, K.; Hayat, T.; Alsaedi, A. Analytic solution for magnetohydrodynamic boundary layer flow of Casson fluid over a stretching/shrinking sheet with wall mass transfer. *Chin. Phys. B* **2013**, *22*, 1–6. [CrossRef]
22. Kumar, G.; Gireesha, K.; Prasannakumara, B.J.; Ramesh, B.C.; Makinde, O.D. Phenomenon of radiation and viscous dissipation on Casson nanoliquid flow past a moving melting surface. *Diffus. Found.* **2017**, *11*, 33–42. [CrossRef]
23. Vijaya, N.; Krishna, H.; Kalyani, K.; Reddy, G.V.R. Soret and radiation effects on an unsteady flow of a Casson fluid through porous vertical channel with expansion and contraction. *Front. Heat Mass Trans.* **2018**, 1–11. [CrossRef]
24. Hayat, T.; Ashraf, M.B.; Shehzad, S.A.; Alsaedi, A. Mixed Convection Flow of Casson Nanofluid over a Stretching Sheet with Convectively Heated Chemical Reaction and Heat Source/Sink. *J. Appl. Fluid Mech.* **2015**, *8*, 803–813. [CrossRef]
25. Ghadikolaei, S.S.; Hosseinzadeh, K.; Ganji, D.D.; Jafari, B. Nonlinear thermal radiation effect on magneto Casson nanofluid flow with Joule heating effect over an inclined porous stretching sheet. *Case Stud. Ther. Eng.* **2018**, *12*, 176–187. [CrossRef]
26. Goodarzi, M.; Tlili, I.; Tian, Z.; Safaei, M. Efficiency assessment of using graphene nanoplatelets-silver/water nanofluids in microchannel heat sinks with different cross-sections for electronics cooling. *Int. J. Numer. Meth. Heat Fluid Flow* **2019**. [CrossRef]
27. Afridi, M.I.; Tlili, I.; Goodarzi, M.; Osman, M.; Khan, N.A. Irreversibility analysis of Hybrid nanofluid flow over a thin needle with effects of energy dissipation. *Symmetry* **2019**, *11*, 663. [CrossRef]
28. Tlili, I.; Khan, W.A.; Ramadan, K. MHD flow of nanofluid across horizontal circular cylinder: Steady forced convection. *J. Nanofluids* **2019**, *8*, 179–186. [CrossRef]
29. Tlili, I.; Khan, W.A.; Ramadan, K. Entropy generation due to MHD stagnation point flow of a nanofluid on a stretching surface in the presence of radiation. *J. Nanofluids* **2018**, *7*, 879–890. [CrossRef]
30. Tlili, I.; Khan, W.A.; Khan, I. Multiple slips effects on MHD SA-Al$_2$O$_3$ and SA-Cu non-Newtonian nanofluids flow over a stretching cylinder in porous medium with radiation and chemical reaction. *Res. Phys.* **2018**, *8*, 213–222. [CrossRef]
31. Tlili, I.; Hamadneh, N.N.; Khan, W.A. Thermodynamic analysis of MHD heat and mass transfer of nanofluids past a static wedge with Navier slip and convective boundary conditions. *Arab. J. Sci. Eng.* **2018**, 1–13. [CrossRef]
32. Tlili, I.; Hamadneh, N.N.; Khan, W.A.; Atawneh, S. Thermodynamic analysis of MHD Couette–Poiseuille flow of water-based nanofluids in a rotating channel with radiation and Hall effects. *J. Therm. Anal. Calorim.* **2018**, *132*, 1899–1912. [CrossRef]
33. Khan, W.A.; Rashad, A.M.; Abdou, M.M.M.; Tlili, I. Natural bioconvection flow of a nanofluid containing gyrotactic microorganisms about a truncated cone. *Eur. J. Mech. B Fluids* **2019**, *75*, 133–142. [CrossRef]
34. Khalid, A.; Khan, I.; Khan, A.; Shafie, S.; Tlili, I. Case study of MHD blood flow in a porous medium with CNTS and thermal analysis. *Case Stud. Therm. Eng.* **2018**, *12*, 374–380. [CrossRef]
35. Hayat, T.; Farooq, M.; Alsaedi, A. Thermally stratified stagnation point flow of Casson fluid with slip conditions. *Int. J. Numer. Methods Heat Fluid Flow* **2015**, *25*, 724–748. [CrossRef]
36. Mahian, O.; Kianifar, A.; Kalogirou, S.A.; Pop, I.; Wongwises, S. A review of the applications of nanofluids in solar energy. *Int. J. Heat Mass Transf.* **2013**, *57*, 582–594. [CrossRef]
37. Wong, K.V.; Leon, O.D. Applications of Nanofluids: Current and Future. *Adv. Mech. Eng.* **2010**. [CrossRef]
38. Hayat, T.; Asad, S. Alsaedi, A. Flow of Casson fluid with nanoparticles. *Appl. Math. Mech.* **2016**, *37*, 459–470. [CrossRef]
39. Naseem, A.; Shafiq, A.; Zhao, L.; Farooq, M.U. Analytical investigation of third grade nanofluidic flow over a Riga plate using Cattaneo-Christov model. *Res. Phys.* **2018**, *9*, 961–969. [CrossRef]
40. Rasool, G.; Shafiq, A.; Khalique, C.M.; Zhang, T. Magneto-hydrodynamic Darcy-Forchheimer nanofluid flow over nonlinear stretching sheet. *Phys. Scr.* **2019**, *94*, 105221. [CrossRef]
41. Rashid, M.; Hayat, T.; Alsaedi, A. Entropy generation in Darcy–Forchheimer flow of nanofluid with five nanoarticles due to stretching cylinder. *Appl. Nanosci.* **2019**, 1–11. [CrossRef]

42. Naseem, F.; Shafiq, A.; Zhao, L.; Naseem, A. MHD biconvective flow of Powell Eyring nanofluid over stretched surface. *AIP Adv.* **2017**, *7*, 1–20. [CrossRef]
43. Rasool, G.; Zhang, T.; Shafiq, A. Second grade nanofluidic flow past a convectively heated vertical Riga plate. *Phys. Scr.* **2019**, *94*, 125212. [CrossRef]
44. Rasool, G.; Zhang, T. Characteristics of chemical reaction and convective boundary conditions in Powell-Eyring nanofluid flow along a radiative Riga plate. *Heliyon* **2019**, *5*, e01479. [CrossRef] [PubMed]
45. Rasool, G.; Zhang, T. Darcy-Forchheimer nanofluidic flow manifested with Cattaneo-Christov theory of heat and mass flux over non-linearly stretching surface. *PLoS ONE* **2019**, *14*, e0221302. [CrossRef] [PubMed]
46. Rasool, G.; Shafiq, A.; Tlili, I. Marangoni convective nano-fluid flow over an electromagnetic actuator in the presence of first order chemical reaction. *Heat Transf. Asian Res.* **2019**, accepted.
47. Rasool, G.; Shafiq, A.; Durur, H. *Darcy-Forchheimer Relation in Magnetohydrodynamic Jeffrey Nanofluid Flow over Stretching Surface*; Discrete and Continuous Dynamical Systems-Series S; American Institute of Mathematical Sciences, 2019. Available Online: https://www.researchgate.net/publication/336373931_Darcy-Forchheimer_relation_in_Magnetohydrodynamic_Jeffrey_nanofluid_flow_over_stretching_surface (accessed on 10 October 2019).
48. Rasool, G.; Shafiq, A.; Khalique, C.M. *Marangoni Forced Convective Casson Type Nanofluid Flow in the Presence of Lorentz Force Generated by Riga Plate*; Discrete and Continuous Dynamical Systems-Series S; American Institute of Mathematical Sciences, 2019. Available Online: https://www.researchgate.net/publication/336373925_Marangoni_forced_convective_Casson_type_nanofluid_flow_in_the_presence_of_Lorentz_force_generated_by_Riga_plate (accessed on 10 October 2019).

© 2019 by the authors. Licensee MDPI, Basel, Switzerland. This article is an open access article distributed under the terms and conditions of the Creative Commons Attribution (CC BY) license (http://creativecommons.org/licenses/by/4.0/).

Article

Three-Dimensional Hydro-Magnetic Flow Arising in a Long Porous Slider and a Circular Porous Slider with Velocity Slip

Naeem Faraz [1,*], Yasir Khan [2], Amna Anjum [1,3,*] and Muhammad Kahshan [4,5]

1   International Cultural Exchange School (ICES), Donghua University, West Yan'an Road 1882, Shanghai 200051, China
2   Department of Mathematics, University of Hafr Al-Batin, Hafr Al-Batin 31991, Saudi Arabia
3   Glorious Sun School of Business and Management, Donghua University, West Yan'an Road 1882, Shanghai 200051, China
4   Faculty of Science, Jiangsu University, Zhenjiang 212013, China
5   Department of Mathematics, COMSATS University Islamabad, Abbottabad 22060, Pakistan
*   Correspondence: nfaraz_math@yahoo.com or naeem@dhu.edu.cn (N.F.); amnaeem14@gmail.com (A.A.); Tel.: +86-150-0075-1065 (N.F.); +86-132-6289-1913 (A.A.)

Received: 21 July 2019; Accepted: 12 August 2019; Published: 16 August 2019

Abstract: The current research explores the injection of a viscous fluid through a moving flat plate with a transverse uniform magneto-hydrodynamic (MHD) flow field to reduce sliding drag. Two cases of velocity slip between the slider and the ground are studied: a long slider and a circular slider. Solving the porous slider problem is applicable to fluid-cushioned porous sliders, which are useful in reducing the frictional resistance of moving bodies. By using a similarity transformation, three dimensional Navier–Stokes equations are converted into coupled nonlinear ordinary differential equations. The resulting nonlinear boundary value problem was solved analytically using the homotopy analysis method (HAM). The HAM provided a fast convergent series solution, showing that this method is efficient, accurate, and has many advantages over the other existing methods. Solutions were obtained for the different values of Reynolds numbers (R), velocity slip, and magnetic fields. It was found that surface slip and Reynolds number had substantial influence on the lift and drag of the long and the circular sliders. Moreover, the effects of the applied magnetic field on the velocity components, load-carrying capacity, and friction force are discussed in detail with the aid of graphs and tables.

Keywords: porous slider; MHD flow; reynolds number; velocity slip; homotopy analysis method

1. Introduction

It is a well-established fact that a moving body reduces drag if it is elevated by a layer of air. This phenomenon is used in air-cushioned vehicles and in air hockey, in which the frictional resistance of moving objects is reduced. Skalak and Wang [1] were the pioneers of studying the three-dimensional flow that arises between a moving porous flat plate and the ground, and they later on wrote an erratum on their own paper [2]. Wang also studied elliptical porous sliders [3]. In the case of Newtonian fluids, past studies have included porous circular, long, inclined, and elliptical sliders. R. C. Bhattacharjee studied a porous slider bearing lubricated with a coupled stress (a magneto-hydrodynamic (MHD) fluid) [4]. Jimit made a comparison of the different porous structures on the performance of a magnetic fluid [5]. Prawal Sinha analyzed the thermal effects of a long porous rough slider bearing [6]. Mohmmadrayian analyzed a rough porous inclined slider bearing lubricated with a ferrofluid in consideration of slip velocity [7]. Ji Lang both theoretically and experimentally investigated the transient squeezing flow in a highly porous film [8]. Similarly, a large amount of literature is available in relation to long porous sliders (LPSs) [1,6,9–13] and circular porous sliders (CPSs) [2,14–18]. Awati investigated the lubrication of a long porous slider by using the homotopy analysis method (HAM) [10]. In a separate study, Khan studied the effects of Reynolds numbers by using different analytical methods [11,12]. Naeem studied the influence of Reynolds numbers (R) on long [13] and circular porous sliders [18]. Ghoreishi studied the circular slider [14]. Madani investigated the circular porous slider by using HPM, and also analyzed its lift and drag [19].

All the above mentioned studies were done without a slip condition on either the immobile ground or slider. However, a slip condition is essential for super-hydrophobic planes, as it is difficult to have a zero mean tangential velocity from where the fluid is injected when there is a slip. Furthermore, in order to minimize adhesion, the fluid could be a rarefied gas, where the compact exterior could be coated with a material, or the ground could be uneven so that an equivalent slip exists or there is a slip flow regime. Wang [16] discussed slip effects, but didn't consider the effects of a transverse magnetic field. Therefore, the goal of the current work is to examine the impact of slip and Reynolds numbers when a transverse magnetic field is affecting the performance of a porous slider. Through the literature survey, it is assumed that a three-dimensional flow with slip and a uniform magnetic field does not exist. Hence, the goal of the current research is to analyze the performance of porous sliders in the presence of slip and a Reynolds number with a constant magnetic field, and to assess their effects on the components of velocity lift and drag.

The structure of the article is as follows: In the introduction, a brief history of the problem of the porous slider and its application is presented. In the second section, the formulation of the problems are given, while in the third section the formulation of a homotopic solution is presented [20]. The fourth section deals with the convergence criteria of the HAM. Results and discussions are given in the fifth section. Finally, the conclusion is given in the sixth section, with a list of nomenclature.

As discussed above, the velocity slip condition is considered in this study. Navier introduced the slip condition for the first time as follows:

$$x_1 = H\varsigma \qquad (1)$$

In Equation (1), tangential velocity u is proportional to the shear stress and H is the constant of proportionality, which is actually a slip coefficient. In order to ignore the end effects, it is assumed that the gap between slider and ground is quite small as compared to the slider's lateral dimension. Both circular and long porous sliders are considered in this study.

2. Problem Formulation of Long and Circular Sliders

In this study, the incompressible and steady flow of a viscous fluid between porous (long and circular) sliders and the ground is considered in the presence of a uniform magnetic field, as shown in Figure 1.

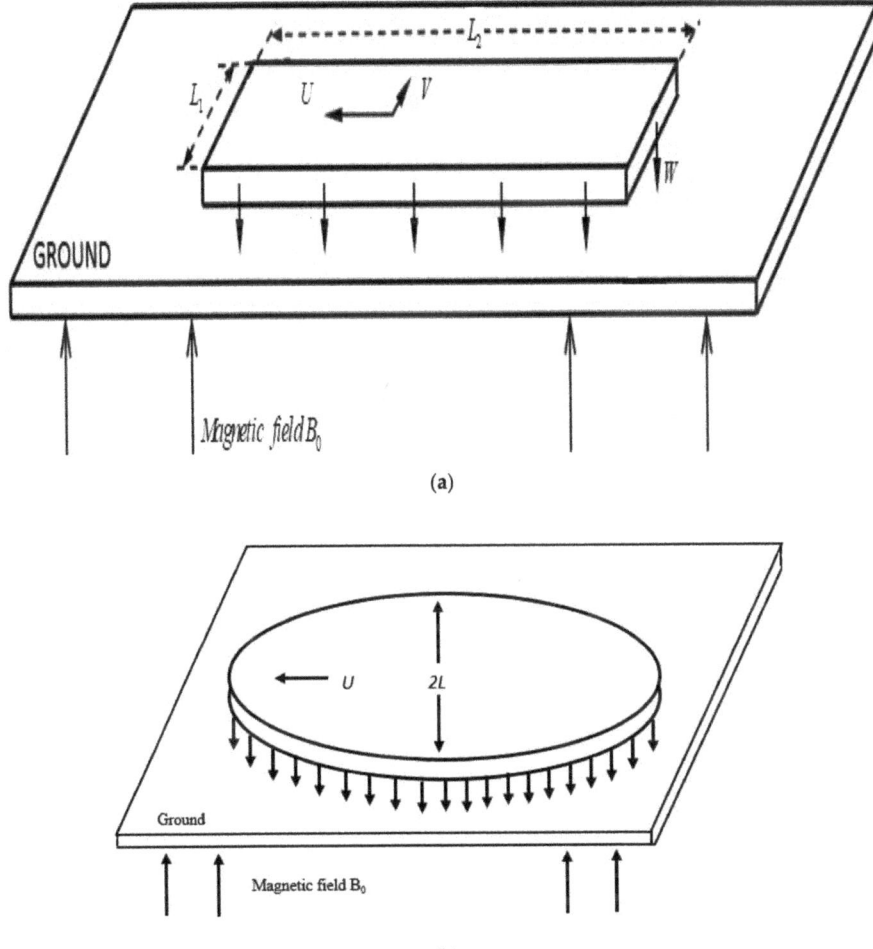

Figure 1. (a) Schematic diagram of the movement of a long porous slider (LPS). (b) Schematic diagram of the movement of a circular porous slider (CPS).

Length and width are quite big compared to height d. The slider moves with the velocity components and is elevated because of the injection of fluid from below with a magnetic field applied externally. In order to avoid the induced magnetic field formed by the movement of the fluid, it is assumed that the magnetic Reynolds number is not very big. Furthermore, the induced and imposed electric field are supposed to be negligible, and therefore the electromagnetic body force per unit volume simplifies $F_{em} = \sigma_0(v \times B) \times B$, where $B = (0, 0, B_0)$ is the magnetic field.

Under the above-stated assumptions and conditions, Navier–Stokes equations take the following form:

$$\phi_1 \frac{\partial \phi_1}{\partial x_1} + \phi_2 \frac{\partial \phi_1}{\partial x_2} + \phi_3 \frac{\partial \phi_1}{\partial x_3} = -\frac{1}{\rho} \frac{\partial p}{\partial x_1} + v\left(\frac{\partial^2 \phi_1}{\partial x_1^2} + \frac{\partial^2 \phi_1}{\partial x_2^2} + \frac{\partial^2 \phi_1}{\partial x_3^2}\right) - \frac{\sigma_0}{\rho} B_0^2 \phi_1 \tag{2}$$

$$\phi_1 \frac{\partial \phi_2}{\partial x_1} + \phi_2 \frac{\partial \phi_2}{\partial x_2} + \phi_3 \frac{\partial \phi_2}{\partial x_3} = -\frac{1}{\rho} \frac{\partial p}{\partial x_2} + v\left(\frac{\partial^2 \phi_2}{\partial x_1^2} + \frac{\partial^2 \phi_2}{\partial x_2^2} + \frac{\partial^2 \phi_2}{\partial x_3^2}\right) - \frac{\sigma_0}{\rho} B_0^2 \phi_2 \tag{3}$$

$$\phi_1\frac{\partial \phi_3}{\partial x_1}+\phi_2\frac{\partial \phi_3}{\partial x_2}+\phi_3\frac{\partial \phi_3}{\partial x_3}=-\frac{1}{\rho}\frac{\partial p}{\partial x_3}+v\left(\frac{\partial^2 \phi_3}{\partial x_1^2}+\frac{\partial^2 \phi_3}{\partial x_2^2}+\frac{\partial^2 \phi_3}{\partial x_3^2}\right) \quad (4)$$

Velocity components are expressed as (ϕ_1, ϕ_2, ϕ_3), where ρ, p, and v are density, pressure and kinematic viscosity, respectively. Law of conservation of mass is as follows:

$$\frac{\partial \phi_1}{\partial x_1}+\frac{\partial \phi_2}{\partial x_2}+\frac{\partial \phi_3}{\partial x_3}=0 \quad (5)$$

According to Naeem [13], the following transform has been used:

$$\phi_1=U\psi_1(\varsigma)+\frac{W}{d}x_1\psi_3{'}(\varsigma), \phi_2=V\psi_2(\varsigma), \phi_3=-W\psi_3(\varsigma). \quad (6)$$

where $\varsigma=\frac{x_3}{d}$. By adding Equation (6) into Equations (2)–(4), the following ordinary differential equations are obtained:

$$\psi_3^{iv}=R\left(\psi_3'\psi_3''-\psi_3\psi_3'''\right)+M^2\psi_3' \quad (7)$$

$$\psi_1''=R\left(\psi_1\psi_3'-\psi_3\psi_1'\right)+M^2\psi_1 \quad (8)$$

$$\psi_2''=-R\left(\psi_1\psi_2'\right)+M^2\psi_2 \quad (9)$$

where R is the Reynolds number ($R=Wd/v$). Boundary conditions at $x_3=0$ and $x_3=d$ are given in Equations (10) and (11), respectively.

$$\phi_1=U+H_1\mu\frac{\partial\phi_1}{\partial x_3}, \quad \phi_2=V+H_1\mu\frac{\partial\phi_2}{\partial x_3}, \quad \phi_3=0 \quad (10)$$

$$\phi_1=-H_2\mu\frac{\partial\phi_1}{\partial x_3}, \quad \phi_2=-H_2\mu\frac{\partial\phi_2}{\partial x_3}=0, \quad \phi_3=-W \quad (11)$$

where H_1, H_2, and $\mu=\rho v$ are slip coefficients and viscosity, respectively. Equations (10) and (11) take the following form:

$$\begin{aligned}&\psi_3'(0)=\beta_1\psi_3''(0)=, \psi_3(0)=0,\\ &\psi_3(1)=1, \psi_3'(1)=-\beta_2\psi_3''(1),\\ &\psi_1(1)=-\beta_2\psi_1'(1), \psi_1(0)-1=\beta_1\psi_1'(0),\\ &\psi_2(1)=-\beta_2\psi_2'(1), \psi_2(0)-1=\beta_1\psi_2'(0).\end{aligned} \quad (12)$$

where $\beta_1=H_1\mu/d, \beta_2=H_2\mu/d$ are slip factors. Equations (7)–(9) and (12) will be solved by the HAM. The expression for pressure can be deduced from Equations (2)–(4) as follows:

$$-\frac{p}{\rho}=\frac{W^2\Lambda x_1^2}{2d}+\frac{1}{2}\phi_3^2-\gamma\phi_{3,x_3}+A \quad (13)$$

where Λ, A are constants and

$$\Lambda=\left(\psi_3'\right)^2-\psi_3\psi_3''-\frac{1}{R}\psi_3'''=\left(\psi_3'(0)\right)^2-\frac{1}{R}\psi_3'''(0). \quad (14)$$

If $2l$ is the width of the slider with ambient pressure p_0, then Equation (13) gives

$$p-p_0=-\rho\frac{\Lambda W^2\left(x_1^2-l^2\right)}{2d^2}. \quad (15)$$

The relationship between depth and lift can be expressed as follows:

$$L = \int_{-1}^{1} (p - p_0) dx = \frac{2\rho W^2 l^3}{3d^2} \Lambda. \tag{16}$$

where $2\rho W^2 l^3 / (3d^2)$ is normalized factor. The relationship between depth and drag in the x_1- direction is

$$D_{x_1} = -\int_{-1}^{1} \mu \frac{\partial \phi_1}{\partial x_3}|_{z=d} dx_1 = -\frac{2\mu Ul}{d} \psi_1'(1). \tag{17}$$

Similarly, $2\mu Ul/d$ is the normalized factor of drag in the x_1- direction, which is $-\psi_1'(1)$, while $-\psi_2'(1)$ is normalized drag for the x_2- direction:

$$D_{x_2} = -\int_{-1}^{1} \mu \frac{\partial \phi_2}{\partial x_3}|_{z=d} dx_1 = -\frac{2\mu Vl}{d} \psi_2'(1). \tag{18}$$

Similarly, from Figure 1b, a circular slider can be seen, where L is the radius of the slider (which can be assumed to be comparatively bigger than the width). Since the slider is levitated, the axes on the slider can be fixed so that the ground is moving with a velocity component in the x_1- direction. For the circular slider, a similar transform [18] helps to reduce the partial differential equations into ordinary differential equations:

$$x_1 = U\psi_5(\varsigma) + \frac{W}{d} x_1 \psi_4'(\varsigma), x_2 = \frac{W}{d} x_2 \psi_4'(\varsigma), x_3 = -2W\psi_4(\varsigma). \tag{19}$$

With the help of Equation (19), Equations (2)–(4) take the following form

$$\psi_4^{iv} - 2R\psi_4 \psi_4''' - M^2 \psi_4' = 0 \tag{20}$$

$$\psi_5'' - R(\psi_5 \psi_4' - 2\psi_4 \psi_5') - M^2 \psi_5 = 0 \tag{21}$$

$$-\frac{p}{\rho} = \frac{W^2 \Lambda (x_1^2 + x_2^2)}{2d} + \frac{1}{2} x_3^2 - \gamma x_{3,x_3} + C \tag{22}$$

in which Λ, C are constants and

$$\Lambda_1 = \left(\psi_4'(0)\right)^2 - \frac{1}{R} \psi_4'''(0). \tag{23}$$

The boundary conditions on $x_3 = 0 \& d$:

$$\begin{aligned}
&\psi_4'(0) = \beta_1 \psi_4''(0) =, \psi_4(0) = 0, \\
&\psi_4(1) = 1/2, \psi_4'(1) = -\beta_2 \psi_4''(1), \\
&\psi_5(1) = -\beta_2 \psi_5'(1), \psi_5(0) - 1 = \beta_1 \psi_5'(0).
\end{aligned} \tag{24}$$

To normalize the lift, integrating the bottom of the slider as a result of the normalized factor can be expressed as $\pi \rho W^2 l^4 / 4d$.

$$L = \frac{4d}{\pi \rho W^2 l^4} \iint_s (p - p_0) ds = \frac{1}{R^3} \Lambda. \tag{25}$$

The relationship between depth and drag in the x_1- direction is

$$D_{x_1} = \frac{d}{\pi \mu U l^2} \iint_s H_{x_3 x_1} ds = -\frac{1}{R^3} \psi_5'(1). \tag{26}$$

3. Homotopic Solution Procedure

To apply HAM [20], the following initial guesses for Equations (7)–(9), (20), and (21) can be chosen as

$$\psi_{3,0}(\varsigma) = \frac{-2\varsigma^3(1+\beta_1+\beta_2)+3\varsigma(1+2\beta_2)(\varsigma+2\beta_1)}{1+4(\beta_1+\beta_2)+12\beta_1\beta_2},$$
$$\psi_{1,0}(\varsigma) = \frac{1-\varsigma+\beta_2}{1+\beta_1+\beta_2}, \psi_{2,0}(\varsigma) = \frac{1-\varsigma+\beta_2}{1+\beta_1+\beta_2}, \quad (27)$$
$$\psi_{4,0}(\varsigma) = \frac{-2\varsigma^3(1+\beta_1+\beta_2)+3\varsigma(1+2\beta_2)(\varsigma+2\beta_1)}{2(1+4(\beta_1+\beta_2)+12\beta_1\beta_2)}, \psi_{5,0}(\varsigma) = \frac{1-\varsigma+\beta_2}{1+\beta_1+\beta_2}$$

For the initial approximation, the following auxiliary linear operators can be chosen:

$$L(\psi_3) = \frac{d^4h}{d\varsigma^4}, L(\psi_1) = \frac{d^2\theta}{d\varsigma^2}, L(\psi_2) = \frac{d^2\theta}{d\varsigma^2}, \\ L(\psi_4) = \frac{d^4\theta}{d\varsigma^4}, L(\psi_5) = \frac{d^2\theta}{d\varsigma^2}. \quad (28)$$

which satisfies

$$L_{\psi_1}[A_5 + A_6\varsigma] = 0, L_{\psi_2}[A_7 + A_8\varsigma] = 0, \\ L_{\psi_3}\left[A_1 + A_2\varsigma + A_3\varsigma^2 + A_4\varsigma^4\right] = 0, \\ L_{\psi_4}\left[A_9 + A_{10}\varsigma + A_{11}\varsigma^2 + A_{12}\varsigma^4\right] = 0, L_{\psi_5}[A_{13} + A_{14}\varsigma] = 0. \quad (29)$$

in which $A_i(i = 1 - 14)$ are constants of integration.

Initial Order Deformation Problem

The deformation equations for the initial order can be viewed as follows:

$$(1-\Phi)I_{\psi_1}\left[\widehat{\psi}_1(\varsigma,\Phi) - \widehat{\psi}_{1,0}(\varsigma,\Phi)\right] = \Phi\hbar_{\psi_1}H_{\psi_1}N_{\psi_1}\left[\widehat{\psi}_1(\varsigma,\Phi)\right] \quad (30)$$

$$(1-\Phi)L_{\psi_2}\left[\widehat{\psi}_2(\varsigma,\Phi) - \widehat{\psi}_{2,0}(\varsigma,\Phi)\right] = \Phi\hbar_{\psi_2}H_{\psi_2}N_{\psi_2}\left[\widehat{\psi}_2(\varsigma,\Phi)\right] \quad (31)$$

$$(1-\Phi)L_{\psi_3}\left[\widehat{\psi}_3(\varsigma,\Phi) - \widehat{\psi}_{3,0}(\varsigma,\Phi)\right] = \Phi\hbar_{\psi_3}H_{\psi_3}N_{\psi_3}\left[\widehat{\psi}_3(\varsigma,\Phi)\right] \quad (32)$$

$$(1-\Phi)L_{\psi_4}\left[\widehat{\psi}_4(\varsigma,\Phi) - \widehat{\psi}_{4,0}(\varsigma,\Phi)\right] = \Phi\hbar_{\psi_4}H_{\psi_4}N_{\psi_4}\left[\widehat{\psi}_4(\varsigma,\Phi)\right] \quad (33)$$

$$(1-\Phi)L_{\psi_5}\left[\widehat{\psi}_5(\varsigma,\Phi) - \widehat{\psi}_{5,0}(\varsigma,\Phi)\right] = \Phi\hbar_{\psi_5}H_{\psi_5}N_{\psi_5}\left[\widehat{\psi}_5(\varsigma,\Phi)\right] \quad (34)$$

and the boundary conditions are

$$\widehat{\psi}_1(1,\Phi) = -\mu_2\widehat{\psi}_1'(1,\Phi), \widehat{\psi}_1(0,\Phi) - 1 = -\mu_1\widehat{\psi}_1'(0,\Phi) \quad (35)$$

$$\widehat{\psi}_2'(0,\Phi) - 1 = -\mu_1\widehat{\psi}_2'(0,\Phi), \widehat{\psi}_2(1,\Phi) = -\mu_2\widehat{\psi}_2'(1,\Phi) \quad (36)$$

$$\widehat{\psi}_3'(0,\Phi) = \mu_1\widehat{\psi}_3''(0,\Phi), \widehat{\psi}_3(0,\Phi) = 0, \widehat{\psi}_3(1,\Phi) = 1, \widehat{\psi}_3'(1,\Phi) = -\mu_2\widehat{\psi}_3''(1,\Phi) \quad (37)$$

$$\widehat{\psi}_4'(0,\Phi) = \mu_1\widehat{\psi}_4''(0,\Phi), \widehat{\psi}_4(0,\Phi) = 0, \widehat{\psi}_4(1,\Phi) = 1/2, \widehat{\psi}_4'(1,\Phi) = -\mu_2\widehat{\psi}_4''(1,\Phi), \quad (38)$$

$$\widehat{\psi}_5(1,\Phi) = -\mu_2\widehat{\psi}_5'(1,\Phi), \widehat{\psi}_5(0,\Phi) - 1 = -\mu_1\widehat{\psi}_5'(0,\Phi) \quad (39)$$

where $N_{\psi_1}, N_{\psi_2}, N_{\psi_3}, N_{\psi_4}$, and N_{ψ_5} are defined as

$$N_{\psi_3}\left[\widehat{\psi}_3(\varsigma;\Phi)\right] = \widehat{\psi}_3'''' - R\left(\widehat{\psi}_3'\widehat{\psi}_3'' - \widehat{\psi}_3\widehat{\psi}_3'''\right) - M^2\psi_3' \quad (40)$$

$$N_{\widehat{\psi}_1}\left[\widehat{\psi}_1(\varsigma;\Phi)\right] = \widehat{\psi}_1'' - R\left(\widehat{\psi}_1\widehat{\psi}_3' - \widehat{\psi}_3\widehat{\psi}_1'\right) - M^2\psi_1, \quad (41)$$

$$N_{\widehat{\psi}_2}\left[\widehat{\psi}_2(\varsigma;\Phi)\right] = \widehat{\psi}_2{}'' + R\left(\widehat{\psi}_3\widehat{\psi}_2{}'\right) - M^2\widehat{\psi}_2, \quad (42)$$

$$N_{\widehat{\psi}_4}\left[\widehat{\psi}_4(\varsigma;\Phi)\right] = \widehat{\psi}_4{}'''' - 2R\widehat{\psi}_4\widehat{\psi}_4{}''' - M^2\widehat{\psi}_4{}' \quad (43)$$

$$N_{\widehat{\psi}_5}\left[\widehat{\psi}_5(\varsigma;\Phi)\right] = \widehat{\psi}_5{}'' - R\left(\widehat{\psi}_5\widehat{\psi}_4{}' - 2\widehat{\psi}_4\widehat{\psi}_5{}'\right) - M^2\widehat{\psi}_5 \quad (44)$$

Here, the auxiliary parameters are $\hbar_{\psi_1} \ne 0, \hbar_{\psi_2} \ne 0, \hbar_{\psi_3} \ne 0, \hbar_{\psi_4} \ne 0$, and $\hbar_{\psi_5} \ne 0$, while the non-zero auxiliary functions are expressed as $H_{\psi_1}, H_{\psi_2}, H_{\psi_3}, H_{\psi_4}$, and H_{ψ_5}, and $\varsigma \in [0,1]$ is the embedding parameter.

From Equations (30)–(34), it is observed that when $\varsigma = 0$ there is

$$\widehat{\psi}_1(\varsigma,0) = \widehat{\psi}_{1,0}(\varsigma), \widehat{\psi}_2(\varsigma,0) = \widehat{\psi}_{2,0}(\varsigma), \widehat{\psi}_3(\varsigma,0) = \widehat{\psi}_{3,0}(\varsigma), \widehat{\psi}_4(\varsigma,0) = \widehat{\psi}_{4,0}(\varsigma), \widehat{\psi}_5(\varsigma,0) = \widehat{\psi}_{5,0}(\varsigma). \quad (45)$$

As $\varsigma = 1$ and $\hbar_{\psi_1} \ne 0, \hbar_{\psi_2} \ne 0, \hbar_{\psi_3} \ne 0, \hbar_{\psi_4} \ne 0, \hbar_{\psi_5} \ne 0$ and $H_{\psi_1} \ne 0, H_{\psi_2} \ne 0, H_{\psi_3} \ne 0, H_{\psi_4} \ne 0, H_{\psi_5} \ne 0$, then Equations (30)–(34) are obtained as

$$\widehat{\psi}_1(\varsigma,1) = \widehat{\psi}_1(\varsigma), \widehat{\psi}_2(\varsigma,1) = \widehat{\psi}_2(\varsigma), \widehat{\psi}_3(\varsigma,1) = \widehat{\psi}_3(\varsigma), \widehat{\psi}_4(\varsigma,1) = \widehat{\psi}_4(\varsigma), \widehat{\psi}_5(\varsigma,1) = \widehat{\psi}_5(\varsigma),$$

In order to get mth-order deformation equations, Equations (30)–(34) are differentiated m-times with respect to ς, after substituting $\varsigma = 0$ and dividing both sides by $m!$. Finally, the mth-order deformation equations take the following forms:

$$L_{\psi_1}[\psi_{1,m}(\varsigma) - \chi_m\psi_{1,m-1}(\varsigma)] = \hbar_{\psi_1}R_{1,m}(\varsigma). \quad (46)$$

$$L_{\psi_2}[\psi_{2,m}(\varsigma) - \chi_m\psi_{2,m-1}(\varsigma)] = \hbar_{\psi_2}R_{3,m}(\varsigma). \quad (47)$$

$$L_{\psi_3}\left[\psi_{3,m}(\varsigma) - \chi_{\psi_3}h_{\psi_3-1}(\varsigma)\right] = \hbar_{\psi_3}R_{3,m}(\varsigma). \quad (48)$$

$$L_{\psi_4}\left[\psi_{4,m}(\varsigma) - \chi_{\psi_4}h_{\psi_4-1}(\varsigma)\right] = \hbar_{\psi_4}R_{4,m}(\varsigma). \quad (49)$$

$$L_{\psi_5}\left[\psi_{5,m}(\varsigma) - \chi_{\psi_5}h_{\psi_5-1}(\varsigma)\right] = \hbar_{\psi_5}R_{5,m}(\varsigma). \quad (50)$$

with boundary conditions

$$\widehat{\psi}_{1,m}(1) = -\mu_2\widehat{\psi}'_{1,m}(1), \widehat{\psi}_{1,m}(0) - 1 = \mu_1\widehat{\psi}'_{1,m}(0) \quad (51)$$

$$\widehat{\psi}_{2,m}(1) = -\mu_2\widehat{\psi}'_{2,m}(1), \widehat{\psi}_{2,m}(0) - 1 = \mu_1\widehat{\psi}'_{2,m}(0) \quad (52)$$

$$\widehat{\psi}'_{3,m}(0) = \mu_1\widehat{\psi}''_{3,m}(0) =, \widehat{\psi}'_{3,m}(0) = 0, \widehat{\psi}_{3,m}(1) = 0, \widehat{\psi}'_{3,m}(1) = -\mu_2\widehat{\psi}''_{3,m}(1) \quad (53)$$

$$\widehat{\psi}'_{4,m}(0) = \mu_1\widehat{\psi}''_{4,m}(0) =, \widehat{\psi}'_{4,m}(0) = 0, \widehat{\psi}_{4,m}(1) = 0, \widehat{\psi}'_{4,m}(1) = -\mu_2\widehat{\psi}''_{4,m}(1) \quad (54)$$

$$\widehat{\psi}_{5,m}(1) = -\mu_2\widehat{\psi}'_{5,m}(1), \widehat{\psi}_{5,m}(0) - 1 = \mu_1\widehat{\psi}'_{5,m}(0) \quad (55)$$

where

$$R_{1,m}(\varsigma) = \psi''_{1,m-1} - R\sum_{k=0}^{m-1}\psi_{1,m-1}\psi'_{3,k} + R\sum_{k=0}^{m-1}\psi_{3,m-1}\psi'_{1,k} \quad (56)$$

$$R_{2,m}(\varsigma) = \psi''_{2,m-1} + R\sum_{k=0}^{m-1}\psi_{3,m-1}\psi'_{2,k} \quad (57)$$

$$R_{3,m}(\varsigma) = \psi_{3,m-1}^{////} - R\sum_{k=0}^{m-1}\psi'_{3,m-1}\psi''_{3,k} + R\sum_{k=0}^{m-1}\psi_{3,m-1}\psi'''_{3,k} \qquad (58)$$

$$R_{4,m}(\varsigma) = \psi_{4,m-1}^{////} - 2R\sum_{k=0}^{m-1}\psi_{4,m-1}\psi'''_{4,k} - M^2\sum_{k=0}^{m-1}\psi'_{4,k} \qquad (59)$$

$$R_{5,m}(\varsigma) = \psi''_{5,m-1} - R\sum_{k=0}^{m-1}\psi'_{5,m-1}\psi'_{4,k} - 2\sum_{k=0}^{m-1}\psi_{4,m-1}\psi''_{5,k} - M^2\psi_5 \qquad (60)$$

A well-known software called MATHEMATICA has been used to solve the modeled problem.

4. Convergence Criteria

HAM was applied to compute the solution of the problems given in Equations (7)–(9), (20), and (21), as HAM contains the non-zero auxiliary parameter $\hbar_i (i = 1-5)$, which ensures the convergence of the solution. To get a suitable value for the $\hbar_i$, $\hbar_i$– curves are displayed. To guarantee the convergence, 20th order $\hbar_i$– curves have been drawn in Figures 2–6. It can easily be seen from the $\hbar_i$ curves that the acceptable values of $\hbar_i$ were $0.5 \leq \hbar_1 \leq 1.5$, $-1.5 \leq \hbar_2 \leq -0.5$, $0 \leq \hbar_3 \leq 6$, for the long slider. Similarly, acceptable values for the circular slider were $0.5 \leq \hbar_4 \leq 1.5$, $-1 \leq \hbar_5 \leq 4$.

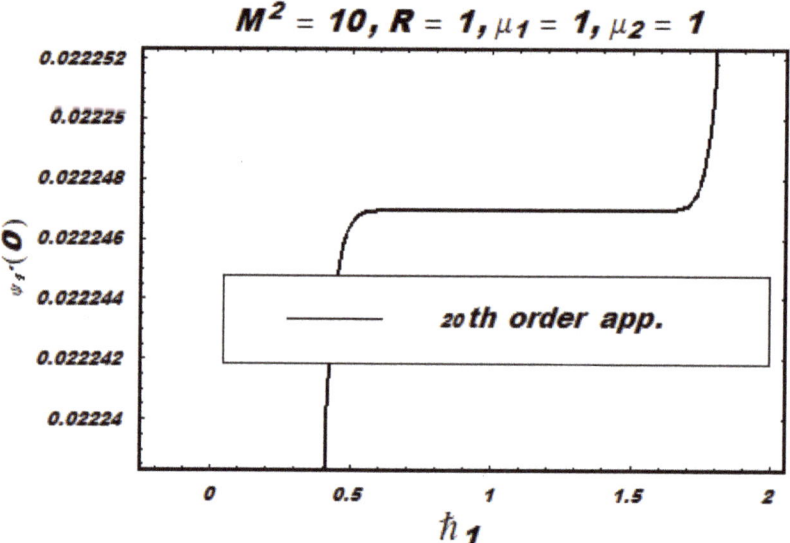

Figure 2. $\hbar_1$ for the strip/long slider.

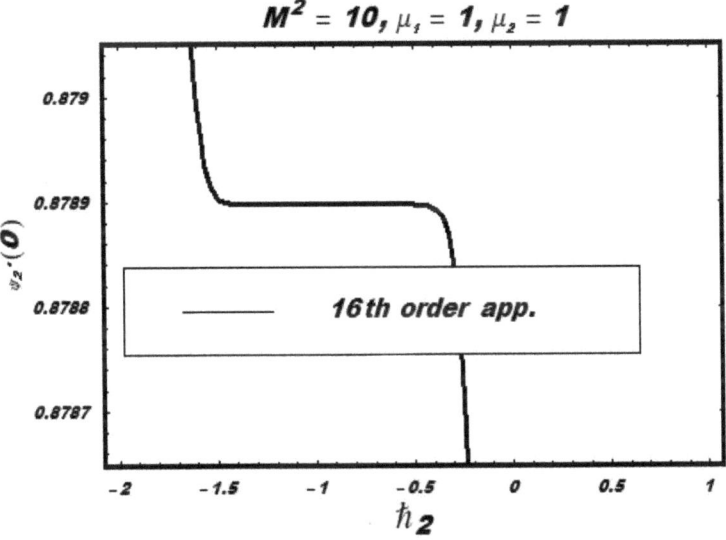

Figure 3. $\hbar_2$ for the strip/long slider.

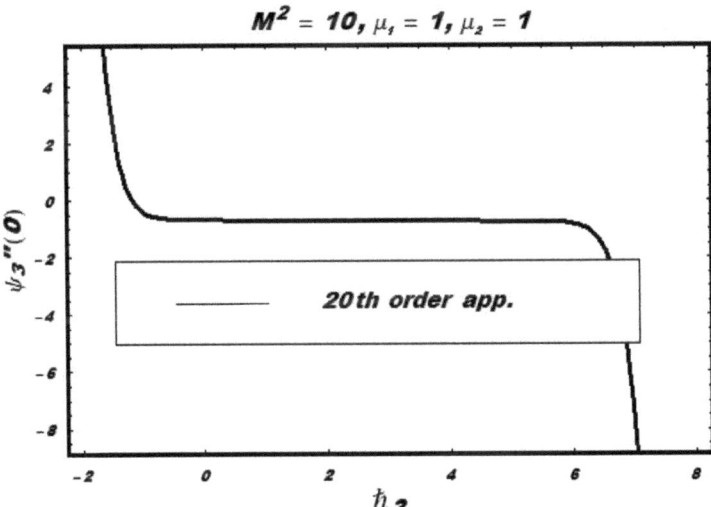

Figure 4. $\hbar_3$ for the strip/long slider.

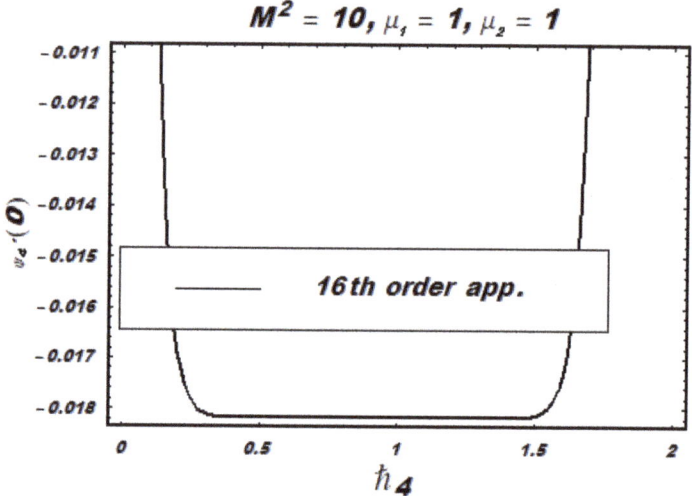

Figure 5. $\hbar_4$ for the circular slider.

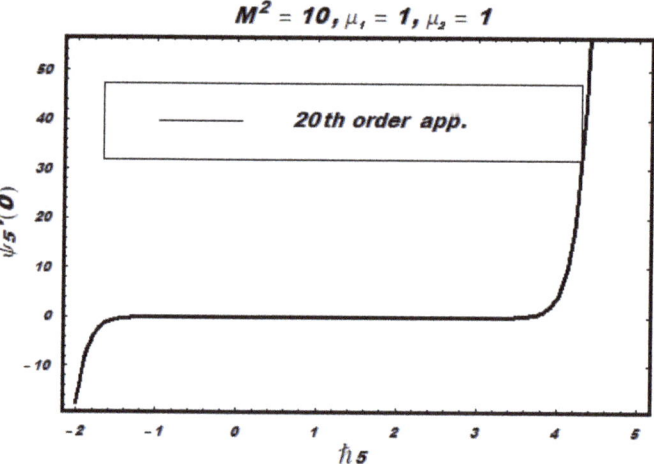

Figure 6. $\hbar_5$ for the circular slider.

5. Results and Discussion

The obtained results from the above-mentioned method (HAM) are presented in the form of tables and graphs. Tables 1 and 2 display the effects of the slip on the dynamic properties of a slider, showing that normalized lift and drag decrease as the slip and/or Reynolds number increases. The lift (per area) of the strip slider was much greater than the circular slider, although the drag remained the same in both cases. The effect of slip could be substantial, affecting the drag much more than the lift.

Table 1. Properties of the long porous slider. Normalized lift Λ, normalized x_1- direction drag, and normalized x_2- direction drag.

β_1, β_2	M^2	R	Λ	$-\psi'_1(1)$	$-\psi'_2(1)$
0, 0	0	0.2	62.33	0.896	0.932
-	-	0.5	26.34	0.760	0.836
-	-	2.0	8.412	0.334	0.467
-	-	5.0	4.917	0.063	0.123
-	-	20	3.267	0	0
-	-	50	2.909	0	0
0.1, 0.1	2	0.2	39.27	0.743	0.780
-	4	0.5	16.78	0.626	0.704
-	6	2.0	6.596	0.4372	0.2536
-	10	5.0	3.436	0.3245	0
-	20	20.0	2.440	0.1520	0
-	50	50.0	2.240	0	0
0.1, 1	2	0.2	20.31	0.424	0.463
-	4	0.5	8.859	0.357	0.436
-	6	2.0	3.159	0.160	0.321
-	10	5.0	2.050	0.035	0.123
-	20	20.0	1.513	0	0.0632
-	50	50.0	1.391	0	0.012
0.1, 10	2	0.2	5.316	0.064	0.082
-	4	0.5	2.702	0.046	0.080
-	6	2.0	1.413	0.013	0
-	10	5.0	1.175	0.002	0
-	20	20.0	1.068	0	0
-	50	50.0	1.047	0	0
1, 1	2	0.2	9.727	0.275	0.315
-	4	0.5	4.591	0.210	0.288
-	6	2.0	2.048	0.068	0.172
-	10	5.0	1.569	0.011	0.047
-	20	20.0	1.355	0	0
-	50	50.0	1.315	0	0

Table 2. Properties of the circular porous slider. Normalized lift Λ, normalized drag $-\psi_1'(1)$.

β_1, β_2	M^2	R	Λ	$-\psi_1'(1)$
0, 0	0	0.2	30.78	0.914
-	-	0.5	12.79	0.797
-	-	2.0	3.833	0.392
-	-	5.0	2.019	0.085
-	-	20	1.349	0
-	-	50	1.194	0
0.1, 0.1	2	0.2	19.33	0.761
-	4	0.5	8.089	0.663
-	6	2.0	2.503	0.310
-	10	5.0	1.445	0.1014
-	20	20.0	0.994	0
-	50	50.0	0.908	0
0.1, 1	2	0.2	9.853	0.441
-	4	0.5	4.130	0.394
-	6	2.0	1.288	0.129
-	10	5.0	0.752	0.0145
-	20	20.0	0.529	0
-	50	50.0	0.483	0
0.1, 10	2	0.2	6.438	0.084
-	4	0.5	2.699	0.076
-	6	2.0	0.841	0.015
-	10	5.0	0.488	0
-	20	20.0	0.338	0
-	50	50.0	0.305	0
1, 1	2	0.2	4.611	0.294
-	4	0.5	2.043	0.244
-	6	2.0	0.776	0.0215
-	10	5.0	0.549	0
-	20	20.0	0.466	0
-	50	50.0	0.453	0

Velocity distributions for the long and circular slider are presented graphically in Figures 7–9.

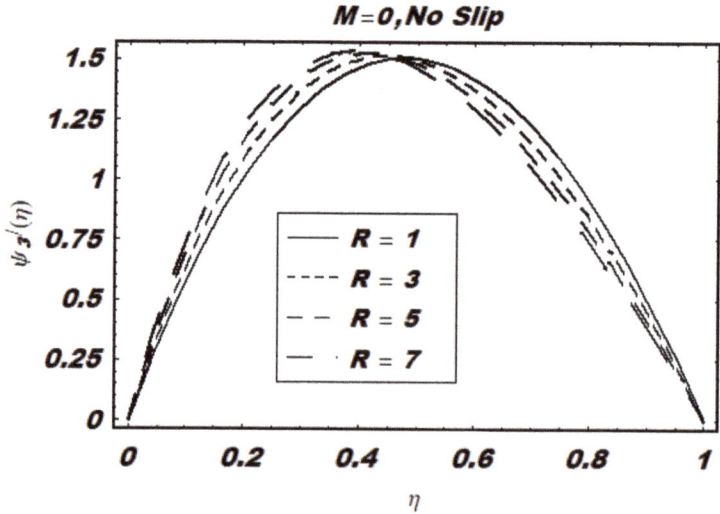

Figure 7. Similarity function ψ_3' for the long slider.

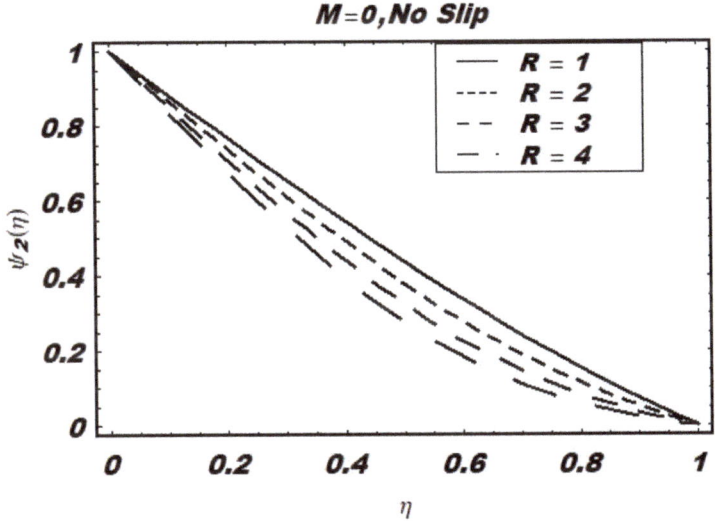

Figure 8. Similarity function ψ_2 for the long slider.

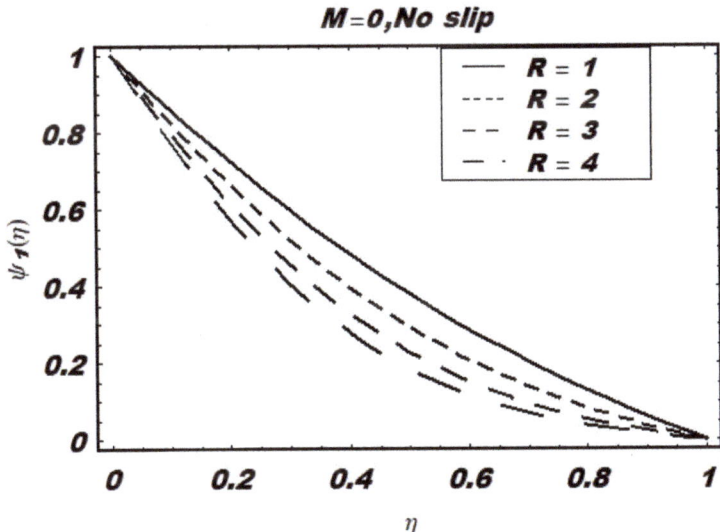

Figure 9. Similarity function ψ_1 for the long slider.

For the long slider, the effect of the Reynolds number in the presence of slip and the magnetic field is shown in Figures 10–18. It is observed that the velocity profile was very much changed. It was seen that slip near the ground reduced the lateral velocity much more than slip on the slider. Moreover, increasing the magnetic parameter decreased the lateral velocity components further (see Figure 12). The effects of the Reynolds number on the typical velocity distribution for the circular slider were similar, as displayed in Figures 19–28. The behavior of velocity profiles was similar for the long and circular sliders in cases of no-slip (see Figures 19 and 20). Further, velocity profiles behaved in a similar fashion in both cases (i.e., parabolic or linear for a low Reynolds number, while a boundary layer formed near the surface in cases of a large Reynolds number). Figures 21–28 demonstrate the effect of the slip parameter on the velocity components corresponding to different Reynolds numbers. These pictorial descriptions demonstrate that velocity profiles decrease with an increase in slip parameters, and that this decrease become even greater after applying the magnetic field. This is due to the fact that slip hinders fluid particles and displaces motion in the vicinity.

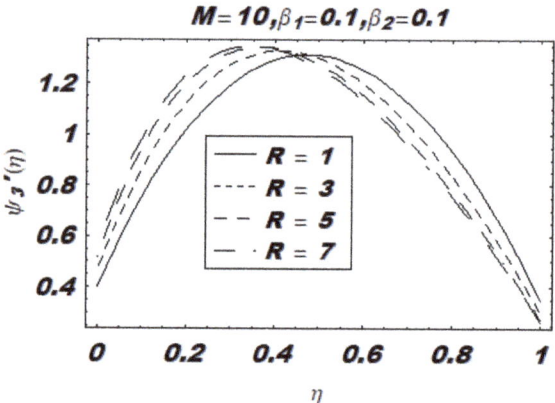

Figure 10. Similarity function ψ_3' for the long slider.

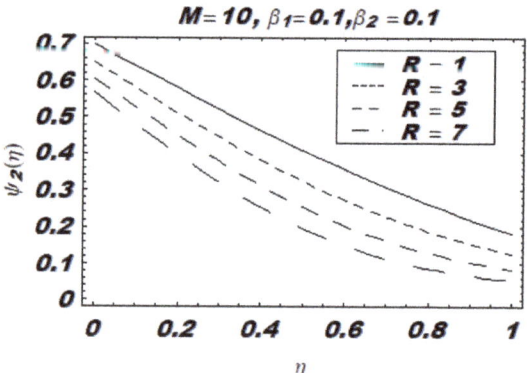

Figure 11. Similarity function ψ_2 for the long slider.

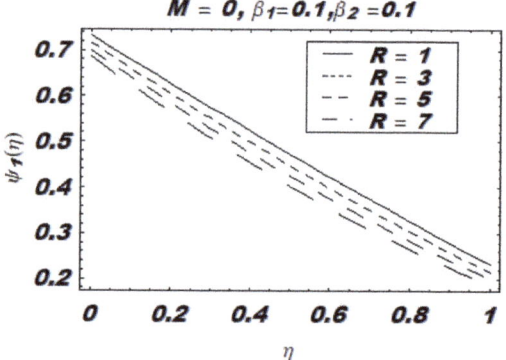

Figure 12. Similarity function ψ_1 for the long slider.

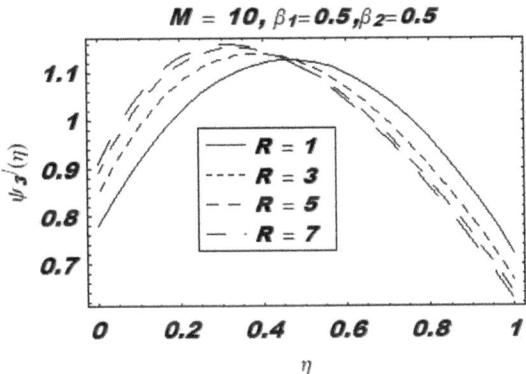

Figure 13. Similarity function ψ_3' for the long slider.

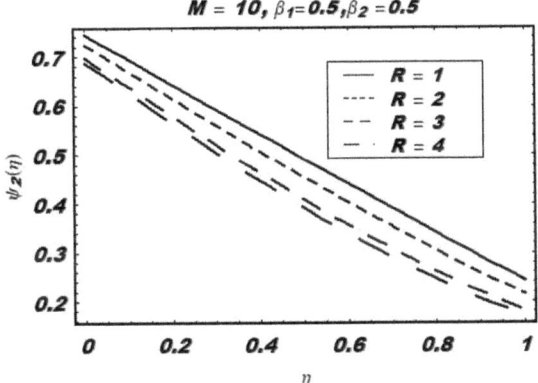

Figure 14. Similarity function ψ_2 for the long slider.

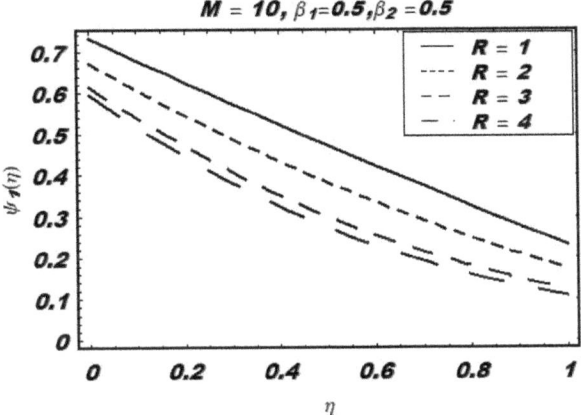

Figure 15. Similarity function ψ_1 for the long slider.

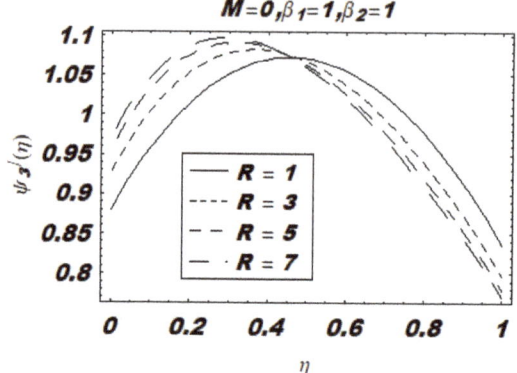

Figure 16. Similarity function ψ_3' for the long slider.

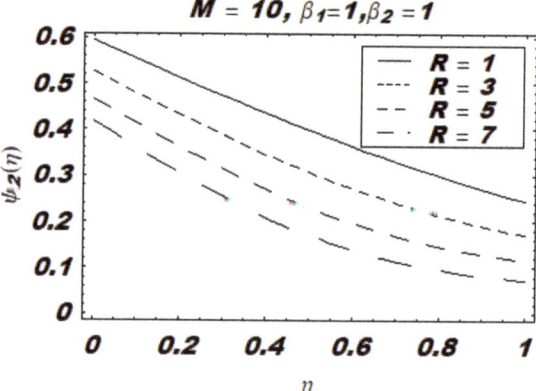

Figure 17. Similarity function ψ_1 for the long slider.

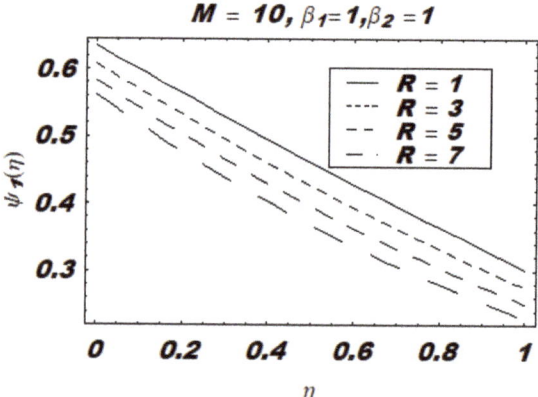

Figure 18. Similarity function ψ_3' for the long slider.

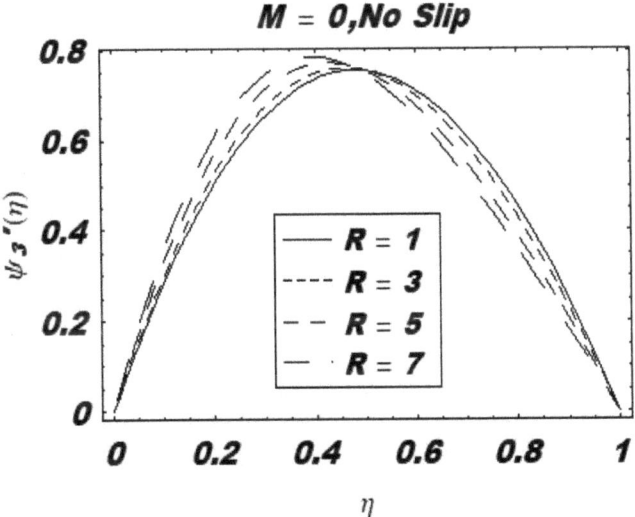

Figure 19. Similarity function ψ_3' for the circular slider.

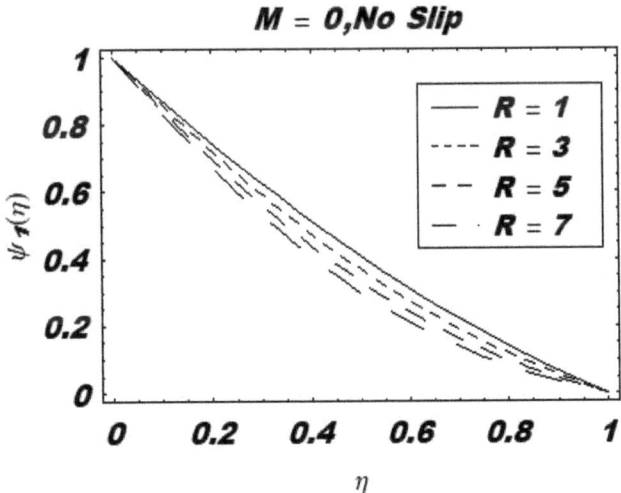

Figure 20. Similarity function ψ_1 for the circular slider.

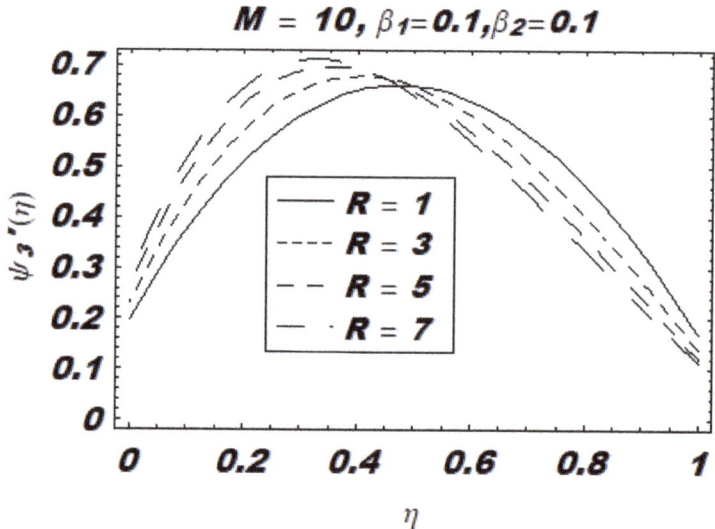

Figure 21. Similarity function ψ_3' for the circular slider.

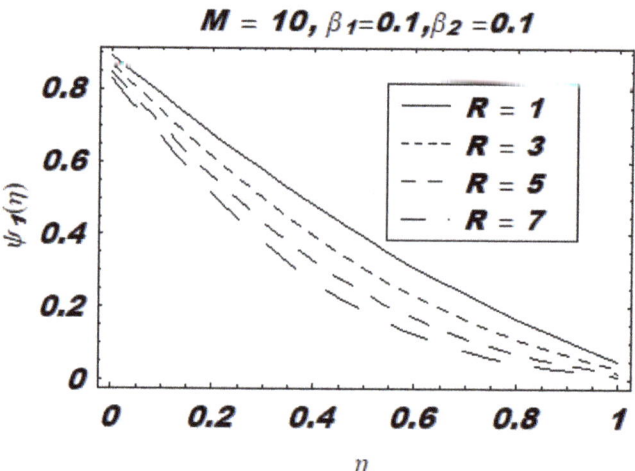

Figure 22. Similarity function ψ_1 for the circular slider.

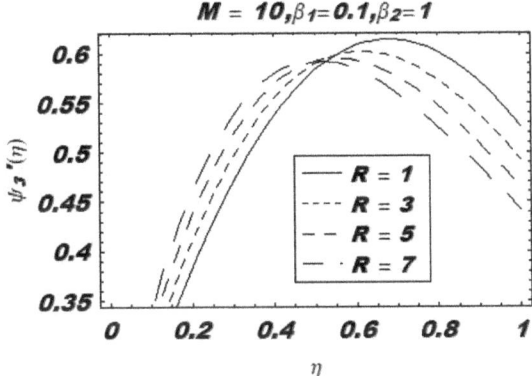

Figure 23. Similarity function ψ_3' for the circular slider.

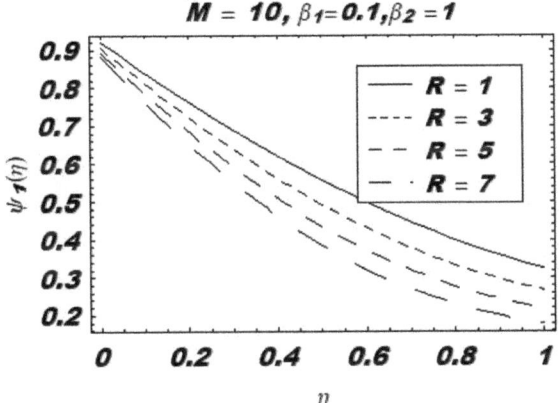

Figure 24. Similarity function ψ_1 for the circular slider.

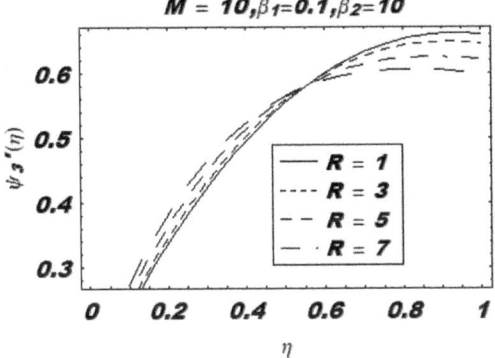

Figure 25. Similarity function ψ_3' for the circular slider.

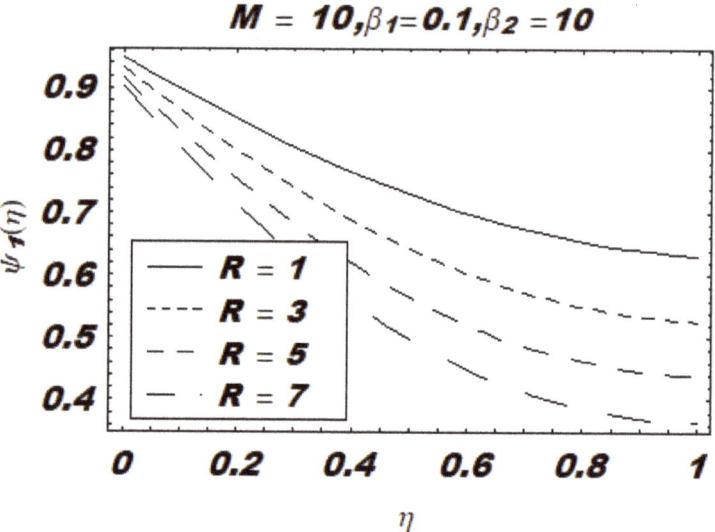

Figure 26. Similarity function ψ_1 for the circular slider.

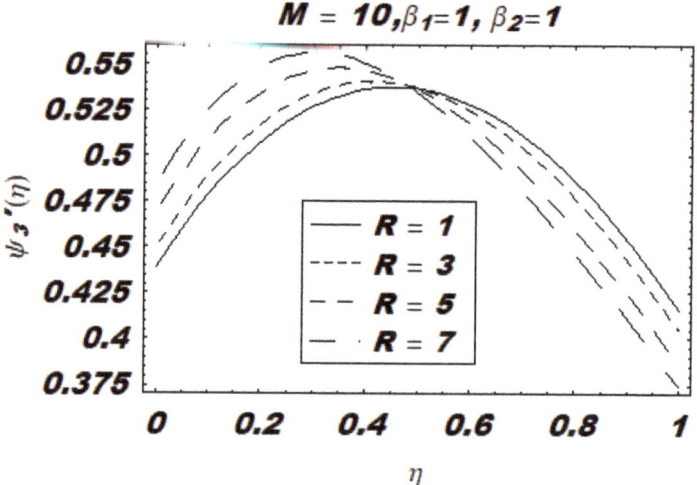

Figure 27. Similarity function ψ_3' for the circular slider.

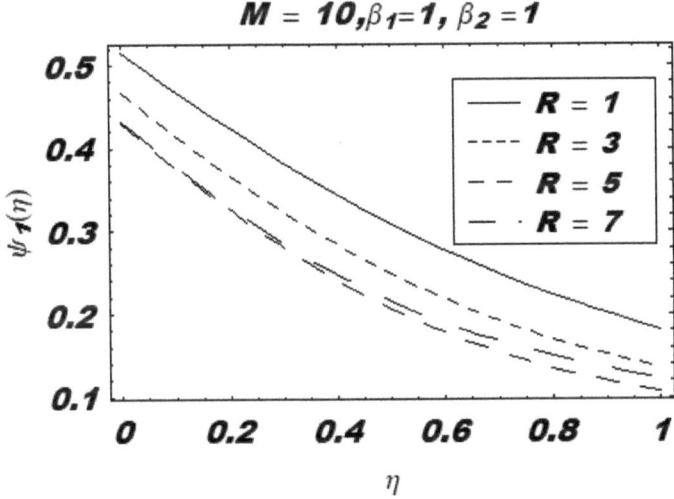

Figure 28. Similarity function ψ_1 for the circular slider.

These results qualitatively confirm the expectation that a drag-like Lorentz force is created by the magnetic field normal to the lateral flow direction, and this force decreases the lateral velocity components. Lift and drag components are important physical quantities for a porous slider. It is interesting to note that the lift is free of translation, but the drag components depend on a cross flow. The effectiveness of a porous slider can be enhanced by making the ratio of friction force to lift smaller. As pointed out by Wang [16], the porous slider should be operated at a cross-flow Reynolds number below unity for optimum efficiency. According to Table 1, porous sliders should be operated at small values that are still valid even when an external uniform magnetic field is applied. Moreover, from the point of view of optimum efficiency, it is more efficient to move a flat slider on a fluid subject than in a high-intensity magnetic field.

6. Conclusions

In this research, different studies have been complied altogether. Different researchers have analyzed fluid flow on a long slider without slip, while others were interested only in a circular slider without slip. Wang presented a comparative study of the both sliders and added velocity slip, but did not cover the effects of a magnetic field. As such, one concern of this study was theoretical investigation of a steady three-dimensional flow of a viscous fluid between a porous slider and the ground in the presence of a transverse uniform magnetic field with velocity slip. The effects of different physical parameter values like Reynolds number and magnetic field on the lateral velocity profiles and lift and drag components were presented in graphs and tables in the presence of velocity slip. It is expected that the results of the present study could be useful for the understanding of various technical problems related to porous sliders where magnetic and velocity slip are the main physical parameters. The main findings are as follows:

- It was shown that normalized lift and drag go down as slip and/or the Reynolds number goes up (see Tables 1 and 2). The lift (per area) of a long slider is much greater than a circular slider. The drag remains the same for both sliders.
- Slip near the ground reduces lateral velocity of the slider much more than slip. By increasing the magnetic parameter, the lateral velocity components decrease further.

- The behavior of velocity profiles is similar for the long and the circular sliders in cases of no-slip (i.e., parabolic or linear for a low Reynolds number).
- In cases of a large Reynolds number, a boundary layer formed near the surface, while velocity profiles decreased with an increase in slip parameters, a decrease which grew more pronounced after applying the magnetic field.

Author Contributions: N.F. and Y.K. conducted whole research such as preparation, creation and/or presentation of the published work by those from the original research group, specifically critical review, commentary or revision – including pre- or post-publication stages. A.A. helped to improve the language of the paper. Drew the schematic diagram of the problem, preparation, creation and/or presentation of the published work, specifically visualization/data presentation. M.K. helped to revise the reviewers' comments.

Funding: This work was supported by the International Cultural Exchange School (ICES), Donghua University, West Yan'an Road 1882, Shanghai 200051, China.

Conflicts of Interest: The authors declare no conflict of interest.

Nomenclature

B_0	Magnetic field	μ	Dynamic viscosity
d	Width	η	Similarity variable
H_1, H_2	Slip coefficient	τ	Extra stress tensor
I	Identity tensor	β_1, β_2	Slip factors
l	Length	ψ_1, ψ_2, ψ_3	Velocity function
p	Pressure	ϕ_1, ϕ_2, ϕ_3	Velocity components
v_0	Constant viscosity	ρ	Fluid density
x_1, x_2, x_3	Space coordinates		

References

1. Skalak, F.; Wang, C.-Y. Fluid Dynamics of a Long Porous Slider. *J. Appl. Mech.* **1975**, *42*, 893–894. [CrossRef]
2. Wang, C.-Y. Erratum: "Fluid Dynamics of the Circular Porous Slider." *J. Appl. Mech.* **1974**, *41*, 343–347. *J. Appl. Mech.* **1978**, *45*, 236. [CrossRef]
3. Wang, C.-Y. The Elliptic Porous Slider at Low Crossflow Reynolds Numbers. *J. Lubr. Technol.* **1978**, *100*, 444. [CrossRef]
4. Bhattacharjee, R.C.; Das, N.C. Porous slider bearing lubricated with couple stress mhd fluids. *Trans. Can. Soc. Mech. Eng.* **1994**, *18*, 317–331. [CrossRef]
5. Patel, J.R.; Deheri, G. A study of thin film lubrication at nanoscale for a ferrofluid based infinitely long rough porous slider bearing. *Facta Univ. Ser. Mech. Eng.* **2016**, *14*, 89–99. [CrossRef]
6. Sinha, P.; Adamu, G. Analysis of Thermal Effects in a Long Porous Rough Slider Bearing. *Proc. Natl. Acad. Sci. India Sectoin A Phys. Sci.* **2017**, *87*, 279–290. [CrossRef]
7. Munshi, M.M.; Patel, A.R.; Deheri, G. Analysis of Rough Porous Inclined Slider Bearing Lubricated with a Ferrofluid Considering Slip Velocity. *Int. J. Res. Advent Technol.* **2019**, *7*, 387–396.
8. Lang, J.; Nathan, R.; Wu, Q. Theoretical and experimental study of transient squeezing flow in a highly porous film. *Tribol. Int.* **2019**, *135*, 259–268. [CrossRef]
9. Madalli, V.S.; Bujurke, N.M.; Mulimani, B.G. Lubrication of a long porous slider. *Tribol. Int.* **1995**, *28*, 225–232. [CrossRef]
10. Awati, V.B.; Jyoti, M. Homotopy analysis method for the solution of lubrication of a long porous slider. *Appl. Math. Nonlinear Sci.* **2016**, *1*, 507–516. [CrossRef]
11. Khan, Y.; Faraz, N.; Yildirim, A.; Wu, Q. A Series Solution of the Long Porous Slider. *Tribol. Trans.* **2011**, *54*, 187–191. [CrossRef]
12. Khan, Y.; Wu, Q.; Faraz, N.; Mohyud-Dind, S.T.; Yıldırım, A. Three-Dimensional Flow Arising in the Long Porous Slider: An Analytic Solution. *Z. Nat. A* **2011**, *66*, 507–511.
13. Faraz, N.; Khan, Y.; Lu, D.C.; Goodarzi, M. Integral transform method to solve the problem of porous slider without velocity slip. *Symmetry* **2019**, *11*, 791. [CrossRef]

14. Ghoreishi, M.; Ismail, A.I.B.M.; Rashid, A. The One Step Optimal Homotopy Analysis Method to Circular Porous Slider. *Math. Probl. Eng.* **2012**, *2012*, 135472. [CrossRef]
15. Shukla, S.D.; Deheri, G.M. Rough porous circular convex pad slider bearing lubricated with a magnetic fluid. In *Lecture Notes in Mechanical Engineering*; Springer: Berlin/Heidelberg, Germany, 2014.
16. Wang, C.Y. A porous slider with velocity slip. *Fluid Dyn. Res.* **2012**, *44*, 065505. [CrossRef]
17. Wang, C.-Y. Fluid Dynamics of the Circular Porous Slider. *J. Appl. Mech.* **1974**, *41*, 343. [CrossRef]
18. Faraz, N. Study of the effects of the Reynolds number on circular porous slider via variational iteration algorithm-II. *Comput. Math. Appl.* **2011**, *61*, 1991–1994. [CrossRef]
19. Madani, M.; Khan, Y.; Mahmodi, G.; Faraz, N.; Yildirim, A.; Nasernejad, B. Application of homotopy perturbation and numerical methods to the circular porous slider. *Int. J. Numer. Methods Heat Fluid Flow* **2012**, *22*, 705–717. [CrossRef]
20. Naeem, F.; Yasir, K. Thin film flow of an unsteady Maxwell fluid over a shrinking/stretching sheet with variable fluid properties. *Int. J. Numer. Methods Heat Fluid Flow* **2018**, *28*, 1596–1612.

© 2019 by the authors. Licensee MDPI, Basel, Switzerland. This article is an open access article distributed under the terms and conditions of the Creative Commons Attribution (CC BY) license (http://creativecommons.org/licenses/by/4.0/).

Article

MHD Flow of Nanofluid with Homogeneous-Heterogeneous Reactions in a Porous Medium under the Influence of Second-Order Velocity Slip

Fahd Almutairi [1], S.M. Khaled [2,3,*] and Abdelhalim Ebaid [4]

1. Department of Chemical Engineering, Faculty of Engineering University of Tabuk, Tabuk 71491, Saudi Arabia; falmutairi@ut.edu.sa
2. Department of Mathematics, Faculty of Sciences, Helwan University, Cairo 11795, Egypt
3. Department of Studies and Basic Sciences, Faculty of Community, University of Tabuk, Tabuk 71491, Saudi Arabia
4. Department of Mathematics, Faculty of Sciences, University of Tabuk, P.O. Box 741, Tabuk 71491, Saudi Arabia; aebaid@ut.edu.sa or halimgamil@yahoo.com
* Correspondence: ksmahmoud@ut.edu.sa; Tel.: +966-56-765-4109

Received: 25 January 2019; Accepted: 21 February 2019; Published: 26 February 2019

Abstract: The influence of second-order velocity slip on the MHD flow of nanofluid in a porous medium under the effects of homogeneous-heterogeneous reactions has been analyzed. The governing flow equation is exactly solved and compared with those in the literature for the skin friction coefficient in the absence of the second slip, where great differences have been observed. In addition, the effects of the permanent parameters on the skin friction coefficient, the velocity, and the concentration have been discussed in the presence of the second slip. As an important result, the behavior of the skin friction coefficient at various values of the porosity and volume fraction is changed from increasing (in the absence of the second slip) to decreasing (in the presence of the second slip), which confirms the importance of the second slip in modeling the boundary layer flow of nanofluids. In addition, five kinds of nanofluids have been investigated for the velocity profiles and it is found that the Ag-water nanofluid is the lowest. For only the heterogeneous reaction, the concentration equation has been exactly solved, while the numerical solution is applied in the general case. Accordingly, a reduction in the concentration occurs with the strengthening of the heterogenous reaction and also with the increase in the Schmidt parameter. Moreover, the Ag-water nanofluid is of lower concentration than the Cu-water nanofluid. This is also true for the general case, when both of the homogenous and heterogenous reactions take place.

Keywords: homogeneous-heterogeneous reactions; porous medium; first slip; second slip; exact solution

1. Introduction

The main characteristic of nanofluid is the significant enhancement of the thermal properties of the base fluid. The term nanofluid comes back to a pioneering experimental research by Choi [1] in which a conclusion had been reached that the thermal conductivity of a base fluid is enhanced up to two times by adding a small amount of nanoparticles. In addition, some authors [2,3] found that the dispersion of a small amount of copper nanoparticles led to 40% of the thermal conductivity of the fluid, while adding a small amount of carbon nanotubes in ethylene glycol or oil led to 50%. Aly and Ebaid [4] considered five metallic and nonmetallic nanoparticles in a base of water, where an effective approach was introduced to derive the exact solution. One of the important results in the field of nanofluid flow has been presented by Majumder [5], in which it was experimentally proven that nanofluidic flow exhibits partial slip against the solid surface, which can be characterized by the so-called slip length. Accordingly, the authors in [6] discussed the effect of partial slip boundary condition on the flow and heat transfer of nanofluids past stretching sheet at constant wall temperature. Furthermore, the no-slip condition is no longer valid for fluid flows at the micro- and nanoscale and, instead, a certain degree of tangential slip must be allowed [7,8]. Very recently, Sharma and Ishak [9] studied the second-order velocity slip effect on the boundary layer flow of Cu-water-based nanofluid with heat transfer over a stretching sheet. Their numerical results were based on the finite element method (FEM). A model for isothermal homogeneous-heterogeneous reactions in boundary layer flow of viscous fluid past a flat plate was studied by Merkin [10]. He presented the homogeneous reaction by cubic autocatalysis and the heterogeneous reaction by a first-order process and showed that the surface reaction is the dominant mechanism near the leading edge of the plate. Chaudhary and Merkin [11] studied the homogenous-heterogeneous reactions in boundary layer flow of viscous fluid. They found the numerical solution near the leading edge of a flat plate. Bachok et al. [12] focused on the stagnation-point flow towards a stretching sheet with homogeneous-heterogeneous reactions effects. Effects of homogeneous-heterogeneous reactions on the flow of viscoelastic fluid towards a stretching sheet were investigated by Khan and Pop [13]. Kameswaran et al. [14] extended the work of [13] for nanofluid over a porous stretching sheet. In general, porous medium is used for transport and storage of energy. Analysis of flow through a porous medium has become the core of several scientific and engineering applications. These applications include the utilization of geothermal energy, the migration of moisture in fibrous insulation, food processing, casting and welding in manufacturing processes, the dispersion of chemical contaminants in different industrial processes, the design of nuclear reactors, chemical catalytic reactors, compact heat exchangers, solar power, and many others. Further, the use of micro/nano electromechanical systems (MEMS/NEMS) has been increased in many industries. Such systems have association with velocity slip [15–19]. Very recently, Hayat et al. [20] studied the MHD flow of nanofluid with homogeneous-heterogeneous reactions of two chemical species and velocity slip. In this field of research, some pioneer works were introduced in [21–24] in which several non-Newtonian models have been analyzed. In [21], a novel radiation MHD activation energy Carreau and nanofluid effects of thermal energy systems have been investigated. The combined electrical MHD Ohmic dissipation forced and free convection of an incompressible Maxwell fluid on a stagnation point heat and mass transfer energy conversion problem have been studied in [22]. In addition, an applied thermal system for heat and mass transfer and energy management problem of hydromagnetic flow with magnetic and viscous dissipation effects of micropolar nanofluids towards a stretching sheet has been investigated by [23]. Moreover, the effect of the slip boundary condition on the stagnation electrical MHD nanofluid mixed convection on a stretching sheet was introduced in [24].

The objective of this work is to extend the model investigated by Hayat et al. [20] by considering the second-order slip velocity. Therefore, the extended model is given as

$$f'''(\eta) = \left(\lambda + (1-\phi)^{2.5}M\right)f'(\eta) - \phi_1\left(f(\eta)f''(\eta) - (f'(\eta))^2\right), \tag{1}$$

$$\frac{1}{Sc}g''(\eta) = Kg(\eta)(h(\eta))^2 - f(\eta)g'(\eta), \tag{2}$$

$$\frac{\delta}{Sc}h''(\eta) = -Kg(\eta)(h(\eta))^2 - f(\eta)h'(\eta), \tag{3}$$

subject to

$$f(0) = 0, f'(0) = 1 + \gamma f''(0) + \mu f'''(0), f'(\infty) = 0, \tag{4}$$

$$g'(0) = K_s g(0), g(\infty) = 1, \tag{5}$$

$$\delta h'(0) = -K_s g(0), h(\infty) = 0, \tag{6}$$

where

$$\phi_1 = (1-\phi)^{2.5}\left(1 - \phi + \phi\frac{(\rho C_p)_s}{(\rho C_p)_f}\right), \quad (7)$$

and ϕ is the solid volume fraction of the nanoparticles, λ is the porosity parameter, M is the Hartman number, Sc is the Schmidt parameter, K is the measure of the strength of the homogeneous reaction, K_s is the measure of the strength of the heterogeneous reaction, δ is the ratio of the diffusion coefficient, ρ_s and ρ_f are respectively the densities of nanoparticles and base fluid, γ and μ are respectively the first and the second velocity slip parameters, and $f'(\eta)$, $g(\eta)$ and $h(\eta)$ are respectively the fluid velocity and the concentrations of the two chemical species.

Following [20], the parameter δ can be taken as unity especially when the diffusion coefficients of two chemical species are the same. In this case, we have [20]

$$h(\eta) + g(\eta) = 1, \quad (8)$$

and hence Equations (2) and (3) reduce to

$$\frac{1}{Sc}g''(\eta) = Kg(\eta)(1 - g(\eta))^2 - f(\eta)g'(\eta), \quad (9)$$

subject to the same boundary conditions given in Equation (5). In [20], the authors applied the homotopy analysis method to solve the set of boundary value problems (1)–(6) in the absence of the second slip parameter (i.e., when $\mu = 0$). However, Equation (1) with the boundary conditions (3) can be exactly solved, even in the presence of the second slip parameter μ, as will be introduced in the next section. This exact solution for $f(\eta)$ will be then compared with the results obtained by [20] at a special case. Further, this exact formula for $f(\eta)$ is to be inserted into Equation (9) to form with the boundary conditions (5) a single nonlinear differential equation in the unknown $g(\eta)$. Details of the suggested procedure are presented in the next section.

2. Methodology

Following [25,26], $f(\eta)$ can be obtained as

$$f(\eta) = \frac{1}{\beta(1 + \gamma\beta - \mu\beta^2)}\left(1 - e^{-\beta\eta}\right), \quad (10)$$

where β is the positive root of the following nonlinear equation:

$$\mu\beta^4 - \gamma\beta^3 - \left(1 + \lambda\mu + \mu M(1-\phi)^{2.5}\right)\beta^2 + \left(\gamma\lambda + M\gamma(1-\phi)^{2.5}\right)\beta + \left(\phi_1 + \lambda + M(1-\phi)^{2.5}\right) = 0. \quad (11)$$

On inserting Equation (11) into Equation (9), we obtain the following nonlinear ordinary differential equation (ODE) for $g(\eta)$:

$$g''(\eta) + \Omega\left(1 - e^{-\beta\eta}\right)g'(\eta) - KScg(\eta)(1 - g(\eta))^2 = 0, \quad (12)$$

where Ω is defined as

$$\Omega = \frac{Sc}{\beta(1 + \gamma\beta - \mu\beta^2)}. \quad (13)$$

The skin friction coefficient is defined in [20] by Equation (14) and hence Equation (15) is obtained by using the exact expression for $f(\eta)$ in Equation (10).

$$\text{Skin friction coefficient} = -\frac{2}{(1-\phi)^{2.5}}f''(0), \quad (14)$$

$$\text{Skin friction coefficient} = \frac{2\beta}{(1-\phi)^{2.5}(1+\gamma\beta-\mu\beta^2)}. \tag{15}$$

At the special case, $K \to 0$, the analytic solution of Equation (12) is given as

$$g(\eta) = \frac{1 + \varepsilon\Gamma(\Omega/\beta, \Omega/\beta e^{-\beta\eta}, \Omega/\beta)}{1 + \varepsilon\Gamma(\Omega/\beta, 0, \Omega/\beta)}, \tag{16}$$

where ε is defined by

$$\varepsilon = K_s(\beta)^{\Omega/\beta-1}\left(e^{\beta^2}/\Omega\right)^{\Omega/\beta}. \tag{17}$$

This case may be of a physical meaning when only the heterogenous reactions occur. The concentration is therefore given as

$$g(0) = \frac{1}{1 + \varepsilon\Gamma(\Omega/\beta, 0, \Omega/\beta)}. \tag{18}$$

3. Discussion

In the beginning, it should be noted that the exact formula for the skin friction coefficient given by Equation (15) will be invested here and used to validate the numerical results obtained in [20] by applying the homotopy analysis method (HAM) when the second slip vanishes (i.e., at $\mu = 0$). The thermophysical properties of water and nanoparticles are introduced in Table 1. These properties have been implemented to conduct the numerical results in Table 2. In view of these comparisons, it may be concluded that the outputs of [20] need some revisions, especially since the differences between the current exact values and the approximate ones seem to be obvious. Besides, the same values of the physical parameters [20] have been selected to hold these comparisons.

Table 1. Properties of water and nanoparticles.

	C_p (J/kg·K)	ρ (kg/m³)	K (W/m·K)
Pure Water	4179	4179	0.613
Copper (Cu)	385	8933	401
Silver (Ag)	235	10500	429
Alumina (Al$_2$O$_3$)	765	3970	40
Titanium Oxide(TiO$_2$)	686.2	4250	8.9538
Silicon Dioxide (SiO$_2$)	765	3970	36

Table 2. Comparisons between the numerical results of skin friction coefficient [20] and the present exact values for copper and silver at $\mu = 0$ and $\lambda = 0.4$.

ϕ	M	γ	Cu HAM [20]	Cu Exact (Present)	Ag HAM [20]	Ag Exact (Present)
0.05	0.5	1.0	1.278	1.23278	1.284	1.23843
0.1			1.465	1.41390	1.475	1.42485
0.2			1.955	1.88407	1.973	1.90636
0.2	0.1		1.897	1.81545	1.917	1.84209
	0.3		1.928	1.85097	1.945	1.87529
	0.7		1.981	1.91500	1.996	1.93552
	0.5	0.1	4.542	4.31610	4.672	4.45659
		0.5	2.827	2.70491	2.865	2.75469
		0.9	2.079	2.00347	2.098	2.02905

In the presence of the second slip, exact values for the skin friction coefficient for the Ag-water and the Cu-water nanofluids are listed in Table 3 at various values of ϕ, M, and γ when $\lambda = 0.4$.

The results reveal that the skin friction coefficient for both nanofluids increases with an increase in the volume fraction ϕ and the Hartman number M; however, it decreases with the increase in the first slip γ and with the decrease in the second slip μ. Further, the variation of the skin friction coefficient is depicted in Figure 1 against the porosity parameter λ at various values of the solid volume fraction ϕ when $\mu = 0$. It is clear from this figure that the skin friction increases with increases in both λ and ϕ. However, in [20], it was found that this behavior is different than the current one. This also confirms the conclusion made above that the method applied in [20] needs further improvement. In addition, the results in Figure 2 indicate that the skin friction decreases with increases in both λ and ϕ in the presence of the second slip parameter. Therefore, the behavior is changed from increasing in Figure 1 ($\mu = 0$) to decreasing in Figure 2 ($\mu = -0.5$), which confirms the importance of the second slip in modeling the boundary layer flow of nanofluids.

Table 3. Values of skin friction coefficient for copper and silver at various values of ϕ, M, γ and μ at $\lambda = 0.4$.

ϕ	M	γ	μ	Cu	Ag
0.01	0.5	1.0	−0.5	0.83677	0.83698
0.03				0.85850	0.85891
0.05				0.92832	0.92930
0.02	0.1			0.83658	0.83755
	0.3			0.84949	0.85013
	0.7			0.86457	0.86483
	0.5	0.1		1.37395	1.37485
		0.5		1.08412	1.08475
		0.9		0.89573	0.89618
		1.0	−0.1	1.06481	1.06653
			−0.5	0.85850	0.85891
			−0.9	0.72553	0.72545

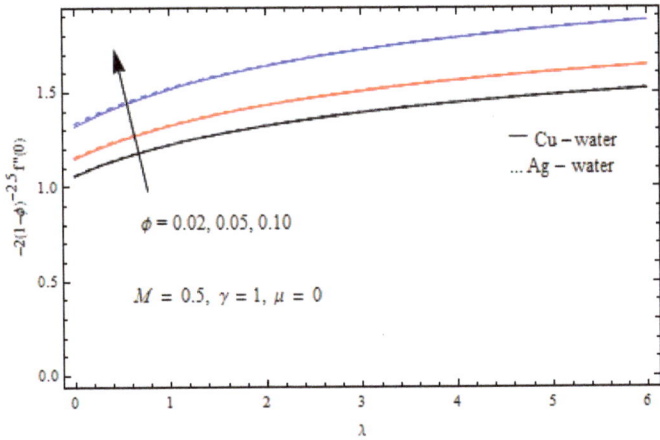

Figure 1. Effects of ϕ on skin friction coefficient when the second slip vanishes.

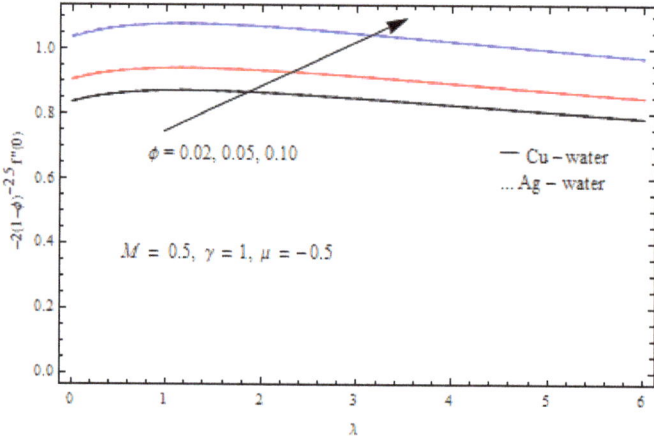

Figure 2. Effects of ϕ on skin friction coefficient in the presence of second slip.

The effect of the first slip parameter γ on the velocity of the nanofluids suspended with five nanoparticles is displayed through Figures 3–5. Figures 3 and 4 show that the velocities of the Ag/Cu/TiO$_2$-water nanofluids satisfy $f'(\eta)|_{Ag} < f'(\eta)|_{Cu} < f'(\eta)|_{TiO_2}$. Figure 5 indicates that $f'(\eta)|_{SiO_2} \approx f'(\eta)|_{Al_2O_3} \approx f'(\eta)|_{TiO_2}$. Therefore, it can be concluded from Figures 3–5 that the Ag-water nanofluid is of lower velocity than any of the four other types. This later conclusion is also observed and confirmed through Figures 6–8 for the effect of the second slip μ on the velocity of the present five types of nanofluids.

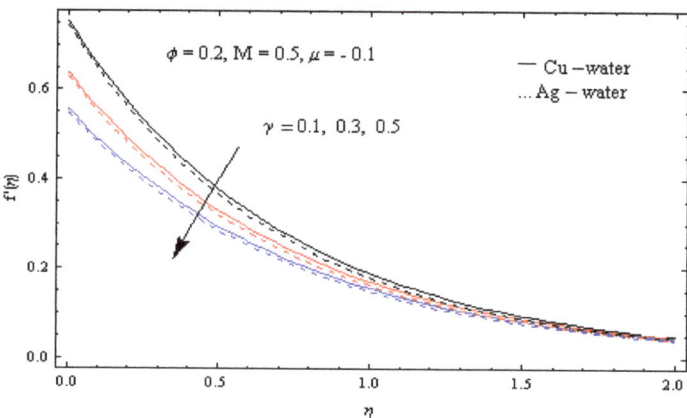

Figure 3. Effect of first slip γ on velocity of Cu-water and Ag-water nanofluids.

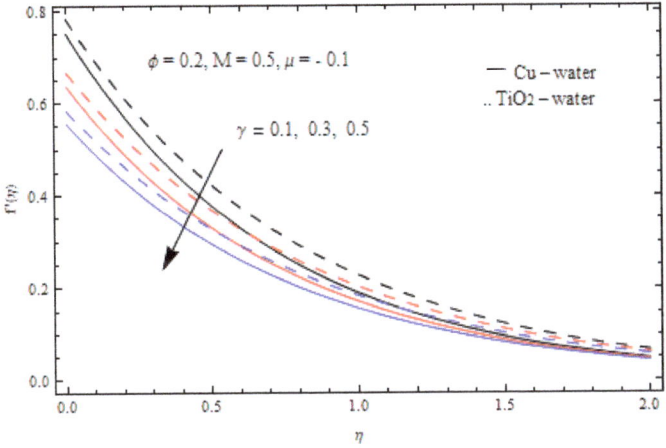

Figure 4. Effect of first slip γ on velocity of Cu-water and TiO$_2$-water nanofluids.

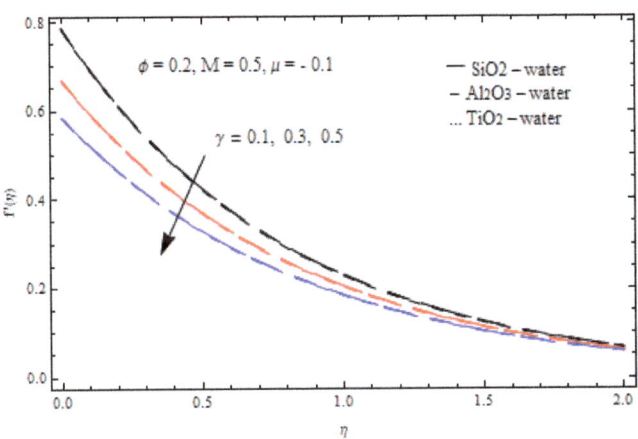

Figure 5. Effect of first slip γ on velocity of SiO$_2$-water, Al$_2$O$_3$-water, and TiO$_2$-water nanofluids.

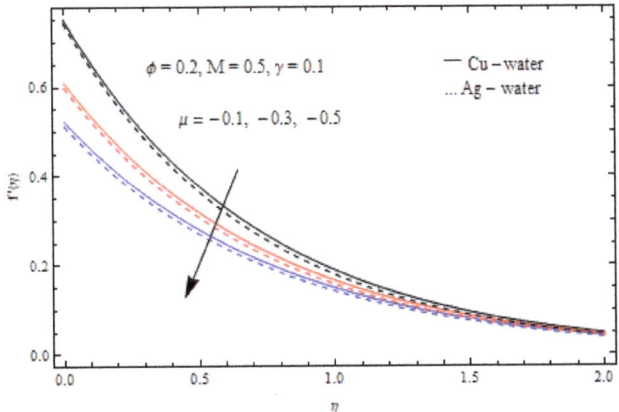

Figure 6. Effect of second slip μ on velocity of Cu-water and Ag-water nanofluids.

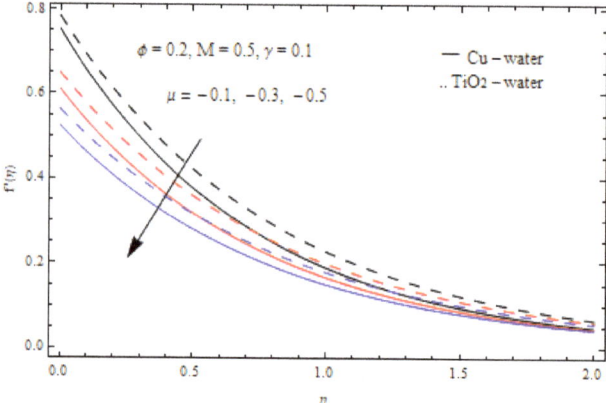

Figure 7. Effect of second slip μ on velocity of Cu-water and TiO$_2$-water nanofluids.

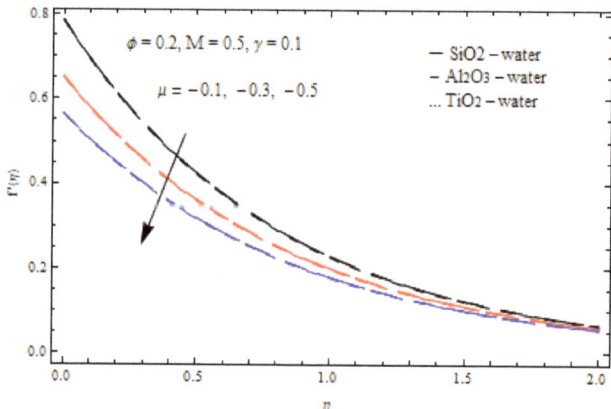

Figure 8. Effect of second slip μ on velocity of SiO$_2$-water, Al$_2$O$_3$-water, and TiO$_2$-water nanofluids.

In the absence of the homogenous reaction (i.e., at $K = 0$), the exact solution for the concentration $g(\eta)$ is available and given by Equation (16). In that case, the effects of K_s and Sc on $g(\eta)$ are plotted in Figures 9 and 10, respectively. It is shown that a reduction in the concentration occurs with the strengthening of the heterogenous reaction K_s and also with the increase in the Schmidt parameter Sc. Moreover, the Ag-water nanofluid is of lower concentration than the Cu-water nanofluid. This is also true for the general case, when both of the homogenous and heterogenous reactions take place in Figure 11, where the NDSolve command in Mathematica 7.0 (Wolfram Research, Champaign, IL, USA) has been used to solve the systems (5) and (12).

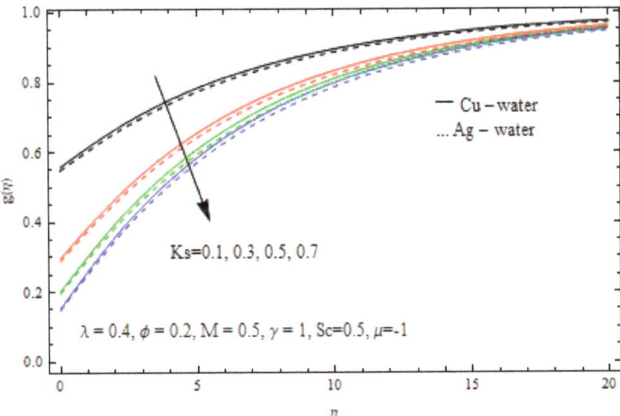

Figure 9. Effects of Ks on g at $K = 0$.

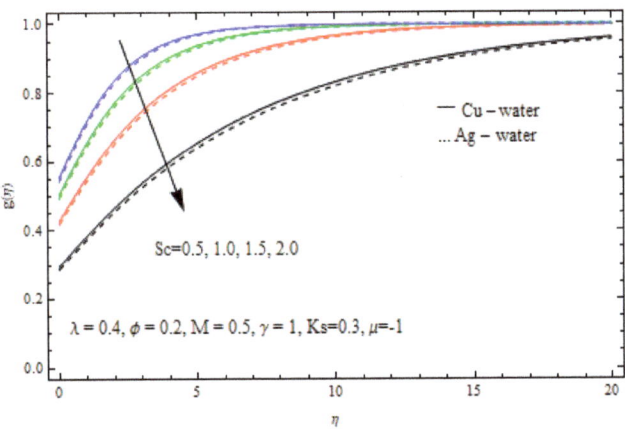

Figure 10. Effects of Sc on g at $K = 0$.

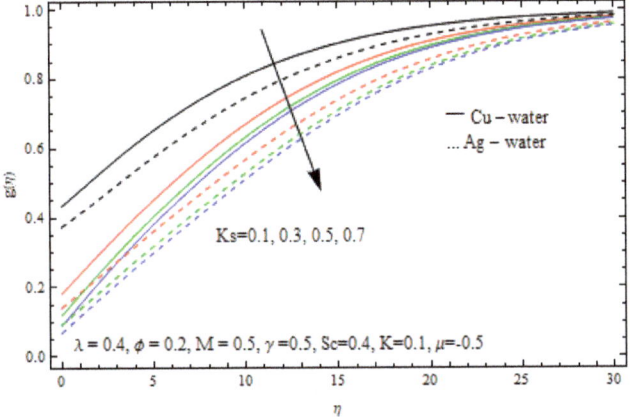

Figure 11. Effects of Ks on g.

4. Conclusions

In this paper, the effect of second velocity slip on the MHD flow of nanofluid in a porous medium with homogeneous-heterogeneous reactions has been analyzed. In the absence of the second slip, remarkable differences have been detected between the current exact results and those in the literature for the skin friction coefficient. For velocity, it has been found that the Ag-water nanofluid is lower than the other four kinds of nanofluids. For concentration, the exact solution has been given when only the heterogeneous reaction occurs. When both of the homogenous and heterogenous reactions take place, the concentration equation has been numerically solved. The concentration reduces with the strengthening of the heterogenous reaction and also with the increase in the Schmidt parameter, where the Ag-water nanofluid is of lower concentration than the Cu-water nanofluid.

Author Contributions: Conceptualization, F.A. and S.M.K.; methodology, F.A.; software, A.E.; validation, A.E.; project administration, F.A.; funding acquisition, F.A.

Funding: This research was funded by the Deanship of Scientific Research (DSR), University of Tabuk, Tabuk, Saudi Arabia, grant number 0074-1439-S.

Conflicts of Interest: The authors declare no conflict of interest.

References

1. Choi, S.U.S. Enhancing thermal conductivity of fluids with nanoparticles. In *Developments and Applications of Non-Newtonian Flows*; Siginer, D.A., Wang, H.P., Eds.; ASME: New York, NY, USA, 1995; Volume 66, pp. 99–105.
2. Eastman, J.A.; Choi, S.U.S.; Li, S.; Yu, W.; Thompson, L.J. Anomalously increased effective thermal conductivity of ethylene glycol-based nanofluids containing copper nanoparticles. *Appl. Phys. Lett.* **2001**, *78*, 718–720. [CrossRef]
3. Choi, S.U.S.; Zhang, Z.G.; Yu, W.; Lockwoow, F.E.; Grulke, E.A. Anomalous thermal conductivities enhancement on nanotube suspension. *Appl. Phys. Lett.* **2001**, *79*, 2252–2254. [CrossRef]
4. Aly, E.H.; Ebaid, A. Exact analytical solution for suction and injection flow with thermal enhancement of five nanofluids over an isothermal stretching sheet with effect of the slip model: A comparative study. *Abstr. Appl. Anal.* **2013**, *2013*, 721578. [CrossRef]
5. Majumder, M.; Chopra, N.; Andrews, R.; Hinds, B.J. Nanoscale hydrodynamics: Enhanced flow in carbon nanotubes. *Nature* **2005**, *438*, 44–46. [CrossRef] [PubMed]
6. Noghrehabadi, A.; Pourrajab, R.; Ghalambaz, M. Effect of partial slip boundary condition on the flow and heat transfer of nanofluids past stretching sheet prescribed constant wall temperature. *Int. J. Sci.* **2012**, *54*, 253–261. [CrossRef]
7. Gad-el-Hak, M. The fluid mechanics of macrodevices-the Freeman scholar lecture. *J. Fluids Eng.* **1999**, *121*, 5–33. [CrossRef]
8. Van Gorder, R.A.; Sweet, E.; Vajravelu, K. Nano boundary layers over stretching surfaces. *Commun. Nonlinear Sci. Numer. Simulat.* **2010**, *15*, 1494–1500. [CrossRef]
9. Sharma, R.; Ishak, A. Second order slip flow of Cu-waternanofluid over a stretching sheet with heat transfer. *Wseas Trans. Fluid Mech.* **2014**, *9*, 26–33.
10. Merkin, J.H. A model for isothermal homogeneous-heterogeneous reactions in boundary layer flow. *Math. Comput. Model.* **1996**, *24*, 125–136. [CrossRef]
11. Chaudhary, M.A.; Merkin, J.H. A simple isothermal model for homogeneous-heterogeneous reactions in boundary layer flow: I. Equal diffusivities. *Fluid Dyn. Res.* **1995**, *16*, 311–333. [CrossRef]
12. Bachok, N.; Ishak, A.; Pop, I. On the stagnation-point flow towards a stretching sheet with homogeneous-heterogeneous reactions effects. *Commun. Nonlinear Sci. Numer. Simul.* **2011**, *16*, 4296–4302. [CrossRef]
13. Khan, W.A.; Pop, I. Effects of homogeneous-heterogeneous reactions on the viscoelastic fluid towards a stretching sheet. *J. Heat Transf.* **2012**, *134*, 1–5. [CrossRef]
14. Kameswaran, P.K.; Shaw, S.; Sibanda, P.; Murthy, P.V.S.N. Homogeneous-heterogeneous reactions in a nanofluid flow due to porous stretching sheet. *Int. J. Heat Mass Transf.* **2013**, *57*, 465–472. [CrossRef]

15. Rashidi, M.M.; Kavyani, M.; Abelman, S. Investigation of entropy generation in MHD and slip flow over a rotating porous disk with variable properties. *Int. J. Heat Mass Transf.* **2014**, *70*, 892–917. [CrossRef]
16. Mahmoud, M.A.A.; Waheed, S.E. MHD flow and heat transfer of a micropolar fluid over a stretching surface with heat generation (absorption) and slip velocity. *J. Egypt. Math. Soc.* **2012**, *20*, 20–27. [CrossRef]
17. Ibrahim, W.; Shankar, B. MHD boundary layer flow and heat transfer of a nanofluid past a permeable stretching sheet with velocity, thermal and solutal slip boundary conditions. *Comput. Fluids* **2013**, *75*, 1–10. [CrossRef]
18. Rooholghdos, S.A.; Roohi, E. Extension of a second order velocity slip/temperature jump boundary condition to simulate high speed micro/nanoflows. *Comput. Math. Appl.* **2014**, *67*, 2029–2040. [CrossRef]
19. Malvandi, A.; Ganji, D.D. Brownian motion and thermophoresis effects on slip flow of alumina/water nanofluid inside a circular microchannel in the presence of a magnetic field. *Int. J. Therm. Sci.* **2014**, *84*, 196–206. [CrossRef]
20. Hayat, T.; Imtiaz, M.; Alsaedi, A. MHD flow of nanofluid with homogeneous-heterogeneous reactions and velocity slip. *Therm. Sci.* **2017**, in press. [CrossRef]
21. Hsiao, K.-L. To Promote Radiation Electrical MHD Activation Energy Thermal Extrusion Manufacturing System Efficiency by Using Carreau-Nanofluid with Parameters Control Method. *Energy* **2017**, *130*, 486–499. [CrossRef]
22. Hsiao, K.-L. Combined Electrical MHD Heat Transfer Thermal Extrusion System Using Maxwell Fluid with Radiative and Viscous Dissipation Effects. *Appl. Therm. Eng.* **2016**. [CrossRef]
23. Hsiao, K.-L. Micropolar Nanofluid Flow with MHD and Viscous Dissipation Effects towards a Stretching Sheet with Multimedia Feature. *Int. J. Heat Mass Transf.* **2017**, *112*, 983–990. [CrossRef]
24. Hsiao, K.-L. Stagnation Electrical MHD Nanofluid Mixed Convection with Slip Boundary on a Stretching Sheet. *Appl. Therm. Eng.* **2016**, *98*, 850–861. [CrossRef]
25. Ebaid, A.; al Mutairi, F.; Khaled, S.M. Effect of velocity slip boundary condition on the flow and heat transfer of Cu-Water and TiO_2-Water Nanofluids in the Presence of a Magnetic Field. *Adv. Math. Phys.* **2014**, *2014*, 538950. [CrossRef]
26. Ebaid, A.; al Sharif, M.A. Application of Laplace Transform for the Exact Effect of a Magnetic Field on Heat Transfer of Carbon Nanotubes-Suspended Nanofluids. *Z. Für Nat. A* **2015**, *70*, 471–475. [CrossRef]

© 2019 by the authors. Licensee MDPI, Basel, Switzerland. This article is an open access article distributed under the terms and conditions of the Creative Commons Attribution (CC BY) license (http://creativecommons.org/licenses/by/4.0/).

Article

Optimal Design of Isothermal Sloshing Vessels by Entropy Generation Minimization Method

Mohammad Yaghoub Abdollahzadeh Jamalabadi [1,2]

[1] Department for Management of Science and Technology Development, Ton Duc Thang University, Ho Chi Minh City 700000, Vietnam; abdollahzadeh@tdtu.edu.vn
[2] Faculty of Civil Engineering, Ton Duc Thang University, Ho Chi Minh City 700000, Vietnam

Received: 20 March 2019; Accepted: 22 April 2019; Published: 26 April 2019

Abstract: In this manuscript, the optimal design of geometry for a forced sloshing in a rigid container based on the entropy generation minimization (EGM) method is presented. The geometry of the vessel considered here is two dimensional rectangular. Incompressible inviscid fluid undergoes horizontal harmonic motion by interaction with a rigid tank. The analytical solution of a fluid stream function is obtained and benchmarked by Finite element results. A parameter study of the aspect ratio, amplitude, and frequency of the horizontal harmonic motion is performed. As well, an analytical solution for the total entropy generation in the volume is presented and discussed. The total entropy generation is compared with the results of the Reynolds-Averaged Navier–Stokes (RANS) solver and the Volume-of-Fluid (VOF) method). Then, the effect of parameters is studied on the total entropy generated by sway motion. Finally, the results show that, based on the excitation frequency, an optimal design of the tank could be found.

Keywords: fluid structure-interaction; vibration suppression; entropy generation minimization; sloshing; damping factor

1. Introduction

The ship maneuver-induced motion in the partially-filled tanks by liquid, sloshing poses a thoughtful danger to the controllability and stability of this phenomenon. The entropy generation minimization method is used for the design of fluid flow motion system [1] as well as thermal systems [2–4] in recent years. Although the method is applied to the thermodynamic optimization of many finite-size systems and finite-time processes [5], the application in isothermal fluid flow is rare [6]. For the specific case of sloshing, as such systems are used to damping the solid motion [7], the minimization of entropy could not be a true objective function for optimization. Even if a new engineering application has emerged in the future where the minimization of the entropy in sloshing fluid is the aim, the fluid cannot consider as a complete thermodynamic system. The fluid motion is caused by a solid structure consists of internal damping which causes entropy generation. The entropy generation in an isothermal wall container could be a measure of viscous dissipation which produces heat and could cause to danger in flammable liquids.

The analytical solution of a similar problem was presented by Ibrahim [8]. The liquid sloshing dynamics of a liquid in a vessel with horizontal excitation was presented Ibrahim [8] while the entropy generation was not discussed. Ikegawa studied the fluid flow problem motion of a rigid container excited by a horizontal harmonic acceleration with Finite Element Methods [9]. His results used in many texts as a benchmark [1]. Damping of surface waves in an incompressible liquid is studied by Case and Parkinson [10].

Jamalabadi et al. [11] found the optimal design of circular baffles in the sloshing problem occurred in a rectangular tank which is horizontally coupled by a one-story structure. Their method was pure numeric, and the optimization was based on the vibration suppression of the liquid motion. Although the problem is a classic case [12–25], the study of its exergy is discussed comprehensively in rare studies [1,26]. The entropy analysis of the flow systems is performed in many flow motions [27–34], as well and the recent developments in fluid modeling [35–39].

The aim of the current paper is to derive an analytical expression for entropy generation isothermal sloshing phenomenon and discuss the use of entropy generation minimization for such systems. The analytical expression for entropy generation in the rectangular tanks is obtained for the first time in the rectangular storage tank.

2. Mathematical Modeling

Consider a rigid rectangular tank as the physical domain of this research with length L, base at $y = -h$, free surface $y = 0$. Figure 1 shows the schematic of the problem with Coordinate system. As a first approximation the fluid motion can considered by the use of velocity potential. The replace of velocity potential in the continuity Equation ($\nabla . V = \frac{\partial u}{\partial x} + \frac{\partial v}{\partial y} = 0$) leads to Laplace equation as, (see Equation 1.23 in [8])

$$\frac{\partial^2 \phi}{\partial x^2} + \frac{\partial^2 \phi}{\partial y^2} = 0. \tag{1}$$

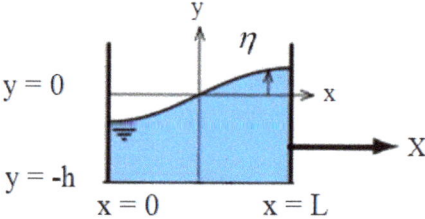

Figure 1. Diagram of fluid-vessel interaction with its cross section.

The boundary condition of the fluid domain in the right wall is the no-slip condition.

$$\left. \frac{\partial \phi}{\partial x} \right|_{x=L} = 0 \tag{2}$$

where L is the tank length. The no-slip condition at the left wall is

$$\left. \frac{\partial \phi}{\partial x} \right|_{x=0} = 0 \tag{3}$$

and the no-slip condition at the bottom wall is

$$\left. \frac{\partial \phi}{\partial y} \right|_{y=-h} = 0 \tag{4}$$

where h is the fluid height. At the free surface, the kinematic boundary is

$$\left. \frac{\partial \phi}{\partial y} \right|_{y=\eta} = \frac{\partial \eta}{\partial t} + \frac{\partial \eta}{\partial x} \left. \frac{\partial \phi}{\partial x} \right|_{y=\eta} \tag{5}$$

and the total pressure equation (neglecting the surface tension) from the Bernoulli equation is

$$P = -\rho\left(\frac{\partial \phi}{\partial t} + \frac{1}{2}\left\{\left(\frac{\partial \phi}{\partial x}\right)^2 + \left(\frac{\partial \phi}{\partial y}\right)^2\right\} + gy + \ddot{X}x\right) \quad (6)$$

where g is the gravity acceleration. The pressure at the free surface can be derived from the Equation of the motion ($\rho\left(\frac{\partial V}{\partial t} + (V.\nabla)V\right) = -\nabla p + \rho(\vec{g} - \vec{a}) + \nabla.\left(\mu(\nabla V + \nabla V^T)\right)$) by the aid of fluid density (ρ) and viscosity (μ) as well. The linearized surface conditions (leads to linear wave theory) are

$$\phi_y(y = 0) = \eta_t, \quad (7)$$

which is the kinematic condition for free surface elevation (η) and

$$\phi_t(y = 0) + g\eta + x\ddot{X} = 0 \quad (8)$$

for kinetic condition. Combining the kinematic and dynamic free-surface conditions leads to the equation

$$\phi_{tt}(y = 0) + g\phi_y(y = 0) = x\ddot{X}. \quad (9)$$

The solution satisfying Equation (1) with the rigid wall boundary conditions, Equations (2)–(4) is obtained in a general form as a sum of infinite sloshing modes as

$$\phi = \sum_{i=1}^{\infty} a_i(t) \cos\left(\frac{i\pi x}{L}\right) \frac{\cosh\left(\frac{i\pi(y+h)}{L}\right)}{\frac{i\pi}{L}\sinh\left(\frac{i\pi h}{L}\right)} \quad (10)$$

where $a_i(t)$ is an arbitrary time function and its related spatial function characterizes the velocity potential function of the nth sloshing mode and the dot notation ($\cdot$) represents $d(\ )/dt$. The free surface profile associated with Equation (10) with the boundary condition of Equation (7) is

$$\eta = \sum_{i=1}^{\infty} a_i(t) \cos\left(\frac{i\pi x}{L}\right). \quad (11)$$

The surface condition of Equation (9) can be used to determine the coefficients $a_i(t)$, which appears in Equation (10) and Equation (11) for the external acceleration of $\ddot{X}$ as

$$\ddot{a}_i(t) + g\frac{i\pi}{L}\tanh\left(\frac{i\pi h}{L}\right)a_i(t) + \frac{4}{i\pi}\tanh\left(\frac{i\pi h}{L}\right)\ddot{X} = 0 \quad (12)$$

where the cosine expansion of the x is used as

$$x = \frac{L}{2} + 2L\sum_{i=1}^{\infty} \cos\left(\frac{i\pi x}{L}\right)\frac{(-1)^i - 1}{(i\pi)^2} \quad (13)$$

to derive Equation (11). The fundamental sloshing frequency (i = 1) of the liquid inside the rectangular tank could be obtained by considering the free oscillation in Equation (11) as

$$f_w = \frac{1}{2\pi}\sqrt{\frac{\pi g}{L}\tanh\left(\frac{\pi h}{L}\right)} \quad (14)$$

and replacing the $X = X_{\max}\cos(\omega t)$ in Equation (11) for the external motion gives

$$\ddot{a}_i(t) + g\frac{i\pi}{L}\tanh\left(\frac{i\pi h}{L}\right)a_i(t) = -\frac{4}{i\pi}\tanh\left(\frac{i\pi h}{L}\right)\omega^2 X_{\max}\sin(\omega t) \quad (15)$$

where $\omega_i^2 = g\frac{i\pi}{L}\tanh(\frac{i\pi h}{L})$. The steady-state solution of Equation (15) is

$$a_i(t) = \tanh(\frac{i\pi h}{L})\frac{4X_{max}}{i\pi}\frac{\omega^2}{\omega^2 - \omega_i^2}\sin(\omega t). \tag{16}$$

The final linearize solutions are

$$\eta = X_{max}\sum_{i=1}^{\infty}\tanh(\frac{i\pi h}{L})\frac{4}{i\pi}\frac{\omega^2}{\omega^2 - \omega_i^2}\sin(\omega t)\cos(\frac{i\pi x}{L}) \tag{17}$$

$$\phi = LX_{max}\omega\sum_{i=1}^{\infty}\frac{\left(\frac{2\omega}{i\pi}\right)^2}{\omega^2 - \omega_i^2}\cos(\omega t)\cos(\frac{i\pi x}{L})\frac{\cosh\left(\frac{i\pi(y+h)}{L}\right)}{\cosh\left(\frac{i\pi h}{L}\right)}. \tag{18}$$

The entropy generated can be calculated by [5]

$$\dot{S}'''_g = \frac{\mu}{T}\varphi + \frac{k}{T^2}(\nabla T)^2 \tag{19}$$

where the dot notation ($'''$) represents the value per volume. The dissipation function in Equation (19) is calculated from

$$\varphi = 2\left[\left(\frac{\partial u}{\partial x}\right)^2 + \left(\frac{\partial v}{\partial y}\right)^2\right] + \left(\frac{\partial u}{\partial y} + \frac{\partial v}{\partial x}\right)^2. \tag{20}$$

The total entropy generated in the volume of the fluid in the case of an isothermal condition ($\nabla T = 0$) is calculated from Equation (18) as

$$S_g = \iint \frac{\mu}{T}\varphi\, dxdy. \tag{21}$$

By substituting the analytical solution in the definition of entropy generation we get:

$$S_g = \frac{16\pi\mu X_{max}^2\omega^2}{T}\sum_{i=1}^{\infty}\left(\frac{\omega^2}{\omega^2 - \omega_i^2}\cos(\omega t)\right)^2\frac{\tanh\left(\frac{i\pi h}{L}\right)}{i}. \tag{22}$$

The entropy appearing in Equation (22) is the total entropy generated by the fluid, and since the energy exchanged with the moving wall has been considered as a thermodynamic system, the entropy of the working fluid is well established and can be used as an objective function. The entropy generation in an isothermal wall container could be a representation of viscous dissipation that could lead to explosion in liquids with flammable materials.

3. Results and Discussion

The analytical solution of Equation (17) is benchmarked with a finite element method (FEM) solution obtained by Ikegawa [9], whose dimensions of the liquid container are $h = 0.6$ m and $L = 0.9$ m. The vessel is exposed to the forced horizontal motion as given by $X = 0.002\sin(5.5t)$. Figure 2 presents the time history of η ($x = +L$, t). The numerical result is denoted by circles and the analytical solution is denoted by a solid line. As shown, there is a good agreement between the analytical solution and numerical result with FEM.

An inspection of the analytical solution of Equation (22) is performed in Figure 3. The maximum distribution of the dimensionless entropy generation rate through the volume, with respect to time, is a plot for various aspect ratios in that figure, according to ($\bar{S}_g = \frac{(\omega^2 - \omega_i^2)^2 TS_g}{16\pi\mu X_{max}^2\omega^4}$). The axes in Figure 3 are Cartesian coordinate system of the vessel in accordance with Figure 1. The dimensions of rectangular storage tanks for each aspect ratio are $L = (0.54a)\,0.5$ and $h = (0.54/a)\,0.5$. As shown by the increase of the aspect ratio, the amount of entropy generated through the volume decreased. Further, the position of maximum entropy changed from the free surface to the side walls. It was expected that by an increase of the aspect ratio, the length of the tanks would increase and the dimensionless penetration length ($\sqrt{\frac{v}{L^2\omega}}$) would decrease. An increase of the aspect ratio made the dampening effects of sidewalls and bottom dominant in comparison with the free surface effects [10].

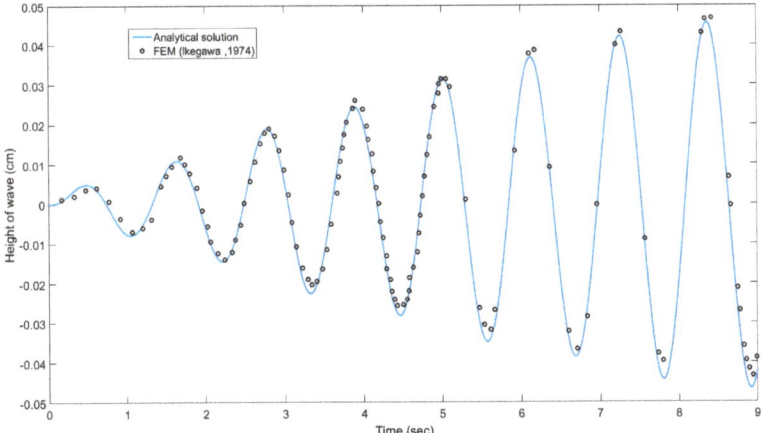

Figure 2. Time history of η ($x = +L, t$).

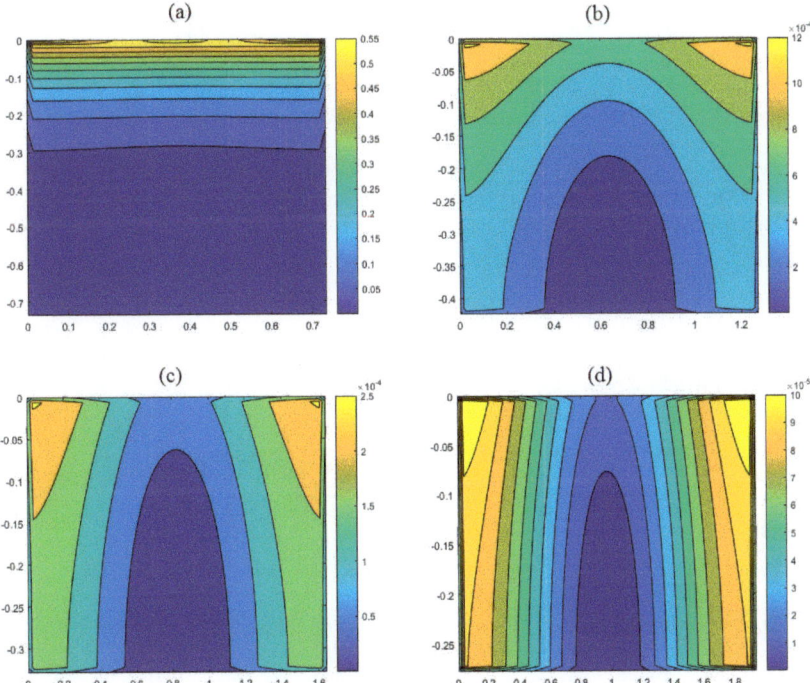

Figure 3. Distribution of dimensionless entropy generation for various aspect ratios (**a**) $\alpha = 1$, (**b**) $\alpha = 3$, (**c**) $\alpha = 5$, and (**d**) $\alpha = 7$.

Horizontal periodic sway motions as $X = A_m \sin(\omega t)$ were applied to the rectangular storage tanks with different aspect ratios, namely the ratios of height to length of the rectangular storage tank (AR). Then, the effect of A_m and ω was studied on the results. The oscillations of total entropy generation rate in the volume for the aspect ratio of $\alpha = 2.05$ are plotted in Figure 4a. Similar to the time history of the wave, the entropy generation rate reaches its maximum after 10 periods of motion and decreases. The beating behavior of the entropy profile repeats as times goes on. To demonstrate the capability and accuracy of the present method, the results of the generated waves are compared with the available numerical calculations. Figure 4b takes from the results of [1]. Results from [1] are opposed to those stemming from this study, where Figure 4a should be compared with AR = 2.05 Figure 4b. The true unit for Sg obtained by the surface integration of volumetric entropy generation is (W/Km). However, in reference [1], as they considered the two-dimensional case with a 1 m depth, the unit appeared as (W/K). Moreover, the results show that an increase in the AR causes a decrease in the total entropy generation rate in the volume.

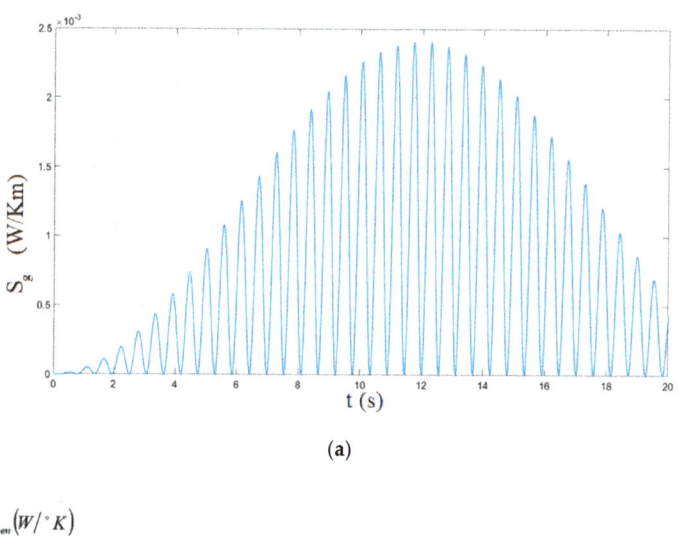

(a)

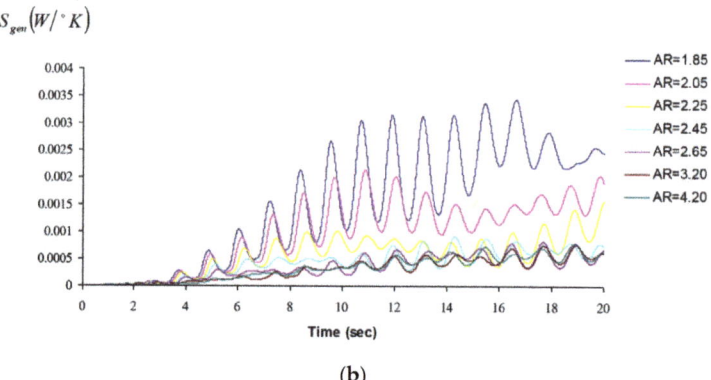

(b)

Figure 4. Entropy generation versus time for $\alpha = 2.05$ (a) current study, (b) Reference [1].

Figure 4b was obtained by using the Reynolds-Averaged Navier–Stokes (RANS) and the Volume-of-Fluid (VOF) methods, together, in a commercial software solver [1]. The RANS equations were discretized and solved using the staggered grid finite difference and simplified marker and cell (SMAC) methods, and the available data were used for the model validation. By comparing the case of

$\alpha = 2.05$, it is clear that the trend and the order of magnitude of a maximum of entropy (2×10^{-2}) are the same. Since the current analytical solution is a suitable measure to decide on optimization based on the entropy generation, the entropy generation distribution offers designers with valuable data about the reasons for the energy loss.

Finally, Figure 5 reveals the value of the total entropy generation rate versus aspect ratio for $\alpha = 3$. As shown, the trend of maximum entropy generation versus aspect ratio decreasing expects a peak point, which is caused by approaching the natural frequency of the system to the external forced frequency. Such phenomena lead to a local minimum point before the resonance, since the $\alpha = 1.4$–1.5 is a candidate for the entropy minimization point. Generally, the overall function has no optimum and a higher aspect ratio leads to lower values of entropy generation.

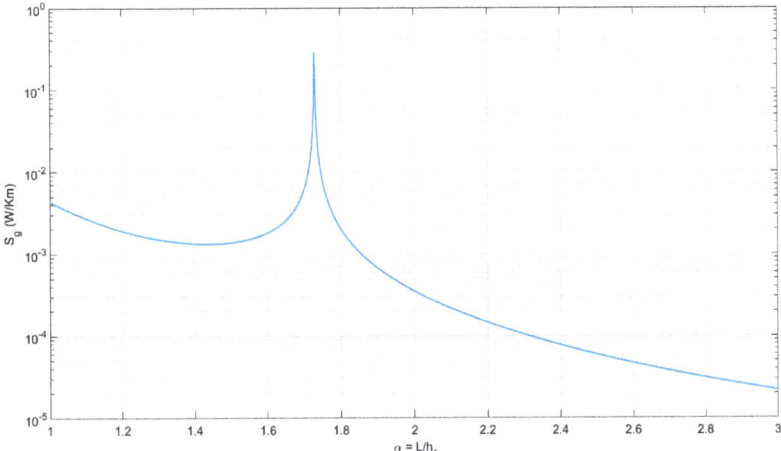

Figure 5. Total entropy generation rate versus aspect ratio for $\alpha = 3$.

As shown in reference [7] (see Figure 3c), 80 percent of energy of the fluid could be dissipated for the dimensionless frequencies in the range 0.95–1.05 ($f' = f/fn$), since the engineers try to design the sloshing vessels with the frequencies near to the structure frequency for highest energy absorbance rate. If the value of the energy of the fluid is symbolized by

$$E_f = \frac{1}{2} m_f (2\xi_f \omega) X_{\max}^2 \omega^2 \tag{23}$$

and the work of no-conservative damping of the coupling structure are

$$E_s = \frac{1}{2} m_s (2\xi_s \omega) X_{\max}^2 \omega^2 \tag{24}$$

then the ratio of structure energy loss to the fluid loss is

$$\gamma = \frac{\xi_s m_s}{\xi_f m_f}. \tag{25}$$

The damping of the fluid could be estimated by the inverse of square root of the Galileo number (ratio of gravity forces and viscous forces)

$$\xi_f = \frac{v^{1/2}}{L^{3/4} g^{1/4}} \tag{26}$$

where v is the kinematic viscosity. The damping factor of 1–2% is predicted for fluids [10] (logarithmic decrement $\approx 6\xi$) and 0.32% for solids [7], since the ratio of structural energy loss to fluid loss is approximated by 0.1–1 of the ratio of structural mass to fluid mass. As an example for engineering applications, the mass ratio of the tuned liquid damper to the solid structure is 1.05% [7], and then the amount of structure energy loss to the fluid loss is about 10–100. Subsequently, most of the energy dissipated in the solid part, which is not considered for optimization.

Since, as stated in the introduction section, for the specific case of sloshing, as such systems are used for the damping of the solid motion entropy of only fluid, they could not be a true objective function, and the energy dissipation in the structure should be considered, too. Today's practical meaning of EGM is very low. Although today engineers in the field of large vessels are mostly focused on frequency response design and exergy efficiency is not considered in engineering code, the entropy minimization method is a growing topic in literature. In the current study, fluid entropy generation used as a measure of optimization of the sloshing phenomenon that is classified among free surface flows. The current research proposes future studies performing experiments for coupled cases with the sum energy dissipation of fluid and structure as an objective function.

4. Conclusions

In this manuscript, the entropy generation rate in a forced sloshing rigid tank was studied analytically. The analytical solution of the fluid was obtained and benchmarked. The following points were concluded:

- By the increase of the aspect ratio, the amount of entropy generated through the volume decreased.
- By the increase of the aspect ratio, the position of maximum entropy is changed from the free surface to the side walls.
- As the order of magnitude of the maximum of entropy for the analytical case and numerical results are the same, the analytical solution is a suitable measure for entropy generation minimization.
- The minimum entropy generation point for the sloshing problem is local and general; the entropy generation has no optimum as a function of aspect ratio.
- The ratio of structural energy loss to fluid loss is approximated by the ratio of structural mass to fluid mass.
- The energy dissipation in the structure coupled with sloshing fluids should be considered for entropy generation minimization.
- The current research proposes to do experiments for coupled cases with total dissipation function (i.e., sum energy dissipation of fluids and structures) as an objective function in future studies.

Funding: This research received no external funding.

Conflicts of Interest: The author declares no conflict of interest.

References

1. Saghi, H.; Lakzian, E. Optimization of the rectangular storage tanks for the sloshing phenomena based on the entropy generation minimization. *Energy* **2017**, *128*, 564–574. [CrossRef]
2. Gholamalizadeh, E.; Jamalabadi, M.Y.A.; Oveisi, M. Optimal design of thermal radiative heating of horizontal thin plates using the entropy generation minimization method. *Energies* **2017**, *10*, 1921. [CrossRef]
3. Jamalabadi, M.Y.A.; Safaei, M.R.; Alrashed, A.A.A.A.; Nguyen, T.K.; Filho, E.P.B. Entropy generation in thermal radiative loading of structures with distinct heaters. *Entropy* **2017**, *19*, 506. [CrossRef]
4. Jamalabadi, M.Y.A.; Hooshmand, P.; Bagheri, N.; KhakRah, H.; Dousti, M. Numerical simulation of Williamson combined natural and forced convective fluid flow between parallel vertical walls with slip effects and radiative heat transfer in a porous medium. *Entropy* **2016**, *18*, 147. [CrossRef]
5. Bejan, A. *Entropy Generation Minimization: The Method of Thermodynamic Optimization of Finite-Size Systems and Finite-Time Processes*; CRC Press: New York, NY, USA, 1996.

6. Chen, L.; Xia, S.; Sun, F. Entropy generation minimization for isothermal crystallization processes with a generalized mass diffusion law. *Int. J. Heat Mass Transf.* **2018**, *116*, 1–8. [CrossRef]
7. Sun, L.M.; Fujino, Y.; Pacheco, B.M.; Chaiseri, P. Modelling of tuned liquid damper (TLD). *J. Wind Eng. Ind. Aerody* **1992**, *43*, 1883–1894. [CrossRef]
8. Ibrahim, R.A. *Liquid Sloshing Dynamics*; Cambridge University Press: Cambridge, UK, 2005.
9. Ikegawa, M. Finite element analysis of fluid motion in a container. In *Finite Element Methods in Flow Problems*; Oden, J.T., Zienkiewicz, 0.C., Gallagher, R.H., Taylor, C., Eds.; UAH Press: Huntsville, AL, USA, 1974; pp. 737–738.
10. Case, K.; Parkinson, W. Damping of surface waves in an incompressible liquid. *J. Fluid Mech.* **1957**, *2*, 172–184. [CrossRef]
11. Jamalabadi, M.Y.A.; Ho-Huu, V.; Nguyen, T.K. Optimal design of circular baffles on sloshing in a rectangular tank horizontally coupled by structure. *Water* **2018**, *10*, 1504. [CrossRef]
12. Abramson, H.N. *The Dynamic Behavior of Liquids in Moving Containers*; National Aeronautics and Space Administration: Washington, DC, USA, 1996.
13. Frandsen, J.B. Sloshing motions in excited tank. *J. Comput. Phys.* **2004**, *106*, 53–87. [CrossRef]
14. Chen, B.F.; Nokes, R. Time-independent finite difference analysis of fully nonlinear and viscous fluid sloshing in a rectangular tank. *J. Comput. Phys.* **2005**, *209*, 47–81. [CrossRef]
15. Wu, C.H.; Faltinsen, O.M.; Chen, B.F. Numerical study of sloshing liquid in tanks with baffles by time-independent finite difference and fictitious cell method. *Comput. Fluids* **2012**, *63*, 9–26. [CrossRef]
16. Huang, S.; Duan, W.Y.; Zhu, X. Time-domain simulation of tank sloshing pressure and experimental validation. *J. Hydrodyn. Ser. B* **2010**, *22*, 556–563. [CrossRef]
17. Pirker, S.; Aigner, A.; Wimmer, G. Experimental and numerical investigation of sloshing resonance phenomena in a spring-mounted rectangular tank. *Chem. Eng. Sci.* **2012**, *68*, 143–150. [CrossRef]
18. Hasheminejad, S.M.; Aghabeigi, M. Sloshing characteristics in half-full horizontal elliptical tanks with vertical baffles. *Appl. Math. Model.* **2012**, *36*, 57–71. [CrossRef]
19. Papaspyrou, S.; Karamanos, S.A.; Valougeorgis, D. Response of half-full horizontal cylinders under transverse excitation. *J. Fluids Struct.* **2004**, *19*, 985–1003. [CrossRef]
20. Shekari, M.R.; Khaji, N.; Ahmadi, M.T. A couple BE-FE study for evaluation of seismically isolated cylindrical liquid storage tanks considering fluid–structure interaction. *J. Fluids Struct.* **2009**, *25*, 567–585. [CrossRef]
21. Gavrilyuk, I.P.; Lukovsky, I.A.; Timokha, A.N. Linear and nonlinear sloshing in a circular conical tank. *Fluid Dyn. Res.* **2005**, *37*, 399–429. [CrossRef]
22. Yue, B.Z. Nonlinear coupling dynamics of liquid filled spherical container in microgravity. *Appl. Math. Mech.* **2008**, *29*, 1085–1092. [CrossRef]
23. Sarreshtehdari, A.; Shahmardan, M.M.; Gharaei, R. Numerical simulation and experimental validation of free surface sloshing in a rectangular tank. *J. Solid Fluid Mech.* **2011**, *1*, 89–95.
24. Mirzabozorg, H.; Hariri-Ardebili, M.; Nateghi, R. Free surface sloshing effect on dynamic response of rectangular storage tanks. *Am. J. Fluid* **2012**, *2*, 23–30. [CrossRef]
25. Rajagounder, R.; Vignesh Mohanasundaram, G.; Kalakkath, P. A study of liquid sloshing in an automotive fuel tank under uniform acceleration. *Eng. J.* **2015**, *20*, 72–85. [CrossRef]
26. Ketabdari, M.J.; Saghi, H. Parametric study for optimization of storage tanks considering sloshing phenomenon using coupled BEM-FEM. *Appl. Math. Comput.* **2013**, *224*, 123–139. [CrossRef]
27. Bejan, A. A study of entropy generation in fundamental convective heat transfer. *J. Heat Transf.* **1979**, *101*, 718–725. [CrossRef]
28. Bejan, A. The thermodynamic design of heat and mass transfer processes and devices. *Int. J. Heat Fluid Flow* **1987**, *8*, 258–276. [CrossRef]
29. Lakzian, E.; Masjedi, A. Slip effects on the exergy loss due to irreversible heat transfer in a condensing flow. *Int. J. Exergy* **2014**, *14*, 22–37. [CrossRef]
30. Lakzian, E.; Shabani, A. Analytical investigation of coalescence effects on the exergy loss in a spontaneously condensing wet-steam flow. *Int. J. Exergy* **2015**, *4*, 383–403. [CrossRef]
31. Lotfi, A.; Lakzian, E. Entropy generation analysis for film boiling: A simple model of quenching. *Eur. Phys. J. Plus* **2016**, *131*, 1–10. [CrossRef]
32. Soltanmohammadi, R.; Lakzian, E. Improved design of wells turbine for wave energy conversion using entropy generation. *Meccanica* **2016**, *51*, 1713–1722. [CrossRef]

33. Lakzian, E.; Soltanmohammadi, R.; Nazeryan, M. A comparison between entropy generation analysis and first law efficiency in a monoplane wells turbine. *Sci. Iran. B* **2016**, *23*, 2673–2681.
34. Wang, T.; Huang, Z.; Xi, G. Entropy generation for mix convection in a square cavity containing a rotating circular cylinder using a local radial basis function method. *Int. J. Heat Master Transf.* **2017**, *106*, 1063–1073. [CrossRef]
35. Shadloo, M.S. Numerical Simulation of Compressible Flows by Lattice Boltzmann Method. *Numer. Heat Transf. Part A* **2019**. [CrossRef]
36. Hopp-Hirschler, M.; Shadloo, M.S.; Nieken, U. Viscous Fingering Phenomena in the Early Stage of Polymer Membrane Formation. *J. Fluid Mech.* **2019**, *864*, 97–140. [CrossRef]
37. Nguyen, M.Q.; Shadloo, M.S.; Hadjadj, A.; Lebon, B.; Peixinho, J. Perturbation threshold and hysteresis associated with the transition to turbulence in sudden expansion pipe flow. *Int. J. Heat Fluid Flow* **2019**, *76*, 187–196. [CrossRef]
38. Shenoy, D.V.; Shadloo, M.S.; Hadjadj, A.; Peixinho, J. Direct numerical simulations of laminar and transitional flows in diverging pipes. *Int. J. Numer. Methods Heat Fluid Flow* **2019**. [CrossRef]
39. Shadloo, M.S.; Weiss, R.; Yildiz, M.; Dalrymple, R.A. Numerical Simulations of the Breaking and Non-breaking Long Waves. *Int. J. Offshore Polar Eng.* **2015**, *25*, 1–7.

© 2019 by the author. Licensee MDPI, Basel, Switzerland. This article is an open access article distributed under the terms and conditions of the Creative Commons Attribution (CC BY) license (http://creativecommons.org/licenses/by/4.0/).

MDPI
St. Alban-Anlage 66
4052 Basel
Switzerland
Tel. +41 61 683 77 34
Fax +41 61 302 89 18
www.mdpi.com

Mathematics Editorial Office
E-mail: mathematics@mdpi.com
www.mdpi.com/journal/mathematics

www.ingramcontent.com/pod-product-compliance
Lightning Source LLC
LaVergne TN
LVHW070250100526
838202LV00015B/2204